The scientific and production quality of the Geological Society's books matches that of its journals. Since 1997, all book proposals are reviewed by two individual experts and the Society's Books Editorial Committee. Proposals are only accepted once any identified weaknesses are addressed.

The Geological Society of London is signed up to the Committee on Publication Ethics (COPE) and follows the highest standards of publication ethics. Once a book has been accepted, the volume editors agree to follow the Society's Code of Publication Ethics and facilitate a peer review process involving two independent reviewers. This is overseen by the Society Book Editors who ensure these standards are adhered to.

Geological Society books are timely volumes in topics of current interest. Proposals are often devised by editors around a specific theme or they may arise from meetings. Irrespective of origin, editors seek additional contributions throughout the editing process to ensure that the volume is balanced and representative of the current state of the field.

Submitting a book proposal

More information about submitting a proposal and producing a book for the Society can be found at https://www.lyellcollection.org/publishing-hub/proposals

It is recommended that reference to all or part of this book should be made in one of the following ways:

Griffiths, J. S., Giles, D. P., Davis, J. A. and Dunlop, J. J. (eds) 2025. *The Peter Fookes Engineering Geological Legacy in Geomodels, Geomaterials and Geomorphology*. Geological Society, London, Engineering Geology Special Publications, **30**, https://doi.org/10.1144/egsp30-000

Lee, E.M., Sweeney, M. and Clarke, J. 2025. Arabia Felix: Fookes in the desert. *Geological Society, London, Engineering Geology Special Publications*, **30**, 191–207, https://doi.org/10.1144/egsp30-2023-33

The Peter Fookes Engineering Geological Legacy in Geomodel Geomaterials and Geomorphology

Geological Society Engineering Geology Special Publication No. 30

The Peter Fookes Engineering Geological Legacy in Geomodels, Geomaterials and Geomorphology

Edited by

J. S. Griffiths
University of Plymouth, UK

D. P. Giles
Geotechnical Consulting Group LLP, UK

J. A. Davis
Geotechnical Consulting Group LLP, UK

and

J. J. Dunlop
Mott MacDonald, UK

2025
Published by
The Geological Society
London

The Geological Society of London

The Geological Society of London is a not-for-profit organisation, and a registered charity (no. 210161). Our aims are to improve knowledge and understanding of the Earth, to promote Earth science education and awareness, and to promote professional excellence and ethical standards in the work of Earth scientists, for the public good. Founded in 1807, we are the oldest geological society in the world. Today, we are a world-leading communicator of Earth science – through scholarly publishing, library and information services, cutting-edge scientific conferences, education activities and outreach to the general public. We also provide impartial scientific information and evidence to support policy-making and public debate about the challenges facing humanity. For more about the Society, please go to https://www.geolsoc.org.uk/

The Geological Society Publishing House (Bath, UK) produces the Society's international journals and books, and acts as European distributor for selected publications of the American Association of Petroleum Geologists (AAPG), the Geological Society of America (GSA), the Society for Sedimentary Geology (SEPM) and the Geologists' Association (GA). GSL Fellows may purchase these societies' publications at a discount. The Society's online bookshop is at www.geolsoc.org.uk/bookshop

Membership

Become part of the Geological Society community and unlock a wealth of benefits. Enjoy access to over 500 books and one journal option online in the Lyell Collection, discounts on printed publications, events, training courses and more. Learn more: www.geolsoc.org.uk

For information about the Society's events, training courses, Corporate Patrons and more go to www.geolsoc.org.uk

Proposing a book

If you are interested in proposing a book then please visit: www.lyellcollection.org/publishing-hub/proposals

Published and manufactured by The Geological Society from:
The Geological Society Publishing House, Unit 7 Brassmill Enterprise Centre, Brassmill Lane, Bath BA1 3JN, UK
www.geolsoc.org.uk and sales@geolsoc.org.uk

For Product Safety concerns and information please contact our EU representative:
Authorised Rep Compliance Ltd, Ground Floor, 71 Lower Baggot Street, Dublin D02 P593, Ireland
www.arccompliance.com

To purchase copies:

Visit the Lyell Collection: www.lyellcollection.org
Online bookshop: www.geolsoc.org.uk/bookshop
Orders: Tel. +44 (0)1225 445046, Email: sales@geolsoc.org.uk

British Library Cataloguing in Publication Data

A catalogue record for this book is available from the British Library.
ISBN 978-1-78620-690-9
ISSN 0267-9914

Printed books

For details of North American and other distributors see: www.geolsoc.org.uk/publishing/contact-us/

Typeset by Nova Techset Private Limited, Bengaluru & Chennai, India
Printed and bound by CPI Group (UK) Ltd, Croydon CR0 4YY, UK

Contents

The Peter Fookes Engineering Geological Legacy in Geomodels, Geomaterials and Geomorphology: an introduction

J. S. Griffiths

School of Geography, Earth and Environmental Sciences (SoGEES), University of Plymouth, Drake Circus, Plymouth PL4 8AA, UK

0000-0002-6431-3992

Correspondence: jim.griffiths@plymouth.ac.uk

Abstract: This introduction provides background to the development of engineering geology as a recognized branch of applied geology from antiquity through to the present day. In the development of the subject in the modern era, Professor Peter Fookes had a leading role in the UK and this is identified and celebrated. Specifically, the legacy Peter Fookes has left to engineering geology in the fields of ground models, geomaterials and geomorphology is briefly reviewed as this forms the focus of the book.

> Engineering geology is neither more nor less than the application of geological knowledge to engineering art and development, and that part of this knowledge which is applicable should increasingly be sought out and used. Otherwise, geology will partly fail in its mission, and the art of engineering will limp
>
> (Paige 1950*b*, p. vi).

This Special Publication is about the legacy of Professor Peter George Fookes (1933–2020), whose life was spent applying geology to the art and development of civil engineering. Peter Fookes has rightly been described as the 'Father of modern British engineering geology' (Charman 2008). In recognition of this, the theme of the Annual Conference of the Engineering Group of the Geological Society of London (Cambridge University 26–28 March 2023) was 'The Legacy of Peter Fookes'. At that meeting it became apparent that Peter's contribution to engineering geology was intertwined with how the discipline had advanced in the twentieth and twenty-first centuries. Therefore, the decision was taken to produce a book that not only commemorated the work of Peter Fookes but went on to illustrate how the discipline had progressed as an applied science based on his work. The thrust of the book would be on looking at how the concepts and papers of Peter have been taken forward and applied to projcets by engineering geologists and geomorphologists.

In this introduction the emergence of engineering geology as a substantive branch of geology from both a global historical and UK perspective is briefly described. The introduction goes on to recognize the pre-eminent role that Professor Fookes played in its development, notably in the fields of geomodels, geomaterials and engineering geomorphology, before briefly outlining the papers that make up the main part of the volume.

History of engineering geology: a global perspective

The first book on engineering geology was published in London in the late nineteenth century (Penning 1880) and there have been many notable papers and books on the historical development of engineering geology, including Paige (1950*a*, *b*), Kiersch (1991*a*), VanDine *et al.* (2006), Culshaw *et al.* (2008) and Griffiths (2014). In the UK, the subject only really started to become mainstream as an identifiable geology subdiscipline after World War II. This is in contrast to the USA and some European countries where the potential value of geology in engineering practice received earlier recognition. For example, Kiersch (1991*a*) edited a book from the Geological Society of America entitled, *The Heritage of Engineering Geology; the First Hundred Years.* This book was dedicated to William Otis Crosby (1850–1925) who was described as 'The Father of Engineering Geology in North America'. In Europe one of the co-authors of a 1929 book on engineering geology was the founder of the discipline of geotechnics, the incomparable Karl Terzaghi (Redlich *et al.* 1929). This book was published in Vienna and indicates that engineering geology was already recognized in Austria at that time.

A broader perspective on how engineering geology developed from antiquity was provided by Turner (2008) who looked at the historical record as a basis for assessing interactions between geology and civil engineering. However, arguably the best treatise on the heritage of engineering geology was by Kiersch (1991*b*) who, prior to modern times (taken as 1940 onwards), identified four periods of development of the subject as described below:

From: Griffiths, J. S., Giles, D. P., Davis, J. A. and Dunlop, J. J. (eds) 2025. *The Peter Fookes Engineering Geological Legacy in Geomodels, Geomaterials and Geomorphology*. Geological Society, London, Engineering Geology Special Publications, **30**, 1–7,
First published online 25 July 2025, https://doi.org/10.1144/egsp30-2024-72

(1) Antiquity through to the Middle Ages prior to AD 1450: during antiquity there is evidence of subsurface mining for copper in the Sinai Peninsula at least 15 000 years ago. There are the Egyptian pyramids built from quarried stone between 2700 and 2300 BC. In Mesopotamia there are indications that bitumen was used to surface streets from around 2800 BC. In Iran (Persia) infiltration galleries or qanats were built to harness groundwater for irrigation more than 2500 years ago. Roman roads, such as that built in 57 BC on the Brest St Bernard's Pass, were remarkable constructions that were based on a real understanding of materials and shallow foundations. During the post-Roman period in Europe the most magnificent structures were those associated with the Christian Church, such as the Patriarchal Cathedral Basilica of St Marks in Rome (opened in AD 1094) and Hagia Sophia in Istanbul completed in AD 537. One of the most remarkable feats of engineering in China during this period was the Forbidden City constructed between 1406 and 1420. All these structures required at least some understanding of engineering geology and geotechnics, even if the disciplines themselves were not recognized.

(2) Renaissance and Baroque periods (*c.* 1450–1750): Kiersch (1991*b*) noted that as the Renaissance developed, French, English and Italian scientists emerged who envisioned new geological concepts, of which some became important for engineering works. Perhaps it was Leonardo da Vinci (1452–1519) who best epitomizes this period. An unsurpassed polymath, Leonardo da Vinci could be regarded as the first applied geologist through his work on canals and other water projects (Clements 1981). He also postulated the notion of mountain building and erosion in a way that was prescient of the concepts of uplift and isostatic adjustment. Elsewhere, Zaruba and Mencl (1963) highlighted the construction of the 1 km-long aqueduct in the King Garden of Prague known as the Rudolf Galley (1581–93) where the original drawings indicated how the designers and builders took account of the geological conditions. Many more examples of the applications of geology in all aspects of societal development during this period can be found in Faul and Faul (1983).

(3) Industrial Revolution to the nineteenth century (1750–1900): this period saw the publication of the *Theory of the Earth* by Hutton (1788), Lyell's (1830, 1832, 1833) seminal books on the *Principles of Geology*, and *On the Origin of Species by Means of Natural Selection* by Charles Darwin (Darwin 1859). Against this background was the massive increase in canal construction in the UK that led to William Smith (1769–1839) creating a geological map of the whole country, the first such map in the world (Winchester 2001). William Smith is arguably the world's first true engineering geologist and hence it is his profile that graces the logo of the Engineering Group of the Geological Society of London, which is the world's oldest geological society, formed in 1807. Apart from canals, this period also saw an unprecedented increase in the construction of railways, roads, tunnels and bridges throughout Europe and North America (Radbruch-Hall 1987). There was also a growing demand for geological information on the soil and rock conditions below major cities in Europe, such as the 300 page study by Professor Suess of Vienna University on *The Foundation Materials and Soils of Vienna* (Suess 1862). It was during the latter part of this period that a few genuine engineering geologists and hydrogeologists started to emerge in publications, notably William Otis Crosby and James F. Kemp in the USA (Kiersch 1991*b*), Penning (1880) and Lucas (1880) in the UK, and Daubree (1887) in France.

(4) Twentieth century (1900–40): this period saw a massive expansion in the application of geology to engineering projects, particularly in North America. This came about both because of notable catastrophes (see the discussion in Kiersch and James 1991) such as the landslides that affected the construction of the Panama Canal (completed in 1914 by US engineers), and the influence of prominent engineers like Karl Tzerghai. Kiersch (1991*b*) suggested that the dominant factor behind the increased interest in geology in the USA was the government and state participation in irrigation, flood control and hydroelectric projects. In both North America and Europe, improvements in the internal combustion engine and automobile manufacture led to a massive expansion in road construction, such as the first superhighway through mountains from Genoa to the Po Plain (Maddalena 1935) and the German autobahns. Also, during World War I the German military applied geology to terrain classification and the construction of underground fortifications.

Modern practice of engineering geology: 1940 to the modern day

Kiersch (1991*c*) details the extensive developments in engineering geology that took place between the 1940s and the 1980s, and Kiersch and Hatheway (1991) review the work of the engineering geology division of the Geological Society of America over this time period. However, the stage for the start of this period was set by the book *Application of Geology to Engineering Practice: Berkey Volume* edited by Paige (1950*a*). This work was produced to celebrate the life and work of Dr Charles P. Berkey who was in the vanguard in emphasizing the critical role that geology had to play in engineering practice in the USA. In this volume Burwell and Roberts (1950, p. 4) identified the primary functions of the geologist in the engineering organization in seven categories that are still valid more than 70 years later:

(1) Those functions related to providing the background of general geological knowledge against which the whole plan of development should be studied and conceived.
(2) Those functions related to the selection of sites for the proposed engineering works.
(3) Those functions related to the discovery of the geological details that bear on the design and construction problems.
(4) Those functions related to finding suitable materials with which to build the required structures.
(5) Those functions related to the preparation of contract plans and specifications.
(6) Those functions related to construction operations.
(7) Those functions related to the effects of the completed works on the natural conditions of the locality.

The Berkey volume (Paige 1950*a*) whilst containing papers on geology in highway, dam and tunnel construction, has a much broader reach as it deals with the application of geology to the petroleum industry, mineral exploitation and the military. The volume also contains a paper by Karl Terzaghi on the mechanics of landslides.

Throughout the period since the 1940s there have been many developments in engineering geology, but the present state of the art is best demonstrated by the *Encyclopedia of Engineering Geology* edited by Bobrowsky and Marker (2018). With over 280 definitions, plus short and long papers by some 200 authors and co-authors, this volume, along with its extensive reference lists, presents a comprehensive review of engineering geology as it is practised in the twenty-first century. Many of the developments described in Bobrowsky and Marker (2018) that have taken place since the Berkey volume was published are in the techniques of site investigation, including engineering geological mapping (Culshaw 2018), remote sensing (Wasowski *et al.* 2018), geophysics (Tuckwell 2018), drilling (Erguler 2018), field testing (Erguler 2018) and instrumentation (Elwood 2018). However, Tepel (2018) also identifies the increasing importance of accreditation of the Earth science degrees offered by institutions of higher education. In addition, Allington (2018) emphasizes how the concept of professional practice along with registration and codes of conduct required by various professional bodies has now become central to practising as an engineering geologist throughout the world.

Engineering geology in the UK

The history of engineering geology in the UK was summarized in 2008 in a series of papers in Volume 41, Issue 2 of *Quarterly Journal of Engineering Geology and Hydrogeology*. This volume included Charman (2008), de Freitas and Rosenbaum (2008), Forster and Reeves (2008) and Reeves (2008). Griffiths (2019) provides a synopsis of the advances in engineering geology in the UK from 1950 (i.e. since the publication of the Berkey volume) up to 2018. As a follow-up to the Charman (2008) paper, Fookes and Lee (2017) documented Peter Fookes' personal views on the early days of engineering geology in the UK.

At the forefront of the development of engineering geology in the UK was a group from Imperial College during the 1950s and 1960s that included Sir Alec Skempton, Sir John Knill, Professor Evert Hoek, Professor John Hutchinson, Dr Mike de Freitas and Professor Peter Fookes, plus Professor William (Bill) Dearman at Newcastle University, supported by some leading figures from industry, notably Dr Rudolph Glossop, Dr Alan Weekes, Ian Higginbottom, Dr Graham West and Ken Early. From this talented group, it was, arguably, Peter Fookes, when he moved out from Imperial College into industry in 1971, who led the development of engineering geology as an applied subject in the UK and who ensured it was a discipline that drew on all aspects of geology.

Engineering geology and Professor Peter G. Fookes (1933–2020)

Peter's life in engineering geology is dealt with in detail in **Charman (2025)**. During his lifetime he worked in over 100 countries, produced some 200 publications, and helped to set up MSc programmes at Imperial College and Queen Mary College, London. From the Geological Society he was awarded the William Smith Medal (in 1985) and was the first recipient of the Glossop Medal in 1996. It was his Glossop Medal lecture (Fookes 1997) that set engineering geology along the path of developing ground models or 'geomodels' for use in understanding the ground conditions pertaining to engineering construction. Although a mainstream engineering geologist, his background in chemistry ensured that he was at the forefront in concrete research and the use of 'geomaterials' (see Smith and Collis 2001; Reeves *et al.* 2006). In addition, working with academic geomorphologists (specifically Professor Denys Brunsden, Professor David Jones, Dr John Doornkamp and Professor Sir Ron Cooke), he led the development of engineering geomorphology as a genuine adjunct subject to engineering geology (Brunsden *et al.* 1975; Lee and Fookes 2015). It is the development of geomodels, geomaterials and geomorphology that represent a key part of Peter Fookes' enduring legacy in engineering geology, and these form the core of this book.

The Fookes legacy – this volume

Ground models

Not surprisingly, both because of its importance to engineering geology and Peter Fookes' role in developing the framework for their creation, six of the papers in this volume deal with ground models. **Griffiths (2025)** explores how the concept of ground models has been adopted in the engineering geological profession following the First Glossop Lecture by Peter Fookes (Fookes 1997) and his subsequent work on the subject. This paper looks at different studies that have used the phrase 'Ground Model', notably the second iteration of Eurocode 7, the IAEG Commission 22 and the CIRIA Guidance on Geotechnical Baseline Reports. Given the different interpretations as to what constitutes a ground model, the paper emphasizes the importance of agreeing a definition in contract documents, as this may have legal ramifications.

Menzies (2025) opens up an interesting area, namely the use of ground models for tailings management, which was a subject not covered in Peter Fookes' Glossop Lecture. The paper refers to the 2020 Global Industry Standard on Tailings Management and its requirement for site characterization. The term 'site characterization' is used extensively in the USA (e.g. Tepel 2004) and has been argued as being broadly equivalent to the UK concept of 'ground model' (Griffiths and Stokes 2008). This paper provides a framework for tailings management through the complete lifespan of tailings and tailings storage facilities with the ground model at its core.

Hearn and Martin (2025) look back on Peter Fookes' work on mountain roads, which they regard as providing his longest lasting legacy. They start with his pioneering work on the Dharan to Dhankuta Road (DDR) in Nepal where he brought in a team of geomorphologists (see Brunsden *et al.* 1975) as he set about creating a geological model that accommodated the complexity and dynamic nature of the terrain in an intensely dissected, seismically active, mountain environment. They note that this provided the basis for Peter Fookes' *Total Geology Approach* (Fookes *et al.* 2000) that they subsequently used on a road across the Blue Nile gorge in Ethiopia.

Pavey *et al.* (2025) demonstrate how modelling has moved on since the 1990s by presenting a computerized three-dimensional geological model in support of an environmental safety case in the nuclear sector. They explain in detail how they were able to draw upon more than 60 years of legacy data, which were collected using different methods and different formats, to create a geological model to support the hydrogeological understanding and future development of a site.

Saroglou and Katsigiannis (2025) draw upon a range of techniques in the creation of a ground model for chalk, with an example from part of the UK's high-speed railway network (known as HS2) where it crossed the North Chilterns. In the case study they used geomorphological mapping, LiDAR surveys, and intrusive and non-intrusive ground investigations to understand the geological environment and the hazards associated with dissolution features in the chalk. A very interesting facet of this study was that they were able to compare the ground conditions exposed during construction with those predicted by the site investigation ground model.

Davis (2025) offers a personal perspective on how and why construction problems and disputes related to unforeseen ground conditions occur. He offers mitigation strategies that include structuring the geotechnical deliverables around the IAEG Commission 25 concept of Engineering Geological Models and an increased awareness of what the engineering geological input to construction projects should be.

Geomaterials

As described in **Charman (2025)**, Peter Fookes' work on geomaterials is normally associated with concrete and road aggregates, the use of clay in construction, borrow materials for earthworks, and rip-rap for coastal protection. However, **Hart (2025)** discusses his involvement with Peter Fookes on the investigation, design and construction of the Thames Barrier in London where use was made of biostratigraphy on the project. The paper describes how this led to the use of micropalaeontology as a key component of the site investigations for the Channel Tunnel, notably in connection with ensuring the tunnel alignment kept to a specific lithology within the chalk. Hart concludes with a view on the role that biostratigraphy can play in the development of ground models.

A particular facet of Peter Fookes' work on geomaterials involves the issues around sulfate-rich soils in deserts and their effects on engineering structures, as described by **Lee *et al.* (2025)** (see below). **Czerewko and Cripps (2025)** expand on this through an investigation on the way in which significant engineering problems may be encountered when landforms and geological materials in temperate and equatorial zones undergo volumetric and mineralogical changes following exposure to surface weathering environments. They provide guidance on the identification of potential problems and suggest ways of avoiding difficulties.

Berry *et al.* (2025) take the concept of geomaterials into a very different direction by presenting a technique for the visualization and semi-quantification of the geotechnical properties of soil and rock. The aim is to provide a visualization tool that communicates ground conditions and potential 'ground risk' to technical and non-technical audiences. Given that one of Peter Fookes' greatest assets was his ability to communicate with all parties on a project, this is exactly the type of development that would have interested him.

Geomorphology

Peter Fookes was not a trained geomorphologist but became a leader in its application to engineering and, through his contacts with many outstanding geomorphologists, developed into a highly skilled practitioner. This is best exemplified by his work in deserts, as described by **Lee *et al.* (2025)**. The paper starts with a description of his work on desert concrete technology but goes on to his leadership that opened the door to a whole generation of desert geomorphologists to work on the development boom in the Middle East. This included the Baharain Surface Materials Resources Survey (Brunsden *et al.* 1979), the Great Man-Made River project in Libya (Fookes *et al.* 1993; Fookes and Gahir 1994) and the Al Yamamah airfield in Saudi Arabia (Griffiths *et al.* 1994). **Lee *et al.* (2025)** note how, in later years, Peter worked on the geohazard challenges posed by routing major pipelines through the Saharan desert. The **Lee *et al.* (2025)** paper describes how, when he was awarded an Honorary Fellowship of the Royal Geographical Society in 2003, Sir Ron Cooke described Peter as a 'true father of engineering geomorphology'.

The legacy of Peter to geomorphology is covered in the references listed in **Charman (2025)** but an indication of the way that geomorphology has embraced his work is brought out in the papers by **Giles (2025)** and **Moses (2025)**.

Moses (2025) provides an insight into coastal geomorphology and the creation of the rock slope geomodel in the study of cliffs. This paper notes how Peter Fookes emphasized the role of climate as a weathering agent that influenced material properties and required consideration in engineering projects. It is recognized that a changing climate leading to increased winter salt and frost weathering, and summer salt and wetting–drying weathering contribute to more frequent coastal rockfalls, whilst increasing storminess and rising sea levels are enhancing marine and subaerial process. The need to improve monitoring, communicate coastal dynamics effectively to policy-makers, and understand the hazard, risk, and resilience and vulnerability of rock coasts is a key recommendation.

Giles (2025) provides case studies on the use of LiDAR for geomorphological investigations in relict periglacial terrain. The first case study is a re-examination of the seminal work on features described as 'solifluction lobes' that caused a realignment of the A21 Sevenoaks Bypass in Kent (Skempton and Weeks 1976), and the second deals with the cambering, gulls and valley budges in the Maidstone district. Using LiDAR, **Giles (2025)** has been able to reassess the nature of the features and demonstrate the efficacy of LiDAR as a rapid geomorphological assessment tool.

Endpiece

The final paper in the volume is by **Shilston *et al.* (2025)**. In this paper the nature of Peter Fookes' legacy is reflected through the experiences of scientists and engineers who worked with him. This paper emphasizes his concentration on the publication of case studies, and how Peter Fookes' ground model concept espoused in his 1997 Glossop Lecture can, and has been, used to accommodate landscape changes that may occur during the lifetime of a project.

Conclusions

Professor Peter G. Fookes was a man of his time and throughout the latter part of the twentieth and the first two decades of the twenty-first centuries he held a unique position in UK engineering geology. The breadth of his knowledge in the Earth sciences, civil engineering and concrete technology allows us, perhaps, to label him a polymath. However, his greatest achievement was to be an inspiration to a whole generation of engineering geologists, particularly those of us who had the privilege to work with him on site. We will all remember his humour, field sketches and dedication. When asked what his favourite project was, he would always reply, 'the one I am working on at the moment'. He had a barrister's ability to absorb enormous amounts of information on a project and was able to ask those questions about the ground conditions that we had either ignored or did not realize needed asking. Long may his legacy continue.

Acknowledgements This volume is dedicated to Professor Peter G. Fookes. All those who worked with him, or have read his published work, recognize the debt that they owe him. Thanks go to Randell A. Stephenson for constructive comments during the reviewing process.

Competing interests The author declares that they have no known competing financial interests or personal relationships that could have appeared to influence the work reported in this paper.

Author contributions **JSG**: writing – original draft (lead).

Funding This research received no specific grant from any funding agency in the public, commercial, or not-for-profit sectors.

Data availability Data sharing is not applicable to this article as no datasets were generated or analysed during the current study.

References

Allington, R. 2018. Professional practice. *In*: Bobrowsky, P.T. and Marker, B. (eds) *Encyclopedia of Engineering Geology*. Springer, Cham, Switzerland, 737–738.

Berry, T., Barratt, E. and Paraskevopoulou, C. 2025. A semi-quantitative geotechnical visualization tool for ground engineering projects. *Geological Society, London, Engineering Geology Special Publications*, **30**, https://doi.org/10.1144/egsp30-2024-2

Bobrowsky, P.T. and Marker, B. (eds). 2018. *Encyclopedia of Engineering Geology*. Springer, Cham, Switzerland.

Brunsden, D., Doornkamp, J.C., Fookes, P.G., Jones, D.K.C. and Kelly, J.M.H. 1975. Large-scale geomorphological mapping and highway engineering design. *Quarterly Journal of Engineering Geology*, **8**, 227–253, https://doi.org/10.1144/GSL.QJEG.1975.008.04.01

Brunsden, D., Doornkamp, J.C. and Jones, D.K.C. 1979. The Bahrain Surface Materials Resources Survey and its application to regional planning. *The Geographical Journal*, **145**, 1–35, https://doi.org/10.2307/633073

Burwell, E.B. and Roberts, G.D. 1950. The geologist in the engineering organisation. *In*: Paige, S. (ed.) *Application of Geology to Engineering Practice: Berkey Volume*. Geological Society of America, New York, 1–10.

Charman, J.H. 2008. Peter George Fookes: father of modern British engineering geology? *Quarterly Journal of Engineering Geology*

and Hydrogeology, **41**, 201–216, https://doi.org/10.1144/1470-9236/07-061
Charman, J.H. 2025. The life and times of Peter George Fookes: 1933–2020. *Geological Society, London, Engineering Geology Special Publications*, **30**, https://doi.org/10.1144/egsp30-2024-4
Clements, T. 1981. Leonardo da Vinci as a geologist. *In*: Rhodes, F.H.T. and Stone, R.O. (eds) *Language of the Earth*. Pergamon Press, New York, 310–314.
Culshaw, M.G. 2018. Engineering geological maps. *In*: Bobrowsky, P.T. and Marker, B. (eds) *Encyclopedia of Engineering Geology*. Springer, Cham, Switzerland, 265–277.
Culshaw, M.G., Reeves, H.J. and Rosenbaum, M. 2008. Two hundred years of engineering geology. *Quarterly Journal of Engineering Geology and Hydrogeology*, **41**, 137–142, https://doi.org/10.1144/1470-9236/07-301
Czerewko, M. and Cripps, J.C. 2025. Implications for ground engineering through deterioration of landforms and geological material caused by atmospheric reactions involving sulfide and sulfate materials. *Geological Society, London, Engineering Geology Special Publications*, **30**, https://doi.org/10.1144/egsp30-2024-22
Darwin, C.R. 1859. *On the Origin of Species by Means of Natural Selection, or the Preservation of Favoured Races in the Struggle for Life*. John Murray, London.
Daubree, A. 1887. *Les eaux souterraines, aux epoques anciennes et a l'epoques actuelle*. Dunod, Paris (3 vols).
Davis, J.A. 2025. Desk studies, geological models and avoiding construction disputes – your records will be used against you. *Geological Society, London, Engineering Geology Special Publications*, **30**, https://doi.org/10.1144/egsp30-2025-2
de Freitas, M. and Rosenbaum, M.S. 2008. Engineering geology at Imperial College London; 1907–2007. *Quarterly Journal of Engineering Geology and Hydrogeology*, **41**, 223–228, https://doi.org/10.1144/1470-9236/07-060
Elwood, D.E.Y. 2018. Instrumentation. *In*: Bobrowsky, P.T. and Marker, B. (eds) *Encyclopedia of Engineering Geology*. Springer, Cham, Switzerland, 518–534.
Erguler, Z.A. 2018. Field testing. *In*: Bobrowsky, P.T. and Marker, B. (eds) *Encyclopedia of Engineering Geology*. Springer, Cham, Switzerland, 336–343.
Faul, H. and Faul, C. 1983. *It Began with a Stone; History of Geology from Stone Age to the Age of Plate Tectonics*. John Wiley & Sons, New York.
Fookes, P.G. 1997. The First Glossop Lecture: Geology for engineers: the geological model, prediction, and performance. *Quarterly Journal of Engineering Geology*, **30**, 293–424, https://doi.org/10.1144/GSL.QJEG.1997.030.P4.02
Fookes, P.G. and Gahir, J.S. 1994. Geotechnical characteristics of some carbonate soils. *In*: Fookes, P.G. and Parry, R.H.G. (eds) *Engineering Characteristics of Arid Soils*. Balkema, Rotterdam, The Netherlands, 313–329.
Fookes, P.G. and Lee, E.M. 2017. The early days of engineering geology in the UK – a personal view by Peter Fookes F.R. Eng. *Quarterly Journal of Engineering Geology and Hydrogeology*, **50**, 472–477, https://doi.org/10.1144/qjegh2017-102
Fookes, P.G., Stoner, J.R. and Mackintosh, J. 1993. Great Man Made River Project, Libya, Phase I: a case study on the influence of climate and geology on concrete technology. *Quarterly Journal of Engineering Geology*, **26**, 25–60, https://doi.org/10.1144/GSL.QJEG.1993.026.01.04
Fookes, P.G., Baynes, F.J. and Hutchinson, J.N. 2000. Total geological history: a model approach to the anticipation, observation and understanding of site conditions. *In*: *Proceedings of GeoEng 2000, An International Conference on Geotechnical and Geological Engineering, 19–24 November, Volume 1*, Technomic Publishing, Lancaster, PA, 370–460.
Forster, A. and Reeves, H.J. 2008. William Smith and the development of engineering geology in England. *Quarterly Journal of Engineering Geology and Hydrogeology*, **41**, 165–170, https://doi.org/10.1144/1470-9236/07-063
Giles, D.P. 2025. LiDAR for rapid geomorphological assessment in relict periglacial landsystems. *Geological Society, London, Engineering Geology Special Publications*, **30**, https://doi.org/10.1144/egsp30-2024-46
Griffiths, J.S. 2014. Feet on the ground: engineering geology past, present and future. *Quarterly Journal of Engineering Geology and Hydrogeology*, **47**, 116–143, https://doi.org/10.1144/qjegh2013-087
Griffiths, J.S. 2019. Advances in engineering geology in the UK 1950–2018. *Quarterly Journal of Engineering Geology and Hydrogeology*, **52**, 401–413, https://doi.org/10.1144/qjegh2018-171
Griffiths, J.S. 2025. What do we mean by a ground model – the Fookes geomodel legacy. *Geological Society, London, Engineering Geology Special Publications*, **30**, https://doi.org/10.1144/egsp30-2023-11
Griffiths, J.S. and Stokes, M. 2008. Engineering geomorphological input to ground models: an approach based on Earth systems. *Quarterly Journal of Engineering Geology and Hydrogeology*, **41**, 73–91, https://doi.org/10.1144/1470-9236/07-010
Griffiths, J.S., Fookes, P.G., Hardingham, A.D. and Barsby, R.D. 1994. Geotechnical soils mapping for construction purposes in Central Saudi Arabia. *In*: Fookes, P.G. and Parry, R.H.G. (eds) *Engineering Characteristics of Arid Soils*. A.A. Balkema, Rotterdam The Netherlands, 69–85.
Hart, M.B. 2025. The value of biostratigraphy in site investigation and its application to the development of 'ground models' (*sensu* Peter Fookes). *Geological Society, London, Engineering Geology Special Publications*, **30**, https://doi.org/10.1144/egsp30-2024-3
Hearn, G.J. and Martin, R.P. 2025. Mountain road engineering: the legacy of Peter Fookes across the continents. *Geological Society, London, Engineering Geology Special Publications*, **30**, https://doi.org/10.1144/egsp30-2024-20
Hutton, J. 1788. Theory of the Earth; or an investigation of the laws observable in the composition, dissolution, and restoration of land upon the globe. *Transactions of the Royal Society of Edinburgh*, **1**, 209–304, https://doi.org/10.1017/S0080456800029227
Kiersch, G.A. (ed.) 1991*a*. *The Heritage of Engineering Geology; the First Hundred Years*. Geological Society of America Centennial Special Volume, **3**.
Kiersch, G.A. 1991*b*. The heritage of engineering geology: changes through time. *Geological Society of America Centennial Special Volume*, **3**, 1–50.
Kiersch, G.A. 1991*c*. Modern practice, training, and academic endeavors, 1940s to 1980s. *Geological Society of America Centennial Special Volume*, **3**, 51–86.
Kiersch, G.A. and Hatheway, A.W. 1991. History and heritage of the engineering geology division, Geological Society of America 1940 to 1990. *Geological Society of America Centennial Special Volume*, **3**, 109–128.
Kiersch, G.A. and James, L.B. 1991. Errors of geologic judgment and the impact on engineering works. *Geological Society of America Centennial Special Volume*, **3**, 517–558.

Lee, E.M. and Fookes, P.G. 2015. A note on the origins of engineering geomorphology in the UK. *Quarterly Journal of Engineering Geology and Hydrogeology*, **48**, 147–156, https://doi.org/10.1144/qjegh2014-048

Lee, E.M., Sweeney, M. and Clarke, J. 2025. Arabia Felix: Fookes in the desert. *Geological Society, London, Engineering Geology Special Publications*, **30**, https://doi.org/10.1144/egsp30-2023-33

Lucas, J. 1880. The hydrogeology of the lower greensands of Surrey and Hampshire. *Minutes of the Proceedings of the Institute of Civil Engineers*, **61**, 200–227, https://doi.org/10.1680/imotp.1880.22114

Lyell, C. 1830. *Principles of Geology: Being an Attempt to Explain the Former Changes to the Earth's Surface by Reference to Causes Now in Operation, Volume 1*. John Murray, London.

Lyell, C. 1832. *Principles of Geology: Being an Attempt to Explain the Former Changes to the Earth's Surface by Reference to Causes Now in Operation, Volume 2*. John Murray, London.

Lyell, C. 1833. *Principles of Geology: Being an Attempt to Explain the Former Changes to the Earth's Surface by Reference to Causes Now in Operation, Volume 3*. John Murray, London.

Maddalena, I. 1935. Geologie appliquee aux autostrades modernes. *Congres International des Mines, de la Métallurgie et de la Geologie Appliquee*, **2**, 569–573.

Menzies, T.J. 2025. Ground models for tailings management. *Geological Society, London. Engineering Geology Special Publications*, **30**, https://doi.org/10.1144/egsp30-2024-21

Moses, C.A. 2025. Rock coast geomodel: process-response dynamics, resilience and vulnerability. *Geological Society, London, Engineering Geology Special Publications*, **30**, https://doi.org/10.1144/egsp30-2023-32

Paige, S. (ed.) 1950*a*. *Application of Geology to Engineering Practice: Berkey Volume*. Geological Society of America, New York.

Paige, S. 1950*b*. Introduction. *In*: Paige, S. (ed.) *Application of Geology to Engineering Practice: Berkey Volume*. Geological Society of America, New York, v–vii.

Pavey, A., Timms, R., Hampson, M., Jones, N., Shevelan, J. and Coleman, C. 2025. 3D geological modelling to support an environmental safety case in the nuclear sector. *Geological Society, London, Engineering Geology Special Publications*, **30**, https://doi.org/10.1144/egsp30-2024-19

Penning, W.H. 1880. *Engineering Geology*. Ballière, Tindall and Cox, London.

Radbruch-Hall, D.H. 1987. The role of engineering – geologic factors in the early settlement and expansions of the conterminous United States. *Bulletin of the International Association of Engineering Geology*, **35**, 9–30, https://doi.org/10.1007/BF02590473

Redlich, K.A., Terzaghi, K. and Kampe, R. 1929. *Ingenieurgeologie*. Springer, Vienna.

Reeves, G.M. 2008. William Robert Dearman: Britain's first professor of engineering geology. *Quarterly Journal of Engineering Geology and Hydrogeology*, **41**, 217–221, https://doi.org/10.1144/1470-9236/07/076

Reeves, G.M., Sims, I. and Cripps, J.C. 2006. *Clay Materials Used in Construction*. Geological Society, London, Engineering Geology Special Publications, **21**, https://doi.org/10.1144/GSL.ENG.2006.021.01.21

Saroglou, C. and Katsigiannis, G. 2025. Understanding the geological model in a chalk environment: an example from the High Speed Two (HS2) railway. *Geological Society, London, Engineering Geology Special Publications*, **30**, https://doi.org/10.1144/egsp30-2024-12

Shilston, D.T., Lee, E.M. and Griffiths, J.S. 2025. Endpiece: reflections on Peter Fookes' legacy. *Geological Society, London, Engineering Geology Special Publications*, **30**, https://doi.org/10.1144/egsp30-2024-18

Skempton, A.W. and Weeks, A.G. 1976. The Quaternary history of the Lower Greensand Escarpment and Weald Clay vale near Sevenoaks, Kent. *Philosophical Transactions of the Royal Society Series A, Mathematical and Physical Sciences*, **A283**, 493–526, https://doi.org/10.1098/rsta.1976.0091

Smith, M.R. and Collis, L. (eds) 2001. *Aggregates: Sand, Gravel, and Crushed Rock Aggregates for Construction Purposes*, 3rd edn (revised by Fookes, P.G., Lay, J., Sims, I., Smith, M.R. and West, G.). Geological Society Engineering Geology, London, Special Publications, **17**, https://doi.org/10.1144/GSL.ENG.2001.017.01.00

Suess, E. 1862. *Der Boden der Stadt Wien nach seiner Bildungsweise, Beschaffenheit und seinen Beziehungen zum bürgerlichen Leben: Eine geologische Studie*. Wilhelm Braumuller, Vienna.

Tepel, R.E. 2004. The engineering geology site characterization as the core function of engineering geology practice. *In*: Tepel, R.E. (ed.) *Visioning the Future of Engineering Geology: Site Characterisation, Volume 3*. Association of Engineering Geologists Special Publications, **19** [CD-ROM].

Tepel, R.E. 2018. Accreditation. *In*: Bobrowsky, P.T. and Marker, B. (eds) *Encyclopedia of Engineering Geology*. Springer, Cham, Switzerland, 4–5.

Tuckwell, G.W. 2018. Geophysical methods. *In*: Bobrowsky, P.T. and Marker, B. (eds) *Encyclopedia of Engineering Geology*. Springer, Cham, Switzerland, 398–406.

Turner, A.K. 2008. The historical record as a basis for assessing interactions between geology and civil engineering. *Quarterly Journal of Engineering Geology and Hydrogeology*, **41**, 143–164, https://doi.org/10.1144/1470-9236/07/064

VanDine, F.F., Nasmith, H.W. and Ripley, C.F. 2006. The emergence of engineering geology in British Columbia: 'An engineering geologist knows a dam site better'. British Columbia Geological Survey Open File, **1992-19**.

Wasowski, J., Giordan, D. and Singhroy, V. 2018. Remote sensing. *In*: Bobrowsky, P.T. and Marker, B. (eds) *Encyclopedia of Engineering Geology*. Springer, Cham, Switzerland, 743–746.

Winchester, S. 2001. *The Map that Changed the World*. Penguin Group, London.

Zaruba, Q. and Mencl, V. 1963. *Engineering Geology*. Developments in Geotechnical Engineering, **10**. Elsevier, Amsterdam.

The life and times of Peter George Fookes: 1933–2020

John Charman

Maltsters, 32 Long Street, Cerne Abbas, Dorchester DT2 7JG, UK
0009-0006-0266-2050
Correspondence: rockjohn@btinternet.com

Abstract: This paper is a personal review of the life and career of Peter Fookes, probably the most influential engineering geologist of his time. It includes personal recollections and a review of his 200 technical publications, his professional career and his influence on how geotechnology is viewed and used by the civil engineering profession. Perhaps his greatest epitaph is the development and use of the ground model.

I feel very lucky to have known Peter, first as his student, then as a working colleague from time to time throughout my career, and also as a friend. I can't think of anyone in the field of engineering geology who did so much to help develop the subject. He worked tirelessly throughout his career to educate and publicize the benefits of geology and geomorphology to the engineering and construction industry. His charisma and enthusiasm enabled him to network continuously, bringing together academics, professional institutions, commercial organizations and his students, and he was always spot on in knowing at what level to pitch his interactions, either as a team member or in presentations to clients. As well as running his own consulting practice, which was in continual demand, he managed to publish some 200 academic papers and several books, and he found time to lecture and run field courses. Although clearly an accomplished academic and researcher, he placed equal – if not more – emphasis on fieldwork (Fig. 1).

Peter Fookes' career began in the 1950s, not in geology, but in chemistry. This soon changed, and the world of applied geology has much to be thankful for. When he began there was a loosely coined 'geology for engineers' but no formal career structure in engineering geology. Later there was a move towards practitioners specifically trained in understanding the ground problems that arise in civil engineering and in the techniques of ground description, classification and investigation for civil engineering development. Peter played a leading role in developing the first engineering geology master's degree course at Imperial College London, helping to establish the subject as an essential component of civil engineering. He was also heavily involved in championing the use of engineering geomorphology, less well known to geologists, and in developing aspects of concrete technology. In addition, he made major contributions to the science, especially focusing on the development of geological models. He published more than 200 technical papers (including some 70 after what many would consider to be a normal retirement age!) and many books, and he participated in numerous national and international committees. This continued virtually up to his death. A full chronological list of his publications is given in the Bibliography. He was a Chairman of the Engineering Group and initiated and chaired many working party reports in the UK and overseas, which were often milestones in the advancement of the subject. He was the recipient of major awards in recognition of his contribution to engineering geology, applied geomorphology, civil engineering, concrete technology and geotechnical engineering.

Throughout his career, he preferred the role of practitioner to becoming involved in geopolitics. This paper summarizes his considerable contribution to the science and the profession.

Career

Peter Fookes was born in 1933 near Fingringhoe on the Essex–Suffolk border, a family stronghold for several centuries. By the time of his secondary schooling, during the 1939–1945 war, his family had moved to Reigate, Surrey, and he gained a scholarship to Reigate Grammar School before continuing his studies at Croydon Polytechnic for the Higher Schools Certificate in chemistry, physics and botany. He then worked as a technician in the geology department at Chelsea Polytechnic, London, while studying part-time for an A-level in geology under the supervision of A. J. (Tony) Barber. Sometime during this earlier period, he damaged his left hand in a laboratory chemical implosion. He says this event diminished his interest in chemistry and turned him to geology, but chemistry clearly had an influence on his later work; for example, on karst, tropical soils and concrete in the Middle East.

In 1957 he won a scholarship to study geology at Queen Mary College, London, under Professor J. F. (John) Kirkaldy,

From: Griffiths, J. S., Giles, D. P., Davis, J. A. and Dunlop, J. J. (eds) 2025. *The Peter Fookes Engineering Geological Legacy in Geomodels, Geomaterials and Geomorphology*. Geological Society, London, Engineering Geology Special Publications, **30**, 9–29,
First published online 12 February 2025, https://doi.org/10.1144/egsp30-2024-4

(a)

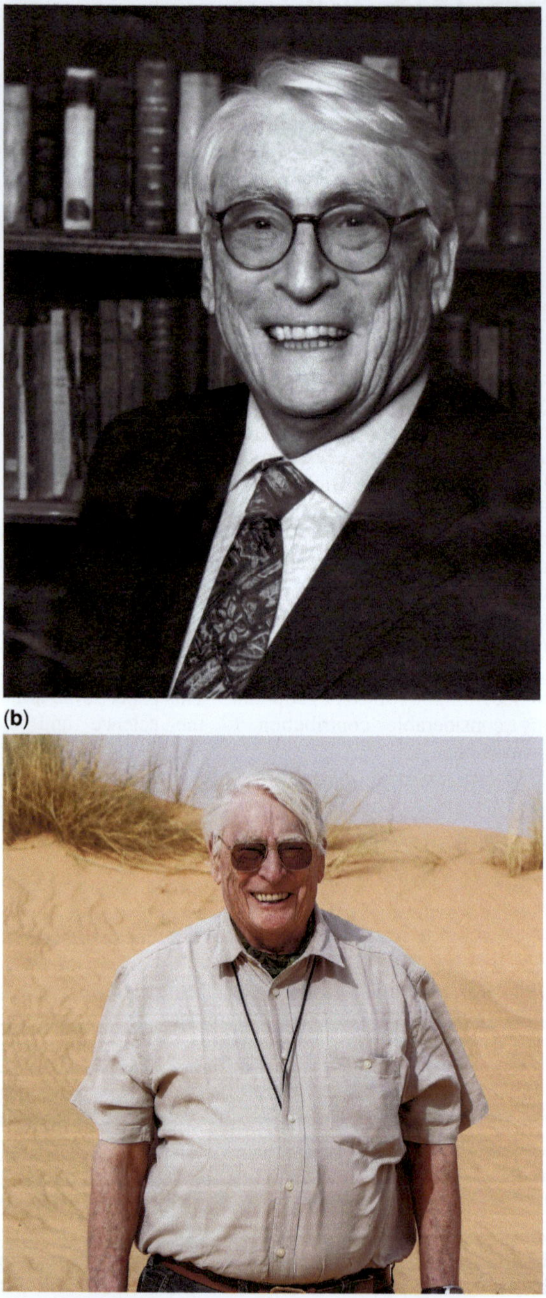

(b)

Fig. 1. Peter Fookes in (**a**) the office and (**b**) the field. Source: Geological Society and Mark Lee.

and he graduated with Special Honours in mineralogy and petrology. On graduating, he spent a short period with a sand and gravel company before joining Binnie, Deacon and Gourley (now part of the RSK Group). At Binnie, Head of Soil Mechanics A. L. (Alan) Little sent Peter for a brief period of practical training at the commercial laboratory of Soil Mechanics Limited in Chelsea. Later Peter was posted as a field geologist to West Pakistan, working on the Mangla, Jari and Sukian earth dams, which are still among the world's largest. These were formative years in terms of the influence of engineering on Peter's geological training. Perhaps the greatest influence on his career (described by Peter as one of his 'Eureka' moments) was when Professor Sir A. W. (Alec) Skempton, consultant to Binnie on the aforementioned projects, convinced Peter to start an external PhD at Queen Mary College, London, on the shear strength characteristics of the overconsolidated Siwalik clay (Fookes 1967*a*), under Skempton's supervision.

In 1965, with Skempton's encouragement, Peter accepted the offer of a lecturing post from J. L. (John) Knill, who was developing an engineering geology group at Imperial College. Later, John and Peter, together with M. (Mike) de Freitas, built up the world's first MSc Engineering Geology course. He was the only one who possessed the commercial and field experience to provide the necessary practical aspects of the course. The influence of this group at that time on the development of engineering geology as a discipline cannot be overestimated. As well as the master's course, Peter's teaching extended to undergraduate civil and mining engineers and undergraduate geologists.

In that heady period of the late 1960s, much was happening in the geotechnical world at Imperial College. E. Hoek was developing a rock mechanics group, and the soil mechanics group was at the forefront of the subject, with research being developed by Skempton and A. W. Bishop, and later also with R. J. Chandler, D. J. Henkel, J. N. Hutchinson, N. Morgenstern, A. Skinner and P. R. Vaughan. The engineering geology master's course was attended by W. R. (Bill) Dearman and T. P. (Peter) Crimes as mature students before they started their own courses at the Universities of Newcastle and Liverpool, respectively.

It was a great experience for me to be a student of that period (Text Box 1), and I particularly remember the practical approach that Peter brought to the teaching. This included many case studies from his direct experience, detailed worksheets and workshops in industry, such as a period of core logging in the Wimpey Laboratories core store with I. E. (Ian) Higginbottom and D. (David) Price.

Many of Peter's lecture notes from that period were indicative of his underlying 'total geology' approach (which he said he learnt from his work on the Mangla Dam), and they were continually honed and developed to appear in later publications. Within this approach he emphasized in his teaching the importance of the interrelationships between the various geotechnical disciplines; Figure 2, reproduced from the Glossop Lecture (Fookes 1997*a*), is typical of a continually developed theme that was initiated in these earlier years. He was also interested in developing the philosophy of site investigation and was one of the first in the UK to advocate a staged approach (Fookes 1967*b*).

Text Box 1. Personal recollection

In 1968 I was studying geology at Imperial College, but the careers focus in geology at that time was heavily on academia, mining or oil, none of which particularly attracted me. Towards the end of my second academic year, I was interested to hear about the rather new engineering geology that was being offered as a third-year specialization. I knocked on Peter's door to find out a little more about it. That was my first meeting with him, and my immediate impression was of his enthusiasm for the subject and his apparent delight that I had made the effort to seek him out. The other impression was his upper row of false teeth, which he constantly pushed in and out of his mouth with his tongue! That started my career in engineering geology and my association with Peter, and I graduated the following year with a geology degree and a specialization in engineering geology. This was followed by researching for a PhD under Peter's supervision.

Peter often dished out little paid jobs or errands to his PhD students. I can remember an occasion when he had booked one of the departmental Land Rovers and asked me and a technician to drive up to a Derbyshire quarry to collect some samples of Darley Dale sandstone. 'Some samples' turned out to be about half a tonne of stone. It was a windy day of gale force winds, and the soft top of the Land Rover was in a bad state. While driving on the M1, an existing rip developed quickly and half of the soft top disappeared somewhere north of Watford. Peter made light of this and said he would sort it out with the departmental administrator. Nevertheless, I was summoned to explain my version of events. My recollection of the administrator is of a steely-eyed senior lecturer in palaeontology with little apparent sense of humour. He was surprised when I said the cover was lost somewhere north of London, and he was less than impressed with Peter, who had booked the vehicle to move an exhibit from Imperial to Chelsea College for a Geologists' Association open day (a 1-mile journey)!

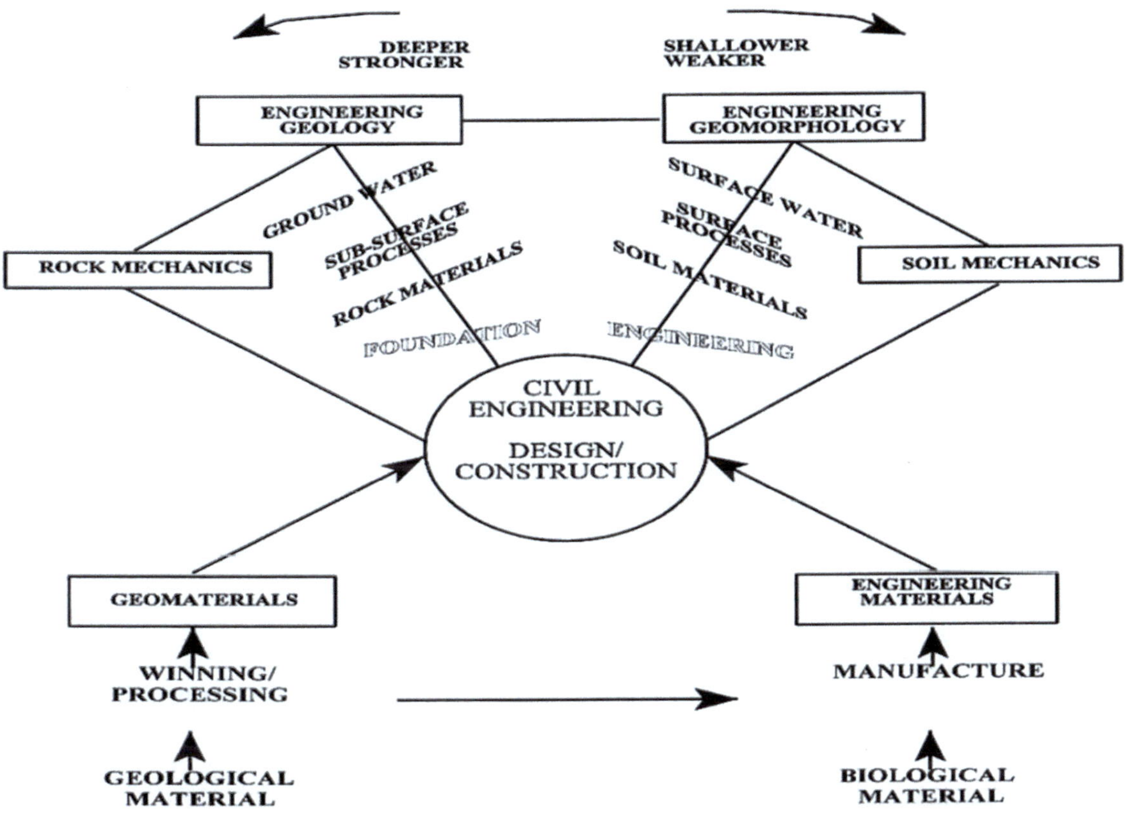

Fig. 2. A schematic relationship between civil engineering, geotechnology and geomaterials. Source: Fookes (1997*a*).

Peter was a prolific writer, and up to 1998, at what would be considered a perfectly respectable retirement age, he had published some 135 papers and five books. After 'retirement' and up to his death he published nearly 70 more, averaging about three per year. His papers on the geotechnical aspects of engineering geology have twice been recognized by the British Geotechnical Society Prize, in 1981 and 2000, and the Australian Engineering Excellence Award in 2004. What distinguishes his work is a highly practical approach and full use of graphics. Almost without fail his publications are both essential references and eminently readable. Examples of landmark book publications are *Geomorphology for Engineers* with E. M. (Mark) Lee and G. (George) Milligan (Fookes *et al.* 2005*b*) and *Engineering Geomorphology, Theory and Practice* with J. S. (Jim) Griffiths and E. M. Lee (Fookes *et al.* 2007).

Peter left teaching at Imperial College in 1971 to begin his career as an independent consultant; he continued until only a few years before his death. During this period, he continued to be a prolific researcher, based on his consulting work, and he published countless papers, many of them keynote. His consulting work was more or less equally divided between consultants and contractors; his appeal came from the practical nature of his advice. Work took him to more than 90 countries: he once estimated that 80% of his projects had been overseas. He worked on literally hundreds of projects in the UK and around the world; for example, the Thames Barrier in the UK, roads and bridges in Bangladesh, Malaysia and Yemen, dams in Indonesia, railways in Australia and mining in Papua New Guinea. In the last 30 years or so he was retained increasingly as a review consultant on many long-term projects, including the Great Man-Made River Project in Libya and mining infrastructure in Australia. More recently he was involved with the Baku–Tbilisi–Ceyhan oil pipeline through Azerbaijan, Georgia and Turkey, and other major pipelines; for example, in Algeria, Egypt, Kovykta and Sakhalin (Siberia) and Papua New Guinea.

Over the years following his departure from Imperial College, he pioneered the application of geology to civil engineering and, using his initial chemistry background, the influence of desert processes and materials on concrete durability. He also championed the role of geomorphology and was a great supporter of the use of geomorphology in engineering projects. When he was awarded an Honorary Fellowship of the Royal Geographical Society in 2005, he was affectionately called the 'father of engineering geomorphology' in the citation by Professor Ron Cooke.

Alongside his busy commercial consultancy, Peter never lost his links to academia, continuing to lecture, lead field courses and initiate groundbreaking research while playing a leading role in the Engineering Group. But it was his busy commercial life that allowed him to pursue the practical research that proved so valuable to the advancement of the profession.

In this book several themes have been chosen to reflect the areas in which he had the most influence in his career. For this paper I have summarized them as:

- the application of geology and geomorphology to civil engineering;
- the role of geomaterials, particularly in the hot desert environment, and their effect on concrete durability;
- and geomodels: his career development of the concept of 'total geology' and the ground model.

Each of these themes will be examined in terms of his publications and academic and industry recognition. He always referred to himself as a generalist, and his publications reflect this. There is arguably no one in this field who can boast so many publications over such a range of geological applications and, because there are so many, they provide an ideal database to allow for some analysis.

Figure 3 shows all of his publications over his 60-year career from 1958 to 2019 and indicates the number of publications per year. It is remarkable that the number of fallow years can be counted on the fingers of one hand and that he published on average more than three publications per year. These publications can be examined in more detail, grouped loosely into the conference themes.

The application of geology and geomorphology to civil engineering

Figure 4 shows my assessment of those publications that cover the role of geological or geomorphological techniques applied to civil engineering. In the figure these are differentiated by colour between engineering geology/geotechnical engineering (below ground) and engineering geomorphology/surface processes and weathering (ground surface). These have been illustrated on the same figure because I have always been of the opinion that geology and geomorphology should have been taught together within the same university department. They were not, but Peter tirelessly championed both geology and geomorphology and sold the benefits of both disciplines to the engineering profession.

Again, it is worth noting that his publications on these themes were spread across the whole of his 60-year career, and it is particularly interesting that his first paper had a distinct geomorphological emphasis. Published in 1958 in *Cosmos* magazine, it was entitled 'Pleistocene sea levels'. Table 1 shows some of what I believe to be his keynote publications in engineering geology and geomorphology.

The link between environment and the engineering characteristics of soils and rocks was a theme that continued throughout Peter's career. From his work in Nepal and the Far East in the late 1970s and 1980s he researched the effects of tropical climates on engineering properties (e.g. Fookes and Marsh 1981*a*, *b*; Fookes 1983*a*, *b*, *c*, *d*, *e*, 1986*a*, *b*, *c*). The value of geomorphology and surface processes was gleaned from interaction with colleagues such as D. (Denys) Brunsden, R. U. (Ron) Cooke, J. C. (John) Doornkamp and D. K. (David) Jones when they worked together from the early 1970s on the Taff Vale trunk road in Wales, the Bahrain Surface Materials Resources Survey

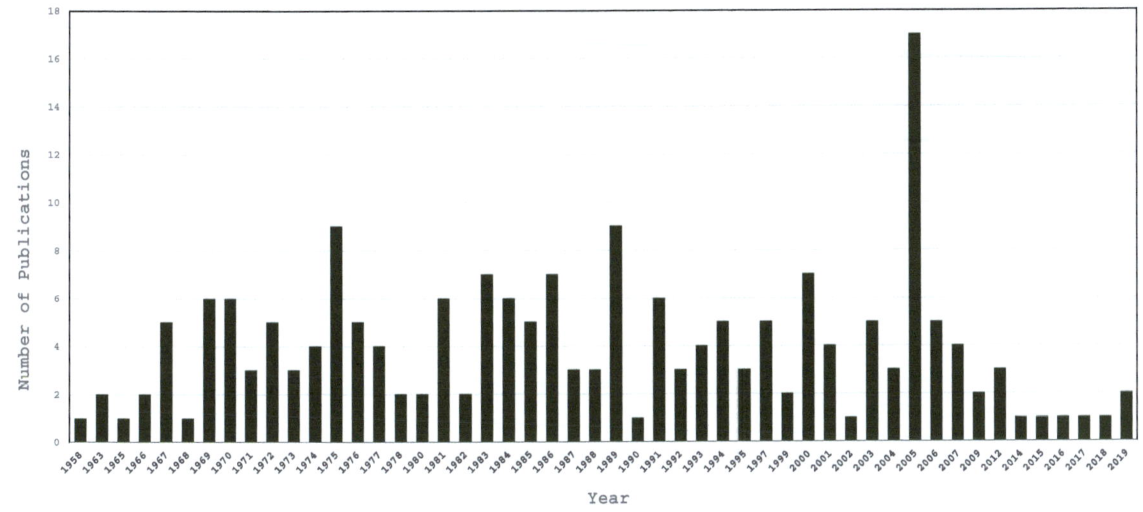

Fig. 3. Number of publications by year.

(Text Box 2) and the Dharan–Dhankuta Road Project in the mountainous environment of eastern Nepal. Authoritative publications inevitably followed (Brunsden *et al.* 1975*a*, *b*; Fookes *et al.* 1985*b*). This work in arid, mountainous, tropical environments and others was drawn together in the publication *A Handbook of Engineering Geomorphology* (Fookes and Vaughan 1986). As far as I am aware this was the first book on the subject and copies are now extremely rare.

On reading these publications in chronological order, it is evident that there is a thread running through them that illustrates Peter's developing thinking around the concept of a ground model. In the early years he was focused on taking a planned approach to ground investigation from an initial literature review, through reconnaissance and mapping of the site to the actual investigation itself. He always maintained that with this approach ground conditions should rarely be unforeseeable. That unforeseen conditions were commonly

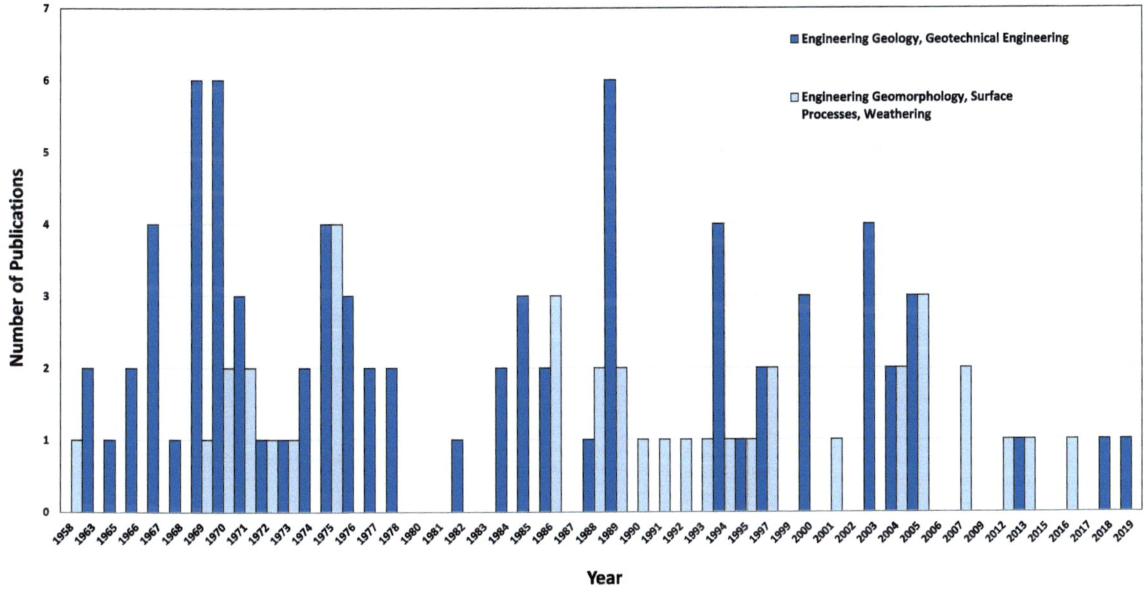

Fig. 4. Publications in engineering geology and geomorphology.

Table 1. Key publications in engineering geology and geomorphology

Date	Title
1958	Pleistocene sea levels. Fookes P. G.
1969	The application of engineering geology in the regional development of north and central Iran. Fookes P. G. and Knill J. L.
1974	A regional engineering geological study of the London Clay in the London and Hampshire Basins. Burnett A. D. and Fookes P. G.
1975	Large scale geomorphological mapping and highway engineering design. Brunsden D., Doornkamp J. C., Fookes P. G., Jones D. K. C. and Kelly J. M. H.
1975	Geomorphological mapping techniques in highway engineering. Brunsden D., Doornkamp J. C., Fookes P. G., Jones D. K. C. and Kelly J. M. H.
1975	Glacial landforms, their deposits and engineering characteristics. Fookes P. G., Gordon D. L. and Higginbottom I. E.
1985	Geological and geotechnical engineering aspects of low-cost roads in mountainous terrain. Fookes P. G., Sweeney M., Manby C. N. D. and Martin R. P.
1986	*A Handbook of Engineering Geomorphology.* Fookes P. G. and Vaughan P. R. (eds)
1990	Tropical Residual Soils. EGGS Working Party. Fookes P. G. (Chairman) *et al.*
2003	Engineering classification of karst ground conditions. Waltham A. C. and Fookes P. G.
2005	*Geomorphology for Engineers*. Fookes P. G., Lee E. M. and Milligan G. (eds)
2007	*Engineering Geomorphology: Theory and Practice.* Fookes P. G., Lee E. M. and Griffiths J. S.

revealed during construction was a result of cost and contractual constraints.

In later papers he includes regional reviews, reviews of different environments and different applications, and he always has a clear focus on the influential factors and their application to engineering behaviour. So, these publications were all part of a developing 'total geology' picture. This was all heading towards the development of a ground model, which he first described in detail in his Glossop Lecture (Fookes 1997*a*).

Geomaterials, concrete and hot deserts

In the 1970s and early 1980s Peter's career was largely influenced by the burgeoning development in the Middle East. He worked on a range of civil engineering projects, such as roads, tunnels, dams and harbours, and the Middle East boom provided some 40% of his workload at that time. Peter became significantly involved in applying his

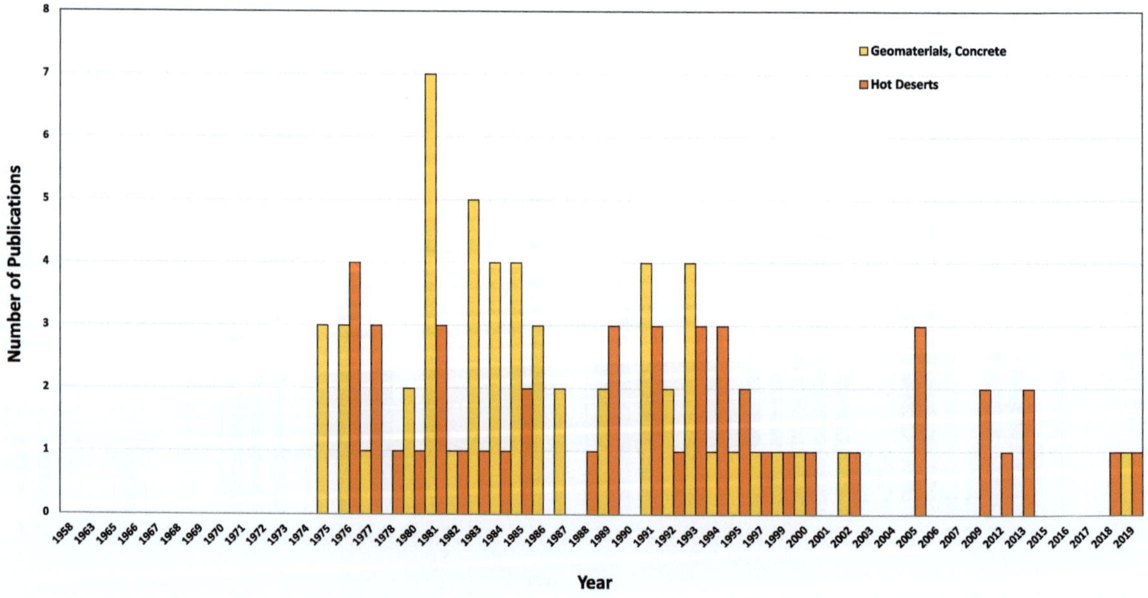

Fig. 5. Publications in hot deserts, geomaterials and concrete.

Table 2. Key publications in hot deserts, geomaterials and concrete

Date	
1975	The classification and description of near-shore carbonate sediments for engineering purposes. Fookes P. G. and Higginbottom I. E.
1976	Road geotechnics in hot deserts. Fookes P. G.
1977	*Concrete in the Middle East*. Fookes, P. G. Reprints of five articles:
	Fookes P. G. (1977*a*) Natural construction materials in the Gulf.
	Fookes P. G. and Collis L. (1975*b*) Problems in the Middle East.
	Fookes P. G. and Collis L. (1975*a*) Aggregates and the Middle East.
	Fookes P. G. (1976a) A plain man's guide to cracking in the Middle East.
	Fookes P. G. and Collis L. (1976) Cracking and the Middle East.
1977	Soluble salt damage to surfaced roads in the Middle East. Fookes P. G. and French W. J.
1978	Middle East – inherent ground problems. Fookes P. G.
1980	Some problems of construction aggregates in desert areas with particular reference to the Arabian Peninsula. 2: Investigation, production and quality control. Fookes P. G. and Higginbottom I. E.
1981	Crack mapping for investigation of Middle East concrete. Pollock D.J., Kay E.A. and Fookes P.G.

Text Box 2. Personal recollections

The Bahrain Surface Materials Resources Survey carried out in 1975 was typical of Peter's networking abilities. While already working in Bahrain with Sandberg and Partners on government quarrying projects, he convinced the government to fund an island-wide resources survey with the help of a team of academics and practitioners he assembled from other contacts.

As a young engineering geologist, this was my first overseas project. The only opportunity Peter had in his busy schedule to give a briefing was the night before my flight, when a small group of us met at a Spanish restaurant under the railway arches at Purley. Peter was in full technical flow, and later in the evening we were the only table left in an otherwise empty restaurant. The owner approached us, said that he had booked a cabaret act and implored us, as the only diners left in the restaurant, to stay to watch. I remember the very surreal experience of having our circular table cleared so that a Spanish lady could climb onto it to perform a belly dance!

A few months later my company employed Peter as a consultant on a project for a new naval dockyard in Kuwait for Brian Colquhoun and Partners. Our role was to seek out potential sources of aggregates. I flew out alone to start a regional reconnaissance of potential resources, which included fine and coarse aggregates and armourstone.

We employed Peter as a consultant, and Figure 6 shows him educating me on sabkha conditions, having flown in for two days to keep me on the straight and narrow. It was a feature of Peter's persona that whoever was involved with him, as a client, colleague or student, and at whatever stage of their career, they came away the wiser.

Fig. 6. A personal photo from 1976 showing Peter in typical teaching mode in the desert in Kuwait.

Text Box 3. Personal recollection

I was involved in route alignment studies for the Baku–Tbilisi–Ceyhan pipeline through Turkey. On one of our field excursions, Mark Lee and I, together with an engineer from our Turkish client, were walking the pipeline route within a few hundred metres of the Georgian border when we found ourselves surrounded by Turkish troops. It was a breakdown of communication between our client and the border force, and we were escorted back to the local village to join the rest of the team, including Peter, who were waiting for us.

It transpired that we needed to go to the border post, where all our identities were to be checked. Not a problem, or so I thought, until I realized that I had committed the cardinal sin of leaving my passport back at the hotel, three hours' drive away. I envisaged spending a night in the cells until the others returned in the morning.

Peter was unruffled and sat quietly looking through his wallet. He eventually pulled out a very old, stained and dog-eared business card of mine that had probably sat in the depths of his wallet for years. So, we all sat in the senior officer's room and presented our passports in turn. I was third or fourth in line and, with as much confidence as I could muster, I calmly handed over the business card. The officer didn't turn a hair, noted down the details from the card and handed it back. Afterwards, I was the subject of much leg-pulling, but Peter had come to the rescue!

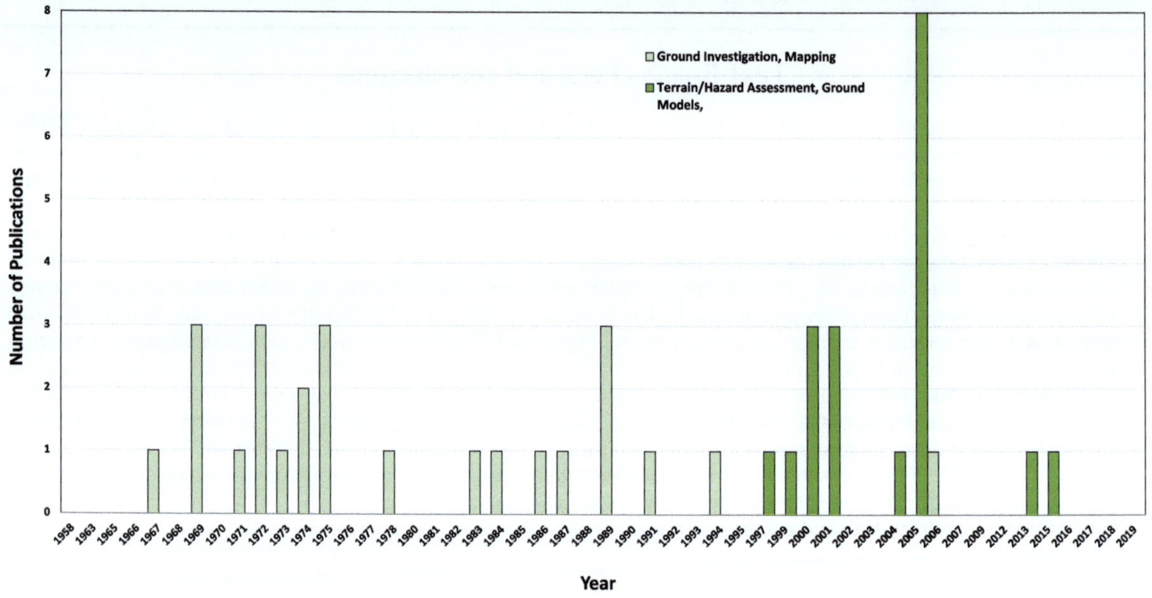

Fig. 7. Publications relating to the development of the ground model.

Table 3. Some key publications in ground investigation and the geological model

Date	Title
1967	Planning and stages of site investigation. Fookes P. G.
1969	Geotechnical mapping of soils and sedimentary rock for engineering purposes with examples of practice from the Mangla Dam project. Fookes P. G.
1971	Engineering geology maps and plans. Fookes P. G.
1972	Preparation of maps and plans in terms of engineering geology. Fookes P. G. (Chairman), Cratchley C. R. and Day J. B. W.
1974	Engineering geological mapping for civil engineering practice in the United Kingdom. Dearman W. R. and Fookes P. G.
1975	Large scale geomorphological mapping and highway engineering design. Brunsden D., Doornkamp J. C., Fookes P. G., Jones D. K. C. and Kelly J. M. H.
1987	Land evaluation and site assessment (hazard and risk). Fookes P. G.
1997	The First Glossop Lecture: Geology for Engineers, the geological model, prediction and performance. Fookes P. G.
2000	Total geological history: a model approach to the anticipation, observation and understanding of site conditions (with Baynes and Hutchinson). Fookes P.G., Baynes F. J. and Hutchinson J. N.

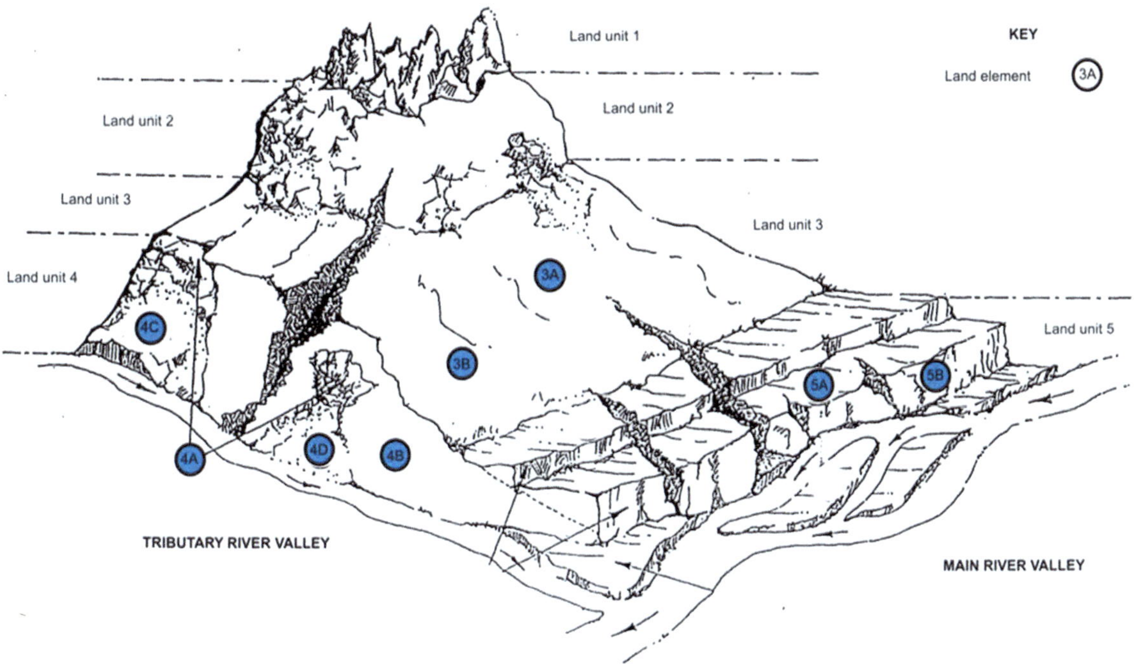

Fig. 8. Adapted from an early example of Fookes' mountain model, showing the key landscape elements in that environment. Source: reprinted from Fookes *et al.* (1985), with permission from Elsevier.

geological knowledge to the practical understanding of the influence of aggressive salts on foundations, roads and concrete durability. He became sought after for this specialist knowledge and was often employed as a 'consultant's consultant'. Peter realized very quickly that much of what was taught to British and US engineers and exemplified in their codes of practice was geared towards conditions in temperate climates and was not necessarily relevant to other environments. There was no more extreme example than the hot and arid conditions of the Middle East, represented by these desert areas.

Using his academic influence, he sowed the seed for academic research into the effects of these climatic conditions on civil engineering design, particularly concrete, locally accessible construction materials and foundations. Many publications relating to geomaterials, concrete and hot deserts followed. Figure 5 shows his publications on these themes.

On closer examination of these publications, it is noteworthy that the subject remained of significant interest to him throughout the remainder of his career, from his initial work in the 1970s through to his very latest works in 2019. Moreover, he continually extended the theme to other regimes, such as tropical environments.

He was at the forefront of identifying hazards to foundations and concrete caused by salts released during high rates of evaporation, and his work did much to describe the processes and develop guidelines since embodied in regional codes and specifications. Table 2 shows some of his key publications from this period. An excellent and still widely referred to summary of the problem is presented in the compendium reprint of five articles comprising *Concrete in the Middle East*, published in *Concrete*.

These publications serve as another example of his movement towards the concept of a ground model. These

Fig. 9. A personal photo from 1989 showing elements of the mountain model in Bhutan.

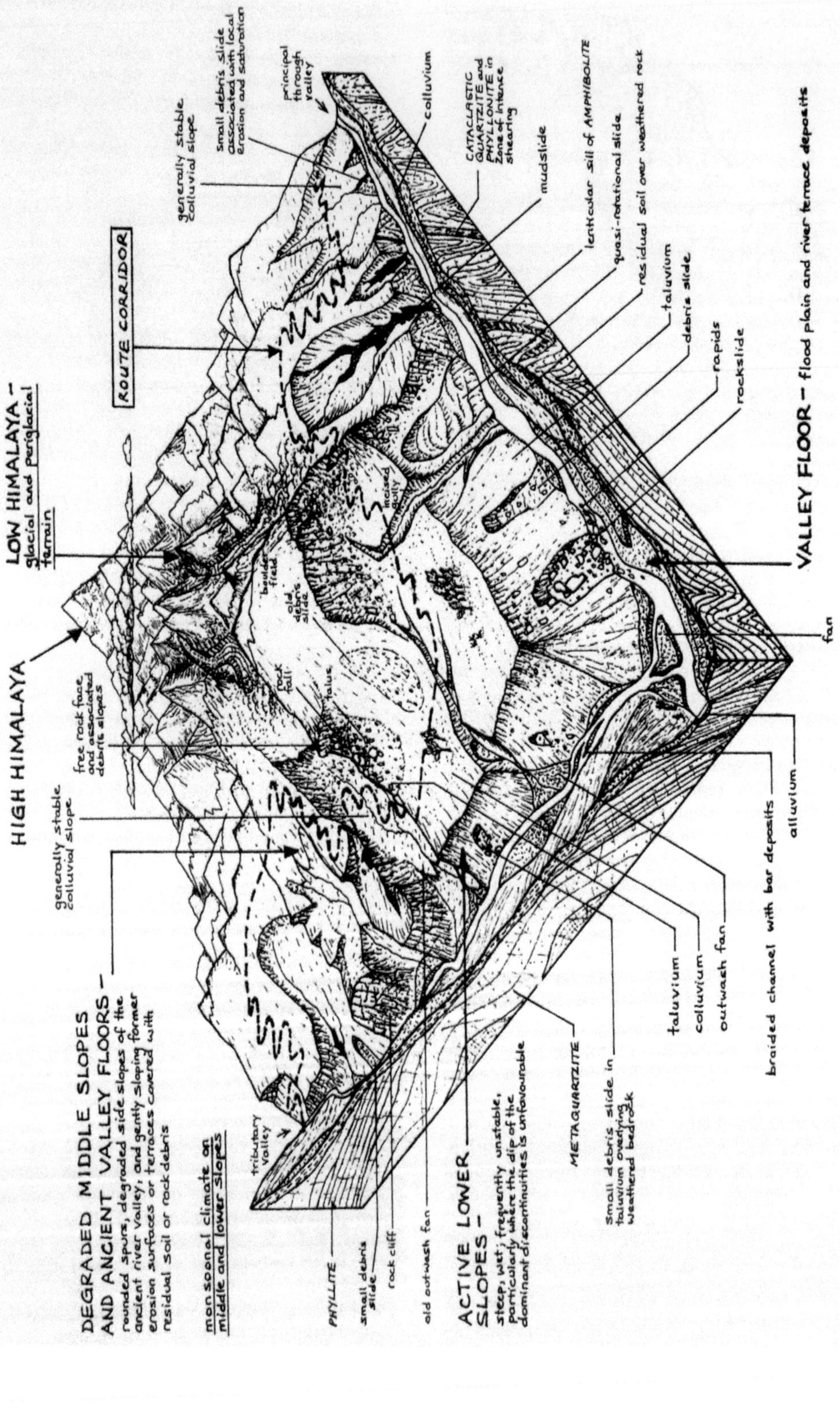

Fig. 10. An evolved form of Fookes' mountain model from his Glossop Lecture. Source: Fookes (1997*d*).

publications encompass a proposed engineering classification of carbonate soils and rocks and apply these principles across the whole civil engineering spectrum. He applied this knowledge of concrete durability to geomaterials and concrete technology in the UK, particularly in SW England, and to other climatic environments in the Far East and in the Nepalese Himalayas.

Geomodels: Peter's career development of the ground model

It was the Glossop Lecture in 1997 (Fookes 1997*a*) that gave Peter the opportunity to demonstrate his approach to site characterization and the tools he used. This was the theme that he had developed throughout his career from his early years, and his whole concept of ground modelling was laid out here. Other publications followed (Fookes *et al.* 2001*a*, *b*; Baynes *et al.* 2005). He considered that he benefited greatly from the work early in his career with Skempton, Hoek, Higginbottom and Brunsden, which emphasized the key links between geology and other disciplines. He likened the characterization of a site to a medical diagnosis in which the general practitioner needs to understand the patient. He recognized the role of geomorphological and geological processes in shaping the characteristics of a site. He had a concern that the multiplicity of geotechnical variables in building even the simplest of models can be daunting. He showed how the ground characteristics of a site are the result of the interaction of processes that need to be put into a historical context: these include a knowledge of the environment of deposition; an appreciation of the plate tectonic processes that have resulted in movement through different climatic zones and environments; and a knowledge of the weathering and surface processes of the Quaternary that have more recently shaped the ground surface.

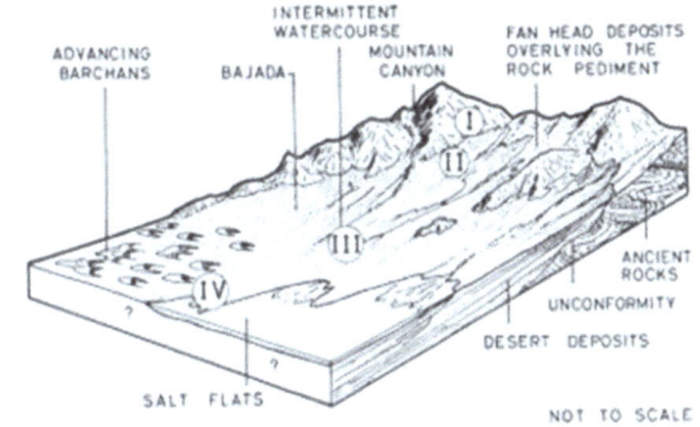

Fig. 11. An early example of Fookes' desert model, showing the key desert elements. Source: reprinted from Fookes and Knill (1969), with permission from Elsevier.

He was clear that the development of a ground model was important right from the earliest stage in the investigation process. Albeit approximate, it provides an initial identification of potential hazards and a basis for refinement as the investigative techniques develop through the project cycle. Peter maintained that by using this approach, ground conditions should only rarely be unforeseeable. Nevertheless, they are often unforeseen through constraints imposed by the contractual process.

I was involved in several linear projects with Peter in the reconnaissance and preliminary investigation stages (Text Box 3). I can single out one client, BP, represented by Mike Sweeney and Jim Clark, who were great supporters of assembling a team of specialists from the relevant disciplines. Invariably Peter would lead the team in an initial brainstorming session that would identify the main geohazards that could potentially influence the project and decide a broad-brush grading system for each one. On the reconnaissance traverse each member of the team was tasked with scoring one of the hazards on the hoof. One lucky member was tasked to be the scribe, while the others shouted out the scores; on one occasion this took place during a flyover of the route by helicopter. This may sound subjective and approximate – and, of course, at an early stage it was – but invariably these preliminary scoring systems helped to focus on areas of concern for more detailed investigation.

A graphical representation of his publications covering mapping, terrain evaluation, site investigation and ground modelling is shown in Figure 7. This again illustrates the continual development of the ground model approach throughout his career and how the focus of the publications shifted through phases of ground investigation and onto the ground model later in his career. A summary of his key publications on these themes is given in Table 3 and also serves to emphasize this continuing development of the modelling theme and

Fig. 12. A personal photo from 1978 showing elements of the desert model in Libya.

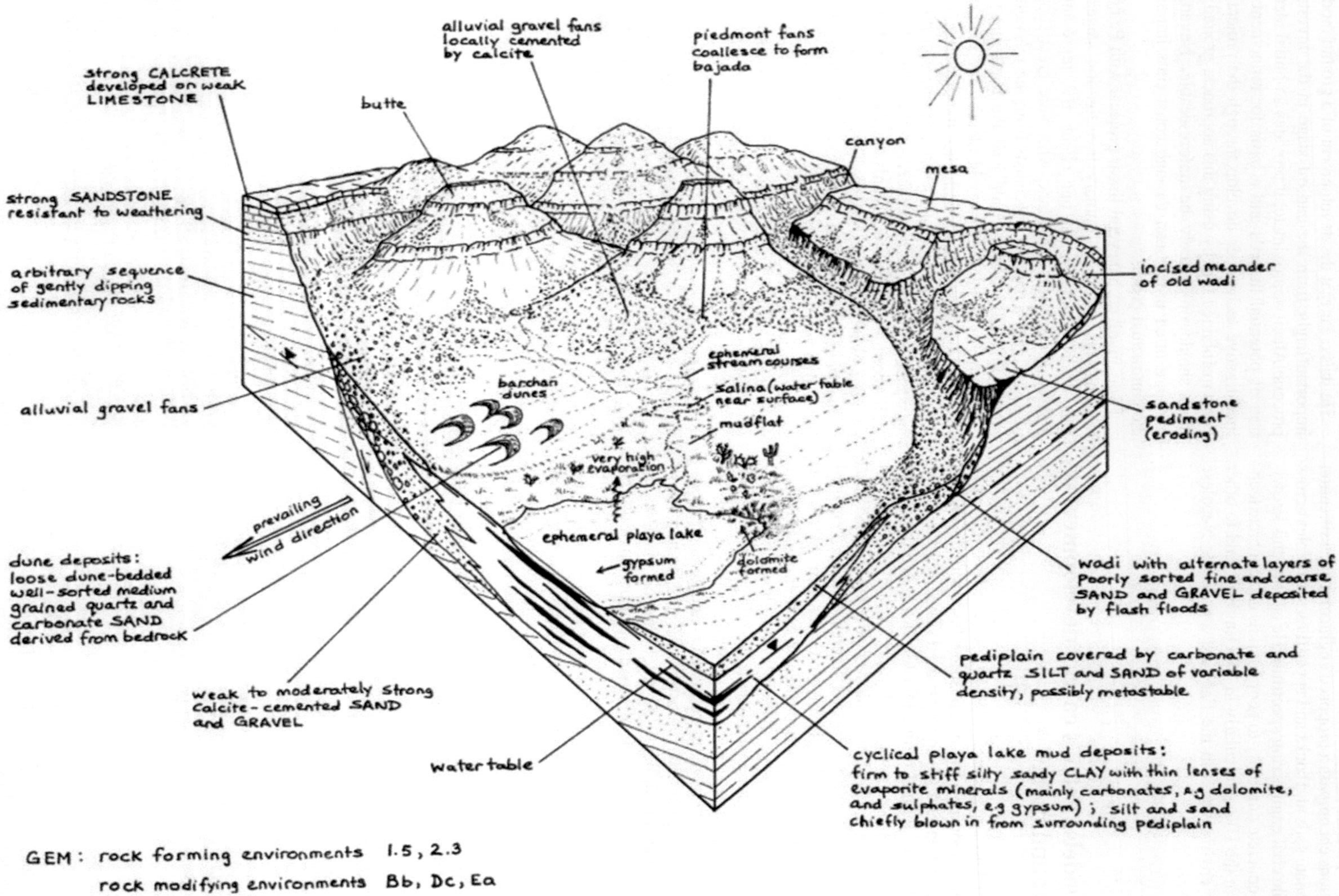

Fig. 13. An evolved form of Fookes' desert model from his Glossop Lecture. Source: Fookes (1997*d*).

how the concept grew through his career, from site investigation techniques and planning, through geological and geomorphological mapping, to hazard assessment and finally modelling.

Some examples of Peter's ground models

Peter's work is distinguished by his highly practical approach and full use of graphics. Almost without fail his publications are both essential references and eminently readable. His illustrative models are tremendous field tools because they represent what you see. Examples of two of the models Peter published are described below and compared to my personal photos of similar terrain to show how representative, clear and useful they are as field companions.

The mountain model (Fig. 8) was first published as a basis for his contributions to the subject of low-cost roads in mountainous regions (e.g. Fookes *et al.* 1985*b*; Fookes 1986*a*, *b*, *c*).

In comparison, Figure 9 is a photo I took while working on a hill irrigation project in eastern Bhutan in the Himalayan foothills. The similarities in the landscape with Peter's mountain model are clear and are an indication of how useful such a model can be if included in a handbook for a field worker. Peter continually evolved his models, building on observations made during his projects. It is interesting to note the differences between his early mountain model (Fig. 8) and the more sophisticated version developed for the Glossop Lecture and shown in Figure 10. Even though the later version is more complex than the earlier model, it is an equally useful tool to identify features that may be encountered in the field.

Similarly, Peter's desert model (Fig. 11), which identifies the basic landscape elements in desert environments, was first published in a paper on the regional development of Iran (Fookes and Knill 1969). Compare this to a personal photo taken in the desert south of Tripoli in western Libya in 1978 (Fig. 12), where the basic desert landscape elements are clearly identifiable. Again, the more sophisticated model (Fig. 13) developed for Peter's Glossop Lecture (Fookes 1997*d*) still retains clarity, even though it is more detailed.

Awards in recognition of his achievements

Peter Fookes was the recipient of many awards in recognition of his seminal contribution to the subject, and particularly his contribution to engineering practice. The key awards are listed in Table 4.

His academic awards include Doctor of Science (Engineering) at Imperial College London (1979), Honorary Professor at Birmingham University (1999), Distinguished Research Associate at Oxford University (2001) and Honorary Doctor of Science from the University of Plymouth (2005). He was particularly proud of the Doctor of Science (Engineering) from Imperial College because, as a geologist, he felt that it demonstrated a respect from his peers in the civil engineering department, acknowledged experts such as Skempton, Bishop, Vaughan, Hutchinson and Chandler.

Other awards he retained particular personal pride in were the William Smith Medal of the Geological Society (1985), which Peter was the first British engineering geologist to receive; Fellow of the Royal Academy of Engineering (1991), the engineering equivalent of the Royal Society;

Table 4. Key awards in Peter Fookes' career

Date	Award
1979	*Doctor of Science (Engineering) at Imperial College London*
1981	Telford Premium of the Institution of Civil Engineers
1981 and 2000	British Geotechnical Society Prize
1982 and 1992	Overseas Premium of the Institution of Civil Engineers
1985	*William Smith Medal of the Geological Society*
1990	George Stephenson Medal of the Institution of Civil Engineers
1990	Webb Prize of the Institution of Civil Engineers
1991	*Fellow of the Royal Academy of Engineering*
1992	Coopers Hill War Memorial Prize of the Institution of Civil Engineers
1992	*Honorary Fellowship of the Institute of Concrete Technology*
1997	*The First Glossop Lecture and Medal: Geology for Engineers: the geological model, prediction and performance*
1999	Honorary Professor at Birmingham University
2001	Distinguished Research Associate at Oxford University
2004	Australian Engineering Excellence Award
2005	Honorary Doctor of Science at University of Plymouth
2005	*Honorary Fellow of the Royal Geographical Society*

Italics denotes those awards that Fookes was particularly proud of.

Honorary Fellowship of the Institute of Concrete Technology (1992) for his contribution to that science; the first Glossop Medal of the Engineering Group of the Geological Society (1997) and presentation of the first Glossop Lecture; and Honorary Fellow of the Royal Geographical Society (2005) for his contribution to geomorphology. These are highlighted in italics in Table 4.

Engineering projects

Peter's consulting work was more or less equally divided between consultants and contractors; his appeal came from the practical nature of his advice. Work took him to more than 90 countries: he estimated that 80% of his projects had been overseas. He worked on literally hundreds of projects; for example, roads and bridges in Bangladesh, Malaysia and Yemen, dams in Indonesia, railways in Australia and mining in Papua New Guinea. In his later years he was increasingly retained as a review consultant on many long-term projects, including the Great Man-Made River Project in Libya and mining infrastructure in Australia. More recently he was involved with the Baku–Tbilisi–Ceyhan oil pipeline through Azerbaijan, Georgia and Turkey, and other major pipelines; for example, in Algeria, Egypt, Kovykta and Sakhalin (Siberia) and Papua New Guinea.

Conclusions

Peter Fookes' contributions to the world of engineering geology were immense. His academic and professional achievements were perfectly summed up in the citation accompanying his award of the William Smith Medal, reproduced in full below.

> With the award of the William Smith Medal, we recognise you as the doyen of British engineering geologists, though your work and reputation are, of course, international. Your taste for application was already developed even before you entered Queen Mary College as an undergraduate in geology. You have been in the forefront of initiation and development of engineering geology and engineering geomorphology. You have made civil engineers aware, for example, of the special conditions of weathering and slope failure that they may meet in arid regions. Your work on the problems of the short-term and long-term deterioration of concrete is of poignant significance. Your activities probably led to the development of the postgraduate school of industrial petrology at Queen Mary College, where you are a most welcome Visiting Professor. You have been involved in such prestigious projects as the Thames Barrier, the Mangla Dam in Pakistan, the Great Man-made River Project in Libya, the Falklands Airport and Himalayan roads.
>
> You have found that geomorphology (a subject which, if I may say so, we should never have allowed to drift away from geology), with its interest in near surface features of the Earth, is a powerful tool with which to tackle problems in civil engineering. You have used petrology (in which you had your initial undergraduate training at Queen Mary College) to resolve problems concerning materials. In spite of your immensely busy life as a consultant engineering geologist, you have found time to lecture widely and to write more than 80 papers, some of which are now standard references for colleagues in your field. You have received two awards from the Institution of Civil Engineers and one from the British Geotechnical Society. You are indeed a most appropriate medallist to follow in the footsteps of William Smith. You are a true practitioner and we salute you as such.
>
> (*Journal of the Geological Society, London*, **143**, 1986, pp. 374–375)

The President at the time was Charles Holland, whose main interests were stratigraphy and palaeontology, particularly cephalopods. I don't know if he prepared this citation himself or if it was written for him, but I think that perfectly sums up Peter's contribution to the profession. It was written in 1985, when he had only just started! Nearly 35 years later those words were still equally applicable.

Finally, I want to say a little about Peter's life beyond the profession. You would be forgiven for believing that he had little time for a private life – but don't you believe it. He lived with his wife Edna in Winchester and then in Wickham in Hampshire, and also spent time in their hideaway property in Swanage. He had five children by his first wife, Gwyneth, and nine grandchildren. He met Gwyneth in Pakistan while she was backpacking and they were married in a Christian chapel in the Khyber Pass.

He met his soulmate Edna while involved in projects in the Falkland Islands in the 1980s. He affectionately referred to her as 'Sir' and that was how she was known to numerous students. She was his wife, secretary and chauffeuse and generally ensured he was in the right place at the right time. She accompanied him on the majority of his UK field trips, but when she remained at home he wrote her a postcard every day. Field trips were meticulously planned so that the party was back in the hotel by 5.30 p.m. for the next episode of his favourite soap, *Neighbours*, and a gin and tonic. Peter and Edna used to own their own custom-designed and -built narrowboat, and over many years they travelled all of the UK's canals with friends and colleagues, including crossing the Wash and the Severn Estuary (both with a pilot), and travelling down the Thames and through the lock at Limehouse Reach into Regent's Canal. He was passionate about steam, regularly visiting the Great Dorset Steam Fair, and he was a life member of the Swanage steam railway in Dorset and the mid-Hants steam railway (Watercress Line) in Hampshire.

Over the last 15 years or so since I retired from commercial work, my wife, Wendy, and I met Peter and Edna regularly for lunch or coffee. When I spoke to Edna a few weeks before the conference she said:

> Geology was his first love, even before me, and it inspired every day of his life. He thought the world of all his students and was very proud of you all. He would be thrilled to bits that the 2023 conference had been organized for him and to see what you are all doing for him.

I think that says it all. He could be exasperating at times, but he was always very supportive and invariably proved to be right. He was the most influential practitioner of his generation and he has left a gap that will be very hard to fill.

Acknowledgements This review is based on my recollections of Peter Fookes' life as his student, from my many professional interactions with him throughout my career and as his friend. I could not have written it without this personal connection during his life. I am indebted in particular to Mark Lee, Jim Griffiths and Mike Sweeney, together with numerous other colleagues who I met on many occasions during the Fookes years. Their discussions and anecdotes have been invaluable. Finally, I must thank Peter's wife, Edna, for her warm personality, friendship and her own insights.

Competing interests The authors declare that they have no known competing financial interests or personal relationships that could have appeared to influence the work reported in this paper.

Author contributions JC: conceptualization (lead), writing – original draft (lead).

Funding This research received no specific grant from any funding agency in the public, commercial or not-for-profit sectors.

Data availability All data generated or analysed during this study are included in this published article (and, if present, its supplementary information files).

References and bibliography

Atkinson, J.H., Pettifer, G.S. and Fookes, P.G. 2000. Mechanical destructuring and disaggregation of Mercia mudstone. *In*: The engineering properties of the Mercia Mudstone Group. Papers presented at a seminar held at Pride Park, Derby, UK on 25 November 1998. *CIRIA Conference Record*, **2000/1**, 5–12.

Atkinson, J.H., Fookes, P.G., Miglio, B.F. and Pettifer, G.S. 2003. Destructuring and disaggregation of Mercia Mudstone during full-face tunnelling. *Quarterly Journal of Engineering Geology and Hydrogeology*, **36**, 293–303. Also in Discussion by West, G. of Destructuring and disaggregation of Mercia Mudstone during full-face tunnelling; reply by J. H. Atkinson, P. G. Fookes, B. F. Miglio and G. S. Pettifer. *Quarterly Journal of Engineering Geology and Hydrology*, **37**, 165–168, https://doi.org/10.1144/1470-9236/03-024

Bamforth, P., Clarke, J., Croft, A., Fookes, P.G., Kay, E. and Walker, M. 2000. *Diagnosis of Deterioration in Concrete Structures. Identification of Defects, Evaluation and Development of Remedial Action.* Concrete Society Technical Report, 54.

Baynes, F., Fookes, P.G. and Hutchinson, J. 2003. Discussion on Atkinson, J. 2002. What is the matter with geotechnical engineering? in *Proceedings of the Institution of Civil Engineers, Geotechnical Engineering* 155, 155–158. *Proceedings of the Institution of Civil Engineers, Geotechnical Engineering*, **156**, 159–162, https://doi.org/10.1680/geng.2003.156.3.159

Baynes, F., Fookes, P.G. and Kennedy, J.F. 2005. Total engineering geology approach to applied railways in the Pilbara, Western Australia. *Bulletin of Engineering Geology and the Environment*, **64**, 67–94, https://doi.org/10.1007/s10064-004-0271-4

Best, R. and Fookes, P.G. 1970. Some geotechnical and sedimentary aspects of ballclays from Devon. *Quarterly Journal of Engineering Geology*, **3**, 207–240, https://doi.org/10.1144/GSL.QJEG.1970.003.04.02

Braybrooke, J.C., Broise, M.Y., Fookes, P.G. and Sweeney, M. 2005. A case study: geohazards risk assessment for a proposed pipeline across diverse geological environments in Papua New Guinea. *Terrain and Geohazard Challenges Facing Onshore Oil and Gas Pipelines Conference, London, 2004.* Institution of Civil Engineers, London, 129–143.

Brunsden, D., Doornkamp, J.C., Fookes, P.G., Jones, D.K.C. and Kelly, J.M.H. 1975*a*. Large scale geomorphological mapping and highway engineering design. *Quarterly Journal of Engineering Geology*, **8**, 227–253, https://doi.org/10.1144/GSL.QJEG.1975.008.04.01

Brunsden, D., Doornkamp, J.C., Fookes, P.G., Jones, D.K.C. and Kelly, J.M.H. 1975*b*. Geomorphological mapping techniques in highway engineering. *Journal of the Institution of Highway Engineers*, **XXII**, 35–41.

Burnett, A.D. and Fookes, P.G. 1974. A regional engineering geological study of the London Clay in the London and Hampshire Basins. *Quarterly Journal of Engineering Geology*, **7**, 257–295, https://doi.org/10.1144/GSL.QJEG.1974.007.03.02

Burnett, A.D., Fookes, P.G. and Robertson, R.H.S. 1972. An engineering soil at Kermanshah, Zagros Mountains, Iran. *Clay Minerals*, **9**, 329–343, https://doi.org/10.1180/claymin.1972.009.3.07

Charman, J.H., Fookes, P.G., Hengesh, J., Lee, E.M., Pollos-Pirallo, S., Shilston, D.T. and Sweeney, M. 2005. Terrain, ground conditions and geohazards: evaluation and implications for pipelines. *Terrain and Geohazard Challenges Facing Onshore Oil and Gas Pipelines Conference, London, 2004.* Institution of Civil Engineers, London, 78–94.

Collis, L. and Fookes, P.G. 1985. Some reflections on concrete deterioration in the Arabian Gulf. *First International Conference on Deterioration & Repair of Reinforced Concrete in the Arabian Gulf.* Bahrain Society of Engineers/Construction Industry Research and Information Association, 71–84.

Davison, L., Fookes, P. and Baynes, F. 2003. Total geological history: a web-based modelling approach to the anticipation, observation and understanding of site conditions. *In*: Rosenbaum, M.S. and Turner, A.K. (eds) *New Paradigms in Subsurface Prediction. Characterization of the Shallow Subsurface Implications for Urban Infrastructure and Environmental Assessment.* Springer, Berlin, 237–252.

Dearman, W.R. and Fookes, P.G. 1972. The influence of weathering on the layout of quarries in south-west England. *Proceedings of the Ussher Society*, **2**, 372– 387.

Dearman, W.R. and Fookes, P.G. 1974. Engineering geological mapping for civil engineering practice in the United Kingdom. *Quarterly Journal of Engineering Geology*, **7**, 223–256, https://doi.org/10.1144/GSL.QJEG.1974.007.03.01

de Freitas, M.H. and Fookes, P.G. 1970. Groundwater problems associated with ground engineering in the reworked alluvium at High Wycombe. *First International Congress of the International Association of Engineering Geology.* Paris, **II**, 1213–1225.

Fookes, P.G. 1958. Pleistocene sea levels. *Cosmos*, **13**, 412–419.

Fookes, P.G. 1965. Orientation of fissures in stiff overconsolidated clay of the Siwalik system. *Géotechnique*, **XV**, 195–206, https://doi.org/10.1680/geot.1965.15.2.195

Fookes, P.G. 1966. London Basin Tertiary sediments (correspondence). *Géotechnique*, **XVI**, 260–263, https://doi.org/10.1680/geot.1966.16.3.260

Fookes, P.G. 1967*a*. Discussion of Mangla. *Proceedings of the Institution of Civil Engineers*, **38**, 337–576.

Fookes, P.G. 1967*b*. London Basin geological structure (correspondence). *Géotechnique*, **XVII**, 168–170, https://doi.org/10.1680/geot.1967.17.2.168

Fookes, P.G. 1967*c*. Planning and stages of site investigation. *Engineering Geology*, **2**, 81–106, https://doi.org/10.1016/0013-7952(67)90026-9

Fookes, P.G. 1967*d*. *The Siwalik Clay Bedrock of the Mangla Dam Project*. PhD thesis, University of London.

Fookes, P.G. 1968. Discussion of Mangla. *Proceedings of the Institution of Civil Engineers*, **38**, 337–576; **41**, 119–203.

Fookes, P.G. 1969. Geotechnical mapping of soils and sedimentary rock for engineering purposes with examples of practice from the Mangla Dam Project. *Géotechnique*, **19**, 52–74, https://doi.org/10.1680/geot.1969.19.1.52

Fookes, P.G. 1970. Welsh water for Wales: visits to Llysyfran and Llyn Brianne Dams. *BNCOLD News & Views (British Section, International Commission on Large Dams)*, Number 8, 23–27.

Fookes, P.G. 1971*a*. Discussion on weathering in dam foundations. *News & Views (British National Committee on Large Dams)*, Number 9, 12.

Fookes, P.G. 1971*b*. Engineering geology maps and plans. *Ground Engineering*, **4**, 24.

Fookes, P.G. 1971*c*. Sample representability. *Quarterly Journal of Engineering Geology*, **4**, 299–307, https://doi.org/10.1144/GSL.QJEG.1971.004.04.03

Fookes, P.G. (ed. & review. 1973*a*. Review and bibliography on rock weathering and residual soil. *Geomechanics Abstracts & International Journal of Rock Mechanics and Mining Sciences*, **4**.

Fookes, P.G. 1973*b*. Geology for engineers. *In*: Blake, L.S. (ed.) *Civil Engineers Reference Book*, **7**, 3rd edn. Butterworths, London, 1–39.

Fookes, P.G. 1973*c*. Site reconnaissance and the engineering geologist. *Transport and Road Research Laboratory (Department of the Environment) Report LR 591 on TRRL Symposium*. Available information for route planning and site investigation by M. J. Dumbleton, 17–22.

Fookes, P.G. 1975. The new engineering geology. *Concrete*, **9**, 40.

Fookes, P.G. 1976*a*. A plain man's guide to cracking in the Middle East. *Concrete*, **10**, 20–22; reprinted in 1977 in *Concrete in the Middle East, Part 1*, as part of reprints of five articles from *Concrete*, 22–24.

Fookes, P.G. 1976*b*. Middle East – inherent ground problems. *Engineering Problems Associated with Ground Conditions in the Middle East, Symposium*. Geological Society, London, 7–23.

Fookes, P.G. 1976*c*. Road geotechnics in hot deserts. *Journal of the Institution of Highway Engineers*, **XXIII**, 11–23.

Fookes, P.G. 1977*a*. Natural construction materials in the Gulf. *In*: *Middle East Construction Conference, Material Specification and Standards*, Dubai, UAE, November 1977, 143–164.

Fookes, P.G. 1977*b*. Some geotechnical aspects of road construction in the Middle East. *Planning and Transport Research and Computation*, London (PTRC), E21, 1–13.

Fookes, P.G. 1978. Middle East – inherent ground problems. *Quarterly Journal of Engineering Geology*, **11**, 33–49, https://doi.org/10.1144/GSL.QJEG.1978.011.01.04

Fookes, P.G. 1980. An introduction to the influence of natural aggregates on the performance and durability of concrete. *Quarterly Journal of Engineering Geology*, **13**, 207–229, https://doi.org/10.1144/GSL.QJEG.1980.013.04.02

Fookes, P.G. 1981. Concrete deterioration and repair – discussion. *Proceedings of the Institution of Civil Engineers*, Part 1, **70**, 330–331.

Fookes, P.G. 1983*a*. Aggregates in Indonesia. *In*: *Concrete Technology Seminar*. Director General of Sea Communications, Jakarta, 1–18.

Fookes, P.G. 1983*b*. Aggregates in the Middle East – an overview. *Conference on Improving Concrete in the Middle East*, Cement & Concrete Association, London, 1–19.

Fookes, P.G. 1983*c*. An overview of aggregates in the Far East. *Conference on Concrete & Construction in the Far East*, Cement and Concrete Association, Wexford Springs, UK, June, 1–19.

Fookes, P.G. 1983*d*. Concrete technology in Indonesia – Part 1. *In*: *Concrete Technology Seminar*. Director General of Sea Communications, Jakarta, 19–23.

Fookes, P.G. 1983*e*. Concrete technology in Indonesia – Part 2. *In*: *Concrete Technology Seminar*. Director General of Sea Communications, Jakarta, 19–23.

Fookes, P.G. 1984*a*. An introduction to concrete in the Far East – Part 1: Aggregates. *Concrete*, **18**, 38–41.

Fookes, P.G. 1984*b*. Arid Lands, Engineering Geology. *In: Encyclopaedia of Applied Geology*, Encyclopaedia of Earth Science Series, **XIII**. Springer, New York, 12–25.

Fookes, P.G. 1984*c*. Discussion on the design and construction of the Thames Barrier cofferdams. *Proceedings of the Institution of Civil Engineers*, Part 1, **76**, 567–568.

Fookes, P.G. 1984*d*. Discussions on rock in breakwaters. *Conference on Breakwaters – Design and Construction*. Institution of Civil Engineers, London, 49–50, 51, 129–130, 162–163.

Fookes, P.G. 1985. An introduction to concrete in the Far East: Part 2. *Concrete*, **19**, 26–30.

Fookes, P.G. 1986*a*. *A Preliminary View on Cut Slope Instability of Roads in Peninsular Malaysia*. Public Works Department, Kuala Lumpur.

Fookes, P.G. 1986*b*. *An Introduction to Weathered Rock and Residual Soils for Engineers*. Public Works Department, Kuala Lumpur (also appeared in *Warta Geologi* – Newsletter of the Geological Society of Malaysia, March–April, 60–71.)

Fookes, P.G. 1986*c*. *Some Aspects of Concrete Technology in Malaysia*. Public Works Department, Kuala Lumpur.

Fookes, P.G. 1987*a*. *A Preliminary View of Stone Construction Materials in Malaysia*. Public Works Department, Kuala Lumpur, 1–39.

Fookes, P.G. 1987*b*. Land evaluation and site assessment (hazard and risk). *Geological Society, London, Engineering Geology Special Publications*, **4**, 273–282, https://doi.org/10.1144/GSL.ENG.1987.004.01.33

Fookes, P.G. 1988. The geology of carbonate soils and rocks and their engineering characteristics and description. State-of-the-art report. *In*: Jewell, R.J. and Khorshid, M.S. (eds) *Engineering for Calcareous Sediments*, **2**. Balkema, Rotterdam, 787–806.

Fookes, P.G. 1989*a*. Additional Discussion on Tappin, R. G. R. and Clark, P. J./Rees, A. and Kim, S. J., Discussion on Jindo and Dolsan bridges: design and construction, *Proceedings of the Institution of Civil Engineers*, Part 1, 1987, 82, June, 663–682. *Proceedings of the Institution of Civil Engineers*, Part 1, **86**, 1233–1235.

Fookes, P.G. 1989*b*. Geology for engineers. *In*: *Civil Engineers Reference Book*, **8**, 4th edn. Butterworths, London, 1–34.

Fookes, P.G. 1989*c*. *Hot Weather Concrete: a Little Theory and a Lot of Practice*. Institute of Engineers, Jakarta [in English and Malay].

Fookes, P.G. 1991*a*. Concrete in the Middle East – past, present and future: a review. *In: The Regional Conference on Damage Assessment, Repair Techniques and Strategies for Reinforced Concrete*, Bahrain, 7–9 December. The Bahrain Society of Engineers, Manama, 9–32.

Fookes, P.G. 1991*b*. Geomaterials. *Quarterly Journal of Engineering Geology*, **24**, 3–15, https://doi.org/10.1144/GSL.QJEG.1991.024.01.02

Fookes, P.G. 1991*c*. Quaternary engineering geology. *In*: Forster, A., Culshaw, M.G., Cripps, J.C., Little, J.A. and Moon, C.F. (eds) *Quaternary Engineering Geology. Geological Society, London, Engineering Geology Special Publications*, **7**, 73–98, https://doi.org/10.1144/GSL.ENG.1991.007.01.04

Fookes, P.G. 1992*a*. Concrete aggregates in the Arabian Peninsula and Gulf: a review. In: *The Regional Conference on Damage Assessment, Repair Techniques and Strategies for Reinforced Concrete*, Bahrain, 7–9 December. The Bahrain Society of Engineers, Manama, 289–311.

Fookes, P.G. 1992*b*. Geology and concrete: a review of British aggregates. The Fifth Sir Frederick Lea Memorial Lecture. *In: Proceedings of 20th Annual Convention of Institute of Concrete Technology*, Coventry, 1–40.

Fookes, P.G. 1993*a*. A simple guide to risk assessment for concrete in hot dry salty environments. *In: Deterioration and Repair of Reinforced Concrete in the Arabian Gulf. Proceedings of 4th International Conference*, Society of Engineers, Bahrain, 1, 161–213.

Fookes, P.G. 1993*b*. Concrete in the Middle East – past, present and future: a brief review. *Concrete*, **27**, 14–20.

Fookes, P.G. 1994. A review: Genesis and classification of tropical residual soils for engineers. *In*: Blight, G.E., Fourie, A.B., Luker, I., Mouton, D.J. and Scheurenberg, R.J. (tech. eds) *Geotechnics in the African Environment, Proceedings of 10th Regional Conference for Africa on Soil Mechanics & Foundation Engineering*, and the *3rd International Conference on Tropical & Residual Soils*, Maseru, 23–27 September 1991, 2, 423–442.

Fookes, P.G. 1995*a*. Concrete in hot dry salty environments. *Concrete*, **29**, 34–39.

Fookes, P.G. 1995*b*. Fighting the flood: river and coastal hazard management. *Geoscientist*, **5**, 17–19.

Fookes, P.G. 1997*a*. Aggregates: a review of prediction and performance. *In*: Glanville, J. and Neville, A. (eds) *Prediction of Concrete Durability. Proceedings of STATS 21st Anniversary Conference, 1996*. The Geological Society, London, 91–170.

Fookes, P.G. 1997*b*. Foreword. *In*: Hearn, G.J. (ed.) *Principles of Low Cost Road Engineering in Mountainous Regions, with Special Reference to the Nepal Himalaya*. Overseas Centre, *Transport Research Laboratory Overseas Road Note*, 16, iii–iv.

Fookes, P.G. 1997*c*. Slip forming across the desert – the construction of the Azraq to Jafr highway project in Jordan by H. S. Mildenhall – Discussion. *Proceedings of the Institution of Civil Engineers – Transport*, May, 147–148.

Fookes, P.G. 1997*d*. The First Glossop Lecture: Geology for engineers: the geological model, prediction and performance. *Quarterly Journal of Engineering Geology*, **30**, 293–424, https://doi.org/10.1144/GSL.QJEG.1997.030.P4.02

Fookes, P.G. (ed.) 1997*e*. *Tropical Residual Soils: A Geological Society Engineering Group Working Party Revised Report*. Geological Society, London, Professional Handbooks.

Fookes, P.G. 1999. Engineering characteristics of weathered rock for use as armourstone. *UK Armourstone Users Group Sixth Annual Meeting*, 4 February. Hydraulic Research Station, Wallingford.

Fookes, P.G. 2000. Eureka moments. *In*: Atkinson, J. and Wheeler, P. (eds). *Living Legends. Ground Engineering, Geotechnical Lives*. Young Geotechnical Engineers Conference 2000 Commemorative Supplement, 8.

Fookes, P.G. 2004. The value of geologists? *Geology Today*, **20**, 131–132.

Fookes, P.G. 2005*a*. Comparing salaries. *Geology Today*, **21**, 130.

Fookes, P.G. 2005*b*. Discussion of the first engineering geological publication in the UK? by M. G. Culshaw, *Quarterly Journal of Engineering Geology and Hydrogeology* 37, 231. *Quarterly Journal of Engineering Geology and Hydrogeology*, **38**, 105–106, https://doi.org/10.1144/1470-9236/04-074

Fookes, P.G. 2005*c*. Rudolph Glossop and the early years of the Engineering Group. *Engineering Group of the Geological Society Newsletter*, May, 6–7.

Fookes, P.G. 2005*d*. Winchester Cathedral underpinned. *Geology Today*, **21**, 18.

Fookes, P.G. 2006*a*. Dorset Ball Clays. *Geology Today*, **22**, 210–212.

Fookes, P.G. 2006*b*. Dorset Clays. *Dorset Geologists Association Newsletter*, December, 2.

Fookes, P.G. 2006*c*. Musings in Durlston Bay. *Geology Today*, **22**, 11.

Fookes, P.G. 2006*d*. Thames Barrier 21st Anniversary. Peter Fookes writes... *Geology Today*, **22**, 48–50, https://doi.org/10.1111/j.1365-2451.2006.00552.x

Fookes, P.G. 2007*a*. Brunel's coastal railway. *Geology Today*, **23**, 24–29, https://doi.org/10.1111/j.1365-2451.2007.00595.x

Fookes, P.G. 2007*b*. Panama Canal in the news. Peter Fookes writes. *Geology Today*, **23**, 52–53.

Fookes, P.G. and Best, R. 1969. Consolidation characteristics of some late Pleistocene periglacial metastable soils of east Kent. *Quarterly Journal of Engineering Geology*, **2**, 103–128, https://doi.org/10.1144/GSL.QJEG.1969.002.02.02

Fookes, P.G. and Collis, L. 1975*a*. Aggregates and the Middle East. *Concrete*, **9**, 14–19; reprinted in 1976 in *Concrete Journal*, Japan Concrete Institute/JCI, No. 130, 76/5,14, No. 7, 50–53 [in Japanese]; 1977 in *Concrete in the Middle East, Part 1*, as part of reprints of five articles from *Concrete*, 7–12.

Fookes, P.G. and Collis, L. 1975*b*. Problems in the Middle East. *Concrete*, **9**, 12–17; reprinted in 1976 in *Concrete Journal*, Japan Concrete Institute/JCI, No. 128, 76/5, 14, No. 5. 67–71 [in Japanese]; 1977 in *Concrete in the Middle East*, Part 1, as part of reprints of five articles from *Concrete*, 1–6.

Fookes, P.G. and Collis, L. 1976. Cracking and the Middle East. *Concrete*, **10**, 14–19; reprinted in 1976 in *Concrete Journal*, Japan Concrete Institute/JCI, No. 133, 76/10, 14, No. 10 [in Japanese]; 1982 in *Concrete in the Middle East, Part 1*, as part of reprints of five articles from *Concrete*, 16–21.

Fookes, P.G. and Dale, S.G. 1992. Comparison of interpretations of a major landslide at an earthfill dam site in Papua New Guinea. *Quarterly Journal of Engineering Geology*, **25**, 313–330, https://doi.org/10.1144/GSL.QJEG.1992.025.04.06

Fookes, P.G. and Denness, B. 1969. Observational studies on fissure patterns in Cretaceous sediments of Southeast England. *Géotechnique*, **19**, 453–477, https://doi.org/10.1680/geot.1969.19.4.453

Fookes, P.G. and French, W.J. 1977. Soluble salt damage to surfaced roads in the Middle East. *Journal of the Institution of Highway Engineers*, **XXIV**, 10–20.

Fookes, P.G. and Gahir, J.S. 1994. Geotechnical characteristics of some carbonate sands in the Fezzan Sahara. *In*: Fookes, P.G. and Parry, R.H.G. (eds) *Engineering Characteristics of Arid Soils, Proceedings of the 1st International Symposium on*

Engineering Characteristics of Arid Soils, London, 1993. AA Balkema, Rotterdam, 313–329.

Fookes, P.G. and Gahir, J.S. 1995. Engineering performance of some coarse-grained arid soils in the Libyan Fezzan. *Quarterly Journal of Engineering Geology*, **28**, 105–130, https://doi.org/10.1144/GSL.QJEGH.1995.028.P2.03

Fookes, P.G. and Gray, J.M. 1986. Geomorphology and civil engineering. *In*: Gardiner, V. (ed.) *International Geomorphology, Part 1*. Wiley, Chichester, 83–105.

Fookes, P.G. and Hawkins, A.B. 1988. Limestone weathering: its engineering significance and a proposed classification scheme. *Quarterly Journal of Engineering Geology*, **21**, 7–31, https://doi.org/10.1144/GSL.QJEG.1988.021.01.02

Fookes, P.G. and Higginbottom, I.E. 1975. The classification and description of near-shore carbonate sediments for engineering purposes. *Géotechnique*, **25**, 406–411, https://doi.org/10.1680/geot.1975.25.2.406

Fookes, P.G. and Higginbottom, I.E. 1980. Some problems of construction aggregates in desert areas, with particular reference to the Arabian Peninsula. 2: Investigation, production and quality control. *Proceedings of the Institution of Civil Engineers*, Part 2, **68**, 69–90.

Fookes, P.G. and Higginbottom, I.E. 1981. Some problems of construction aggregates in desert areas with particular reference to the Arabian Peninsula. *Proceedings of the Institution of Civil Engineers*, Part 1, **70**, 519–522.

Fookes, P.G. and Higginbottom, I.E. 2003. Discussion on Burton, C. J., Waltham, A. C. & McLaren, S. J. 2001. Strength variation in young reef limestones, in *Géotechnique*, 51, 887–889. *Géotechnique*, **53**, 764–766, https://doi.org/10.1680/geot.2003.53.8.764

Fookes, P.G. and Horswill, P. 1969. Discussion on engineering grade zones. *Proceedings, Conference on In Situ Testing Soils Rock*, Institution of Civil Engineers, London, 53–57.

Fookes, P.G. and Knill, J.L. 1969. The application of engineering geology in the regional development of northern and central Iran. *Engineering Geology*, **3**, 81–120, https://doi.org/10.1016/0013-7952(69)90001-5

Fookes, P.G. and Lee, E.M. 2005. Introduction to engineering geomorphology. *In*: Fookes, P.G., Lee, E.M. and Milligan, G. (eds) *Geomorphology for Engineers*. Whittles, Caithness, 1–28.

Fookes, P.G. and Lee, E.M. 2007. Climate variation: a simple geological perspective. *Geology Today*, **23**, 66–73, https://doi.org/10.1111/j.1365-2451.2007.00603.x

Fookes, P.G. and Lee, E.M. 2009*a*. Desert environments of inland Oman. *Geology Today*, **25**, 226–231, https://doi.org/10.1111/j.1365-2451.2009.00735.x

Fookes, P.G. and Lee, E.M. 2009*b*. Desert environments: landscapes and stratigraphy. *Geology Today*, **25**, 172–180, https://doi.org/10.1111/j.1365-2451.2009.00722.x

Fookes, P.G. and Lee, E.M. 2017. The early days of engineering geology in the UK – a personal view by Peter Fookes F.R.Eng. *Quarterly Journal of Engineering Geology and Hydrogeology*, **50**, 472–477, https://doi.org/10.1144/qjegh2017-102

Fookes, P.G. and Lee, E.M. 2018. The engineering geology of playas, salt playas and Salinas. *Quarterly Journal of Engineering Geology and Hydrogeology*, **51**, 287–298, https://doi.org/10.1144/qjegh2017-084

Fookes, P.G. and Lee, E.M. 2019. The engineering geology of concrete in hot drylands. *Quarterly Journal of Engineering Geology and Hydrogeology*, **52**, 257–259, https://doi.org/10.1144/qjegh2018-185

Fookes, P.G. and Marsh, A.H. 1981*a*. Some characteristics of construction materials in the low to moderate metamorphic grade rocks of the Lower Himalayas of East Nepal. 1: Occurrence and geological features. *Proceedings of the Institution of Civil Engineers*, Part 1, **70**, 123–138, https://doi.org/10.1680/iicep.1981.1961

Fookes, P.G. and Marsh, A.H. 1981*b*. Some characteristics of construction materials in the low to moderate metamorphic grade rocks of the Lower Himalayas of East Nepal. 2: Engineering characteristics. *Proceedings of the Institution of Civil Engineers*, Part 1, **70**, 139–162, https://doi.org/10.1680/iicep.1981.1962

Fookes, P.G. and Martin, P.L. 1978. Site investigation and geotechnical considerations. *Conference on Thames Barrier Design*, Institution of Civil Engineers, London, 29–50.

Fookes, P. and Martin, P. 2006. Discussion on 'What lies beneath: surveying the Thames at Woolwich'. Lenham, J., Beyer, V., Edmonds, H., Harris, D., Mortimore, R., Reynolds, J. & Black, M. *Proceedings Institution of Civil Engineers: Civil Engineering*, **159**, 32–41, https://doi.org/10.1680/cien.2006.159.1.32

Fookes, P.G. and McKenna, J.M. 1963. An approach to rock mechanics by Klaus W. John – a discussion. *Journal of the Soil Mechanics and Foundations Division*, Proceedings of the American Society of Civil Engineers, Part 1, **89**, 139–140.

Fookes, P.G. and Mellors, T.W. 1970. Discussion on soil anchors. *Ground Engineering, Proceedings of the Conference*. Institution of Civil Engineers, London, 103–105.

Fookes, P.G. and Parrish, D.G. 1969. Observations on small-scale structural discontinuities in the London Clay and their relationship to regional geology. *Quarterly Journal of Engineering Geology*, **1**, 217–240, https://doi.org/10.1144/GSL.QJEG.1969.001.04.01

Fookes, P.G. and Parry, R.H.G. (eds) 1994. Engineering characteristics of arid soils. *Proceedings of the 1st International Symposium on Engineering Characteristics of Arid Soils*, London, 1993. AA Balkema, Rotterdam.

Fookes, P.G. and Poole, A.B. 1981. Some preliminary considerations on the selection and durability of rock and concrete materials for breakwaters and coastal protection works. *Quarterly Journal of Engineering Geology*, **14**, 97–128, https://doi.org/10.1144/GSL.QJEG.1981.014.02.03

Fookes, P.G. and Revie, W.A. 1982. Mica in concrete – a case history from Eastern Nepal. *Concrete*, **16**, 12–16. Reprinted in *Concrete in the Middle East, Part 2*, as part of reprints of three articles from *Concrete*, 3–33.

Fookes, P.G. and Shilston, D.T. 2001. Building the geological model: case study of a rock tunnel in SW England. *Geological Society, London, Engineering Geology Special Publications*, **18**, 123–128, https://doi.org/10.1144/GSL.ENG.2001.018.01.18

Fookes, P.G. and Sweeney, M. 1976. Stabilization and control of local rock falls and degrading rock slopes. *Quarterly Journal of Engineering Geology*, **9**, 37–55, https://doi.org/10.1144/GSL.QJEG.1976.009.01.03; reprinted in 1977 in *Laboratorio del Transporte y Mecánica del Suelo, Boletin de Informacion*, **119**, 17–31 [in Spanish]

Fookes, P.G. and Thomas, R.S. 1986. Rapid site appraisal of potential breakwater rock at Qeshm, Iran. *Proceedings of the Institution of Civil Engineers*, Part 1, **80**, 1297–1325, https://doi.org/10.1680/iicep.1986.585

Fookes, P.G. and Vaughan, P.R. (eds) 1986. *A Handbook of Engineering Geomorphology*. Surrey University Press, Guildford.

Fookes, P.G. and Weltman, A.J. 1989. Rock slopes: stabilization and remedial measures against degradation in weathered and fresh rock. *Proceedings of the Institution of Civil Engineers*, Part 1, **86**, 359–380; also in 1989 in Discussion, Rock slopes:

stabilization and remedial measures against degradation in weathered and fresh rock. *Proceedings of the Institution of Civil Engineers*, Part 1, 86, 1209–1211.

Fookes, P.G. and Wilson, D.D. 1966. The geometry of discontinuities and slope failures in Siwalik Clay. *Géotechnique*, **XVI**, 305–320, https://doi.org/10.1680/geot.1966.16.4.305

Fookes, P.G., Dearman, W.R. and Franklin, J.A. 1971. Some engineering aspects of rock weathering with field examples from Dartmoor and elsewhere. *Quarterly Journal of Engineering Geology*, **4**, 135–185, https://doi.org/10.1144/GSL.QJEG.1971.004.03.01

Fookes, P.G. (Subcommittee Chairman), Cratchley, C.R., and Day, J.B.W. 1972*a*. Preparation of maps and plans in terms of engineering geology. Report by the Geological Society Engineering Group Working Party. *Quarterly Journal of Engineering Geology*, **5**, 293–381, https://doi.org/10.1144/GSL.QJEG.1972.005.03.05

Fookes, P.G., Hinch, L.W. and Dixon, J. 1972*b*. Geotechnical considerations of the site investigation for Stage IV of the Taff Vale Trunk Road in South Wales. *Permanent International Association of Road Congresses*. Second British Regional Congress, Cardiff, 1–25.

Fookes, P.G., Gordon, D.L. and Higginbottom, I.E. 1975*a*. Glacial landforms, their deposits and engineering characteristics. *The Engineering Behaviour of Glacial Materials, Symposium Proceedings Midland Society of Soil Mechanics and Foundation Engineering*, Birmingham, 18–51.

Fookes, P.G., Hinch, L.W., Huxley, M.A. and Simons, N.E. 1975*b*. Some soil properties in glacial terrain – the Taff Valley, South Wales. *The Engineering Behaviour of Glacial Materials, Symposium Proceedings Midland Society of Soil Mechanics and Foundation Engineering*, Birmingham, 100–123.

Fookes, P.G., Reeves, B.J. and Dearman, W.R. 1977. The design and construction of a rock slope in weathered slate at Fowey, Southwest England. *Géotechnique*, **27**, 533–556, https://doi.org/10.1680/geot.1977.27.4.533

Fookes, P.G., Pollock, D.J. and Kay, E.A. 1981. Middle East concrete (2) – Rates of deterioration. *Concrete*, **15**, 12–19; and reprinted in 1982 in *Concrete in the Middle East, Part 2*, as part of reprints of three articles from *Concrete*, 12–22.

Fookes, P.G., Comberbach, C.D. and Cann, J. 1983*a*. Field investigation of concrete structures in south-west England – Part 1. *Concrete*, **17**, 54–56.

Fookes, P.G., Comberbach, C.D. and Cann, J. 1983*b*. Field investigation of concrete structures in south-west England – Part 2. *Concrete*, **17**, 60–65.

Fookes, P.G., Cann, J. and Comberbach, C.D. 1984. Field investigation of concrete structures in south-west England Part 3. *Concrete*, **18**, 12–16.

Fookes, P.G. (Chairman), Smith, M.R. *et al.* 1985*a*. *Aggregates: Sand, gravel and Crushed Rock Aggregates for Construction Purposes*. Engineering Group Working Party Report, Geological Society, London. (Summarized by I. J. Sims, 1986 in *Quarterly Journal of Engineering Geology*, **19**, 325–338.)

Fookes, P.G., French, W.J. and Rice, S.M.M. 1985*b*. The influence of ground and groundwater geochemistry on construction in the Middle East. *Quarterly Journal of Engineering Geology*, **18**, 101–128, https://doi.org/10.1144/GSL.QJEG.1985.018.02.01

Fookes, P.G., Sweeney, M., Manby, C.N.D. and Martin, R.P. 1985*c*. Geological and geotechnical engineering aspects of low-cost roads in mountainous terrain. *Engineering Geology*, **21**, 1–152, https://doi.org/10.1016/0013-7952(85)90002-X

Fookes, P.G., West, R., Mettam, J.D. and Berry, J.G. 1985*d*. Concrete deterioration and corrosion of steel in marine environments. *26th International Navigation Congress*, Permanent International Association of Navigation Congresses, Brussels, June, Section II, Subject 4, 99–111.

Fookes, P.G., Simm, J.D. and Barr, J.M. 1986. Marine concrete performances in different climatic environments. *Marine Concrete 86 – International Conference on Concrete in the Marine Environment, London, September*. The Concrete Society, London, 115–130.

Fookes, P.G., Barr, J.M. and Simm, J.D. 1987. Performance of marine concrete and characteristics of component materials in different climatic environments. *Proceedings of Conference on Improving Concrete in Marine Environments*. Institution for International Research, Hong Kong, June, 7.1–7.38.

Fookes, P.G., Gourley, C.S. and Ohikere, C. 1988. Rock weathering in engineering time. *Quarterly Journal of Engineering Geology*, **21**, 33–57, https://doi.org/10.1144/GSL.QJEG.1988.021.01.03

Fookes, P.G. (Chairman) 1990. Tropical Residual Soils: Geological Society Engineering Group Working Party Report. *Quarterly Journal of Engineering Geology*, **23**, 1–101, https://doi.org/10.1144/GSL.QJEG.1990.023.001.01

Fookes, P.G., Dale, S.G. and Land, J.M. 1991*a*. Some observations on a comparative aerial photography interpretation of a landslipped area. *Quarterly Journal of Engineering Geology*, **24**, 249–265, https://doi.org/10.1144/GSL.QJEG.1991.024.03.01

Fookes, P.G., Stoner, J.R. and Mackintosh, J.H. 1991*b*. The Libyan Great Man-made River Project Phase 1 Part 2: concrete technology of prestressed concrete pipe manufacture. *Proceedings of the Institution of Civil Engineers*, Part 1, **90**, 881–919. Also in 1992, Discussion, The Libyan Great Man-made River Project Phase 1; Parts 1 and 2. *Proceedings of the Institution of Civil Engineers*, **94**, 501–502.

Fookes, P.G. (Chairman), Smith, M.R. *et al.* 1993*a*. *Aggregates: Sand, Gravel and Crushed Rock Aggregates for Construction Purposes*, 2nd edn. Geological Society, London, Engineering Group Working Party Report.

Fookes, P.G., Stoner, J.R. and Mackintosh, J. 1993*b*. Great Manmade River Project, Libya, Phase 1: a case study on the influence of climate and geology on concrete technology. *Quarterly Journal of Engineering Geology*, **26**, 25–60, https://doi.org/10.1144/GSL.QJEG.1993.026.01.04

Fookes, P.G., Lee, E.M. and Sweeney, M. 1999. Engineering terrain evaluation for a desert pipeline. *Onshore Pipelines Conference, IBC 3rd International Event*, 18–19 November, InterContinental Hotel, Berlin, 1–42.

Fookes, P.G., Baynes, F.J. and Hutchinson, J.N. 2000*a*. Total geological history: a model approach to the anticipation, observation and understanding of site conditions. *In: GeoEng2000, an International Conference on Geotechnical & Geological Engineering*, 19–24 November 2000, Melbourne, Australia. Conference Proceedings, Technomic Publishing, Lancaster, USA, 370–460.

Fookes, P.G., Lee, E.M. and Sweeney, M. 2000*b*. Engineering terrain evaluation: a cost-cutting route – investigation method. *In: Onshore Pipelines Cost Reduction Conference*, organized by Pipes and Pipelines International UK and Clarion Technical Conferences, US, 3–5 April, Marriott Hotel, Amsterdam.

Fookes, P.G., Lee, E.M. and Sweeney, M. 2000*c*. Engineering terrain evaluation – with a desert pipeline case record. *In: First BP AMOCO Forum*, 27 February–1 March, Woodlands Resort, Houston, TX.

Fookes, P.G., Baynes, F.J. and Hutchinson, J.N. 2001*a*. Total geological history: a model approach to understanding site conditions (Technical Note). *Ground Engineering*, **34**, 42–47.

Fookes, P.G., Lee, E.M. and Sweeney, M. 2001*b*. Pipeline route selection and ground characterization, Algeria. *In*: Griffiths, J.S. (ed.) *Land Surface Evaluation for Engineering Practice*. Geological Society, London, Engineering Geology Special Publications, **18**, 115–121, https://doi.org/10.1144/GSL.ENG.2001.018.01.17

Fookes, P.G., Hearn, G.J. and Tollin, B. 2005*a*. Aligning oil and gas pipelines through the Makarov Mountains, Sakhalin Island, Russia. *Terrain and Geohazard Challenges Facing Onshore Oil and Gas Pipelines Conference, London, 2004*. Institution of Civil Engineers, London, 243–258.

Fookes, F.G., Lee, E.M. and Milligan, G. (eds) 2005*b*. *Geomorphology for Engineers*. Whittles, Caithness.

Fookes, P.G., Lee, E.M. and Sweeney, M. 2005*c*. In Sala Gas Project, Algeria – Part 1: Terrain evaluation for desert pipeline routing. *Terrain and Geohazard Challenges Facing Onshore Oil and Gas Pipelines Conference, London, 2004*. Institution of Civil Engineers, London, 144–161.

Fookes, P.G., Lee, E.M. and Griffiths, J.S. 2007. *Engineering Geomorphology: Theory and Practice*. Whittles, Caithness.

Fookes, P.G., Hart, A.B. and Lee, E.M. 2012. Some near-surface desert features of significance in engineering geology evaluations. *Quarterly Journal of Engineering Geology and Hydrogeology*, **46**, 259–266, https://doi.org/10.1144/qjegh2012-083

Fookes, P.G., Pettifer, G.S. and Waltham, A.C. 2015. *Geomodels in Engineering Geology: An Introduction*. Whittles, Caithness.

Fookes, P.G., Waring, D. and Lee, M. 2019. Wilfred Thesiger and the Umm as Samim, Oman. *Geology Today*, **35**, 221–227.

French, W.J., Pettifer, G.S. and Fookes, P.G. 2000. The occurrence of smectite in the Mercia mudstone of Leicestershire. *In:* The engineering properties of the Mercia Mudstone Group. Papers presented at a seminar held at Pride Park, Derby, UK on 25 November 1998. *CIRIA Conference Record*, **2000/1**, 137–141.

Griffiths, J.S., Fookes, P.G., Hardingham, A.D. and Barsby, R.D. 1994. Geotechnical soils mapping for construction purposes in Central Saudi Arabia. *In*: Fookes, P.G. and Parry, R.H.G. (eds) *Engineering Characteristics of Arid Soils, Proceedings of the 1st International Symposium on Engineering Characteristics of Arid Soils*, London, AA Balkema, Rotterdam, 1993, 69–85.

Griffiths, J.S., Hutchinson, J.N., Brunsden, D., Petley, D.J. and Fookes, P.G. 2004. The reactivation of a landslide during the construction of the Ok Ma tailings dam, Papua New Guinea. *Quarterly Journal of Engineering Geology and Hydrogeology*, **37**, 173–186, https://doi.org/10.1144/1470-9236/02-034

Griffiths, J.S., Fookes, P.G., Goudie, A.S., Stokes, M. and Engineering Group Working Party 2012. Processes and Landforms in Deserts. *Geological Society, London, Engineering Geology Special Publications*, **25**, 33–95, https://doi.org/10.1144/EGSP25.03

Higginbottom, I.E. and Fookes, P.G. 1970. Engineering aspects of periglacial features in Britain. *Quarterly Journal of Engineering Geology*, **3**, 85–117, https://doi.org/10.1144/GSL.QJEG.1970.003.02.02

Higginbottom, E.A. and Fookes, P.G. 1982. Geotechnical considerations for construction in Saudi Arabia – Discussion. *American Society of Civil Engineers. Geotechnical Division*, June, 900–902.

Hinch, L.W. and Fookes, P.G. 1989*a*. Taff Vale trunk road stage 4, South Wales. 1: location and geology. *Proceedings of the Institution of Civil Engineers*, Part 1, **86**, 139–160.

Hinch, L.W. and Fookes, P.G. 1989*b*. Taff Vale trunk road stage 4, South Wales. 2: geotechnical design. *Proceedings of the Institution of Civil Engineers*, Part 1, **86**, 161. Also in 1989 in Discussion, Taff Vale trunk road stage 4, South Wales. 2: geotechnical design. *Proceedings of the Institution of Civil Engineers*, Part 1, **86**, 1021–1022.

Jackson, J.O. and Fookes, P.G. 1974. The relationship of the estimated former burial depth of the Lower Oxford Clay to some soil properties. *Quarterly Journal of Engineering Geology*, **7**, 137–179, https://doi.org/10.1144/GSL.QJEG.1974.007.02.03

Kay, E.A., Fookes, P.G. and Pollock, D.J. 1981. Middle East concrete (3) – Deterioration related to chloride ingress. *Concrete*, **15**, 22–28; reprinted in 1982 in *Concrete in the Middle East, Part 2*, as part of reprints of three articles from *Concrete*, 23–33.

Knill, J.L. and Fookes, P.G. 1967. Preliminary studies on the geotechnical properties of recent wash sediments. *7th International Sedimentological Congress*, Reading and Edinburgh, 1–4.

Lee, E.M. and Fookes, P.G. 2005*a*. Climate and weathering. *In*: Fookes, P.G., Lee, E.M. and Milligan, G. (eds) *Geomorphology for Engineers*. Whittles, Caithness, 31–56.

Lee, E.M. and Fookes, P.G. 2005*b*. Hot drylands. *In*: Fookes, P.G., Lee, E.M. and Milligan, G. (eds) *Geomorphology for Engineers*. Whittles, 419–453.

Lee, E.M. and Fookes, P.G. 2014. A note on the origins of engineering geomorphology in the UK. *Quarterly Journal of Engineering Geology and Hydrogeology*, **48**, 147–156, https://doi.org/10.1144/qjegh2014-048

Lee, E.M., Griffiths, J.S. and Fookes, P.G. 2004. Engineering geology/geomorphology. *In*: Selley, R.C., Cocks, R. and Plimer, I.R. (eds) *Encyclopaedia of Geology*. Elsevier, Amsterdam, 474–481.

Lee, E.M., Fookes, P.G. and Hart, A.B. 2012. Observations on the impact of climate change on landform behaviour and geohazards in the Algerian Sahara. *Quarterly Journal of Engineering Geology and Hydrogeology*, **46**, 107–116 https://doi.org/10.1144/qjegh2012-058

Lee, E.M., Fookes, P.G. and Hart, A.B. 2016. Landslide issues associated with oil and gas pipelines in mountainous terrain. *Quarterly Journal of Engineering Geology and Hydrogeology*, **49**, 125–131, https://doi.org/10.1144/qjegh2016-020

Little, A.L., Stewart, J.C. and Fookes, P.G. 1963. Bedrock grouting tests at Mangla Dam, West Pakistan. *Grouts & Drilling Muds in Engineering Practice, Symposium*, Institution of Civil Engineers, London, 91–97.

Lovegrove, G.W. and Fookes, P.G. 1972. The planning and implementation of a site investigation for a highway in tropical conditions in Fiji. *Quarterly Journal of Engineering Geology*, **5**, 43–68, https://doi.org/10.1144/GSL.QJEG.1972.005.01.05

Milligan, G., Fookes, P.G. and Lee, E.M. 2005. Engineering behaviour of soils and rocks. *In*: Fookes, P.G., Lee, E.M. and Milligan, G. (eds) *Geomorphology for Engineers*. Whittles, Caithness, 137–169.

Norman, J.W., Leibowitcz, T.H. and Fookes, P.G. 1975. Factors affecting the detection of slope instability with air photographs in an area near Sevenoaks, Kent. *Quarterly Journal of Engineering Geology*, **8**, 159–176, https://doi.org/10.1144/GSL.QJEG.1975.008.03.01

Obika, B., Freer-Hewish, R.J. and Fookes, P.G. 1989. Soluble salt damage to thin bituminous road and runway surfaces. *Quarterly Journal of Engineering Geology*, **22**, 59–73, https://doi.org/10.1144/GSL.QJEG.1989.022.01.05

Pettifer, G.S. and Fookes, P.G. 1994. A revision of the graphical method for assessing the excavatability of rock. *Quarterly Journal of Engineering Geology*, **27**, 145–164, https://doi.org/10.1144/GSL.QJEGH.1994.027.P2.05

Pollock, D.J., Kay, E.A. and Fookes, P.G. 1981. Crack mapping for investigation of Middle East concrete. *Concrete*, **15**, 12–18; reprinted in 1982 in *Concrete in the Middle East, Part 2*, as part of reprints of three articles from *Concrete*, 3–11.

Poole, A.B., Fookes, P.G., Dibb, T.E. and Hughes, D.W. 1984. Durability of rock in breakwaters. *Conference on Breakwaters – Design and Construction*. Institution of Civil Engineers, London, 31–42.

Sanders, M.K. and Fookes, P.G. 1970. A review of the relationship of rock weathering and climate and its significance to foundation engineering. *Engineering Geology*, **4**, 289–325, https://doi.org/10.1016/0013-7952(70)90021-9

Sanders, M.K. and Fookes, P.G. 1974. The performance of a trial embankment on Mangrove swamp, Fiji. *Proceedings of Symposium on Field Instrumentation in Geotechnical Engineering, Part 2*. Institution of Civil Engineers, London, 633–636.

Shilston, D.T., Lee, E.M., Fookes, P.G. and Standish, I. 2005*a*. In Sala Gas Project, Algeria – Part 2: Detailed terrain evaluation for site design. *Terrain and Geohazard Challenges Facing Onshore Oil and Gas Pipelines Conference, London, 2004*. Institution of Civil Engineers, London, 162–174.

Shilston, D.T., Lee, E.M., Pollos-Pirallo, S., Morgan, D., Clarke, J., Fookes, P.G. and Brunsden, D. 2005*b*. Terrain evaluation and site investigations for design of the Trans-Caucasus oil and gas pipelines in Georgia. *Terrain and Geohazard Challenges Facing Onshore Oil and Gas Pipelines Conference, London, 2004*. Institution of Civil Engineers, London, 283–300.

Simm, J.D. and Fookes, P.G. 1989. Improving reinforced concrete durability in the Middle East during the period 1960–1985: an analytical review. *Proceedings of the Institution of Civil Engineers*, Part 1, **86**, 333–358. Also in 1990 in Discussion, Improving reinforced concrete durability in the Middle East during the period 1960–1985: an analytical review. *Proceedings of the Institution of Civil Engineers*, Part 1, **86**, 165–178.

Stoner, J.R. and Fookes, P.G. 1989. The Great Man-made River Project (Photographic Feature). *Quarterly Journal of Engineering Geology*, **22**, 253–256, https://doi.org/10.1144/GSL.QJEG.1989.022.04.01

Stoner, J.R. and Fookes, P.G. 1991. The Libyan Great Man-made River Project Phase 1 Part 1: manufacture of prestressed concrete cylinder pipe. *Proceedings of the Institution of Civil Engineers*, Part 1, **90**, 85–879. Also in 1992 in Discussion, The Libyan Great Man-made River Project Phase 1; Parts 1 and 2. *Proceedings of the Institution of Civil Engineers*, **94**, 501–502.

Walker, M., Bamforth, P. *et al.* 2002. *Guide to the Construction of Reinforced Concrete in the Arabian Peninsula*. CIRIA and The Concrete Society, London.

Waltham, T. and Fookes, P.G. 2001. Ice wedges of the Dalton Highway, Alaska (Photographic Feature). *Quarterly Journal of Engineering Geology and Hydrogeology*, **34**, 65–70, https://doi.org/10.1144/qjegh.34.1.65

Waltham, A.C. and Fookes, P.G. 2003. Engineering classification of karst ground conditions. *Quarterly Journal of Engineering Geology and Hydrogeology*, **36**, 101–118, https://doi.org/10.1144/1470-9236/2002-33

Waltham, A.C. and Fookes, P.G. 2005. Engineering classification of karst ground conditions. *Speleogenesis and Evolution of Karst Aquifers. Online Scientific Journal*, **3**, Issue 1.

What do we mean by a ground model – the Fookes geomodel legacy

James S. Griffiths

SoGEES, University of Plymouth, Drake Circus, Plymouth PL4 8AA, UK
0000-0002-6431-3992
Correspondence: jim.griffiths@plymouth.ac.uk

Abstract: The first Glossop lecture by Professor Fookes in 1996 established the importance of the geological model in predicting ground conditions in engineering geology. The concept of the geological model was expanded in his 2000 paper on 'total geology', and this becomes the 'geomodel' in a 2015 book. Fookes was clear that the *a priori* geological model/geomodel developed from existing knowledge and site observations and was fundamental to planning and interpreting the ground conditions data that emerged from investigations.

A geological model was referred to in the first iteration (2007) of Eurocode 7 and fed into a geotechnical design model. This usage was reaffirmed in 2022 within an overall Engineering Geological Model Framework by IAEG Commission 25. In Hong Kong the 2007 engineering geology manual identified a geological model as feeding into a ground model and then into a geotechnical model.

The expression 'ground model' has subsequently emerged as providing a synopsis of the ground conditions that feeds into geotechnical design. This is recognized in the second iteration of Eurocode 7 and is in the 2023 CIRIA guidance on Geotechnical Baseline reports. The Society for Underwater Technology 2022 Report refers to the ground model and defines it as a database of available information that encompasses written reports, the database of collated information and the geotechnical risk register.

Given these different views it is necessary to agree what constitutes a ground model, as this has important implications for contract documentation and establishing the legal responsibilities of all ground engineering professionals.

In memoriam Peter G. Fookes (1933–2020)

In the Oxford English Dictionary, the term 'model' is defined as:

(1) A three-dimensional representation of a person or thing or of a proposed structure, typically on a smaller scale than the original.
(2) A thing used as an example to follow or imitate.

These definitions fit with the accepted use of the term in engineering geology and geotechnics where 'models' of various types are used to provide conceptualizations of the ground conditions in order to facilitate safe and economic design of the foundations to structures of all types. However, whilst the term 'model' is used extensively, the accepted definitions for the various types of models associated with engineering geology and geotechnics are still subject to debate. For example, Norbury (2021) pointed out that there was no universally agreed definition to the widely used term 'ground model'. Similarly, Griffiths and Stokes (2008) showed that the terms geotechnical model (Site Investigation Steering Group 1993), engineering geological model (Culshaw 2005), geological model (Fookes 1997) and site characterization (Mayne *et al.* 2001; Hatheway 2002; Tepel 2004) have all been used in similar and overlapping ways. This has important implications for engineering works because contract documents will be produced for a project where the requirement may be to produce 'models' of the site conditions and the nature, extent, accuracy, and suitability of these models could be a matter of legal contention in the event of claims or disputes arising during construction or after the works are complete. For example, the occurrence of a landslide during the construction of the Ok Ma tailings dam in Papua New Guinea in 1985 led to a £575 million insurance claim and the inadequacy of the geomorphological, engineering geological and geotechnical models was cited as contributing to the failure (Griffiths *et al.* 2004).

Given this background it is advantageous to see how the term 'model' in engineering geology came into use, has developed to mean different things to different groups, and to note the fundamental role that Professor Fookes played in establishing the approach. This paper reviews the geological model described by Fookes (1997) in the first Glossop Lecture, that developed into a 'total geology' model (Fookes *et al.* 2000), and then a 'geomodel' (Fookes *et al.* 2015). The manner in which the term 'model' has progressed through other engineering geological and geotechnical publications is then explored, namely in: Hong Kong (Geotechnical Engineering Office 2007); Commission 25 of the IAEG on engineering geological models (Parry *et al.* 2014; Baynes and Parry 2022); the CIRIA report on baseline conditions (Davis *et al.* 2023); Europe through Eurocode 7 Geotechnical

From: Griffiths, J. S., Giles, D. P., Davis, J. A. and Dunlop, J. J. (eds) 2025. *The Peter Fookes Engineering Geological Legacy in Geomodels, Geomaterials and Geomorphology*. Geological Society, London, Engineering Geology Special Publications, **30**, 31–46,
First published online 29 January 2025, https://doi.org/10.1144/egsp30-2023-11

Design, with the first iteration in 2007 (EN 1997-2 2007), the second issued in draft form in late 2022 (prEN 1997-1 2022; prEN 1997-2 2022; prEN 1997-3 2022) and parts in published form in September 2024 (BS EN 1997-2 2024; Menteth 2024); and in the offshore site investigation industry (Thomas 2017; Cook 2022). Finally, drawing on this review, some recommendations are made about what the 'model' in engineering geology actually is and how engineering geologists are key to ensuring it is created and developed in a way that allows for safe geotechnical design and construction.

Professor Peter Fookes and geomodels

First Glossop Lecture 1996 (published in 1997)

The First Glossop Lecture (Fookes 1997) was entitled 'Geology for Engineers: The geological model, prediction and performance'. In this landmark paper Fookes (Fig. 1) stated:

> I believe that in the context of site investigation and in the understanding of the site, application of geological education and experience should be able to make a moderately close approximation of the actual geological conditions from the desk study and that when this is supplemented by ground and laboratory investigations there should be, ideally, nothing that has not been discovered (p. 295).

Fookes emphasized that *a priori* knowledge forms the basis for the initial understanding of site conditions, and he went on to say:

> I do believe that it should be a realistic goal to be able to make a very close approximation to the actual site conditions, particularly if you know what you are looking for and what questions to ask. I have called this approximation the **Geological Model** (p. 295).

To illustrate the geological model, Fookes (1997) contained numerous three-dimensional block diagrams illustrating various geological and geomorphological situations. Some of these block diagrams were then reproduced in colour in Fookes *et al.* (2007) and an example is provided in Figure 2 of a mountain terrain model that evolved from the work on the Dharan–Dankuta road constructed in the Nepalese Himalayas (Fookes *et al.* 1985).

Fig. 1. Professor Peter Fookes in Liberia in 1983 providing the scale beside a dolerite dyke subject to spheroidal weathering in a wet tropical environment.

Having defined the geological model and provided many examples of these, Fookes (1997, p. 300) identified the responsibility of the engineering geologist: 'The main role of the engineering geologist is to get the geology right'.

Fookes (1997, p. 405) went on to say:

> I believe any near-surface geological feature should be anticipatable by a well-devised model supported by a an equally well-designed site investigation. I think that all potential geological situations on a civil engineering site should be foreseeable but in the context of the site investigation carried out may not actually have been foreseen.

An area that Fookes was concerned about in 1997 was the potential loss of field observation and mapping skills as technology developed. He stated:

> Despite the steady and successful rise of technology and the development of black boxes to supplement and even replace field observation, the skills of the field geologist and field geomorphologist are still a cornerstone, if not the main strength of engineering geology. There is, as yet no substitute for an informed brain and trained pair of eyes (p. 301).

The loss of core field skills, along with an understanding of Quaternary geology, are concerns that had been raised by a number of practitioners over subsequent years, notably in the Glossop series of lectures (see Hutchinson 2001; Brunsden 2002; DeFreitas 2009; Griffiths 2014).

Total Geology – Fookes *et al.* (2000)

At the 2000 Melbourne International Conference on Geotechnical and Geological Engineering, Professor Fookes joined two other outstanding geoscientists, Professor Hutchinson and Dr Baynes, to present a keynote paper on the concept of 'Total Geology'. This was an extension of his 1997 Glossop paper, with the premise being that the conditions and geotechnical characteristics of the ground are the product of the geological and geomorphological history of the site, including past and present climatic conditions, i.e. its total geological history. The paper showed how knowledge of the geological environment aided anticipation of the site conditions by the development of a preliminary engineering geological model, which guided the planning and interpretation of site ground investigations and fed into the design and construction of the project. The model developed progressively to be more site specific as the understanding of local geology improved during project development.

To predict the regional-scale geology, in their paper Fookes *et al.* (2000) presented ten 2D initial global-scale tectonic models based on the concepts of plate tectonics. These were divided into three groups:

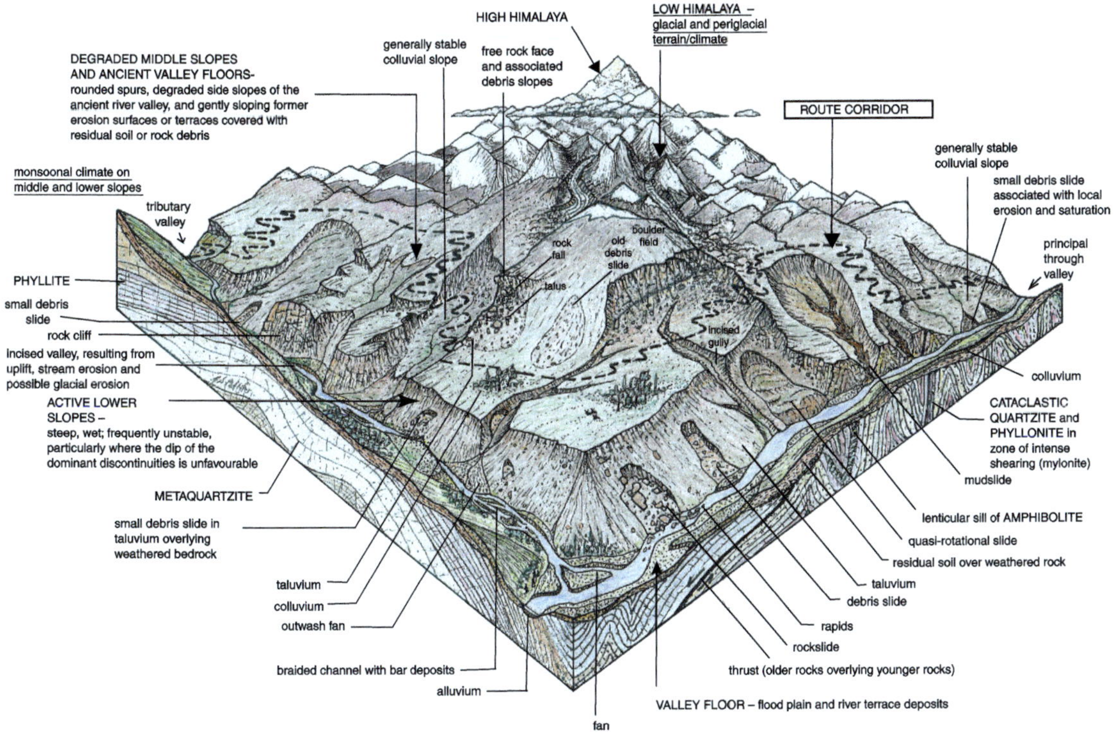

Fig. 2. The Mountain landscape. Source: Fookes *et al.* (2007, fig. 38.9), based on Fookes *et al.* (1985) and reproduced with kind permission of Whittles Publishing. Original drafting by Dr Geoff Pettifer.

(1) divergent plate boundaries – continental rifts, oceanic rifts;
(2) convergent plate boundaries – accretional prisms, fold and thrust belts, magmatic arcs, collision complexes, foreland basins;
(3) intraplate cratons, mobile belts, platform sediments and basins.

To help anticipate the local-scale geology there were 17 3D models. Each model had, what was described as, an anticipation list encompassing the rock-forming environments and tectonic and diagenetic modifications to these environments:

(1) igneous: basic volcanics, acid volcanic, plutonic intrusions;
(2) sedimentary: continental fluvial, colluvial & lacustrine; continental deltaic & coal measures, shelf carbonates & evaporates, deep marine and continental slopes;
(3) metamorphic: schists & phyllites; gneisses & migmatites;
(4) structural: faults (normal, strike slip, thrust); joints (in undeformed sediments); open folds and joints; plastic fold with cleavage; multiple folds and shears; tectonized melange.

To foresee the local geomorphology eight 3D initial geomorphological landform models were presented, each with an anticipation list related to climate and processes. These were in keeping with the concept of morphogenetic landforms (Chorley *et al.* 1984):

(1) zonal features from the following climates: glacial (valley glaciations), periglacial, temperate, hot dry, hot wet. These were labelled 'zonal features' because they were believed to align approximately with the world's climatic zones;
(2) azonal features: coastal, & soluble rock features. These were labelled azonal features as they were regarded as being independent of global climate. There are other azonal features related to volcanicity and structural geology and these formed some of the local geological models under the relevant rock type.

It should be noted that the concept of morphogenetic landforms has been subject to a great deal of debate in geomorphology. The consensus view is that whilst the broad regional climate conditions will be a major control on the development of landscapes, there needs to be as much emphasis on tectonics. It also must be recognized that climate zones migrated repeatedly during the Quaternary Period leaving relict landforms of these former climates in the present-day environments (Griffiths *et al.* 2012).

Geomodels in Engineering Geology: an introduction (Fookes *et al.* 2015)

The 2015 book on geomodels written with two other very experienced engineering geologists, Dr Pettifer and Dr Waltham, brought together all Fookes' previous work on the subject. In the preamble to the book, it is stated:

> **Geomodel** is the general term for any form of interpretation of a geological situation (p. iii).
>
> The pictorial models [in the book] are the springboard for developing the geomodel approach as a tool for use in ground investigations and to help the development of geotechnical models for design and construction. The development of site geomodels (ground models) during investigation is an important part of this book (p. vii).

The book is divided into five parts and an appendix.

Part 1 – describes the underlying factors of climate and geology and provides the geomorphological characteristics of the major morphoclimatic zones. Also included are the Fookes *et al.* (2000) total geology models (an example of the model for sediments and sedimentary rocks is provided in Fig. 3).

Part 2 – covers the changes to the near-surface ground conditions resulting from weathering together with a basic description of pedological soils and landscapes. Also included are some hillslope models that are found in weathered terrain.

Part 3 – describes the basic geological environments that influence engineering design and construction. In this section nine geomodels are presented based on the morphoclimatic zones discussed in Part 1 (an example of the savanna environment is provided as Fig. 4).

Part 4 – entitled 'Ground Investigations', this part demonstrates how knowledge increases at various stages during a site investigation.

Part 5 – this section provides nine synoptic case histories and some basic ground characteristics and properties.

The Appendix contains information on the geotechnical properties associated with different types of engineering soil.

In the main text of the book (p. 2), it is stated:

> Geomodels created during the desk study of a ground [site] investigation will allow a better understanding of the site and more efficient planning of the investigation.

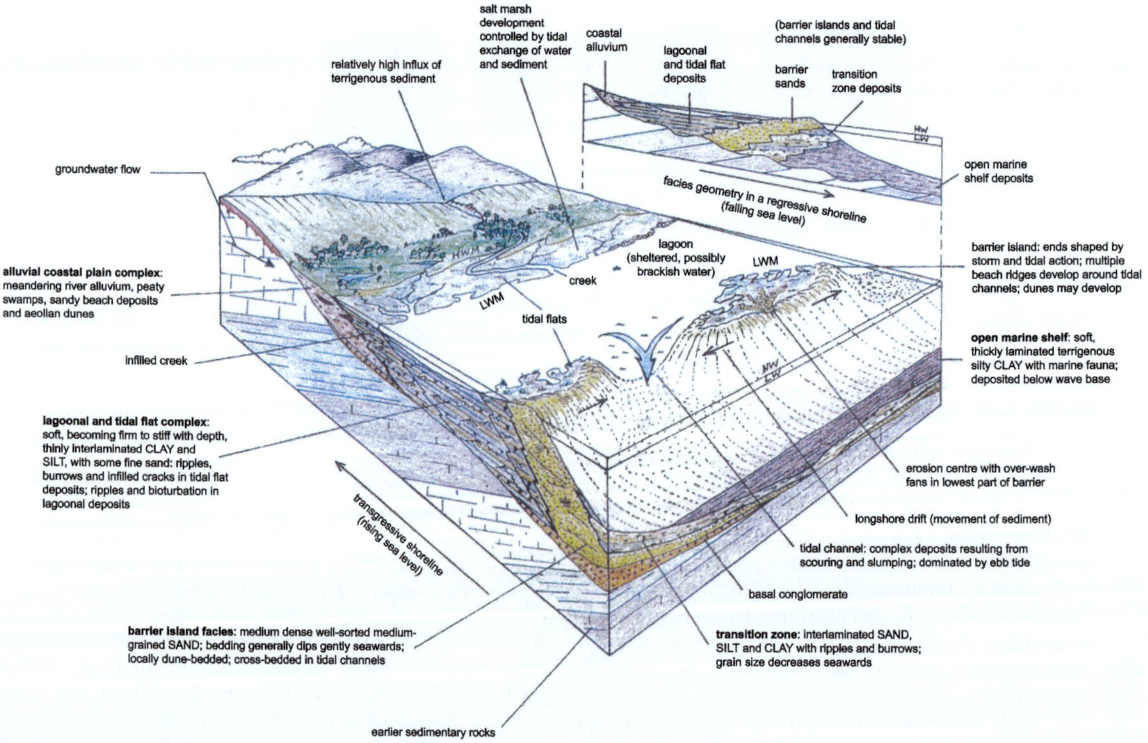

Fig. 3. Sediments and sedimentary rocks. Source: Fookes *et al.* (2015, fig. 1.7, p. 32), reproduced with kind permission of Whittles Publishing. Original drafting by Dr Geoff Pettifer.

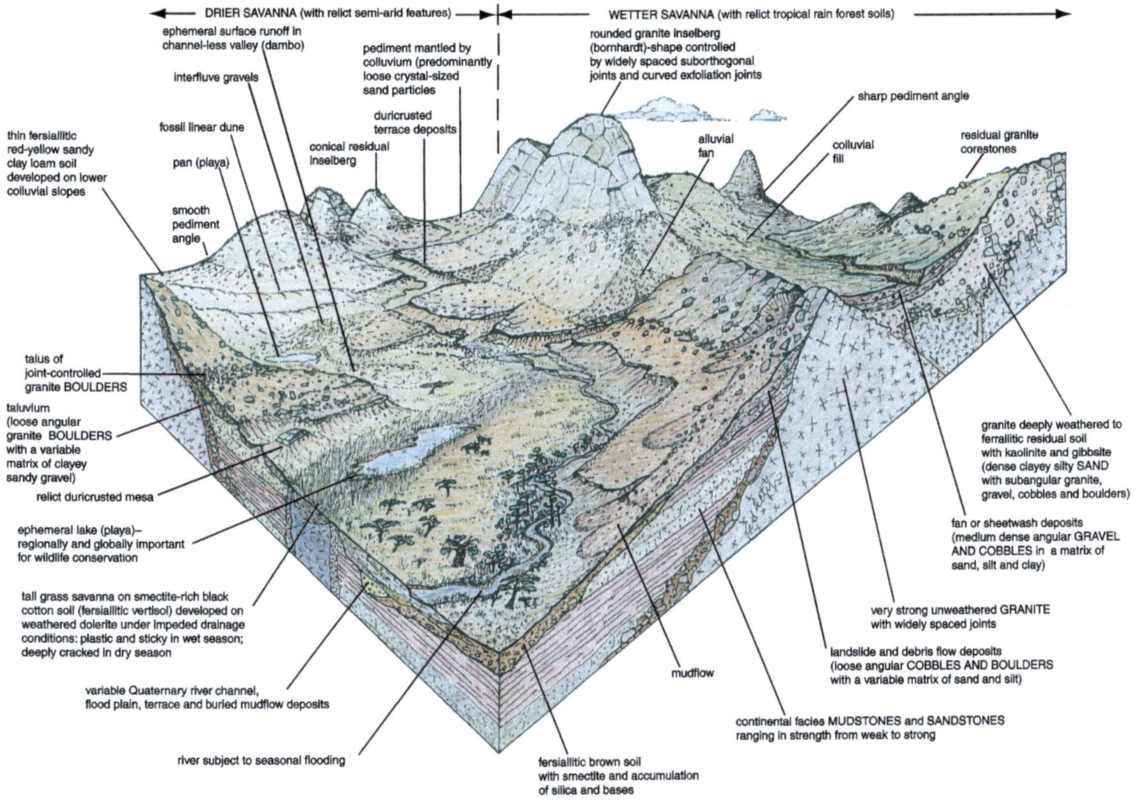

Fig. 4. Savanna Environments. Source: Fookes *et al.* (2015, fig. 3.7, p. 102), reproduced with kind permission of Whittles Publishing. Original drafting by Dr Geoff Pettifer.

> The models are continually reviewed and improved as the investigation evolves. It is essential to continually check, refine, or modify the initial model so that it becomes increasingly site specific.
>
> The total geological concept is required during the ground investigation stages following the early desk study to both develop thoroughly the initial geomodels and to present a comprehensive picture of the ground conditions.

The lessons of the Fookes geomodels' philosophy

In his Glossop lecture, Fookes (1997) quotes Rudolph Glossop from his 1968 Rankine Lecture: 'If you do not know what you should be looking for in a site investigation, you are not likely to find much of value' (Glossop 1968, p. 113).

In this case Fookes, and Glossop, are referring to the aspects of the site investigation (SI) that requires exploratory holes, which in the UK is now called the ground investigation (GI), although that distinction is not used in all countries. However, from this statement it is evident that the Fookes geomodel must be established *a priori* (literal Latin translation 'from what is before'). For Fookes it was axiomatic that no ground investigation should proceed without a clear idea of what was being 'looked for'. On that basis the Fookes geomodels, and how they are formulated, should be part of the toolkit of every engineering geologist. However, geomodels always require ground-truthing and customizing to the actual situation encountered in the field. In this context engineering geomorphology, notably through field mapping (Brunsden *et al.* 1975; Smith *et al.* 2011), emerged as a discipline that was central to producing bespoke observational geomodels models that fed into ground investigation design and interpretation (Lee and Fookes 2015; Fookes *et al.* 2005, 2007). It remains a fundamental requirement that the geomodels are continually updated as further data are obtained during the investigation and construction stages of a project.

Hong Kong Engineering geological practice (Geotechnical Engineering Office 2007)

In 2007 the Geotechnical Engineering Office in Hong Kong produced report GEO 1/2007 on engineering geological practice. This report introduced the principles of engineering geology as applicable to Hong Kong and illustrated their application to civil engineering works using examples and references. The report stated:

> Engineering geology provides the link between geology and engineering through the formation of geological models which can be used to identify geological hazards and uncertainty, plan effective ground investigations, and define blocks of ground and geological structures in an engineering context to facilitate geotechnical risk assessment and design (p. 22).

Within this comprehensive document there is a flow diagram that shows the typical development and application of the model approach for a major project. A simplified version of this flow diagram is presented in Figure 5, and this identifies three types of models: the geological, the ground and the design models.

(1) The **Geological Model** characterizes the site and focuses on the geological, geomorphological and hydrogeological features and characteristics that are relevant to the engineering project. As highlighted in Figure 5 an initial geological model is produced at the planning and feasibility stage of a project. Geotechnical Engineering Office (2007) refers to document PNAP 274 Geotechnical Design Information (Building Department 2003) that states:

> … the ground investigation fieldwork should then be planned with the objective of refining the geological model and the parameters to be used in the design, and identifying the various uncertainties involved as far as possible (p. 28).

(2) The **Ground Model** builds on the geological model and embeds the range of engineering parameters and ground conditions that need to be considered in the design. It is stated that:

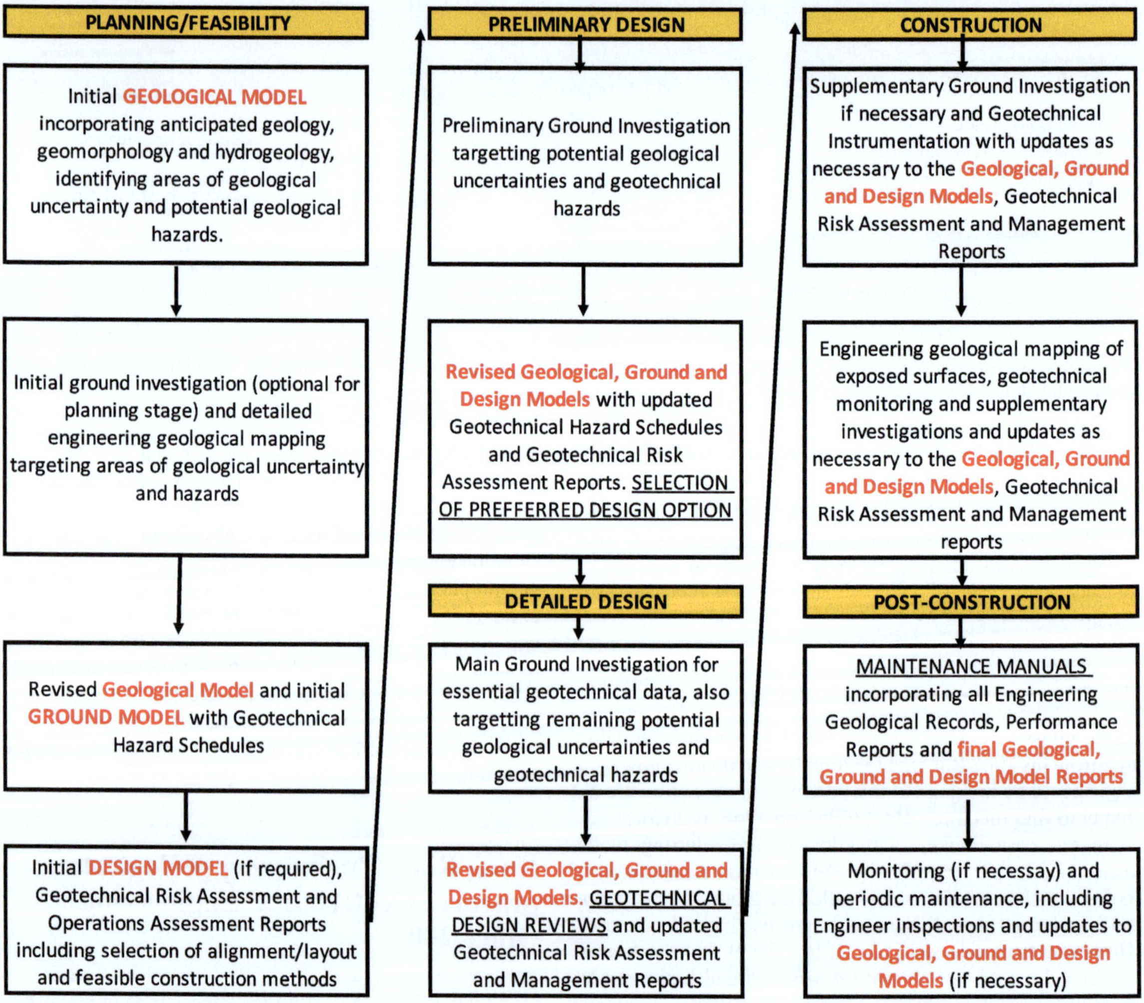

Fig. 5. The model approach for a major project. Source: adapted from GEO (2007).

> The ground model refines the geological model by defining and characterizing bodies of ground with similar engineering properties and identifies boundaries at which changes in geotechnical conditions may occur. Engineering geological input assists in ensuring as far as practicable that the ground model reflects the ground conditions indicated by the geological model (p. 29).

(3) The **Design Model** is primarily concerned with assessing the response of the ground to the proposed works for use in the geotechnical design. Geotechnical Engineering Office (2007) states that design models for empirical, prescriptive, and quantitative designs depend on the engineering applications, degree of conservatism in the empirical/prescriptive models and the level of [acceptable] geotechnical risk.

Reference to Figure 5 illustrates that all three models are continually revised as a project moves from feasibility, through design and construction to post-construction monitoring and maintenance. What is very apparent from this document is that it is not possible to create a ground model without first producing a geological model, a view that supports the recommendations of Fookes (1997). Useful guidance for producing the geological model for engineering projects, based mainly on experience in Hong Kong, can be found in Fletcher (2016). The link between geology and ground models, again mainly concentrating on work in Hong Kong, is presented in the book *Practical Engineering Geology* by Hencher (2012). Both these books provide an excellent insight into the type and nature of data that need to be included in geological models if they are to be used in engineering construction and, whilst mainly referring to Hong Kong, they have global applicability.

IAEG Commission 25 (Baynes and Parry 2022)

Based on an initial earlier report (Parry *et al.* 2014), new guidelines for the development and application of engineering geological models on projects were issued by IAEG Commission 25 in 2022 (Baynes and Parry 2022) and they are briefly introduced in this paper.

The new Commission 25 guidelines describe an Engineering Geological Model (EGM) as a comprehensive knowledge framework that supports the interpretation and assessment of the engineering geological conditions and facilitates the evaluation of the interaction of these conditions with a proposed project. This will allow appropriate engineering decisions to be made throughout the life cycle of a project from inception through to decommissioning. In adopting this definition, Baynes and Parry (2022) state that their intention was to move beyond the concept that a 'model' is a simplified and static three-dimensional representation of the ground conditions, and recognize that the formation and development of the EGM is an on-going process of knowledge accumulation that provides direction and control to the ground engineering throughout a project.

An overview of the EGM knowledge framework is presented in Figure 6. The EGM is the core of the work (the central ellipse) and feeding into this are the nature of the project, engineering and geological knowledge, the results of desk studies, reconnaissance mapping, and various preliminary geological and geomorphological models. Within the EGM the evaluation of the site conditions go through conceptualization to enable on-site investigations to be planned and then initiated. The results of these investigations are interpreted within the EGM knowledge framework that allows the next iteration of the geological model to be produced, plus the geotechnical model and any other models that are needed (e.g. a ground model, digital terrain model, analytical model). Further investigations might be required if the levels of uncertainty about ground conditions are still viewed as producing unacceptable levels of construction risk. However, once feasibility has been demonstrated then design work proceeds, that can then lead to construction and post-construction monitoring.

The EGM knowledge framework approach places an emphasis on the role and importance that geology and geomorphology play in understanding both the evolution of and the extant ground conditions at a site (i.e. the total geology of Fookes *et al.* 2000). It demonstrates that without this understanding the level of risk to construction at a site will remain an undefined or even unknown quantity.

UK engineering geological practice

There are two critically important documents produced in 2023–24 that will affect the creation and application of ground models in the UK over the next decade: the Geotechnical Baseline Report guidance from CIRIA (Davis *et al.* 2023) and the second iteration of Eurocode 7 Geotechnical Design. These are both discussed below.

CIRIA C807, Geotechnical Baseline Reports (GBR) January 2023 (Davis *et al.* 2023)

This guide was produced to promote and improve the application of GBRs in the UK. The GBR is used to allocate ground-related risk and commercial responsibility between the employer and the contractor, and is defined in Clause 1.1 (p. 1) as '… a contract document containing measurable contractual descriptions of the subsurface conditions anticipated, or to be assumed to be anticipated, during construction'. It is noted that the GBR forms '… the basis of dealing with uncertainty in below-ground conditions for construction projects.'

The guide contains a history of the development of GBRs both in the UK and overseas, recognizing that the concept originated in the USA during the mid- to late 1970s when it was acknowledged that there was a need to reduce the growing trend of claims, litigation and adverse publicity in the tunnel construction industry. The guide then goes on to explore

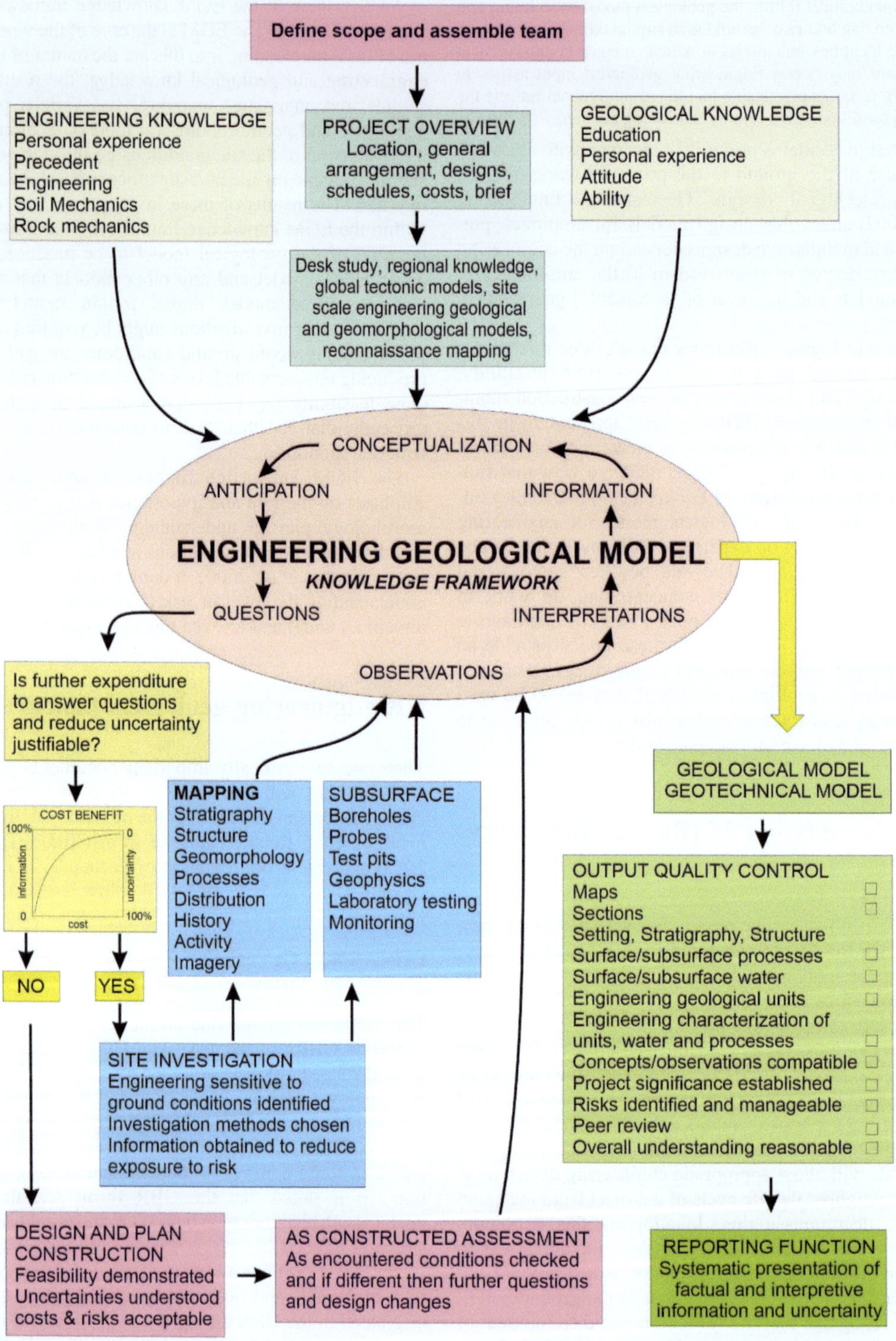

Fig. 6. The Engineering Geological – Model Knowledge Framework. Source: IAEG Commission 25 (Baynes and Parry 2022), reproduced with permission of IAEG.

their required authorship, content, scope, usage and, crucially, how GBRs can be included in contract documents (Clause 4). This is all supported by case studies.

Clause 5.2 (p. 26) in the guide is entitled, 'Commercial ground models as a basis for GBR drafting'. This section states that,

> … knowledge of the ground before construction is always imperfect. The hazards present prior to construction often arise from this imperfect knowledge, which could be considered to be gaps in a project ground model.

It then notes that (p. 27),

> … the best way to address these [i.e. gaps in knowledge] is through the explicit use of ground models as a tool to help formulate baseline statements. The discipline best able to identify the potential content of the gaps is usually engineering geology.

Clause 5.2 draws on the IAEG Commission 25 original guidelines (Parry *et al.* 2014; note that, as described above, these were updated in 2022) on the use of engineering geological models to define two of the four types of models relevant to the formation of the GBR:

Conceptual Model – described as the model created at the desk study stage and based on knowledge, experience and data that is not specific to the project nor necessarily related to a defined locality.

Observational Model – this is based on measured or observed data and is project specific. This model will be derived from all forms of data collection including desk studies, remote sensing, field mapping and ground investigations and draws on the initial conceptual model.

Analytical Model – this is the design-based stage of modelling that takes the observational model and adapts it for use in design by including measured data relevant to the project along with an appreciation of their inherent uncertainty and the range of their values. It is worth noting that the content of an analytical model is usually constrained by inherently cautious design standards (John Davis pers. comm.). Davis *et al.* (2023) state that this model, being focused on design and the requirements of design standards, has no role in the drafting of the GBR.

Commercial Model – this model is not taken from IAEG Commission 25 and emerges from the fact that the GBR should never be used for design. However, it does require the balance of risk to be apportioned through negotiation and hence the GBR can be considered a commercial model.

In their assessment of these four types of model Davis *et al.* (2023, p. 28) state that

> … the GBR is a separate and unique alternative commercial end point of an engineering geological modelling process. It takes the conceptual and observational models and uses them to help predict the range of conditions a contractor might encounter during the works.

This fits perfectly into the concept of the engineering geological model being a conceptual framework, as described by Baynes and Parry (2022) in IAEG Commission 25.

With respect to who should prepare GBRs, Davis *et al.* (2023, p. 16) state that '… within the UK, GBRs have typically been prepared by consultants on behalf of employers; it is often left to engineering geologists and geotechnical engineers to write these documents …'.

Later in the guide, in reference to the competency for site management (Clause 7.6), it is noted that where specialist advice is needed (p. 47), '… the specialist should be an engineering geologist experienced in the use of GBRs, the particular engineering applications, and, preferably, the geological conditions'.

Based on this document it is clear that GBRs have a critical role in improving construction practice in the UK and that ground models, preferably produced by engineering geologists, are central to this process.

Eurocode 7: 1st and 2nd iterations

The 2nd iteration of Eurocode 7 is presently undergoing a phased introduction (Menteth 2024; European Commission 2024) with the general rules and ground properties guidelines published in late 2024 (BS EN 1997-1: 2024; BS EN 1997-2: 2024) and geotechnical structures due in April 2025 (BS EN 1997-3: 2025), although they have been available in final draft form since 2022 (prEN 1997-1 2022; prEN 1997-2 2022; prEN 1997-3 2022). However, it is worth starting with a brief look at components of the 1st iteration that were originally issued in 2007 and will not be officially withdrawn until 30 March 2028. In EN 1997-2 2007 Annex B there was a diagram that illustrated the full scope of site investigation in Europe according to the code (reproduced in Fig. 7). This showed the four stages in the design and construction of a structure. In the strategy for design (Stage 1), geology was seen as a key input to the geotechnical design. In the geotechnical investigations and design (Stage 2), geology and the geological model were aspects of the planning and evaluation of the investigation programme, and these would be expected to be included in the ground investigation report. The ground investigation report would then feed into the design where a geotechnical model would be produced. Geology and the geological model, therefore, were seen as an integral part of the geotechnical investigation during the 1st iteration of Eurocode 7, but there is no reference in this document to a 'ground model'.

The 2nd iteration of Eurocode 7 has been under review for several years (see Norbury 2021). There are three parts to the revised code.

Part 1 – Geotechnical Design: General Rules. In this the terms relating to analysis and models are described in Clause 3.1.6.

Part 2 – Ground Properties. This is where the geological input requirements can be found.

Part 3 – Geotechnical Structures, where description of geological and geotechnical ground conditions is referred to in Clause 6.2.7 entitled 'Ground Investigations'.

A major change from the 1st iteration is that 'geological model' has disappeared from the requirements. Part 1, Clause

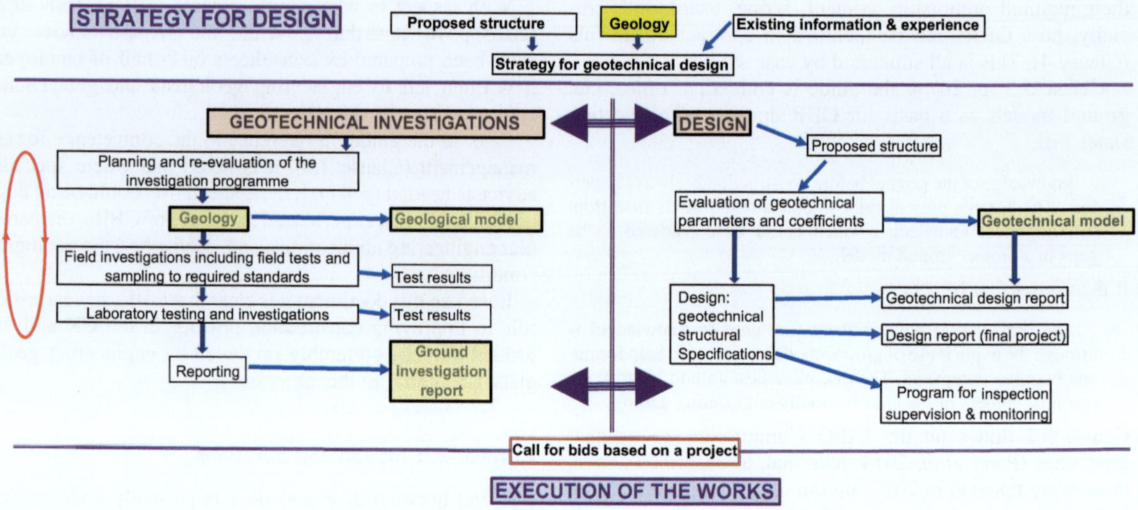

Fig. 7. The scope of site investigation in Europe according to the 1st iteration of Eurocode 7. Source: adapted from Annex B, BS EN 1997-2.

4.1.1, item 3 identifies only two models that are required under the 2nd iteration of the code for the safety, serviceability, robustness and durability of geotechnical structures:

(1) the Ground Model, as specified in Part 2, Clause 4 (see discussion below);
(2) the Geotechnical Design Model, as specified in Part 1, Clause 4.2.3, where it is noted that this should be developed for each design situation. The measures required to validate the model are included in Part 1, table 4.5 of the Eurocode.

In Part 2, Clause 4.1, the required content of the ground model is defined and a number of pertinent notes are provided to clarify the definition:

(1) A Ground Model shall comprise the geological, hydrogeological, and geotechnical conditions at the site based on the ground investigation results.
Note 1 – Geological conditions include, for example, the description of the site geomorphology, the lithology of the geotechnical units, the potential presence and level of a rockhead, geometrical and geotechnical properties of discontinuities and weathered zones.
Note 2 – Hydrogeological conditions address surface, groundwater and piezometric levels, including their potential variation with time, potential water flows and the presence of other fluids or gasses affecting the site.
Note 3 – Geotechnical conditions include, for example, the disposition of the geotechnical units and the mechanical behaviour of the ground described by the properties of the geotechnical units.
(2) Variability and uncertainty of geological, hydrogeological and geotechnical conditions and properties shall be included in the Ground Model.
(3) The detail and extent of the Ground Model shall be consistent with the Geotechnical Category and zone of influence of the structure.
(4) The Ground Model shall be progressively developed and updated based on potential new information.
(5) The Ground Model shall include the derived values of relevant ground properties for all geotechnical units encountered in the zone of influence.
(6) The Ground Model shall be documented in the Ground Investigation Report
Note 1 – The Ground Model is the main output of the Ground Investigation.
Note 2 – The Ground Model forms the basis for development of the Geotechnical Design Model.

The geotechnical design model itself is defined in Part 1, Clause 3.1.6.4 as a 'conceptual representation of the site derived from the ground model for the verification of each appropriate design situation and limit state'. Details on the required content of the geotechnical design model are provided in Part 1, Clause 12 of Eurocode 7.

In the 2nd iteration of Eurocode 7 the specific need to develop a geological model of the ground has disappeared and the requirement is to go direct to a ground model. It could be argued that this ground model should be an amalgam of geological, geomorphological, engineering geological and geotechnical models, but this is not actually stated. The ground model in Eurocode 7 is seen as the output from the ground investigations and its purpose is to provide input to the geotechnical design model. In particular, the level of detail required in the ground model is limited by the Geotechnical Category and its spatial extent to the zone of influence of the structure. This suggests there is no particular requirement that a site would be placed in its broader geological or geomorphological context by looking at its evolution through geological time and the effects of processes beyond the site boundaries. There appears to be no reference to engineering geology in the 2nd iteration, except in Part 2, Annex C, where it is noted that geological engineering geological maps should be consulted in desk studies. In Part 2, Clause 5.3.1, with respect to site inspection techniques, it does refer to the use of geological mapping and geotechnical mapping but not geomorphological mapping. It is also worth noting that there is no mention of engineering geologists in the code.

The 2nd iteration of Eurocode 7 is very clearly focused on ground engineering design and as such, contains a wealth of information and guidance for all aspects of geotechnical engineering. In the UK, drafts of the 2nd iteration have been widely circulated for comment in the engineering geological and geotechnical community over the past few years. It must be assumed, therefore, that the final publication reflects the views of the UK ground engineering industry.

Offshore industries

The offshore construction industry has been formulating its own methodologies for investigating and understanding seabed conditions, and this work has tended to develop independently from the onshore ground investigation industry, although clearly there are many overlaps in techniques. However, offshore there is inevitably a far greater emphasis on the use of geophysical techniques (e.g. multibeam bathymetry, side-scan sonar, seismic techniques) and the use of CPT technology. Nevertheless, the development and use of models to help interpret the shallow seabed for all forms of offshore structure is well established. Two approaches that incorporate models of various forms are highlighted below, namely the guidance notes provided by the Society for Underwater Technology (Cook 2022), and the methodology proposed by Thomas (2017).

Society for Underwater Technology (SUT)

Cook (2022), on behalf of the Offshore Site Investigation and Geotechnics Committee (OSIG), produced the *Guidance notes for the planning and execution of geophysical and geotechnical ground investigation for offshore renewable energy developments.* This was an updated version of the OSIG 2014 guidance notes, and the revised document contains a whole chapter (Chapter 3, pp. 6–10) on the ground model. This chapter describes what a ground model is, its use, and the stages in its development from the initial desk studies, creation of a hazard register, to data collection through reconnaissance and engineering surveys. The final section is on how the nature of the ground model is progressed and used, noting that it should be considered as an evolving model that continually develops and maintains relevance throughout the life of a project.

The definition of a ground model in this document is quite distinct: 'A ground model is a database of available information, such as the structural geology, geomorphology, sedimentology, stratigraphy, geotechnical properties and geohazards of a site' (Cook 2022, p. 6). For OSIG the ground model represents an approach for collating all available site information but specifically, in relation to offshore renewable energy developments, is used '... to identify all relevant unknowns and project hazards, to direct investigations, and to inform the foundation design and installation methods for a field development' (Cook 2022, p. 6).

It is noted that the form of a ground model varies but is influenced by the attitude of a developer to risk and by the complexity of the site and project. However, the typical format for OSIG is described as consisting of three parts (Cook 2022, p. 6):

(1) written reports that detail the phases of development of the ground model;
(2) a database of all the information collated (including raw and interpreted data) in an internationally recognized format;
(3) geotechnical risks register.

For all but the simplest of sites it is expected the ground model should be able to present data as a 3D or 4D interactive model so that it can be used not just by geoscientists but by engineers and other relevant stakeholders. A particularly pertinent observation is made in relation to data acquisition and project risk:

> Due to the iterative nature of the process of gaining an understanding of ground conditions, and in order to reduce the ground risks, a ground model will normally comprise several stages of evolution before these risks are considered to be acceptable (Cook 2022, p. 6).

Figure 8, from the OSIG document, shows how the process of creating a ground model starts at the desk study stage and, based on the project requirements, is used to determine the scope of the geotechnical investigations. After the initial investigations have been completed the ground model is updated and the critical question to be asked is 'are the data enough to meet requirements and manage ground risks?'. If not, then more investigations are carried out. Thus, the focus is on understanding and managing the ground risks.

The guidance notes recognize that there is no ideal model or site investigation type that will suit every development, mainly because every site will have different characteristics that require appropriate investigation and assessment methodologies. Nevertheless, the ground model is regarded as a key input to the geotechnical design parameters and how these vary across a site. Also, whilst the ground model will initially be developed to support the project design, it is emphasized that it should be continually updated as new data emerge and is noted that it will be relevant throughout the life of a project.

This OSIG approach to the use of a ground model is neatly summarized at the end of Chapter 3: 'A well maintained and current ground model will provide confidence to the various parties that the risks are well understood and being well-managed' (Cook 2022, p. 10).

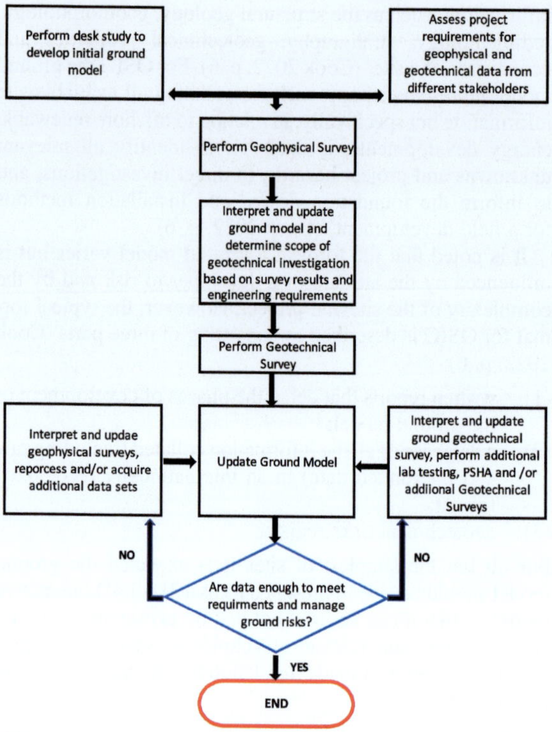

Fig. 8. A typical process for understanding ground risk. Source: adapted from OSIG Guidance Notes September 2022.

Steve Thomas OSIG keynote address 2017

The OSIG document by Cook (2022) concentrated on offshore renewable energy developments. However, Thomas (2017), drawing on his decades of experience at Fugro GB Marine Limited, proposed a method for characterizing the seabed conditions for all types of offshore sites. In this paper he uses the term 'site characterization' and in a marine context this is defined as '… the understanding of geological, geotechnical, environmental and metocean conditions in relation to the planned development of a site' (Thomas 2017, p. 71).

This definition has parallels with the use of the expression 'site characterization' in the US Federal Highways manual on subsurface investigation (Mayne *et al.* 2001) and as the core function of engineering geologists by Tepel (2002). Thomas's (2017) paper provides an overview of the investigative toolkit that is available to characterize large offshore sites, including the application software and geological concepts of classification, deduction and mapping. Thomas (2017, p. 71) states:

> There is a focus on the key engineering geological concepts as they provide the framework to identify which components of the investigative toolkit are likely to be required, where within the site the toolkit needs to be deployed, and when the results are needed based on an understanding of the interdependencies within the interpretation process.

Thomas's approach to offshore site characterization is synthesized in Figure 9. This figure identifies the various stages of the process on the y-axis (system definition, appraise, select, define, execute/operate), with the tasks, deliverables and acquired data on the x-axis. A numbered list of the analytical studies required for a project is presented at the bottom of the figure and where these are undertaken can be found in the flow diagram of interpretative tasks. A key concept within the proposed methodology is the creation of an evolutionary ground model, which is defined as '… the three-dimensional understanding of geological, geomorphological, geotechnical, and metocean features and conditions through time' (Thomas 2017, p. 72).

Thomas (2017) notes that this model provides the epistemological framework to make inferences concerning likely depositional environments and sediment properties using deduction based on geological principles. Thus, underpinning the evolutionary ground model is an understanding of geological processes and history, undertaking terrain evaluation and mapping (Moore *et al.* 2007), and describing geological structures and sediments. Under routine or static conditions interpretation of the evolutionary model will lead to the production of a geotechnical model that will be calibrated by the acquisition of site-specific data through investigations. Under active or non-routine conditions, the evolutionary model will feed into a process of geohazard identification, mapping, susceptibility assessment and, with respect to a specific project, geohazard risk assessment.

Thomas (2017) goes on to demonstrate his integrated approach to site characterization for a hypothetical deep water gas field on a passive non-salt-influenced continental margin with a canyon-prone continental slope. This example illustrates the process of site characterization and highlights the role played by the evolutionary ground model throughout a project.

Discussion and conclusions

In Fookes *et al.* (2015, p. 1) it is stated:

> It must be emphasized that, at any scale, every geomodel is different and must be tailor-made for the current project. There is no ideal model, just relevant approaches, and good practices for the situation.

With respect to good practices, Fookes *et al.* (2015, p. 1) also made it very clear that the role of the engineering geologist was '… to interpret the geology and ground conditions correctly. Creating an initial model of the geology of the site is an excellent start'.

This has some very interesting parallels with the historic work of Dr Charles Berkey (1867–1955), arguably the

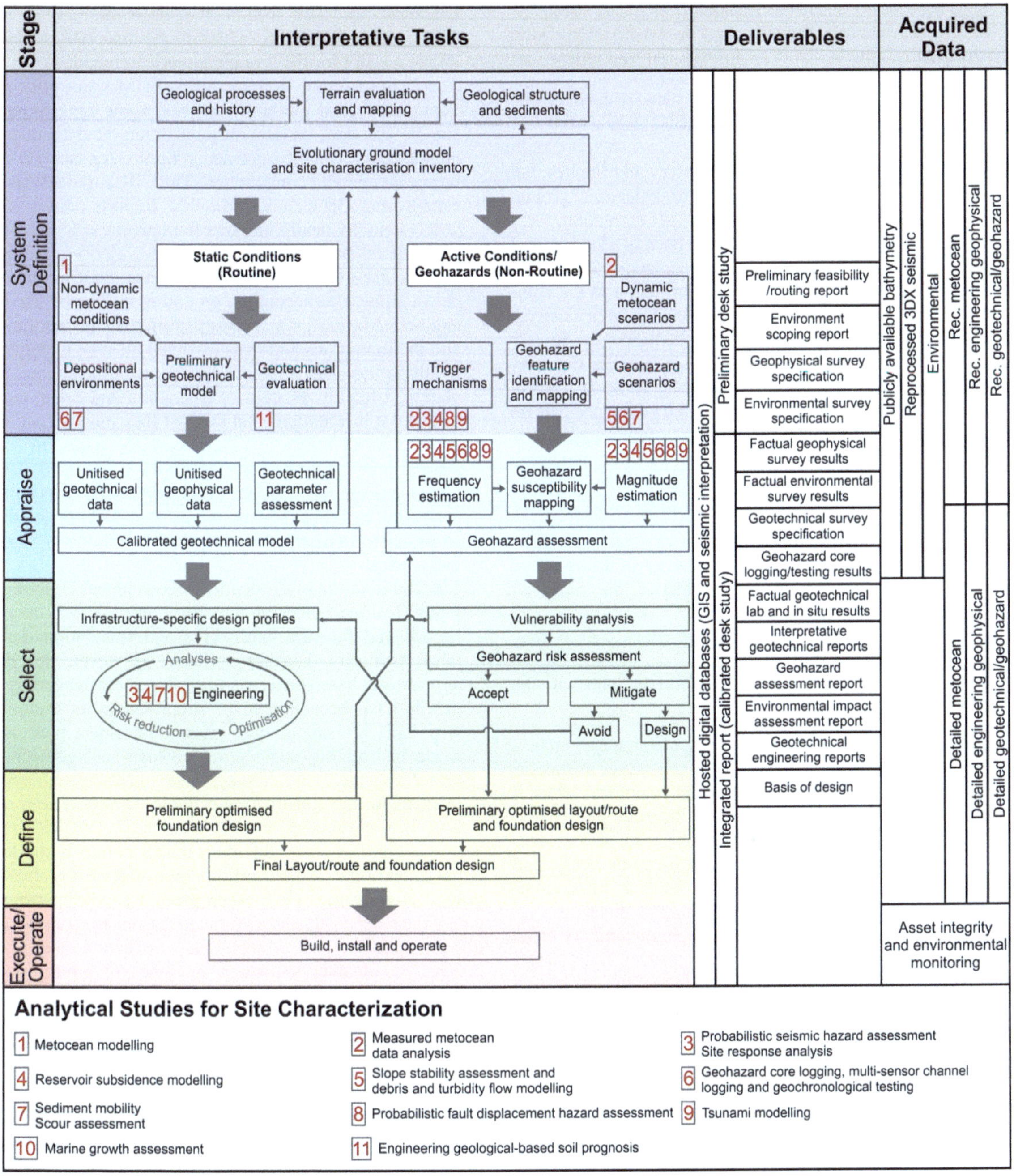

Fig. 9. The Thomas (2017) approach to offshore site characterization. Source: reproduced by kind permission of the Society for Offshore Technology from their 2017 Conference volume.

founder of the modern science of engineering geology. In their biography of his work, Savage and Rhoades (1950, p. xvi) stated about Dr Berkey:

> … contributing to his successful association with engineering works is his conviction that the geology applied to engineering must be *good* geology. He has not originated new geological

principles for the solution of engineering problems, but he has insisted unremittingly that the geology involved must be studied thoroughly, scientifically, in great detail, and with imagination; he has argued and proved that problems in engineering geology, however bewildering on the basis of superficial study, frequently solve themselves when the entire geological situation is understood …'.

Essentially, the same message from Berkey and Fookes, but 65 years apart, is that if you are going to understand ground conditions for any construction project, get the geology and geomorphology right. However, both Berkey (Savage and Rhoades 1950) and Fookes (1997) recognized that it is crucial to be able to communicate this information to the engineers in a way that they are able to utilize in their designs. A development on this theme was identified by Eggers (2016), who wrote on the importance and need for diversity in the science and practice of engineering geology. Eggers (2016, p. 3) stated that 'geologists and engineers can be quite different beasts and their brains are often wired differently'. He referred to the Baynes (1999) communication model for engineering geology on large, complex projects that highlighted three aspects as being critical in the process.

(1) The filters between each phase of the communication path can potentially reduce the transfer of information; for the communications to be effective, the impact of the filters must be limited.
(2) The impact of the filters must be considered as they always exist.
(3) The feedback loop is an important part of the communication path.

Eggers (2016, p. 5) stated that, 'geological language can be a barrier to communication and is a large reason for misunderstanding of the [geological] model by the engineers'. Thus, it is not just the responsibility of the engineering geologist to understand the geology and geomorphology but also to identify why and how this might affect engineering design and construction. However, this can be difficult as many engineering geologists have limited experience in this area.

This leads back to the focus of this paper on what we mean by a 'ground model' and why does this matter? There is still no definitive answer as to what it is, although the work from the Geotechnical Engineering Office (2007) in Hong Kong is probably the least ambiguous. However, based on personal experience and from listening to views expressed by engineering geologists, it is clear that any ground model must be based on sound geological and geomorphological principles, and it should be derived from a geological model. Therefore, it is possible to suggest a possible way forward.

- At the outset it is necessary to draw upon the *a priori* knowledge and skills identified by Fookes (1997) and Fookes *et al.* (2000) under the concept of 'total geology', to ensure ground investigations provide the information required, with the issues affecting a project correctly identified, and the ground conditions established. This will involve the creation of the 'geomodel' (Fookes *et al.* 2015) that initially will be conceptual in nature but becomes more site specific (i.e. observational) as new data emerge. This fits with all the material reviewed above, except for the 2nd iteration of Eurocode 7.
- The IAEG Commission 25 (2022) EGM framework provides an overall guide within which site investigations can be planned, carried out, and interpreted to provide engineers with all the information needed for safe and economic design and construction. The CIRIA guidelines for establishing Geotechnical Baseline Reports (Davis *et al.* 2023) fits very neatly into this framework.
- The 'filters' (terminology from Baynes 1999) or simplifications that are applied to the geological model or geomodel to allow it to become a 'ground model' to be used by engineers for design and construction must be recognized and the limitations and risks involved must be understood by all parties. The responsibility lies with the engineering geologist for this process of evaluation and communication and it is a fundamental skill set they must all acquire.
- As part of all stages in this process the focus on 'Risk Assessment' suggested by the OSIG Committee (Cook 2022) and supported by the work of Thomas (2017), will ensure the end requirements and consequences of any decisions and interpretations on ground conditions remain paramount.
- The 2nd iteration of Eurocode 7 Geotechnical Design provides clear guidelines on the procedures and processes to be adopted for safe earthworks and foundation design when sufficient data are available. However, this does appear to be based on a ground model where the geological details have been filtered or simplified in an undefined way. This is disappointing, given the progress previously made in bringing engineering geology and geotechnics closer together. and it may have unanticipated consequences for the construction of future structures.

The question as to why this does matter concerns the legal requirements and responsibilities of everyone involved in ground engineering. Unforeseen ground conditions remain one of the most costly and significant reasons for construction disputes, claims and project delays (Clayton and Smith 2013). If a dispute transpires and goes to arbitration or court, then the definitions of the geological model, geomodel and ground model in the contract documents may be critical. Engineering geologists, if they are involved, must be fully cognisant with the requirements and their responsibilities under the contract for compiling and presenting the various types of models. They must know what they are expected to produce, along with its limitations, and ensure their clients have been made fully aware of the level of risk they are accepting and liable for in the contract.

It is also pertinent to recognize that the 2nd iteration of Eurocode 7 is set to dominate the UK construction industry for the next decade, and it has a quite specific definition and usage of the term 'ground model'. As stewards of the profession (Tepel 2002), and because we, as engineering geologists, are aware how critical understanding the geology and geomorphology of a site is to ensure safe construction, it is beholden on us to ensure it is engineering geologists and not just geotechnical engineers or civil engineers who are responsible for producing the ground models required.

Acknowledgements The author wishes to thank the following for comments on earlier drafts of this paper: David Norbury (for the section on Eurocode 7); John Davis (for the section on CIRIA C870); and Steve Parry (for the section on IAEG Commission 25). Finally, the author wishes to thank Peter Fookes (Fig. 1) for his long-term support of his career in engineering geology & geomorphology, ever since his interview for the Rendel Palmer and Tritton PhD studentship in 1976.

Competing interests I have no known competing financial interests or personal relationships that could have appeared to influence the work reported in this paper.

Author contributions JSG: writing – original draft (lead).

Funding This research received no specific grant from any funding agency in the public, commercial, or not-for-profit sectors.

Data availability Data sharing is not applicable to this article as no datasets were generated or analysed during the current study

References

Baynes, F.J. 1999. Engineering geological knowledge and quality. *In*: Vitharana, N. and Colman, R. (eds) *Proceedings of the Eighth Australia-New Zealand Conference on Geomechanics*. Institute of Engineers, Hobart, Australia, **1**, 227–234.

Baynes, F.J. and Parry, S. 2022. Guidelines for the development and application of engineering geological models on projects. IAEG Commission 25 Publication No.1, https://www.iaeg.info/commission-25-use-of-engineering-geological-models/ [last accessed 18 July 2023].

Brunsden, D. 2002. The Fifth Glossop Lecture. Geomorphological roulette for engineers and planners: some insights into an old game. *Quarterly Journal of Engineering Geology and Hydrogeology*, **35**, 102–142, https://doi.org/10.1144/1470-9236200 1-40

Brunsden, D., Doornkamp, J.C., Fookes, P.G., Jones, D.K.C. and Kelly, J.M.N. 1975. Large-scale geomorphological mapping and highway engineering design. *Quarterly Journal of Engineering Geology*, **8**, 227–253, https://doi.org/10.1144/GSL.QJEG.1975.008.04.01

BS EN 1997-1: 2024. *Eurocode 7 – Geotechnical design – Part 1: General Rules*. CEN/British Standards Institution.

BS EN 1997-2: 2024. *Eurocode 7 – Geotechnical design – Part 2: Ground properties*. CEN/British Standards Institution.

BS EN 1997-3: 2025. *Eurocode 7 – Geotechnical design – Part 3: Geotechnical structures*. CEN/British Standards Institution.

Building Department 2003. *PNAP 274: Geotechnical Design Information*. Buildings Department, Hong Kong.

Chorley, R.J., Schumm, S.A. and Sugden, D.E. 1984. *Geomorphology*. Methuen, London.

Clayton, C.R.I. and Smith, D.M. 2013. *Effective Site Investigation*. The Site Investigation Steering Group, Institution of Civil Engineers. Thomas Telford, London.

Cook, M. (ed.) 2022. *Guidance Notes for the Planning and Execution of Geophysical and Geotechnical Ground Investigations for Offshore Renewable Energy Developments*. Society for Underwater Technology, London.

Culshaw, M.G. 2005. The Seventh Glossop Lecture. From concept towards reality: developing the attributed 3–D geological model of the shallow subsurface – the Seventh Glossop Lecture. *Quarterly Journal of Engineering Geology and Hydrogeology*, **38**, 231–284, https://doi.org/10.1144/1470-9236/04-072

Davis, J., Essex, R., Farooq, I. and Drake, A. 2023. *Geotechnical baseline reports: a guide to good practice*. **Report C807**, CIRIA, London.

DeFreitas, M.H. 2009. The Ninth Glossop Lecture: geology; its principles, practice, and potential for geotechnics. *Quarterly Journal of Engineering Geology and Hydrogeology*, **42**, 397–441, https://doi.org/10.1144/1470-9236/09-014

Eggers, M.J. 2016. Diversity in the science and practice of engineering geology. *Geological Society, London, Engineering Geology Special Publications*, **27**, 1–18, https://doi.org/10.1144/EGSP27.1

EN 1997-2 2007. *Eurocode 7 – Geotechnical design – Part 2: Ground investigation and testing*. CEN, Brussels.

European Commission: Joint Research Centre, Garin, H., Bladwin, M. *et al.* 2024. *Assembling the ground model and the derived values*. (Garin, H., ed.). Publications Office of the European Union, Luxembourg, https://data.europa.eu/doi/10.2760/6390378, JRC140556.

Fletcher, V.J.N. 2016. *Geology for Ground Engineering Projects*. CRC Press, Boca Raton.

Fookes, P.G. 1997. Geology for engineers: the geological model, prediction, and performance – the First Glossop Lecture. *Quarterly Journal of Engineering Geology and Hydrogeology*, **30**, 293–424, https://doi.org/10.1144/GSL.QJEG.1997.030.P4.02

Fookes, P.G., Baynes, F.J. and Hutchinson, J.N. 2000. Total geological history: a model approach to the anticipation, observation and understanding of site conditions. *GeoEng 2000: an International Conference on Geotechnical and Geological Engineering*, 19–24 November, Volume **1**. Technomic, 370–460.

Fookes, P.G., Lee, E.M. and Milligan, G. (eds) 2005. *Geomorphology for Engineers*. Whittles, Caithness.

Fookes, P.G., Lee, E.M. and Griffiths, J.S. 2007. *Engineering Geomorphology*. Whittles, Caithness.

Fookes, P.G., Pettifer, G. and Waltham, T. 2015. *Geomodels in Engineering Geology: an Introduction*. Whittles, Caithness.

Fookes, P.G., Sweeney, M., Manby, C.N.D. and Martin, R.P. 1985. Geological and geotechnical engineering aspects of low-cost roads in mountainous terrain. *Engineering Geology*, **21**, 1–152, https://doi.org/10.1016/0013-7952(85)90002-X

Geotechnical Engineering Office 2007. *Engineering Geological Practice in Hong Kong*. GEO Publication 1/2007. Civil Engineering and Development Office, Hong Kong.

Glossop, R. 1968. The rise of geotechnology and its influence on engineering practice. *Géotechnique*, **18**, 105–150, https://doi.org/10.1680/geot.1968.18.2.107

Griffiths, J.S. 2014. The Fourteenth Glossop Lecture. Feet on the ground: engineering geology, past present and future. *Quarterly Journal of Engineering Geology and Hydrogeology*, **47**, 116–143, https://doi.org/10.1144/qjegh2013-087

Griffiths, J.S. and Stokes, M. 2008. Engineering geomorphological input to ground models: an approach based on Earth systems. *Quarterly Journal of Engineering Geology and Hydrogeology*, **41**, 73–91, https://doi.org/10.1144/1470-9236/07-010

Griffiths, J.S., Hutchinson, J.N., Brunsden, D., Petley, D. and Fookes, P.G. 2004. The reactivation of a landslide during the construction of the Ok Ma tailings dam, Papua New Guinea. *Quarterly Journal of Engineering Geology and Hydrogeology*, **37**, 173–186, https://doi.org/10.1144/1470-9236/02-034

Griffiths, J.S., Stokes, M., Stead, D. and Giles, D. 2012. Landscape evolution and engineering geology: results from IAEG Commission 22. *Bulletin of Engineering Geology and the Environment*, **71**, 605–636, https://doi.org/10.1007/s10064-012-0434-7

Hatheway, A.W. 2002. Geoenvironmental protocol for site and waster characterization of former manufactured gas plants: worldwide remediation challenges in semi-volatile organic wastes. *Engineering Geology*, **64**, 317–338, https://doi.org/10.1016/S0013-7952(01)00097-7

Hencher, S. 2012. *Practical Engineering Geology*. Spon Press, London.

Hutchinson, J.N. 2001. The Fourth Glossop Lecture: Reading the ground: morphology and geology in site appraisal. *Quarterly Journal of Engineering Geology and Hydrogeology*, **34**, 7–50, https://doi.org/10.1144/qjegh.34.1.7

Lee, E.M. and Fookes, P.G. 2015. A note on the origins of engineering geomorphology in the UK. *Quarterly Journal of Engineering Geology and Hydrogeology*, **48**, 147–156, https://doi.org/10.1144/qjegh2014-048

Mayne, M.W., Christopher, B.R. and Dejong, J. 2001. *Manual on Subsurface Investigation: Geotechnical Site Characterization*. Federal Highway Administration, National Highways Institute Publication, FHWA NHI-01-031.

Menteth, T. 2024. Eurocode 7 guidelines on design implementation and ground models published. *Ground Engineering*, 20 December 2024, on-line news feature.

Moore, R., Usher, N. and Evans, T.G.E. 2007. Integrated multidisciplinary seismic geomorphology assessment of West Nile Delta geohazards. *Proceedings of the 6th International Conference*, Offshore Site Investigation and Geotechnics, SUT, London.

Norbury, D. 2021. Ground models; a brief overview. *Quarterly Journal of Engineering Geology and Hydrogeology*, **54**, 1–6, https://doi.org/10.1144/qjegh2020-018

Parry, S., Baynes, F.J. *et al.* 2014. Engineering geological models: an introduction. IAEG Commission 25. *Bulletin of Engineering Geology and Environment*, **73**, 689–706, https://doi.org/10.1007/s10064-014-0576-x

prEN 1997-1 2022. *Eurocode 7 – Geotechnical design – Part 1: General Rules*. CEN/British Standards Institution.

prEN 1997-2 2022. *Eurocode 7 – Geotechnical design – Part 2: Ground properties*. CEN/British Standards Institution.

prEN 1997-3 2022. *Eurocode 7 – Geotechnical design – Part 3: Geotechnical structures*. CEN/British Standards Institution.

Savage, J.L. and Rhoades, R. 1950. Charles Peter Berkey. *In*: Paige, S. (ed.) *Application of geology to engineering practice*. Geological Society of America, New York, xi–xix.

Site Investigation Steering Group 1993. *Without Site Investigation Ground is a Hazard*. Thomas Telford, London.

Smith, M.J., Paron, P. and Griffiths, J.S. (eds) 2011. *Geomorphological Mapping: Methods and Applications*. Developments in Earth Sciences Series, Elsevier, Amsterdam.

Tepel, R.E. 2002. *Visioning the Future of Engineering Geology: Sustainability and Stewardship*. Association of Engineering Geologists, Special Publication, **19**, Vol. 1 [CD-ROM].

Tepel, R.E. 2004. The engineering geology site characterization as the core function of engineering geology practice. *In*: Tepel, R.E. (ed.) *Visioning the Future of Engineering Geology: Site Characterisation*. Association of Engineering Geologists, Special Publication, **19**, Vol. 3 [CD-ROM].

Thomas, S. 2017. A Phased and integrated data interpretation approach to site characterisation. *Offshore Site Investigation Geotechnics 8th International Conference Proceedings*, held at the Royal Geographical Society, London, Vol. **1**. Society for Underwater Technology, 71–87.

Ground models for tailings management

Tristan Jónsson Menzies

WSP Australia, 5 Spring Street, Perth 6000, Australia
0009-0009-2898-0657

Correspondence: tristan.menzies@wsp.com; tristanmenzies1992@gmail.com

Abstract: Tailings and tailings storage facilities (TSF) are an unavoidable by-product of mining, the improper design, construction and management of which has historically caused significant environmental and societal impacts. The 2020 'Global Industry Standard on Tailings Management' (GISTM) was introduced into tailings management practice to 'achieve the ultimate goal of zero harm to people and the environment and zero tolerance for human fatality' and provides the framework by which facility operators, designers, consultants and other stakeholders approach tailings management. GISTM includes the requirement of a site characterization, considered to be equivalent to an Engineering Geological Model (EGM). The development of robust site characterization(s) for tailings management is considered to require a series of interrelated ground models that make up the EGM and critically include a conceptual geological model, a total geological history approach and a component of digital modelling. This paper provides a framework for the development of ground models for tailings management through the complete TSF lifespan, from concept to closure, by the reinterpretation of existing ground model literature in a mine-waste specific and GISTM compliant context.

The extraction of mineral resources from the Earth generally results in the creation of two byproduct mine waste streams: bulk excavated overburden or non-ore-containing materials placed in a soil and/or rock waste dump; and tailings, the waste residue generated from ore processing. Tailings are generated as slurries, mixtures of crushed rock (rock 'flour') and processing fluids (water, chemical reagents and unrecoverable metals) from mills, washeries or concentrators (Piciullo *et al.* 2022). They typically comprise a contaminated wet mixture of sand and silt that are stored for perpetuity in a Tailings Storage Facility (TSF). While several methods of disposal/storage are possible, tailings are most typically piped and deposited into impoundments behind earth-fill embankments (referred to as tailings dams) (Fig. 1). Often embankments are raised multiple times during the life of tailings deposition to increase storage capacity.

The failure of TSFs, attributed primarily to lack of knowledge on tailings behaviour and the poor performance of monitoring and management processes, has resulted in disastrous environmental and human tragedies (Piciullo *et al.* 2022). The development of safe TSF landforms requires a multidisciplinary approach involving scientists (geologists, geomorphologists, hydrogeologists, hydrologists, geochemists, environmentalists, geophysicists, metallurgists) and engineers (geotechnical, civil, water, process, environmental). In the context of ground models, tailings dams are fundamentally different to other types of dams in that progressive raising of embankments and deposition of tailings during operation means that static stability and seepage conditions are in continuous change (APEGBC 2016). As a result, multiple dynamic, progressive and interrelated ground models are foundational to the development of safe TSFs as steady-state conditions are not typically reached until facility closure.

This paper discusses the requirement of ground models in the design, construction and management of safe TSF, with a focus on the importance of a total geological history approach, collaborative digital model solution and robust geological conceptualization.

The global industry standard on tailings management

Following (in part) the catastrophic failure of Vale S.A.'s TSF at Corrego do Feijão mine in Brumadinho, Brazil (Robertson *et al.* 2019), the Global Industry Standard on Tailings Management (GISTM) (GTR 2020) was developed. This international standard provides the framework by which TSF operators, designers, consultants and other stakeholders approach tailings management. The standard comprises a series of 'Topics' and 'Requirements'. Topic II of GISTM requires the development, maintenance and use of an interdisciplinary knowledge base, which includes a specific requirement (Requirement 2.2) for a site characterization. Requirement 2.2 (GTR 2020, p. 8) states

> Prepare, document and update a detailed site characterisation of the tailings facility site(s) that includes data on climate, geomorphology, geology, geochemistry, hydrology and hydrogeology

From: Griffiths, J. S., Giles, D. P., Davis, J. A. and Dunlop, J. J. (eds) 2025. *The Peter Fookes Engineering Geological Legacy in Geomodels, Geomaterials and Geomorphology*. Geological Society, London, Engineering Geology Special Publications, **30**, 47–53,
First published online 7 February 2025, https://doi.org/10.1144/egsp30-2024-21

Fig. 1. Slurry tailings deposition on a tailings dam surface, Pilbara, Western Australia.

> (surface and groundwater flow and quality), geotechnical, and seismicity. The physical and chemical properties of the tailings shall be characterised and updated regularly to account for variability in ore properties and processing.

Consequently, implementing Requirement 2.2 site characterization is considered in part to describe the development of a ground model.

Ground models and GISTM 'site characterization'

Current best practice for geotechnical investigation and site characterization adopts the concept of development of geological, engineering geological and geotechnical models to represent conceptualizations and available data for a given site. There are various documents that describe model development; for example, notable documents typically utilized by the author in Australia are:

(a) International Society for Engineering Geology and the Environment (IAEG) *Guidelines for the Development and Application of Engineering Geological Models on Projects* (IAEG 2022);
(b) *Guidelines for Geotechnical Investigations of Dams, their Foundations, and Appurtenant Structures* (ANCOLD 2020); and
(c) AS 1726:2017 *Geotechnical Site Investigations* (AS 1726 2017).

Globally, regional differences in standards, guidelines and edicts exist (e.g. APEGBC 2016; ICOLD 2022). However, the IAEG Guidelines are considered to provide the most robust instruction on the development of ground models. The guidelines have been developed by world-wide experts to be applicable globally, thereby providing a single consistent framework that can be applied to tailings management projects worldwide, in addition to regional requirements, guidelines or standards.

The IAEG guidelines utilize the terminology 'Engineering Geological Model (EGM)'. An EGM is defined as

> a comprehensive knowledge framework that allows for the logical evaluation and interpretation of the geological, geomorphological and hydrogeological conditions that could impact a project and their engineering characteristics. The EGM comprises both conceptual and observational components and may consist of several interrelated models and approaches. The Geological Model, the Geotechnical Model and a Geohazard Assessment are outputs from the EGM knowledge framework (IAEG 2022, p. 4).

The following conclusions are drawn from this definition in the context of GISTM (GTR 2020) site characterization requirements:

(a) Requirement 2.2 in part describes an EGM, specifically geology, geomorphology and hydrogeology characterization;
(b) Requirement 2.2 seismic and geochemical characterization are partially informed by a geological model (an output from the EGM knowledge framework);
(c) Requirement 2.2 geotechnical characterization is informed by a geotechnical model (an output from the EGM knowledge framework); and
(d) an EGM must consider the dimension of time (IAEG 2022) and is directly influenced by climate change (Nathanail and Banks 2009); therefore, there should be interaction between the EGM and the Requirement 2.2 climate characterization and Requirement 3.1 climate variation (GTR 2020).

As demonstrated above, TSF site characterization in accordance with IAEG requirements should achieve the majority of GISTM Requirement 2.2 compliance; and/or provide the foundational knowledge from which other interrelated characterization(s) can be achieved (for example, a site-specific seismic hazard assessment).

'Total geological history'

Peter Fookes has influenced the methodology by which TSF sites are investigated and characterized through the concept of 'total geological history' or 'total geology' (Fookes *et al.* 2000; Fookes and Baynes 2008), as recommended for dam sites (ANCOLD 2020).

The total geological history concept is that the 'geotechnical characteristics of the ground are a product of the geological and geomorphological history of the site, including past and present climatic conditions, in short, its total geological history' (Fookes *et al.* 2000, p. 370). Application of the concept involves understanding the site history and 'adopting strategies such as staged investigations, an emphasis on geomapping, establishing reference conditions, the development of models and applying the observational method' (Fookes and Baynes 2008, p. 39).

Geological mapping (considered fundamental to the total geological history approach) is recommended in a variety of literature pertaining to TSF investigation (APEGBC 2016; ICOLD 2022; SME 2022) and EGMs (IAEG 2022). Despite this, project-specific geological mapping at a variety of scales and throughout the life of the facility/mine is often absent. Without a robust conceptual model, supported by geological mapping, intrusive investigations alone are considered insufficient for reliably assessing the spatial distribution of soil and rock materials (particularly surficial soils). Additionally, targeting staged intrusive investigations appropriately and understanding the future evolution of soil and rock (particularly pertaining to changing climate) requires a total geological history approach.

Collaboration through a digital modelling solution

GISTM (GTR 2020) requires a collaborative approach to data management with multi-disciplinary knowledge disseminated amongst all stakeholders to support informed decision-making. This means ground models must be accessible and dynamic, set up in a way that allows data within the framework to be continually interrogated and updated. Additionally, data collaboration with mine-based geoscientists (resource context geologists, hydrogeologists, etc.), has in recent years provided an overload of information that requires technological solutions to be adequately reviewed and incorporated in a TSF context site characterization. For these reasons, a digital modelling solution is considered a fundamental component of a robust ground model.

Current typical requests for information to help inform a ground model may equate to tens of thousands of data points and include resource definition drillholes, exploration drillholes, sterilization drillholes, hydrogeological boreholes, geotechnical boreholes (of both mined pits and workings as well as civil infrastructure and processing plants), surface mapping data, face mapping data, groundwater monitoring data, geophysical data, geotechnical/environmental/geochemical laboratory testing results, current and historical topographic surveys and reports. Resource-context digital models may also be available which generally provide valuable stratigraphy and structural geology information.

A digital modelling solution centres around collaborative data repositories (such as Bentley Systems OpenGround Cloud and Seequent Central), and 3D data visualization and implicit modelling software (for example Seequent Leapfrog). The aforementioned software applications and cloud systems are currently utilized by the author (Menzies and Simjee 2023). Dynamic links are also established between the digital model and seepage, stability and hydrogeological modelling software packages.

Typically, multiple digital models are created for a single site, with varied mesh resolutions, model units and/or model extents depending on the end model use. For example, a hydrogeological digital model may be greater than 100 km^2 in area and comprise relatively simplified hydrogeological stratigraphy, whereas the equivalent geotechnical digital model for the site may contain more refined units, higher-resolution meshes and extend less than a few hundred metres from the facility extent. An example digital geotechnical model is shown in Figure 2. Although multiple models are developed, a single Leapfrog workspace is utilized to maintain dynamic data connections more easily.

The temptation of digital modelling tools is to believe the data will reveal the correct answer to the question of geological unit geometry. However, implicit modelling techniques still require the modeller to input fundamental geological rules to extrapolate model surfaces between data points. This means the modeller needs to understand (in their mind or based on a conceptual model) what the ground model

Fig. 2. Image of a TSF digital geotechnical model. Vertically exaggerated.

should look like and, secondly, that the modeller must in most cases be a competent geologist. The importance of conceptual ground models is discussed further below.

Conceptualization

Conceptual ground models are considered foundational to subsequently generated observational models and to communicating TSF site characterization. Problematic or unreliable ground models will likely be generated where knowledge and experience are lacking, resulting in critical aspects of design or construction omitted in concept and therefore not addressed in later stages of investigation and model development (Baynes *et al.* 2021). A conceptualization is often accompanied by non-explicit visuals (conceptual cross-sections or block diagrams) to assist in effectively communicating ground conditions and geohazards to all stakeholders, not just geoscience professionals.

A conceptual model is developed primarily on pre-existing knowledge and experience. As a result, the assembly of an appropriate project team is the first step in beginning a site characterization (IAEG 2022). It is important to recognize mine site geoscientists not only provide valuable model input data, but that individuals on the mine site are likely to have a more thorough understanding of ground conditions than appointed tailings consultants, albeit in a potentially different context. This means they play an important role in the development of the conceptual geological model.

Conceptual models are often indicated to be an initial assessment of ground conditions as part of the first phase of site characterization (e.g. ANCOLD 2020). However, they should be continually reviewed as part of data validation during the full lifecycle of a TSF and updated if 'unknown unknowns' are encountered or the context of the site or facility changes. Context changes include, for example, design or construction variations, or life of mine production changes.

Evolution of ground models for tailings management

Site investigation and characterization guidelines typically define different model types based on the available geological and geotechnical data at different stages of project development/investigation. While regional differences in model development and definition exist, the pertinent takeaway is that the process of model development is iterative throughout the lifespan of a project. For a TSF site, project lifespan is typically from option-level studies, iterative levels of design, construction, operation (often concurrent with further construction) and closure.

Depending on the evolution of a project or facility, or the presence of multiple facilities at various stages of their life cycle within a single mine site, a simplified flow chart of site characterization may not be appropriate. Mine sites can move millions of tonnes of material annually and this modification of ground condition needs to be reflected in continued updating of the ground model if the context of the subject structure changes. For example, hydrogeological conditions and contaminant transport may change through dewatering or recharge. Additionally, as active TSFs are commonly being raised concurrently with operation, the pertinent geotechnical conditions addressed by parts of the ground model may change. For example, understanding the geometry of a lower strength layer of clay within the foundation may become more critical as the embankment height increases.

While literature pertaining to iterative ground model development exists, tailings-specific literature addressing site characterization (e.g. APEGBC 2016; ICOLD 2022) pre-dates or is inconsistent with the IAEG *Guidelines for the Development and Application of Engineering Geological Models on Projects* (IAEG 2022), considered to provide the most robust instruction on model development. Figure 3 provides an EGM pathway for tailings-specific ground model development utilizing a combination of literature sources. The intent of the figure is to provide a framework by which to combine existing references (APEGBC 2016; ANCOLD 2020; IAEG 2022; ICOLD 2022) into a cohesive ground model solution for tailings projects. This framework is designed to ensure robust conceptualization, a total geological history approach and a digital solution are incorporated in the site characterization, and GISTM compliance is achieved.

An Engineering Geological Model (as defined earlier in this chapter) is a knowledge framework (IAEG 2022). All the ground models that make up the site characterization, the knowledge base and associated reports are part of a site- or TSF-specific EGM. Model type definitions (as shown in Fig. 3) are explained here.

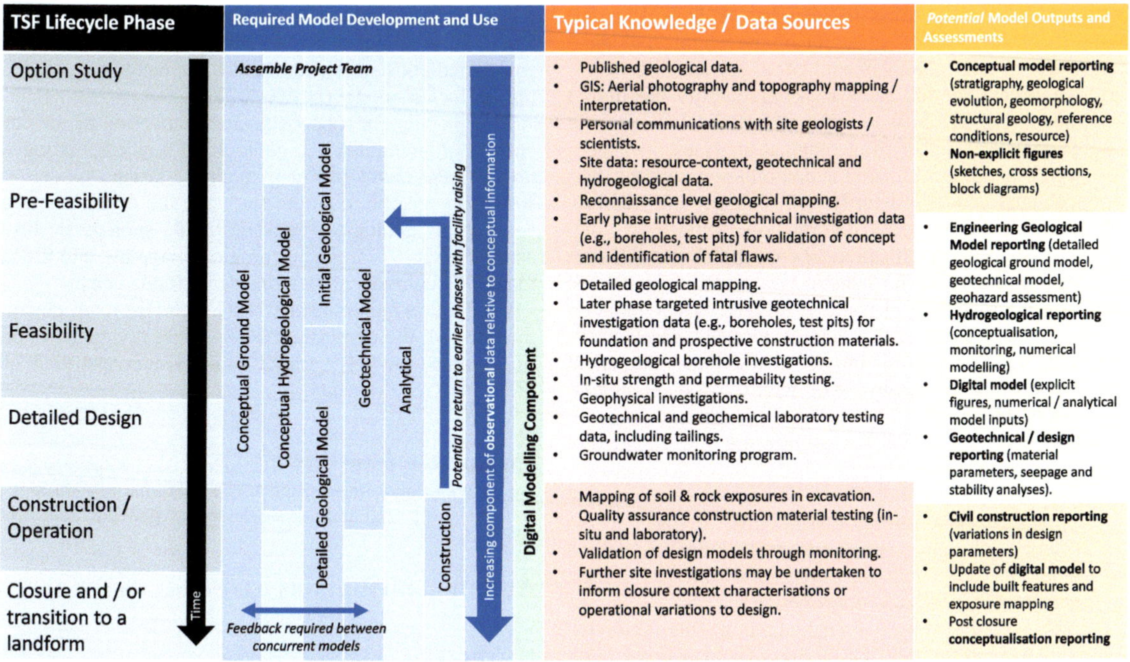

Fig. 3. An Engineering Geological Model (EGM) framework for the development and evolution of ground models for tailings management. Not all sites are the same and investigation and site characterization requirements vary. As a result, the overlapping of ground model types, information sources and outputs is intentional. Boundaries and output requirements should be considered guidance only. The initial conceptual model, developed in the context of the proposed TSF, should reveal project-specific ground model requirements.

(a) **Conceptual ground model.** A model based mainly on geological and geomorphological concepts and interpretations, and the knowledge that certain geological conditions and processes are likely to have certain engineering characteristics (IAEG 2022). The extent of the conceptual ground model should be determined in the context of the proposed TSF structure, in consideration of the area the TSF has the potential to impact. It should extend across the facility footprint, surrounding slopes and likely several kilometres upstream and downstream of the facility. The conceptual ground model should describe the regional geological setting, geological history, resource geology, structural geology and geomorphology.

(b) **Conceptual hydrogeological model.** A conceptual understanding of site stratigraphy and associated aquifers and aquitards, groundwater flow and the consequences of poor groundwater management in the context of the study structure. Although conceptual, the model may eventually be updated to contain hydrogeological observational data.

(c) **Initial geological model.** An observational model. Based on the conceptual ground model(s) and geological observations and measurements that are constrained in space and time. Represents/describes the spatial distribution of soil and rock engineering geological units and hydrogeological units and accounts for geological and geomorphological processes. Accompanied with a digital model(s), geohazard assessment and gap analyses required to inform the detailed geological model.

(d) **Detailed geological model.** An observational model. Based on the conceptual ground model(s) and observational data at an increasing ratio of observational data to conceptualization information, relative to the initial geological model. Represents/describes the spatial distribution of soil and rock engineering geological units and hydrogeological units and accounts for geological and geomorphological processes. Accompanied with a digital model(s) and geohazard assessment.

(e) **Digital model.** Presentation of geological observational data in 3D space and the development of 3D model units utilizing implicit modelling techniques. Units are developed in the context of the TSF structure and relevant subsequent analyses (geotechnical, hydrogeological, etc.); therefore, multiple digital models are typically required.

(f) **Geotechnical model.** Provides engineering characteristics and geotechnical parameters of relevant aspects of the observational geological model(s) and tailings. Feedback is required to inform earlier model types if inconsistencies in material characterization data exist (e.g. lab testing, *in situ* testing). Typically considered to be a follow on/output from the detailed geological model; however, it can be an output of the initial geological model in early phases of study (e.g. pre-feasibility).

(g) **Analytical model.** Relevant outputs from the digital model (e.g. 2D cross-sections, 3D meshes) to facilitate geotechnical and hydrogeological analyses. In a geotechnical context, utilized in combination with the geotechnical model primarily for seepage and stability analyses, numerical deformation modelling and advancement/modification of conceptual embankment design to account for construction material properties. In a hydrogeological context, utilized in combination with the observational geological model(s) and conceptual hydrogeological model to facilitate numerical groundwater modelling and hydrogeological assessments. Geometries of units used for analyses may be modified to account for engineering judgement and design sensitivity.

(h) **Construction model.** The collection of observational geological data during construction, update of the detailed geological model and geotechnical model (as required) to account for revision of unit geometries, unit characteristics and construction material variations. Inclusion of the observational data in the digital model. Review/investigation of tailings characteristics (geotechnical, hydrogeological and chemical) during the life of the facility.

For each model, reliability and uncertainty should be considered. This includes for example, the input assumptions, accuracy of observational data and the potential for multiple realizations of interpreted geometries.

GISTM (GTR 2020) requires the appointment of a facility-specific Engineer of Record (EOR),

> the qualified engineering firm responsible for confirming that the tailings facility is design, constructed, and decommissioned with appropriate concern for integrity of the facility, and that it aligns with and meets applicable regulations, statutes, guidelines, codes, and standards (GTR 2020, p. 27).

The EOR's responsibilities are continuous throughout the life of the facility and therefore involve continued facility review, inspection, monitoring and reporting. Although the EOR is defined as an engineering firm, the technical responsibility within the firm is typically with a single individual, most regularly a tailings engineer with a geotechnical background. It is recommended that review of the TSF ground model(s) and update if required is included as part of the EOR's annual responsibilities and role.

Conclusions

TSFs are an unavoidable requirement for the extraction of metals from the Earth. Poor design, construction and management of TSFs have the potential to result in significant environmental and societal impacts, including loss of life. GISTM (GTR 2020) has improved mine waste practice and the requirement for a site characterization (Requirement 2.2) is considered consistent with the development of an engineering geological model (IAEG 2022).

Best practice for robust site characterization by the development of an engineering geological knowledge framework involves the development of multiple ground models and, while site-specific requirements vary, the development of a conceptual geological model, total geological history approach centred around geological mapping, and a digital modelling component are considered fundamental.

Finally, it is recommended that to fulfil the requirements of the GISTM, the nominated EOR should review and update (if required) the TSF ground model(s) amongst their other predefined requirements.

Competing interests The authors declare that they have no known competing financial interests or personal relationships that could have appeared to influence the work reported in this paper.

Author contributions **TJM**: writing – original draft (lead).

Funding This research received no specific grant from any funding agency in the public, commercial, or not-for-profit sectors.

Data availability Data sharing is not applicable to this article as no datasets were generated or analysed during the current study.

References

ANCOLD (Australian National Committee on Large Dams) 2020. *Guidelines for Geotechnical Investigations of Dams, Their Foundations, and Appurtenant Structures*. ANCOLD, Hobard, Tasmania, Australia.

APEGBC (Association of Professional Engineers and Geoscientists of British Columbia) 2016. *Site Characterization for Dam Foundation in BC*. APEGBC Professional Practice Guidelines V1.2, Burnaby, BC, Canada.

AS (Australian Standard) 1726 2017. *Geotechnical Site Characterisation*. Standards Australia, Sydney, Australia.

Baynes, F.J., Parry, S. and Novotný, J. 2021. Engineering geological models, projects and geotechnical risk. *Quarterly Journal of Engineering Geology and Hydrogeology*, **54**, 2, https://doi.org/10.1144/qjegh2020-080

Fookes, P.G. and Baynes, F.J. 2008. Successful use of the total geology approach, using reference conditions for the design and construction of heavy duty railways in the Pilbara, Western Australia. *Ground Engineering*, March, 39–45, https://www.geplus.co.uk/technical-paper/technical-paper-successful-use-of-the-total-geology-approach-using-reference-conditions-for-the-design-and-construction-of-heavy-duty-railways-in-the-pilbara-western-australia-01-03-2008/

Fookes, P.G., Baynes, F.J. and Hutchinson, J.N. 2000. Total geological history: a model approach to the anticipation, observation and understanding of site conditions. *Proceedings of the GeoEng2000 Conference, an International Conference on*

Geotechnical and Geological Engineering, November 2000, Melbourne, Australia, 370–460.

GTR (Global Railings Review) 2020. *Global Industry Standard on Tailings Management*. https://globaltailingsreview.org/wp-content/uploads/2020/08/global-industry-standard_EN.pdf [last accessed 19 September 2024].

IAEG (International Association for Engineering Geology and the Environment) 2022. *Guidelines for the development and application of engineering geological models development and application of engineering geological models on projects*. IAEG Commission 25 Publication No. 1.

ICOLD (International Commission on Large Dams) 2022. *Tailings Dam Safety*. ICOLD Pre-Print Bulletin 194 Version 1.0.

Menzies, T.J. and Simjee, Y. 2023. Engineering Geological Models and a Digital Modelling Solution for Tailings Management. *Proceedings of Tailings and Mine Waste 2023 Conference*, November 2023, Vancouver, Canada, https://doi.org/10.14288/1.0438117

Nathanail, J. and Banks, V. 2009. Climate change: implications for engineering geology practice. *Geological Society London Engineering Geology Special Publications*, **22**, 65–82, https://doi.org/10.1144/egsp22.5

Piciullo, L., Storrosten, E.B., Liu, Z., Nadim, F. and Lacasse, S. 2022. A new look at the statistics of tailings dam failures. *Engineering Geology*, **303**, 106657, https://doi.org/10.1016/j.enggeo.2022.106657

Robertson, P.L., de Melo, L., Williams, D.J. and Wilson, G.W. 2019. *Report of the Expert Panel on the Technical Causes of the Failure of Feijão Dam 1*, https://bdrb1investigationstacc.z15.web.core.windows.net/assets/Feijao-Dam-I-Expert-Panel-Report-ENG.pdf

SME (Society for Mining, Metallurgy & Exploration Inc) 2022. *Tailings Management Handbook: A Life-Cycle Approach*. Society for Mining, Metallurgy & Exploration Inc, Inglewood, Colorado, USA.

Mountain road engineering: the legacy of Peter Fookes across the continents

Gareth J. Hearn[1]* and Richard P. Martin[2]

[1]Hearn Geoserve Ltd, Worthing, West Sussex BN125NJ, UK
[2]GeoconsultHK Ltd, Hong Kong 123456, Hong Kong
GJH, 0000-0002-9503-8049; RPM, 0009-0002-3476-540X
*Correspondence: Gareth@Hearngeoserve.co.uk

Abstract: Peter Fookes played a pioneering role in many fields of engineering geology, but the legacy of his work in mountain regions, and especially with respect to the design of mountain roads, was among his most lasting. In the 1970s, he took on the challenge of assembling and leading a team of geo-specialists to assist in the design and construction of the Dharan–Dhankuta Road (DDR) through intensely dissected fold mountain terrain in eastern Nepal. Peter did not fail to deliver, and the mountain model he produced to accommodate the complexity and dynamic nature of the terrain helped guide project decision-making and design. Peter's awareness of the need to understand the geological and geomorphological history of any site led to the *Total Geology Approach* that offered a blueprint for engineering geological practitioners thereafter. The authors describe a project to construct a road across the Blue Nile gorge in Ethiopia where this blueprint was applied. Design and construction of the DDR and the road across the Blue Nile gorge were especially demanding. To date, both roads have performed relatively successfully since they were opened to traffic and are fitting tributes to Peter's legacy.

Although Professor Fookes ('Peter') was a stickler for the detail required to resolve geological complexity, he was also a pioneer of simplicity. He combined detail and simplicity in the development of conceptual ground models to aid the interpretation of landscapes, their geology, ground conditions and geohazards (e.g. Fookes 1997; Fookes *et al.* 2015) and to convey their implications to the civil engineering end-user. Through the use of geological models, more generally geomodels, he dedicated most of his career to helping civil engineers understand and accommodate geological complexity and geohazard uncertainty in their design and project management. One of Peter's most notable geological models was the mountain model he and his colleagues developed for highway engineering purposes in young fold mountains, and the Himalayan terrain in particular (Fookes *et al.* 1985; Hearn and Martin 2022).

Peter's first encounter with civil engineering in the mountains was in Pakistan, in connection with the geological structure and slope stability of the Siwalik Clays on the Jari and Mangla Dam projects on the southern margin of the Himalaya (Fookes 1965; Fookes and Wilson 1966). Thereafter, Peter worked on engineering schemes in many mountainous regions, but it was his contribution to road building in Nepal that was perhaps his most profound achievement in this sector.

The 50 km long Dharan to Dhankuta Road (DDR), built between 1977 and 1982, was by no means the first road to be constructed in the Himalayan foothills but it was arguably pioneering in its engineering geological approach to design, and particularly with respect to the role played by geomorphological assessment (Brunsden *et al.* 1975; Fookes *et al.* 1985; Hearn 2001; Lee and Fookes 2015; Hearn and Martin 2022). This was an early example of another of Peter's strengths and major contributions to the engineering geological profession: the creation of multi-skilled geo-teams to resolve complex and dynamic problems.

The engineering geological and geomorphological work undertaken for the design, construction and operation to date of the DDR is among the most notable and referenced work globally of this nature. Professor Megh Dhital, one of the most accomplished engineering geologists in Nepal, stated that Peter's work on the DDR had been pioneering and he was responsible for maps that were the first of their kind in the country (M. Dhital pers. comm. 2023). This chapter summarizes the legacy of Peter's work in ensuring the successful completion and operation of the DDR and describes how his approach was adopted on challenging road projects elsewhere, and particularly in Ethiopia.

Dharan–Dhankuta Road, Nepal

Project background

The DDR is located in the Siwalik and Mahabharat Ranges of the low Himalaya of eastern Nepal (Fig. 1). The Sangure Ridge is the most rugged and southerly ridge within the Mahabharat Lekh, and rises to 1440 m above sea level (asl) to the north of Dharan, itself located on the Terai Plain at

From: Griffiths, J. S., Giles, D. P., Davis, J. A. and Dunlop, J. J. (eds) 2025. *The Peter Fookes Engineering Geological Legacy in Geomodels, Geomaterials and Geomorphology*. Geological Society, London, Engineering Geology Special Publications, **30**, 55–77,
First published online 25 March 2025, https://doi.org/10.1144/egsp30-2024-20

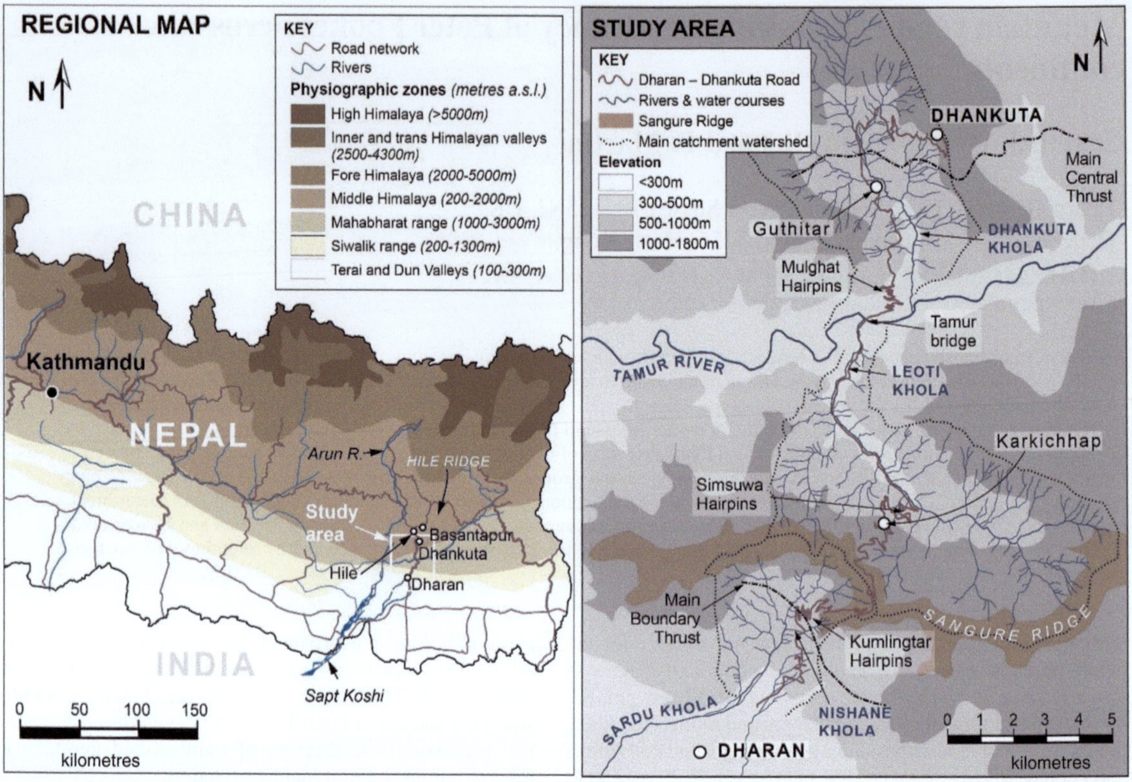

Fig. 1. Location of the Dharan–Dhankuta Road in eastern Nepal. Source: Hearn and Martin (2022), reproduced with permission from Springer Nature.

525 m asl. North of the Sangure Ridge, the terrain occupied by the road corridor descends rapidly to 260 m asl where the road crosses the Tamur River and rises again to 1200 m asl at the town of Dhankuta.

Episodes of geological uplift and drainage incision have created a polycyclic landscape with mid- and high-level terraces and plateaux separated by steep slope units up to 40° in inclination. Slopes are underlain predominantly by closely jointed and frequently sheared metasedimentary rocks, phyllites, schists and gneisses (Fookes and Marsh 1981). Depths of moderately to completely weathered rock can be several metres in the schists and up to 15 m in the gneisses. Annual rainfall exceeds 2000 mm in Dharan and reduces to 950 mm in Dhankuta, falling predominantly during the June–September wet season when 24-hour rainfalls can, exceptionally, reach 300 mm. Earthquakes, landslides and floods are the key hazards that road building and operation are forced to contend with (Brunsden *et al.* 1981; Fookes *et al.* 1985; Hearn and Martin 2022).

Peter the geo-team builder

Peter would have been impressed with the work of Professors Denys Brunsden and David Jones on landslide-related geomorphological investigations in Dorset, UK (e.g. Brunsden and Jones 1972) when, as consultant to Rendel, Palmer and Tritton (RPT, now Rendel Ltd), he was considering how to tackle the prospect of a new road through the Himalaya in the early 1970s. At the same time, Dr John Doornkamp was also applying geomorphological mapping to aid land resource surveys in Uganda (e.g. Doornkamp 1971) and these three accomplished geomorphologists were assembled by Peter to map the geomorphology of alignment alternatives for the DDR and provide insight into slope and drainage hazards for design purposes.

Peter's geo-team assembled for the DDR was accommodated through RPT's recruitment team, and combined geomorphological, engineering geological and materials specialists. An ability to anticipate how a site investigation might unfold, and the engineering questions that will need to be answered as it does, is essential when creating geo-teams. This anticipation is borne from wide-ranging experience and conceptual ground modelling: attributes Peter was able to bring to bear in almost any geo-setting. Furthermore, his ability to 'squeeze' all he could from his assembled specialists (Fig. 2) and turn his conceptual model into observational reality for engineering decision-making and design was probably unrivalled.

Fig. 2. Peter overseeing field investigations with RPT geo-team members and project management contractor's staff at the proposed site of the Tamur Bridge in 1978. From left to right: 1st left, K. Cross, chief engineer at the preliminary design stage; 2nd left, C. Manby, engineering geologist; 5th left, A. Marsh, materials engineer; 6th left, Peter. Source: photo taken by R. Martin, engineering geomorphologist.

Peter the ground modeller

Figure 3 is the conceptual mountain model developed by Peter and his colleagues to advance the engineering geological investigation and design of the DDR. The geomorphological roots of the model were developed by Brunsden *et al.* (1975), but Peter recognized the importance of classifying the terrain according to geomorphological evolution and created the five-zone model that was used to: (1) explain and predict the location of slope and fluvial hazards in the landscape; and (2) summarize anticipated ground conditions and engineering recommendations for accommodating them in design (Table 1). Zone 4 in the model represented the steep lower valley sides where landslides and other geohazards were most severe, while Zone 5 defined fluvially active valley floors where flooding, scour and sediment deposition hazards were to be expected. The principal alignment recommendations from the model were to:

(1) avoid lengthy horizontal traverse in Zone 4;
(2) find stable climbing corridors (hairpin stacks) in Zone 4 to gain access from Zone 5 to Zone 3 and vice versa;
(3) and maximize horizontal traverse in Zones 2 and 3.

Because the north–south road corridor was normal to a west–east topographic grain with extremely high relative relief it was inevitable that all of these terrain zones would need to be crossed (Fig. 4). In total, 43% of the alignment was constructed across terrain Zones 2 and 3. Terrain Zone 4 accounted for 34% of the alignment, mainly through the construction of carefully located and designed multi-hairpin climbing corridors. The remaining 23% was located in terrain Zone 5, a much higher proportion than had been advised from engineering geological considerations during design (see explanation in Hearn and Martin 2022). Figure 5 illustrates how the classification is applied to selected sections of the alignment.

Peter's use of the 'moveable cross-section' was a simple but highly effective technique for ensuring continuity of the model from alignment selection into design and construction. During a reconnaissance survey of the selected DDR alignment he would quickly sketch hillside profiles with his estimated ground conditions, then superimpose a roughly scaled standard carriageway on tracing paper which he moved laterally across the profiles as he summarized the pros and cons for cross-section design in terms of the changing earthworks balance (cut/fill), and implications for

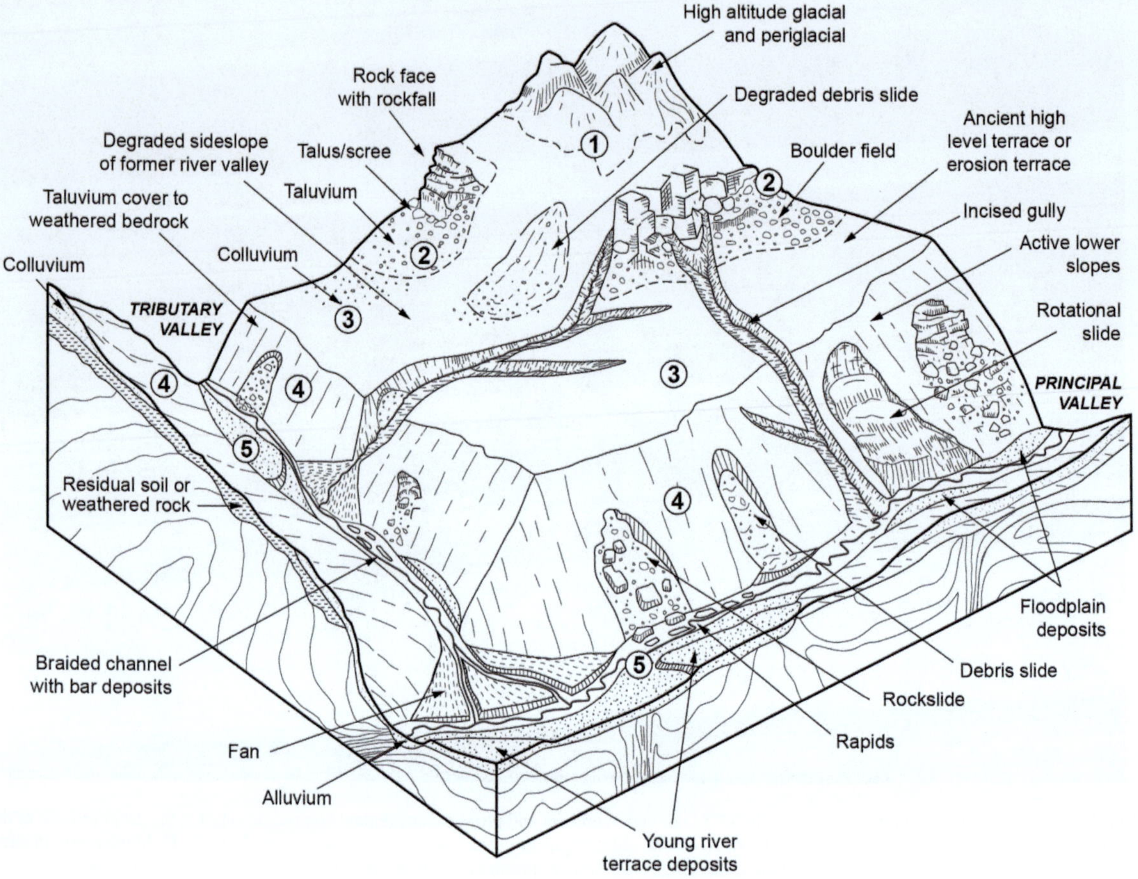

Fig. 3. The Fookes *et al.* (1985) mountain model. Zone 1: glacial & periglacial terrain, not represented in the DDR area; Zone 2: steep ridge lines & rock slopes; Zone 3: mid-slope, gentle topography, reflecting a dormant period of landscape development without uplift or incision; Zone 4: steep, geomorphologically-active zone created by renewed drainage network incision in response to tectonic uplift; Zone 5: modern valley floor, with active river flooding, scour, sediment transport & deposition, including debris fans from unstable tributary valleys. Source: modified with permission from Elsevier.

retained support of the road formation and/or protection or support of cut faces.

Model continuity, from investigation, through design to construction completion, was facilitated by Peter's ongoing role as a visiting specialist, together with a full-time geo-team required to update the model as: (1) ground conditions became revealed through excavation; (2) the evolving geomorphology of the alignment was observed; and (3) the performance of the constructed works was progressively reviewed. Inevitably, some compromises had to be made, with budgets and construction expediency sometimes taking priority over engineering geology (Hearn and Martin 2022).

Outcome

Hearn and Martin (2022) undertook a detailed review of the performance of the DDR 40 years after its completion. During this period the project area was impacted by several extreme rainstorms, hydrological floods, landslide dam outburst floods and two major earthquakes. The authors visited the area on numerous occasions to record the effects of these events on the stability and integrity of the road. They concluded that all the alignment in terrain Zone 2 and practically all of it in Zone 3 had remained secure as far as the stability of the road formation was concerned. Although terrain Zone 4 had proven to be the most geomorphologically unstable over the 40-year period (as had been predicted in Fig. 3), careful route selection and design in accordance with Table 1 meant that only 3–4% of the road constructed within this zone had subsided due to ground movement. Furthermore, less than 1% of the entire alignment had suffered from recurrent ground movement. In such difficult terrain, this can be described as nothing less than a major achievement.

Table 1. Summary of geomorphological observations and engineering recommendations made by Fookes *et al.* (1985), according to mountain model terrain zone

Zone	Description	Geomorphology	Specific hazards	Highway engineering guidelines	
				Alignment	Design and construction
1	High altitude glacial–periglacial	Glacial and periglacial processes, landforms and deposits		Not considered	
2	Free rock face and associated debris slopes	Rock cliffs and steep slopes, mostly talus (scree), taluvium, boulder fields	Rock falls, rock slides, talus creep	(1) Cross talus slopes close to their lower, less-steep, sections where larger-size debris has greater frictional strength and hazard from further rock fall is lower due to increased distance from source	(1) Generally use full-cut cross-sections in rock and talus (2) Talus slopes can be cut 5–12° steeper than existing naturally. Use gabion walls to retain fill across talus. Found high walls on rock beneath talus where practicable (3) Taluvium can be cut to 33–45°, and steeper (up to 53°) when dry and with cohesion (4) Large boulders derived from Zone 2 rock falls may be exposed in cut slopes and will require support (5) Careful balancing needed between open cut (larger volumes) and partial retaining or revetment walls to steeper cut slopes (6) Rock slope support and rock fall control will usually be required (7) Foundation design is critical to below-road retaining walls and other structures (8) Slope drainage generally not necessary, though earth bunds may assist the control of runoff (9) Check dams and cascades unnecessary at stream crossings
3	Degraded middle slopes and ancient valley floors	Low- and moderately angled residual, colluvial and alluvial soil slopes, ancient fans, terraces, and degraded rock failure deposits from Zone 2. Can include: (1) localized rock slope problems similar to Zone 2; (2) reactivation of landslides in	*Relatively rare:* Rock falls, rock slides, rotational slides, debris slides, earth slides, debris flows, earth flows, rills, gullies	(1) The ideal alignment is along the mature, stable terraces and plateaux of this zone	(1) Retaining walls and revetments for cut and fill slopes generally not required (2) Colluvium can be cut to 33–45°, and steeper (up to 53°) when dry and with cohesion (3) Generally suitable for construction of compacted spoil benches (4) Slope drainage generally not necessary unless obvious sources of runoff need to be controlled

(*Continued*)

Table 1. *Continued*

Zone	Description	Geomorphology	Specific hazards	Highway engineering guidelines	
				Alignment	Design and construction
		localized Zone 4 terrain within Zone 3; (3) steep-sided and unstable tributaries extending upslope from Zone 4			(5) Check dams and cascades generally unnecessary at stream crossings (6) The cost of geotechnical works (earthworks, slope works, retaining walls, scour protection) are typically between a quarter and a third of the costs in Zones 2, 4 and 5
4	Active lower slopes	Steep shallow taluvium and colluvium overlying weathered rock, unstable rock cliffs and rock slopes, steep-sided gullies, unstable soil slopes, soil creep, surface and sub-surface erosion. Toe erosion from Zone 5	*Relatively common*: Rock falls, rock slides, debris slides, earth slides, debris flows, rills, gullies. High groundwater *Relatively rare:* Rotational slides, earth flows	(1) Gain or lose elevation via climbing corridors (hairpin stacks) on relatively stable slopes to avoid extended contouring alignments and long continuous climbs across slopes in this zone (2) If space permits use offset stacks so that sub-stacks are self-contained and any instability does not progress to other parts of the stack (3) Hazard ranking surveys may assist in alignment design (4) Design horizontal and vertical alignment to avoid deep cut slopes (5) Choice of bridge sites may have a negative effect on alignment stability (6) Stable crossing points of streams and large gullies required, affecting detailed alignment design	(1) Engineering geological mapping during design and construction needs to be more intensive than in other zones (2) Minimize size of cuts and fills as far as practicable. Retained cut is preferred to retained fill as rock cutting will yield a more stable road (3) Ensure adequate wall foundations on rock (4) Zone 5 river scour is an important design consideration for Zone 4 slopes, also river scour and sediment deposition in Zone 4 side-streams (5) Careful balancing needed between open cut and partial retaining or revetment walls to cut slopes (6) Gabion structures preferred to support deep (6 m +) cuts in colluvium (7) Small slides (<1000 m^3) can be removed and scarps cut to reduced angle (8) Large landslides may require partial removal, drainage and revegetation (9) Foundation design is critical to below-road retaining walls, i.e. on rock or beneath depth of transported soils/ground movement (10) Even limited spoil disposal can trigger instability. Where topography prevents spoil bench construction, tip spoil as far from the road as possible (11) Check dams and cascades commonly required at stream crossings and may need to be extended up and down the channel (12) Cross unstable gullies with high sediment load by bridges or large

					culverts. Consider fords, though these require high maintenance (13) Drop outlets preferred to drop inlets where culverts are constructed through retained fill (14) Channel bed protection required beneath culverts in weak and erodible materials and continuously between culverts on hairpin stacks (15) Rock slope support and rock fall control may be required
5	Valley floors	Low-angle river channel and terraces, tributary fans, coarse debris, river scour, bedrock-cut river channels and cut terraces, unstable terrace banks	Aggrading debris fans from tributary valleys, river scour, sediment transport	(1) Use valley floor margins. Compare potential slope instability against river flooding and scour and design alignment according to whether Zone 4 cut is preferred over Zone 5 fill (2) Cross aggrading tributary fans as close to apex/throat as possible (3) Consider crossing Zone 4 slopes to avoid problematic fan crossings (4) Consider bridging main river to avoid Zone 4 slopes and Zone 5 fans (5) If valley floor terraces are too steep for road gradient, consider 'stepping down' in Zone 4	(1) River scour and sedimentation are important design considerations (2) Preference for road construction on fill to avoid cutting into Zone 4 slopes (3) Consider main river realignment and river training (4) If tributary fans cannot be crossed at their apex/throat, mid-fan bridging will require expensive river training and scour protection. (5) Ford crossings of fans will be cheaper than bridges or vented causeways but will require high maintenance (6) Foundation design is critical to below-road retaining walls and other structures where river scour is a potential hazard (7) Surcharge of embankments may be necessary on compressible alluvial soils (8) Rock slope support and rock fall control may be required in river gorge sections (9) Detailed consideration required for design and construction of road support structures and scour protection

Source: from Hearn and Martin (2022, reproduced with permission from Springer Nature).

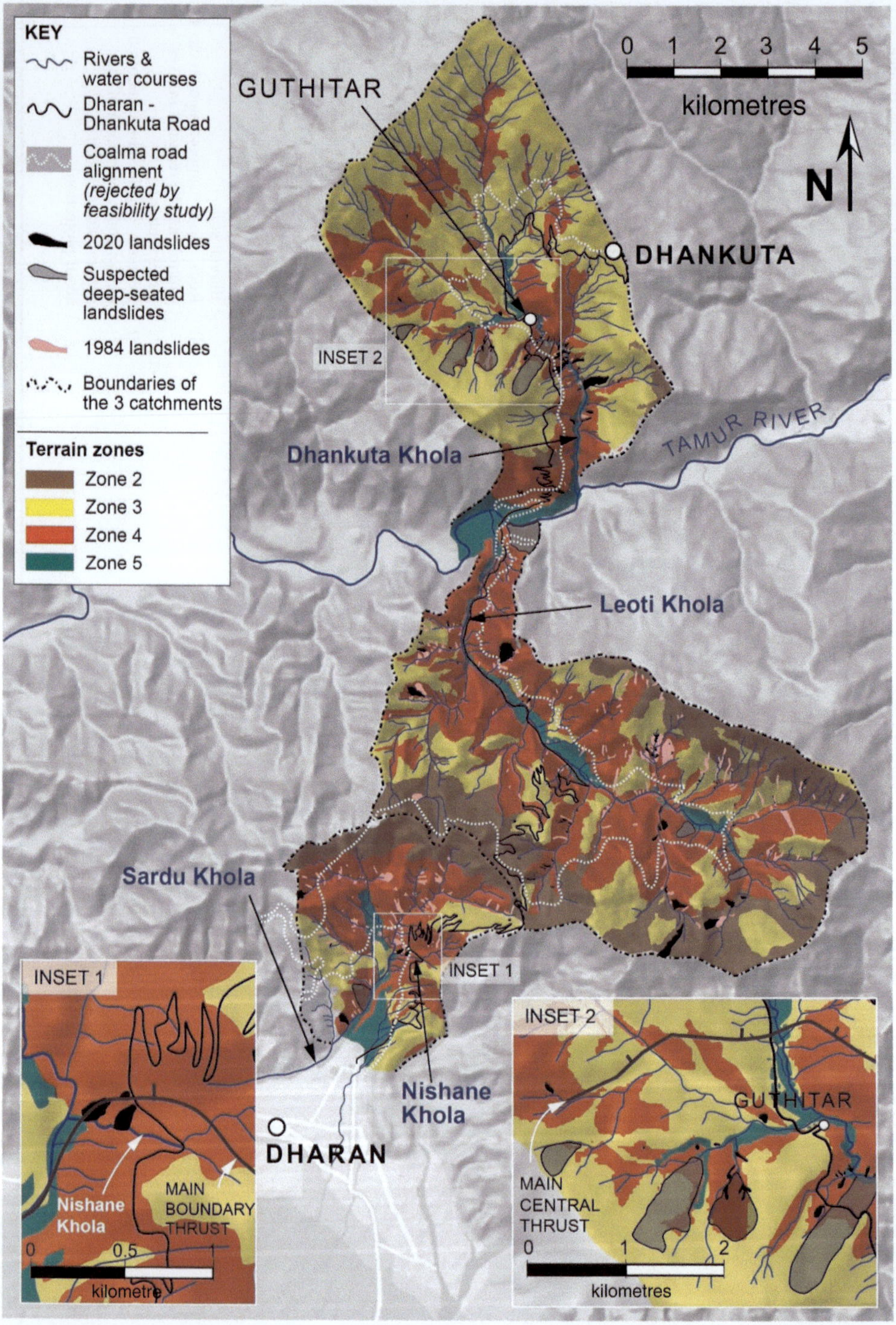

Fig. 4. The Fookes *et al.* mountain model mapped out for the DDR alignment by Hearn and Martin (2022) with larger-scale insets to show specific details. The distribution of landslides, especially large and deep-seated movements, demonstrates the value of the model in classifying the terrain for engineering purposes. These landslides were mapped variously by Brunsden *et al.* (1975), Fookes *et al.* (1985), Hearn (1987, 2017) and Hearn and Martin (2022), and they predominate in Zone 4.

(a) Ascent to the Sangure Ridge across Zones 4 & 3, hairpin stack at Kumlingtar right centre as shown in inset 1, Figure 4

(b) Sangure Ridge to Leoti Khola floor in Zones 2, 3 & 4 with short 'Simsuwa' hairpin stack in the 'bowl' at the toe

(c) Alignment on the floor of the Leoti Khola valley in Zone 5, including river terrace and fan crossings

(d) Hairpin stack at Mulghat in Zone 4 providing high-level access to the Dhankuta Khola valley above the gorge lower right

(e) Side-long traverse across Zone 4 above the Dhankuta Khola gorge. The right-hand (northern) section of this image corresponds to the area shown in inset 2 of Figure 4

Fig. 5. Examples of the mountain model terrain classification in the constructed road corridor. See Figure 4 for the plan view and terrain zone colour codes. All the photographs were taken during or soon after construction, i.e. late 1970s to early 1980s. Source: Hearn and Martin (2022), reproduced with permission from Springer Nature.

For economic and construction expediency reasons, and against engineering geological advice, 7.5 km (15% of the entire alignment) was constructed on the valley floor of the Leoti Khola river in Zone 5 (Figs 4 & 5c) and 13% of this alignment has been subject to recurrent damage by river scour. In total, less than 3% of the entire alignment has suffered from recurrent landslide and fluvial hazards. Hearn and Martin (2022) demonstrated how the stability of the DDR compared favourably with statistics derived for other roads in the country. While Peter must take a sizeable amount of credit for this outcome, the efforts of the RPT geo-team, civil engineers and, in particular, the skill of the main route selection and alignment engineer (John Warren) were equally important.

In addition to his overall contribution to the project in terms of ground modelling and geo-team building, Peter also provided valuable insights and advice on specific design issues during construction. An example was his suggestion to undertake early trial excavations of steep-cut slopes in weathered gneisses close to the Dhankuta end of the road and observe how the slopes degraded naturally for several wet seasons before finalizing the slope designs. Martin and Howell (2024) reviewed the performance of two of these cut slopes and demonstrated how a combination of steep design angles, slope drainage and bio-engineering measures reduced slope degradation rates by over two orders of magnitude over a period of 40 years.

The DDR legacy

Hearn and Martin (2022) describe how the mountain model has been widely used elsewhere, including academic publications, engineering guidelines and engineering applications, especially in Nepal. A variation on the model was developed and applied to the analysis of access road corridors for the Arun III hydroelectric project located in the Middle Himalaya to the north of Dhankuta. This work was undertaken in the 1990s and is reported in Hearn and Lawrance (2000) and Hearn (2001, 2017). The models are shown in Figure 6 and were used to systematically compare route options in terms

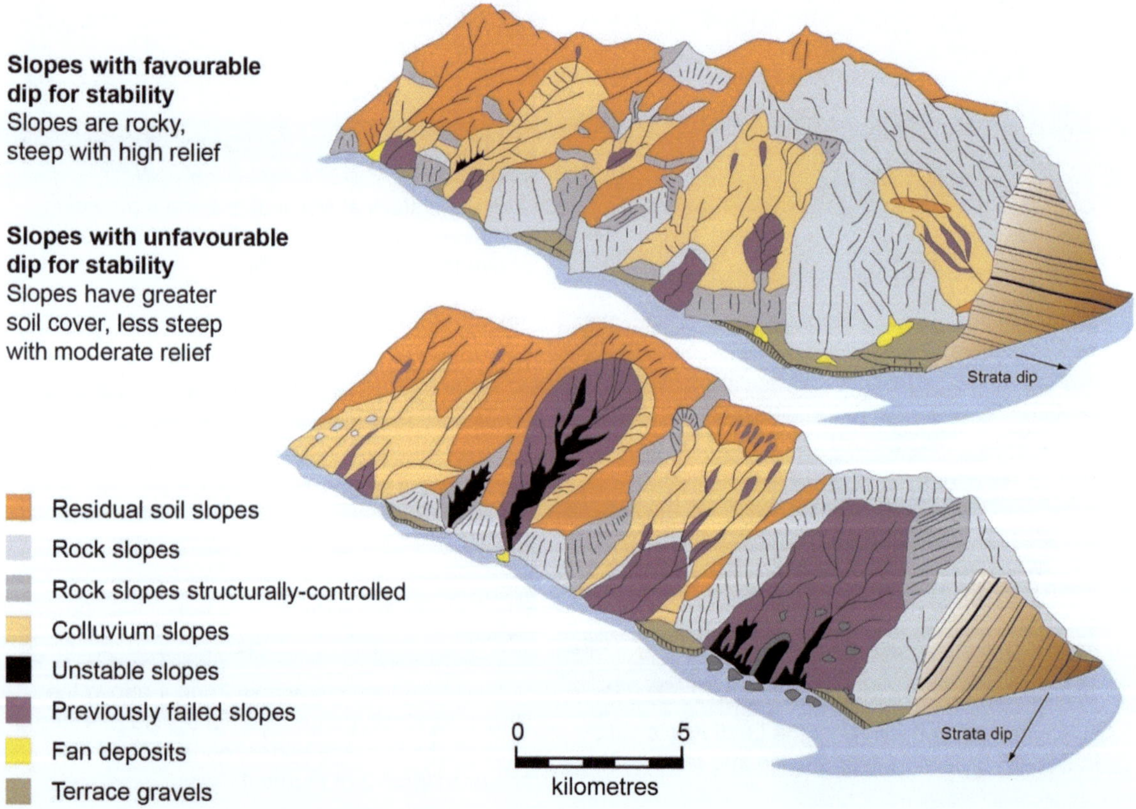

Fig. 6. Terrain models used to review route options for the Arun III hydropower project access road. Source: Hearn (2017), reproduced with permission of the Geological Society, London.

of anticipated average earthworks quantities, frequency and size of stream and river crossings, landslide hazards and anticipated slope stabilization and erosion protection requirements. Hearn and Martin (2022) provided illustrations of other roads in Nepal where the Fookes *et al.* mountain model appears to apply reasonably well and can be used to explain the incidence of favourable and less successful outcomes.

Both authors (Hearn and Martin) gained immensely from their experience working in Nepal and time spent learning from Peter. Martin went on to work for most of his career in the Hong Kong Geotechnical Engineering Office, and in his first 10 years there carried out several studies using field techniques and analytical methods of the type Peter would have recommended. One of these was a regional-scale risk assessment of catchwaters (open reservoir intake channels to capture hillside runoff for local water supply) based on geomorphological hazard mapping. Some of the maps produced in this study (at *c.* 1:5000 scale) were back-analysed 25 years later to assess their ability to predict debris flow run-out from cases of catchwater breach or blockage recorded in the interim (Martin 2009). The original assessment was undertaken by a geo-team comprising a geotechnical engineer, an engineering hydrologist and an engineering geomorphologist (Martin).

Another of his undertakings was the detailed study of a large (*c.* 5000 t), upstanding rock outcrop on the Mid-level cliffs overlooking part of the main urban area of Hong Kong Island. This involved production of plans and sections at 1:200 to 1:20 scale based on precise measurement of joint-bounded block geometry, kinematic assessment of a range of credible failure modes and stability modelling. Both studies were later summarized in a thematic paper on the benefits of bespoke landslide hazard mapping (Martin 2014), a further topic of long-standing interest to Peter throughout his career (e.g. Dearman and Fookes 1974; Fookes 1997; Fookes *et al.* 2007).

As a member of one of his geo-teams, Hearn was able to benefit immensely from Peter's work as a team builder and ground modeller at Ok Tedi in Papua New Guinea. This work was undertaken in the early 1990s and is notable for its intensity of investigation and multi-disciplinary approach. The project was an operating open-cast gold and silver mine at Ok Tedi, located in steep, geologically complex and seismically active mountainous terrain in Papua New Guinea where some of the highest rainfalls in the world are recorded. As for the DDR in Nepal, Peter assembled a wide range of geological, geomorphological and geotechnical expertise to develop ground models to help understand structural geology, engineering geomorphology and ground conditions for landslide hazard assessment.

Some of this Ok Tedi work is described in Hearn (1992, 1995*a*, *b*) and Hearn *et al.* (2001). However, this is just one element of the *total geology approach* that Peter applied through his team of experts to help his client (Ok Tedi Mining Ltd) understand and manage geohazards within its operational area. The total geology approach to geotechnical and geohazard problem-solving was later expounded by Peter and his colleagues in relation to railway projects in Western Australia (Fookes *et al.* 2001; Baynes *et al.* 2005; Fookes and Baynes 2008). It is based on the need to understand the total geological and geomorphological history of a site or project area from an early stage in the site investigation. This enables robust interpretation of geology, climate, geomorphology, ground conditions and geohazards using thorough desk study, aerial and satellite image interpretation, landscape and ground modelling, terrain classification, soil and rock reference condition mapping, specialist field mapping and ground investigation. It forms an important background to the guidelines on conceptual ground models developed by the International Association of Engineering Geology and the Environment (Baynes and Parry 2024).

Mekane Selam–Gundewein Road, Ethiopia

The project

Though Peter was not directly involved in this project, its implementation was based on the same principles he developed and applied on the DDR in Nepal and elsewhere, namely: geo-team development and engineering integration at all project stages (feasibility study, design and construction supervision); landscape evolution and ground modelling; and the total geology approach.

Mekane Selam and Gundewein are two small towns in the Central Highlands of Ethiopia (Fig. 7), separated by the Blue

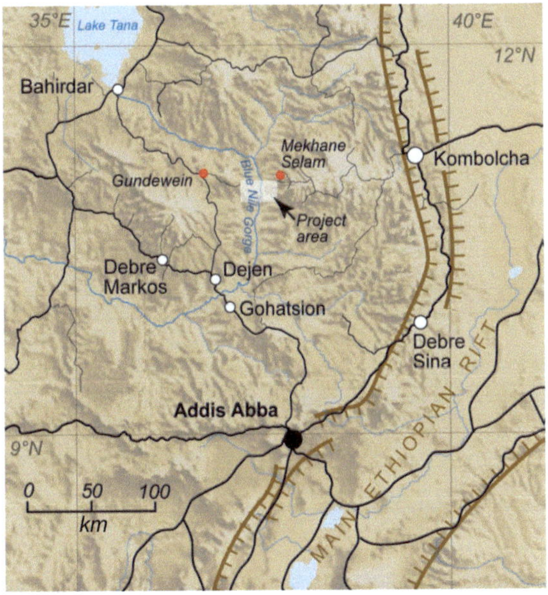

Fig. 7. Location of the Mekane Selam–Gundewein Road across the Blue Nile gorge in Ethiopia.

Nile (Abay) gorge that is up to 1300 m deep and less than 2 km wide in places. Design and construction of a new road across the gorge between these towns was funded by the Ethiopian Roads Authority (ERA) and the road was opened to traffic in 2014. The cumulative rise and fall of the alignment in the gorge is more than 2500 m over a route length of approximately 58 km and a straight-line distance of 21 km.

The gorge has cut into a sequence of lava flows, pyroclastic deposits and sedimentary rocks. Jurassic limestone, with interbedded marl and shale, occupies the base of the gorge and is overlain by Cretaceous sandstone, with subordinate conglomerate, siltstone and shale. The sandstone is in turn overlain by a sequence of Oligocene and Miocene basalt flows and volcaniclastic deposits, which, due to differential weathering and erosion, impart a stepped profile to the side walls of the gorge (Fig. 8). Colluvial deposits and taluvial aprons mantle the lower slopes while deep-seated cliff failures occupy many of the upper benches (Fig. 9). Expansive black cotton soils have developed on the weathered basalts and are unstable on all but the gentlest of slopes. The average annual rainfall in the Blue Nile basin is a little over 1400 mm with 70% falling between June and September. Surface and subsurface erosion, karst cavity collapse, flooding, debris flows and aggradation are active processes in the gorge and its neighbouring valley side slopes, and combine with rockfalls, translational and rotational slides, deep-seated cambering and complex ground conditions to create an extremely challenging environment for road construction.

Total geology approach

A geo-team was assembled capable of providing the following skills:

- remote sensing, and particularly stereo aerial photograph interpretation, to record principal geological and geomorphological features;
- terrain classification;
- landscape evolution modelling, including the ability to interpret the contemporary landscape in relation to its geological and geomorphological history;
- engineering geological and geomorphological field mapping and exposure logging;
- geotechnical reference condition classification and mapping, grouping terrain units with similar landforms,

Fig. 8. Typical side wall of the Blue Nile gorge with stepped profile in limestone (lower-most), sandstone and volcanic rocks (upper-most), forming cliffs and terraces.

Fig. 9. Failed cliffs are the norm rather than the exception in both limestone and sandstone.

geomorphological processes and geological materials and with a defined range of geotechnical parameters;

- design and supervision of sub-surface geotechnical investigations and ground modelling;
- route selection and design of earthworks, slope stabilization, drainage control and erosion protection measures; and
- observation and exposure logging during construction for comparison with anticipated reference condition and adjustment, as required.

The implementation of these skills as part of the total geology approach to design and construction supervision is described and illustrated by Hearn and Pettifer (2016). Desk study commenced in 2006 and focused on aerial photograph interpretation, the production of a simple terrain classification and the manipulation of published geological and topographic data to produce perspective views of the landscape in advance of landscape modelling (Fig. 10) and geotechnical reference condition mapping (see below). Initially, aerial photographs were made available at a scale of 1:50 000, which was considered too small for mapping anything other than apparent fault traces, terrain types and the largest of geomorphological features. Nevertheless, this and the interpretation of larger scale (1:10 000) aerial photography obtained later during the design phase, provided useful background. The outputs contributed to the geotechnical reference condition mapping (see below) and were documented both in tabular and mapping format.

Field reconnaissance commenced on horseback in 2006. At the time this was the only feasible way of reaching the gorge from either direction and had to be undertaken one side at a time. The 1000 km drive from one side of the Blue Nile to the other via Addis Ababa was essential if the potentially treacherous boat journey across the 100 m wide river, known to harbour crocodiles, was to be avoided. Later, fieldwork was facilitated by the establishment of mobile camp sites until vehicular access tracks were established along the approximate centre line of the road corridor. Eventually, a temporary bridge was built across the river to enable access until the permanent structure was operational. Difficulties with access meant that drilling investigations were limited to the two main bridge sites, i.e. the Blue Nile and its main tributary, the Difarsa River. Drilling capability at the time was also below international standards and field investigations were focused primarily on mapping, trial pitting and the production of geological cross-sections.

The geotechnical reference condition classification proved to be the most important tool in conveying ground conditions and design parameters to the construction supervision team. A

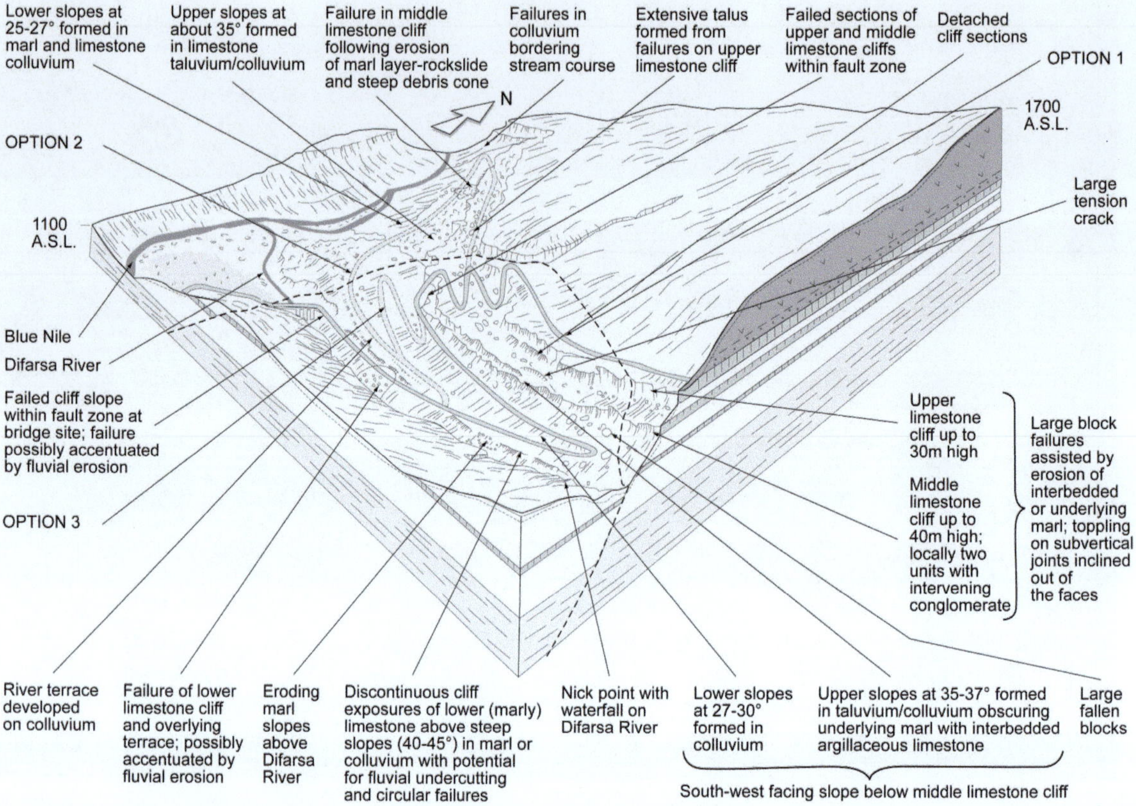

Fig. 10. Landscape models were prepared for the western and eastern side of the Blue Nile gorge. This example shows how the technique helped compare route corridor alternatives in an area of particularly difficult topography and ground conditions. For location, see Figure 15. Source: Hearn and Pettifer (2016), reproduced with permission from Springer Nature.

total of 34 different geotechnical reference conditions were identified: 17 for *in situ* rock, 10 for transported materials, 6 for different landslide types, and 1 for faults and other tectonic features. The main rock and soil reference condition classes are shown in Table 2. They were tabulated according to chainage and shown on perspective alignment drawings along with terrain and key geomorphological features, as illustrated in Figure 11. The details and recommendations listed below were tabulated for each reference condition and an example reference condition dataset is given in Figure 12:

- lithology;
- landform;
- material descriptions;
- engineering geological considerations;
- preferred cross-section;
- embankment foundation stability;
- cut slope stability;
- anticipated maximum cut slope against height; and
- expected excavation techniques.

The geotechnical reference condition mapping was used to help refine the alignment and consider route alternatives. Recommendations included realignments to avoid sections of tunnel, the preferred routes across or through cliffs, route options alongside the Blue Nile river, bridge locations across the Blue Nile and its major tributary and realignments to avoid landslides and difficult side-long ground. The decision to use geotechnical reference conditions was based on the success with which they were used by Peter Fookes and his colleagues in Western Australia, referred to above. The latest version (Baynes and Parry 2024) of the IAEG guidelines on the development of engineering geological models refers to reference condition classes as engineering geological units, though the purpose remains the same.

Outcome

Using observations made by the lead author during construction supervision (the last visit was made during the defects liability period in 2015), Hearn and Pettifer (2016) carried out a detailed comparison of the observed ground conditions with those predicted by the geotechnical reference condition mapping. It was found that a total of 15% of the alignment in the gorge encountered ground conditions that were not

Table 2. Summary of rock and soil reference condition classes derived for the project area

Primary classification	Origin	Rock/soil type			Landscape
In situ rock	Volcanic	Basalt	Slightly to moderately weathered (SW–MW)		Spur and ridge lines
					Cliffs and steep slopes
					Cliffs with turrets or towers
					Knolls
			Highly to completely weathered (HW–CW)		Gentle slopes
			Residually weathered	Granular soils	Plateaux and benches
				Black cotton soil	
		Volcaniclastic (including pyroclastic) rocks with or without interlayered basalt	Tuff/ash		Benches and slopes
			Agglomerate		
			Volcanic breccia		
			Scoriaceous basalt		
			Tuffaceous sandstone/siltstone		
	Sedimentary	Sandstone			Benches and slopes
					Cliffs
		Limestone			Benches and slopes
					Cliffs
		Interbedded limestone and marl			Benches and slopes
					Cliffs
		Marl			Gentle slopes
Transported soils and mass movement deposits	Colluvium/Taluvium	Volcanic material			Benches and slopes
		Limestone/marl material			
	Talus	Volcanic material			
		Limestone material			
	Debris flows	Volcanic material			
		Limestone material			
	Rock avalanches	Large rock fragments			Lower slopes and terraces
	Landslides	Differentiated according to source material and mechanism			Mainly on steep slopes
Fluvial deposits	Terraces/fans	Fine and coarse material			River terraces and valley floors
	Flood plains	Coarse material			
	Point bars				

Source: Hearn and Pettifer (2016, reproduced with permission from Springer Nature).

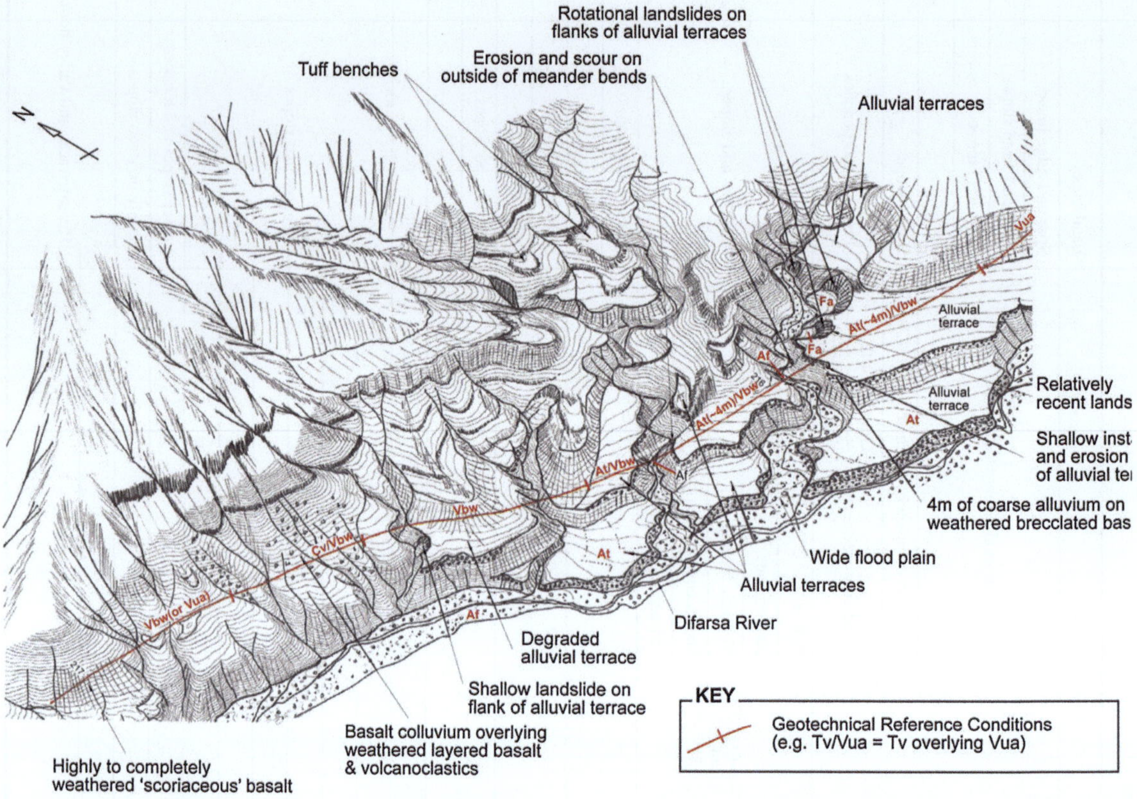

Fig. 11. Extract from the geotechnical reference condition mapping. For location, see Figure 15. Source: Hearn (2011), reproduced with permission of the Geological Society, London.

predicted. In these areas, failures in cut slopes, deep-seated ground movements and cavity collapse have damaged the road over a total length of approximately 1 km, i.e. <2% of the alignment in the gorge. Significantly, the presence of plastic clay at depth within interlayered volcanic sequences was judged to be the cause of ground instability in at least one location that would have been very difficult to foresee without the aid of good quality drill core. Other damage was caused, at least in part, by subgrade erosion and cavity collapse within limestone and marl colluvium caused by excess road runoff.

There has been a general trend for slopes to fail where, for whatever reason, they have been cut to angles steeper than recommended for the reference condition. There were also cases where cut slopes had to be excavated to shallower angles than recommended, primarily due to the unforeseen occurrence of higher weathering grades and weaker horizons within volcanic sequences. Nevertheless, recommended cutting angles were defined in the reference condition tables as maximum and multi-slope profiles were proposed where excavations would need to accommodate varying material strengths with depth. Other problems were encountered where some fill slopes, especially those constructed in rock fill, had to be built steeper than specified in the design because of topographic constraints. This resulted in localized frontal slope instability and cracking to the road shoulder. These defects were rectified through fill reconstruction using a higher grading specification and fill reinforcement using gabion wire mesh and gabion revetment (Fig. 13). Figure 14 shows some selected photographs of the constructed alignment in 2016.

It has not been possible to re-visit the road to update the record of its geotechnical performance because of political unrest in that part of the country. The Ethiopian Roads Authority was requested to provide an update but, for the same reason has been unable to do so, other than by confirming that the road remains open.

Fortunately, Google Earth has high-resolution (sub-metre) imagery for 2024 of the entire road corridor (Fig. 15). From this imagery, the road appears to have remained intact along most, if not all, of its alignment, though there are short sections where the wearing course appears a different shade to that on either side, possibly suggesting that it might have been replaced for whatever reason. In one area there is obvious damage to the road surface along 100 m of its length where it was constructed across steep, dissected terrain formed in

Reference Condition	Lithology	Landforms	Materials
Vbw	Highly to completely weathered basalt	Basalt plateau, moderate slopes	Spheroidal weathering of basalt resulting in basalt gravels, cobbles and small boulders with a clayey and sandy matrix.

Geomorphology
Underlying large areas of the basalt plateau above the gorge and some moderately steep slopes within the volcanic terrain of the gorge.
Stratigraphic origin
Ashangi Basalts (eastern side of gorge) – Eocene to Oligocene (~55 to 24 My)
Blue Nile Basalts (western side of gorge) – Paleocene to Oligocene (~69-23 My)
Geological description
A mass ranging from basalt cobbles and small boulders within a dark brown clayey to sandy matrix, to clayey and sandy basalt cobbles.
Structural geological characteristics
The basalt weathers to a disintegrated rock mass and structures become impersistent.
Degree of weathering
The weathering profile in the basalt grades from a moderately consolidated, but uncemented soil near the surface to a closely fractured rock mass with relict joints recognisable at depth. Rock mass weathering typically involves variable spheroidal weathering around strong, subrounded, cobble and boulder-size basalt corestones.
Material strength and rock mass quality
Basalt boulders/clasts are usually strong but the overall weathered rock mass is generally weak and very closely fractured – i.e. behaves as a dense soil.

Reference Condition	Vbw
Bulk density	2.25 Mg/m³
Material strength, UCS	1.25-50 MPa
RQD estimate	<20% though considered a soil
Rock mass quality	Very poor to poor
GSI	25-40

Embankment foundation suitability: Variable and unreliable embankment foundation conditions depending on weathering profile and depth.

Expected excavation techniques: Conventional.

Comments: Potentially loose rocks and boulders should be scaled from cut slopes and rock catch ditches constructed where necessary

Anticipated maximum cut slope angles

	Anticipated maximum cut slope angles vertical distance : horizontal distance					Comments
Cut slope height, H	<5m	5-10m	10-15m	15-20m	20-30m	
Vbw	1:1	1:1.75	Weathering grade not expected beyond 10m depth			

Fig. 12. Geotechnical reference condition table for Vbw – highly–completely weathered basalt.

Fig. 13. Rock fill embankments, supported by masonry retaining walls and protected with gabion revetment. For location, see Figure 15.

Side-long ground cut & fill. Note the long masonry channel across farmland to safely convey roadside drainage

Hairpin stack from the Difarsa R bridge to an upper limestone plateau. The boxed area is underlain by in situ and failed limestone cliffs within a fault zone with limestone and marl colluvium containing cavities caused by natural and road-related seepage erosion

Road constructed in box cut through detached and subsided blocks of limestone cliff

Disturbed basalt exposed in 13m deep box cut. Local heave to the road pavement is thought to be due to movement along a clay seam within a volcaniclastic sequence at depth beneath road, as exposed in a nearby trial pit. 2023 imagery on Google Earth suggests manageable deformation to the road surface continues

Limestone and marl taluvium beneath 40m high cliffs. Slopes below road too steep (35°) for fill and unsuitable for wall foundations. Gabion, RC retaining walls and revetments were considered but a 1:1 open cut design was selected with planting and low gabion catch walls

20-25° climbing corridor between high-level plateau & mid-slope bench. The 7-bend stack is built in balanced cut and fill using gabion walls to support slopes excavated through basalt colluvium overlying weathered tuff

Fig. 14. Examples of the constructed road with summaries of the ground conditions encountered and their accommodation in design. For locations, see Figure 15.

Fig. 15. Looking approximately east, showing the constructed road in the Blue Nile gorge. Source: image date 2024 Google Earth/CNES/ Airbus.

weathered, scoriaceous basalt and volcaniclastic breccia. The lead author cannot find any record of there being a problem in this area during design or construction and it is difficult to determine from the imagery if this damage has been due to ground movement or the effects of post-construction quarrying and quarry waste disposal. The steepness of the topography, the weakness of the underlying rock, the presence of groundwater seepage within the volcanic and underlying sedimentary sequences and the destabilizing effects of gully erosion combine to create landslide potential throughout much of the landscape.

There are several other locations where cut slope instability has deposited debris onto the road, probably blocking it from time to time, and one or two instances where ephemeral flows in stream channels have deposited fans across the road. The critical sections of alignment, including all the hairpin stacks, appear to have remained stable (Fig. 16).

Concluding remarks

The climate, geology, topography and ground conditions found in Nepal and Ethiopia are different but the projects posed similar scales of engineering challenge and demanded an inclusive and integrated approach to geological, geomorphological and geotechnical assessment. Although both case studies concern road design and construction, the same total geology approach is equally applicable to the resolution of complex ground conditions and geohazards on any engineering project in mountainous terrain, as briefly outlined in relation to open-cast mining in Papua New Guinea. All sources of data must be closely examined and verified, and every feasible method of investigation pursued to yield as robust an interpretation as possible. By combining the observations and interpretations of geo-specialists with different backgrounds and experience, the opportunities to overlook or misinterpret important details are minimized.

Peter Fookes was a key developer of the geo-team approach and was one of the most accomplished geo-practitioners in the mental assimilation of complex detail and its classification and translation into clear summaries and recommendations for his engineering clients. His use of conceptual and observational ground models, or 'geomodels' (e.g. Fookes 1997; Fookes *et al.* 2015) facilitated combined interpretation of structural geology, landscape evolution, landforms, materials and geomorphological processes and was fundamental to his approach. It has formed a blueprint for so much engineering geological work that has followed, and has been the principal goal on all projects carried out by the authors since Peter's pioneering studies in the Himalaya.

In advance of his presentation at the Geological Society of London Engineering Group Annual Conference in Cambridge in March 2023, the lead author collected tributes

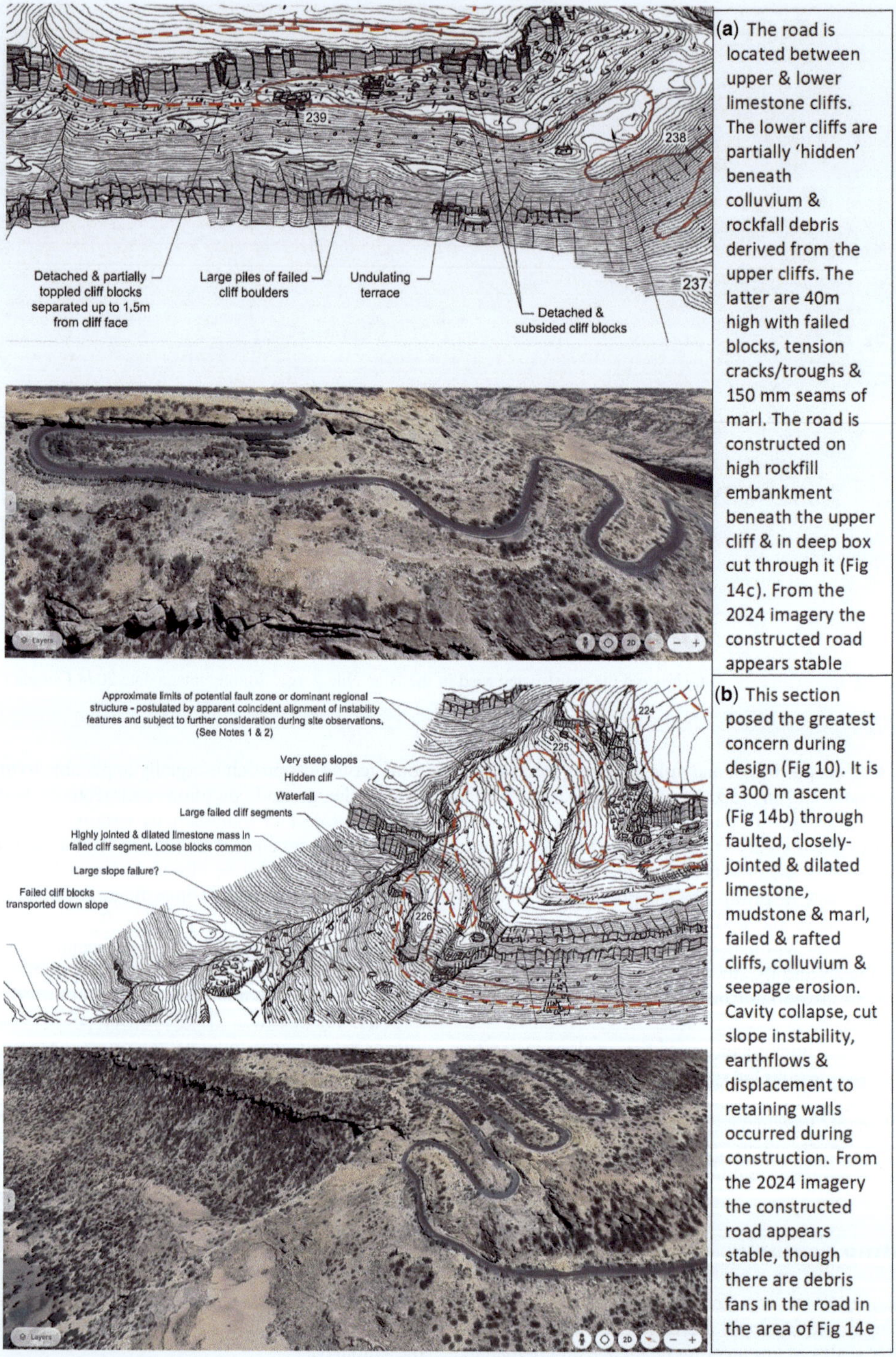

Fig. 16. Geomorphological interpretation (2007) and constructed road (2024) for two critical sections of alignment. Note that the geomorphological interpretation was primarily from aerial photograph interpretation and field reconnaissance plotted onto plans and perspective views using digitized contours from published topographic mapping projected in Autocad. Source: imagery Google Earth/CNES/Airbus.

from selected colleagues and clients of Peter's. These are presented in the textbox below and provide fitting confirmation of the role Peter played in their work and his enormous contributions in helping to resolve engineering problems.

Professor Denys Brunsden *Peter was an enthusiast, a supporter, a constant stimulus and certainly the prime supporter of our subject. Peter wanted ideas, problems, solutions. He could persuade you to do anything and go anywhere. We did it but he thought of it.*

Professor Russell Blong *(Peter) made a terrific job of giving all the (Ok Tedi) consultants and others freedom to do their sub-projects the way they wanted but at the same time keeping the whole show focused on the project aims and the big picture ... Peter did a superb job of pulling the whole project together, keeping everyone focused on their part of the project, and keeping everyone pretty happy most of the time ... I also worked with Peter on a smaller project in PNG for BP Development ... Again Peter did a great job of maintaining focus, keeping everyone happy, and taking everyone's contributions into account, with a lot of humour (and red wine).*

Paul Maconochie (superintendent geotechnical engineer and client, Ok Tedi Mining Ltd) *Peter opened up to me a wonderful world of interaction between geology and engineering. At Ok Tedi he brought to bear his ideas formalized later in the Total Geological History Approach. Beyond that was his ability to communicate geo-concepts to lay people. I learnt a lot through observing how he organized his thoughts and marshalled them into coherent reports with very little editing. His use of tables to organize his thoughts and present information was a revelation.*

Brian Tollin (geotechnical engineer and client, Sakhalin Energy Investment Company Ltd, Russia) *I first encountered Peter at a conference in the early 1980s. His reputation for excellence in engineering geology became fully apparent to me. Little did I know then that we would work together 30 years later when, as Principal Geotechnical Engineer on the Sakhalin II project, I engaged Peter as geohazard peer review consultant for the pipelines alignment through the Makarov Mountains. Peter was firmly rooted in the practical aspects of engineering geology and how best to derive and convey information to aid geotechnical and civil engineers. The central role he played in stimulating debate at the ICE/BP onshore oil and gas pipeline geohazard conference in 2005 reflected the respect that he was afforded by the profession.*

Geoff Pettifer *I was one of Peter's students on the MSc Geomaterials course at QMC in 1981. His lectures were packed full of interesting information. He always stressed the importance of geology, hydrology and geomorphology in the interpretation of ground conditions for excavation design and construction engineering. I worked with Peter in Lesotho, Hong Kong, North Wales, Indonesia and lastly Russia (Sakhalin). In Lesotho, he persuaded me to construct a painted cardboard model to illustrate ground conditions in a way that could be easily understood by non-specialists. Peter saw I had a knack for presenting relevant ground information in block diagrams, and he got me to draw several of these for publication, leading to work with Gareth Hearn producing sets for linear construction projects.*

Dr Andrew Hart *After first working with Peter in Romania, he quicky became a mentor. He was always curious about things, asking questions, challenging preconceived ideas. He was willing to take advice from others (as long as he trusted their opinion). He liked classification schemes for assessing ground conditions, engineering constraints and geohazards. These were a simplification, but allowed rapid assessment of the problems and helped focus on the key issues.*

Acknowledgements The authors would like to acknowledge the invaluable contributions made by the following colleagues on the projects described:

Nepal – Denys Brunsden; David Jones; John Doornkamp; Mike Sweeney; Chris Manby; Adrian Marsh; Cliff Lawrance; John Dowling; and many others.

Ethiopia – Geoff Pettifer; Jonathan Hart; Colin Chant; Darryn Wise; Lulseged Ayalew; Sarah D'Agostino; Phil Talbot; Solomon Kuliche; Mesfin Teshome; Seyoum Belew, Kirti BahadurThapa; Solomon Assefa; and many others.

Papua New Guinea – Russell Blong; Geoff Humphries; Russell Mason; and many others.

Competing interests The authors declare that they have no known competing financial interests or personal relationships that could have appeared to influence the work reported in this paper.

Author contributions **GJH**: conceptualization (lead), formal analysis (lead), writing – original draft (lead), writing – review & editing (supporting); **RPM**: conceptualization (supporting), formal analysis (supporting), writing – review & editing (lead).

Funding This research received no specific grant from any funding agency in the public, commercial, or not-for-profit sectors.

Data availability All data generated or analysed during this study are included in this published article (and, if present, its supplementary information files).

References

Baynes, F.J. and Parry, S. 2024. *Guidelines for the Development and Application of Engineering Geological Models on Projects.*

International Association for Engineering Geology and the Environment (IAEG) Commission, **25** Publication No. 1, https://www.iaeg.info/C25EGMGuidelines/

Baynes, F.J., Fookes, P.G. and Kennedy, J.F. 2005. The total engineering geology approach applied to railways in the Pilbara, Western Australia. *Bulletin of Engineering Geology and the Environment*, **64**, 67–94, https://doi.org/10.1007/s10064-004-0271-4

Brunsden, D. and Jones, D.K.C. 1972. The morphology of degraded landslide slopes in South West Dorset. *Quarterly Journal of Engineering Geology*, **5**, 205–222, https://doi.org/10.1144/gsl.qjeg.1972.005.03.01

Brunsden, D., Doornkamp, J.C., Fookes, P.G., Jones, D.K.C. and Kelly, J.M.N. 1975. Large scale geomorphological mapping and highway engineering design. *Quarterly Journal of Engineering Geology*, **8**, 227–253, https://doi.org/10.1144/GSL.QJEG.1975.008.04.01

Brunsden, D., Jones, D.K.C., Martin, R.P. and Doornkamp, J.C. 1981. The geomorphological character of part of the Low Himalaya of eastern Nepal. *Zeitschrift für Geomorphologie, Neue Folge, Supplementband*, **37**, 25–72.

Dearman, W.R. and Fookes, P.G. 1974. Engineering geological mapping for civil engineering practice in the United Kingdom. *Quarterly Journal of Engineering Geology*, **7**, 223–256, https://doi.org/10.1144/GSL.QJEG.1974.007.03.01

Doornkamp, J.C. 1971. Geomorphological mapping. Chapter 2. *In*: Ominde, S.H. (ed.) *Studies in East African Geography and Development*. Heinemann, London, 9–28.

Fookes, P.G. 1965. Orientation of fissures in stiff over-consolidated clay of the Siwalik System. *Géotechnique*, **15**, 195–206, https://doi.org/10.1680/geot.1965.15.2.195

Fookes, P.G. 1997. Geology for engineers: the geological model, prediction and performance. *Quarterly Journal of Engineering Geology and Hydrogeology*, **38**, 231–284, https://doi.org/10.1144/GSL.QJEG.1997.030.P4.02

Fookes, P.G. and Baynes, F.J. 2008. Successful use of the Total Geology Approach, using Reference Conditions for the design and construction of heavy duty railways in the Pilbara, Western Australia. *Ground Engineering*, 1 March, 39–45, https://www.geplus.co.uk/technical-paper/technical-paper-successful-use-of-the-total-geology-approach-using-reference-conditions-for-the-design-and-construction-of-heavy-duty-railways-in-the-pilbara-western-australia-01-03-2008/

Fookes, P.G. and Marsh, A.H. 1981. Some characteristics of construction materials in the low to moderate grade metamorphic rocks of the Lower Himalayas of east Nepal, 1: occurrence and geological features. *Proceedings of the Institution of Civil Engineers Part 1*, **70**, 123–138, https://doi.org/10.1680/iicep.1981.1961

Fookes, P.G. and Wilson, D.D. 1966. The geometry of discontinuities and slope failures in Siwalik Clay. *Géotechnique*, **16**, 305–320, https://doi.org/10.1680/geot.1966.16.4.305

Fookes, P.G., Sweeney, M., Manby, C.N.D. and Martin, R.P. 1985. Geological and geotechnical engineering aspects of low cost roads in mountainous terrain. *Engineering Geology*, **21**, 1–152, https://doi.org/10.1016/0013-7952(85)90002-X

Fookes, P.G., Baynes, F.J. and Hutchinson, J.N. 2001. Total geological history: a model approach to the anticipation, observation and understanding of site conditions. *GeoEng 2000 Conference, An International Conference on Geotechnical and Geological Engineering*, 19–24 November, 2000, Melbourne Australia. Technomic, Lancaster, Pennsylvania, **1**, 370–460.

Fookes, P.G., Lee, E.M. and Griffiths, J.S. 2007. *Engineering Geomorphology: Theory and Practice*. Whittles, Caithness.

Fookes, P.G., Pettifer, G. and Waltham, T. 2015. *Geomodels in Engineering Geology: An Introduction*. Whittles, Caithness.

Hearn, G.J. 1987. *An Evaluation of Geomorphological Contributions to Mountain Highway Design with Particular Reference to the Lower Himalaya*. PhD thesis, University of London.

Hearn, G.J. 1992. Terrain hazard mapping at Ok Tedi, Papua New Guinea. *In*: Bell, D.H. (ed.) *Proceedings of the 6th International Symposium on Landslides*, Christchurch, New Zealand. Balkema, Rotterdam, **2**, 971–976.

Hearn, G.J. 1995*a*. Landslide and erosion hazard mapping at OK Tedi copper mine, Papua New Guinea. *Quarterly Journal of Engineering Geology*, **28**, 47–60, https://doi.org/10.1144/GSL.QJEGH.1995.028.P1.05

Hearn, G.J. 1995*b*. Engineering geomorphological mapping and opencast mining in unstable mountains a case study. *Transactions of the Institution of Mining and Metallurgy*, **104**, A1–A18.

Hearn, G.J. 2001. Low-cost road construction and rehabilitation in unstable mountain areas. *Geological Society, London, Engineering Geology Special Publications*, **18**, 135–141, https://doi.org/10.1144/GSL.ENG.2001.018.01.20

Hearn, G.J. 2011. Field mapping. Chapter B3. *Geological Society, London, Engineering Geology Special Publication*, **24**, https://doi.org/10.1144/EGSP24.7

Hearn, G.J. 2017. Engineering geomorphology of the Koshi Highway, east Nepal. *Quarterly Journal of Engineering Geology and Hydrogeology*, **50**, 354–367, https://doi.org/10.1144/qjegh2016-112

Hearn, G.J. and Lawrance, C.J. 2000. *Geomorphological Processes in Eastern Nepal as a Basis for Road Design*. Transport Research Laboratory, Unpublished Project Report, **PR/INT/186/00**. Department for International Development, Crowthorne, https://www.gov.uk/research-for-development-outputs/geomorphological-processes-in-eastern-nepal-as-a-basis-for-road-design

Hearn, G.J. and Martin, R.P. 2022. The performance of the Dharan-Dhankuta Road, east Nepal, in the context of the Fookes *et al.* (1985) mountain model: a tribute to the late Professor PG Fookes. *Bulletin of Engineering Geology and the Environment*, **81**, 4, https://doi.org/10.1007/s10064-021-02514-w

Hearn, G.J. and Pettifer, G.S. 2016. The role of engineering geology in the route selection, design and construction of a road across the Blue Nile gorge, Ethiopia. *Bulletin of Engineering Geology and the Environment*, **75**, 1, https://doi.org/10.1007/s10064-015-0724-y

Hearn, G.J., Blong, R. and Humphries, G. 2001. Terrain hazard around the Ok Tedi copper mine, Papua New Guinea. *Geological Society, London, Engineering Geology Special Publications*, **18**, 143–149, https://doi.org/10.1144/GSL.ENG.2001.018.01.21

Lee, E.M. and Fookes, P.G. 2015. A note on the origins of engineering geomorphology in the UK. *Quarterly Journal of Engineering Geology and Hydrogeology*, **48**, 147–156, https://doi.org/10.1144/qjegh2014-048

Martin, R.P. 2009. Qualitative risk assessment of catchwaters in Hong Kong based on simplified geomorphological hazard mapping. *Quarterly Journal of Engineering Geology and Hydrogeology*, **42**, 459–471. (Preprint published in the *Proceedings of the*

Conference on Engineering Geology in Geotechnical Risk Management, Hong Kong, 2007, 102–113), https://doi.org/10.1144/1470-9236/08-002

Martin, R.P. 2014. Geotechnical insights from flexible field mapping of landslide hazards. *Proceedings of the Conference on Advances in Terrain Mapping and Analysis for Landslide Hazard Assessment*, Hong Kong. Geological Society of London, Hong Kong Regional Group, 145–166, GSL_Conference_2014_350dpi_GeotechnicalInsightsfromFlexibleFieldMapping of Landslide Hazards.pdf

Martin, R.P. and Howell, J.H. 2024. Degradation and protection of cut slopes in weathered gneiss; a 40 year case study. *Quarterly Journal of Engineering Geology and Hydrogeology*, **57**, 2, https://doi.org/10.1144/qjegh2023-073

3D geological modelling to support an environmental safety case in the nuclear sector

Amber Pavey[1*], Rhys Timms[1], Matthew Hampson[1], Neil Jones[1], John Shevelan[2] and Christopher Coleman[1]

[1]Amentum, HQ Building, Thomson Avenue, Harwell Campus, Harwell OX11 0GD, UK
[2]Nuclear Waste Services, Old Shore Road, Holmrook, Cumbria CA19 1XP, UK
AP, 0009-0000-5780-2539; RT, 0000-0002-0894-9080; MH, 0000-0002-9194-1353; NJ, 0000-0002-7273-1457; JS, 0000-0002-4314-7593
*Correspondence: amber.pavey@global.amentum.com

Abstract: The Low Level Waste Repository in west Cumbria requires, as part of their environmental safety case, a geological model to support hydrogeological understanding and future development of the site. A range of legacy data accrued over the last 60 years exists which is characterized by variable data collection methods, formats, and quality. The geological model development process is atypical due to this large body of existing data and the requirements of the environmental safety case (ESC), resulting in an initial, primarily observational digital geological model. Outlined here is the approach undertaken for geological model development and includes terminology and data management practices, and considers the role of conceptualization and its direct application to the observational 3D digital model when there is significant existing observational data. Sites with significant legacy data and atypical requirements may be more common in the future, as more site investigations are completed and data are shared. The process applied is a development of Peter Fookes' work, highlighting the importance of the conceptual model, but that application of conceptual ideas to the development of 3D digital geological models may look different in a world of increasing observational data.

A key part of Peter Fookes' legacy, and one epitomized by his inaugural Glossop Lecture, is the importance of the geological model to engineering geology (Fookes 1997), particularly the development of a conceptual model that can provide a basis for investigation, be objectively tested, and then developed into an observational geological model. Furthermore, Fookes advocated the premise of the 'total geological history', that is a thorough understanding of the complete geological and geomorphological history of the site (Fookes *et al.* 2000). This is crucial, and an understanding of the balance of conceptual ideas v. observational data of a ground model needs to be at the forefront of modelling practices, particularly as it is now commonplace to utilize 3D digital software to communicate the ground model. The International Association for Engineering Geology and the Environment (IAEG) Commission 25 (C25) guidance released in late 2022 for the development of an engineering ground model (EGM) (Baynes and Parry 2022), integrates much of this original thinking and provides a standardized framework from which modelling studies can be undertaken.

The Low-Level Waste Repository (LLWR) in west Cumbria is the UK's principal disposal facility for low-level radioactive waste. For the continued operation of the facility, the Environment Agency (EA) requires submission of an environmental safety case (ESC), a part of which requires the development of geological and subsequent hydrogeological models. There have been several models built to describe the LLWR site, based on a large volume of investigative data collected as the site developed from a Royal Ordnance Factory, to the facility today where waste is stored in highly engineered vaults. An updated model was commissioned in 2021, following additional site investigations including cored and comprehensively logged boreholes, refraction and reflection data.

A geological model is an output of an EGM. A typical EGM for engineering purposes progresses from a conceptual model to observational (see later section on terminology), with the EGM facilitating design of further site investigation. A typical EGM is often driven by engineering geological units with similar engineering characteristics, and the requirement to support a geotechnical model. In contrast, the LLWR geological model for the ESC can begin largely observational due to existing data, is largely driven by geological units with shared geological history and potential hydrogeological characteristics rather than geotechnical characteristics, and where data gaps exist, this is primarily managed through uncertainty studies rather than further site investigation at this stage.

As a result, the process for geological model development at LLWR is atypical. The significant body of legacy data must be carefully interpreted and integrated, with consideration to

From: Griffiths, J. S., Giles, D. P., Davis, J. A. and Dunlop, J. J. (eds) 2025. *The Peter Fookes Engineering Geological Legacy in Geomodels, Geomaterials and Geomorphology*. Geological Society, London, Engineering Geology Special Publications, **30**, 79–92,
First published online 31 January 2025, https://doi.org/10.1144/egsp30-2024-19

the variable data quality, range of data collection methods and changeable standards through time. Given this data availability and abundance, the initial geological models in the process are already observationally driven. However, the region is defined by a complex Quaternary stratigraphy (Merritt and Auton 2000; Coleman *et al.* 2021; Smith *et al.* 2023) meaning that conceptualization is applied directly to the 3D digital geological model later in the process. This later application of conceptual thinking is to refine the models, and to consider what is geologically reasonable and most likely, a process that can only really occur once the observational data are integrated and digested. We present a case study of this geological model development process, in the context of the LLWR in west Cumbria. This atypical approach may offer lessons for similar sites with a large catalogue of legacy data on which to base initial models.

Low-Level Waste Repository: a case study

The LLWR is the UK's low-level radioactive waste disposal facility situated in west Cumbria (Fig. 1.), which has disposed of low-level waste for over 50 years. The site is underlain by a variable thickness of complex Quaternary deposits resting on Sherwood Sandstone bedrock (Akhurst *et al.* 1997; Merritt and Auton 2000; Ambrose *et al.* 2014; Coleman *et al.* 2021; Smith *et al.* 2023).

As a condition of the LLWR's operational licence and disposal of radioactive waste, an environmental permit is required from the regulator, the EA. As part of this process, an ESC is required, at intervals as requested by the EA, which seeks to demonstrate the required safety standard for people and the environment, both at the time of disposal and in the future (for example, as a result of climate change) (Environment Agency 2015). The ESC includes safety assessment calculations which consider the movement of groundwater through the repository and subsequent transport of contaminants to potential receptors. A 3D geological model is required as the basis for a 3D groundwater flow model to support the assessment calculations of groundwater flow through the underlying geology (LLW Repository Ltd 2011). A detailed understanding of the geology also supports LLWR in the management and future development of the site.

As a result of this process, the LLWR has been the subject of several model builds, with multiple models produced over the years and in different formats, from 2D hand-drawn sections to 3D digital explicit and implicit models, to support environmental safety cases (e.g. Michie *et al.* 2010; Smith 2011). A wealth of data has been, and continues to be, collected to support the evolution of this understanding and the ongoing development of the site, including groundwater monitoring wells, geotechnical boreholes and non-intrusive investigation. The wider region to the north has also been subject to a thorough characterization from 1989 to 1997 as part of the Nirex geological investigations and associated updates to British Geological Survey (BGS) mapping (BGS 1993; Nirex 1995, 1997; Chaplow 1996, 2001; Akhurst *et al.* 1997; Merritt and Auton 1997, 2000).

In the last decade, updated geological and hydrogeological models were produced by Wood and Fugro (consultants), which are summarized in Coleman *et al.* (2021). Coleman *et al.* (2021) developed the interpretation of the Quaternary deposits and integrated a range of datasets. The same work highlights the importance of understanding the depositional processes which have resulted in the Quaternary units, and their influence on the extent of the associated deposits to support the geological model. The 2020 paper also provides conceptual ideas to consider in ongoing and future modelling work, which can be tested, reassessed and developed with the addition of new geoscience data.

Since the development of these conceptual ideas and the previous geological modelling efforts for the LLWR, additional geoscience data have been collected as part of ongoing site characterization, including regional geomorphological mapping conducted by Sinead Birks, Imperial College London (S. Birks pers. comm. 2022). In late 2021 Jacobs were commissioned to undertake a development of the geological and hydrogeological models to support the forthcoming ESC and to consider newer and wider regionally focused information. The resultant initial model produced at the end of 2022 integrated and utilized:

- nearly 600 intrusive records, which are predominantly shallow boreholes, some cone penetration tests (CPTs), and a number of fully cored boreholes dating from 2021 with detailed sedimentological logs. Regional deep boreholes from the Nirex investigations penetrate the bedrock sufficiently to differentiate bedrock units, but few others achieve this;
- sedimentological logging of exposed cliff lines and outcrops, including at Drigg and Seascale;
- geomorphological mapping of Quaternary landforms;
- geophysical data, including reflection, refraction, electrical resistivity tomography (ERT), horizontal to vertical spectral ratio (HVSR) and ground penetrating radar (GPR), with a total length of over 3000 km, including offshore seismic reflection, were used to support regional modelling;
- geotechnical data from boreholes on the LLWR site, including *in situ* or field methods, for example standard penetration tests (SPTs), hand vanes and sample testing, for example, particle size distributions (PSDs), liquid and plastic limits, oedometer tests and triaxial tests;
- targeted micro-sedimentological thin section investigations on deeper Quaternary deposits from boreholes and cliff section samples to support understanding of depositional environment, supplemented by microfauna/flora assessments;
- radiocarbon dating of macroflora remains to support correlation of the event stratigraphy.

This model was a new construct compared to the previous, built independently of the existing with a reassessment of geological interpretations, while utilizing existing conceptual

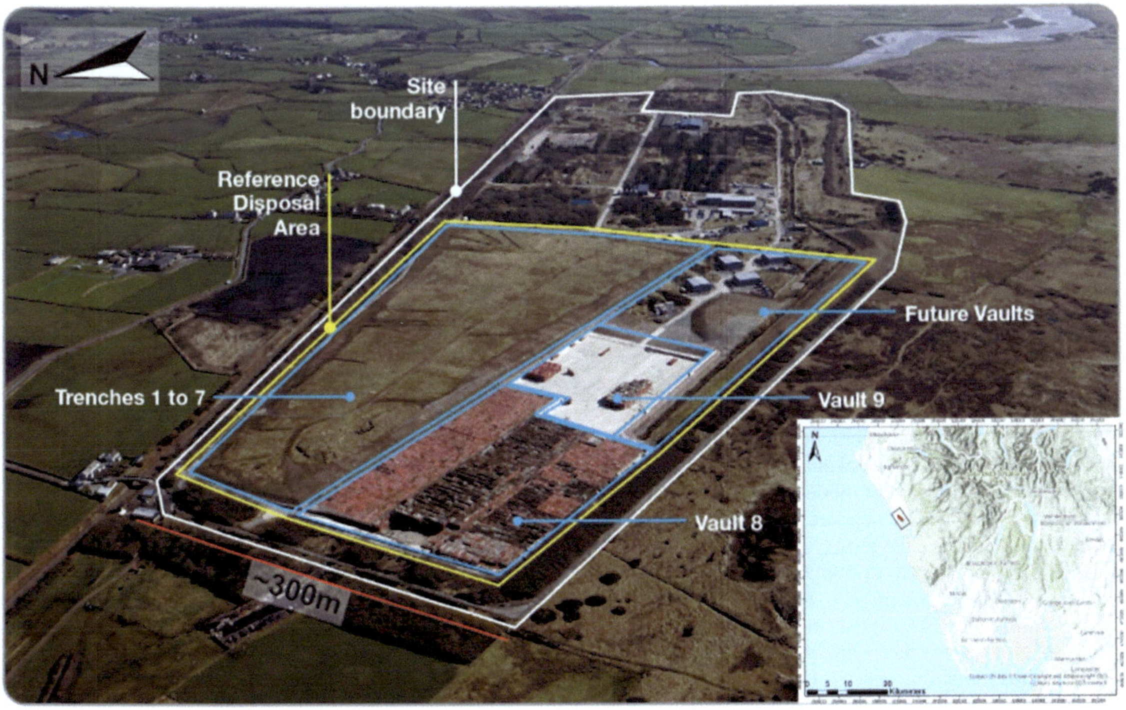

Fig. 1. Annotated aerial imagery of the Low-Level Waste Repository, covering just over 1 km^2. Source: aerial image with credit to LLW Repository Ltd, copyright Nuclear Decommissioning Authority.

ideas and understanding as a basis. This approach was taken due to new data collection on the site in 2021, including boreholes to bedrock with full-diameter core of Quaternary and bedrock, downhole geophysical logs and refraction and reflection geophysical data. The data provided additional information to develop the conceptualization and add detail and confidence to the model in the vicinity of LLWR. Regionally, models of this detail had not previously been produced but were now recognized as of value to support the geological model. The model continued to evolve, developing through 2023 into 2024 with additional data obtained and ongoing validation and quality control, to produce a fully quality-checked and validated version with the additional data by April 2024. It is in the nature of the site, with ongoing data collection and the requirements of the site licensing to continually develop site understanding, that models require regular review and development. This paper focuses on the initial version of the model at the end of 2022 as aligned to the presentation at the Engineering Group of the Geological Society Annual Conference in March 2023, with reference to changes since made where relevant.

Due to the requirements of the subsequent hydrogeological modelling, the complexity of the geology and limitations in the modelling software discussed below, four component 3D digital models were produced to represent the geological model for the LLWR: a regional-scale bedrock surface model; a complex bedrock model representing subdivisions of the bedrock; a regional-scale Quaternary deposits model; and a site-scale Quaternary deposits model. Subdividing the model allowed for sufficient resolution at different scales which varied depending on end users/uses. To provide sufficient representation of the complex sequence and adequately support hydrogeological modelling, 26 Quaternary stratigraphic units were explicitly represented in the 3D digital regional-scale Quaternary deposits model, and seven bedrock stratigraphic units in the bedrock model. Unit divisions were determined by depositional/deformational history and material properties to align with shared hydrogeological parameters. It is important to consider that the stratigraphy is tailored to the needs of the end use of the model. While the 3D digital model endeavours to provide a representation of the geological understanding it is through this lens, which may differ for a different end purpose such as for construction.

Terminology

An engineering geological model is defined in the C25 guidance (p. 4) as 'a comprehensive knowledge framework that allows for the logical evaluation and interpretation of the geological, geomorphological and hydrogeological conditions that could impact a project and their engineering characteristics'. The geological model is defined (p. 4) as an output of the framework, representing 'the distribution in 3D space of the engineering geological units, hydrogeological conditions and geological and geomorphological processes'. Within this paper the term geological model is used, as the process described relates to the development of the 3D digital geological model, as opposed to the wider EGM. There are multiple ways the EGM may be represented in a model; Table 1 provides supporting definitions.

To manage the integration of data into a 3D digital geological model, a consistent and clearly defined terminology specifically within the geological model development process was required. The C25 guidelines discuss observational data and conceptual ideas associated with observational and conceptual models, and utilize 'input data' to describe inputs to the model. However, specific to the 3D digital model development process, additional detail was required to differentiate input data at different model iterations, its use in the 3D digital model, and its support from observational data or conceptual ideas. The terms and definitions listed in Table 2 were developed to support data integration and may be useful for other projects to support differentiation of data during the 3D digital modelling process. These terms are also in parallel with data management stages throughout the modelling process (Fig. 2).

Modelling process applied at LLWR

3D digital geological modelling for the LLWR ESC utilized Leapfrog Works (version 2022.1). Leapfrog Works was chosen over other software due to its visualization capability, easy user manipulation of modelled outputs, and compatibility with the existing and preferred hydrogeological modelling software (ConnectFlow, Jacobs 2023). Leapfrog Works was also chosen as it is an implicit modelling software, which is of benefit for the volume of data available and allows for the application of control to explicitly define modelled surfaces where required.

The hydrogeological modelling requires a regional-scale model to reach assumed no flow boundaries. These boundaries are physical boundary conditions to groundwater flow such as rivers and extend about 5 km from the LLWR onshore and 2 km offshore for the hydrogeological model. The geological model needed to be larger than this, to support definition of the hydrogeological model boundaries where geologically defined (e.g. faulting against impermeable lithologies, such as the Triassic Mercia Mudstone bedrock unit offshore) and to allow incorporation of regional observational data which could be used to inform the model within the area of interest. The resultant area of the regional geological models was roughly 25 × 23 km (Fig. 3).

In addition, a site-scale Quaternary deposits-only model (approximately 5 km^2, Fig. 3) was produced to allow representation of more complex features within a higher-resolution model. This allowed a 5 m resolution of modelled geological bounding surfaces at site scale compared to 100 m in regional scale. Further limitations of the software, to require fault terminations to extend to the bounds of the model or other faults, meant the decision was made to split the Quaternary and bedrock units into separate models, and separately model the bedrock surface to serve a common boundary between the two. As a result of this and of early interpretation, faults could not cut the bedrock surface but allowed the bedrock

Table 1. Definitions of key model terms

Term	Definition	Typical role/considerations of use in EGM development
Conceptual model	A model based largely on geological information and created to anticipate what might be encountered on site.*	Should be developed early in the project, facilitates design of site investigation. Observational model assessed for consistency with conceptual model, with iterative development of each.
Observational model	A model based on a mixture of geological and engineering information collected during a site investigation which constrains the model in time and space. *	Developed following collection of observational data, typically after a conceptual model. May be 2D cross-sections or, now frequently, 3D digital models. May be geological model or geotechnical model.
Explicit modelling	'Traditional' manual method of wire framing and digitizing, where the modeller defines the surfaces that constrain units.†	May be more suited to smaller or simpler projects given time consuming nature to produce and update. Although complex features may benefit from this approach.
Implicit modelling	Algorithms interpolate and extrapolate between data points to create a model, which can be a continuum or defined by boundaries.†	Allow rapid production and update, so may be more suited to larger or complex projects. Geological judgement must assess reasonableness of resultant model and may need explicit editing.

*Adapted from Parry *et al.* (2014).
†Adapted from Baynes and Parry (2022).

Table 2. Terms and their use in the 3D digital model development process

Term	Use
Reference data	All data available that may be able to inform the geological model, whether it is applied directly to the 3D digital model or not. This may include observational data or conceptual ideas, such as borehole records, LiDAR data, previous reports including interpretations, publications, maps, geophysics data, interpretations, geomorphological mapping outputs, photographs, cliff section photographs or logs.
Direct input data (observational or conceptual)	Direct input data are applied to the 3D digital model within the modelling software to produce the model outputs. Examples of direct input data include borehole interpretations or geophysics interpretations used as contacts for modelled geological boundaries. Input data, as shown in the examples given, are more typically observational data driven; however, they may be conceptual. For example, conceptual cross-sections developed to represent the regional setting could be conceptual input data directly added to the 3D digital model.
Indirect input data	Indirect input data support the geological model but are not directly applied to the 3D digital model. Examples of indirect input data include previous interpretations and published literature which inform interpretations and the stratigraphic framework but are not directly applied in the modelling software themselves.
Model control	Controls applied to the 3D digital model after an initial stage of implicit modelling utilizing input data. This control utilizes geological understanding, is dependent on the initial output and must be carefully reviewed when model updates are made to ensure they are still appropriate to the developing geological understanding with additional data availability. These may include polylines or points which limit the extent of geological units within the model where there is a lack of observational data.

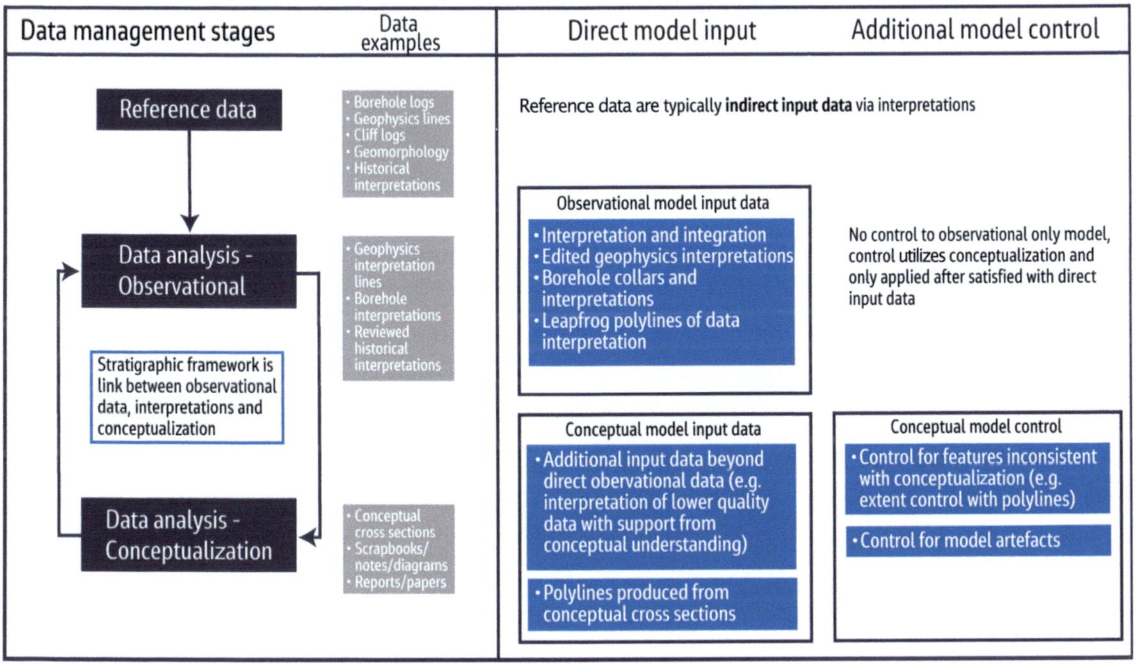

Fig. 2. Data terminology used and alignment to data management approach. Reference data require data analysis within a stratigraphic framework to produce input data to the model, and model input data are divided into observational and conceptual, with control separating data applied that are dependent on the initial model produced.

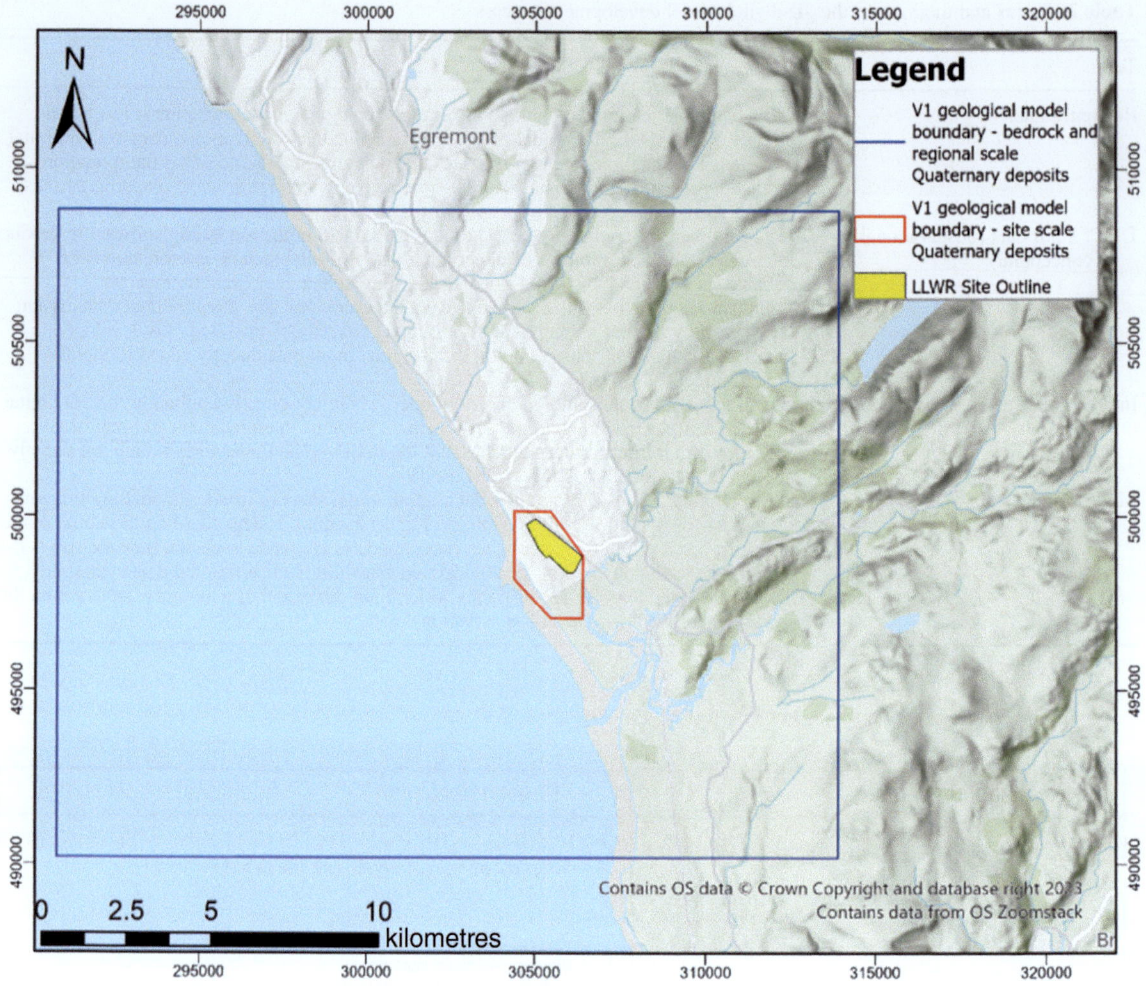

Fig. 3. Geological model boundaries produced for the geological models in 2022, relative to the west Cumbria region and LLWR site outline. The regional geological model extends beyond the requirements of the hydrogeological modelling expected, to support the definition of the hydrogeological model boundaries.

surface to be simply modelled, avoiding the issue of multiple bedrock units and multiple Quaternary units contacting at this unconformity. The resultant four models we term 'component models' were then merged to produce surfaces for the hydrogeological model (Fig. 4).

Each component model followed the same fundamental modelling process (to be described herein), tailored to the input data available for the specific model. Figure 5 shows an overview of the process and the balance of observational data v. conceptual ideas and level of conceptualization directly applied to the model. While this is the overarching process applied, focusing on the scenario where significant observational data are available (the site-scale Quaternary deposits model), the approach applied and balance between conceptual ideas and observational data used in the 3D digital modelling process varies dependent on the data available and, as a result, the scale of the models, as highlighted as a fundamental principle in the C25 guidelines (1.1.2.3; Baynes and Parry 2022).

For the LLWR, creating a data catalogue was a first step in collating the legacy data collected over numerous decades. For intrusive data additional information catalogued included locations, type of intrusive data collection (e.g. boreholes, trial pit, CPT), drilling/excavation method (e.g. rotary, sonic, cable percussion), date, log formats available and depth of penetration. This information, including an assessment of the log descriptions, served to inform quality assessments.

A quality assessment exercise prioritized boreholes with detailed and comprehensive logging descriptions, depth of

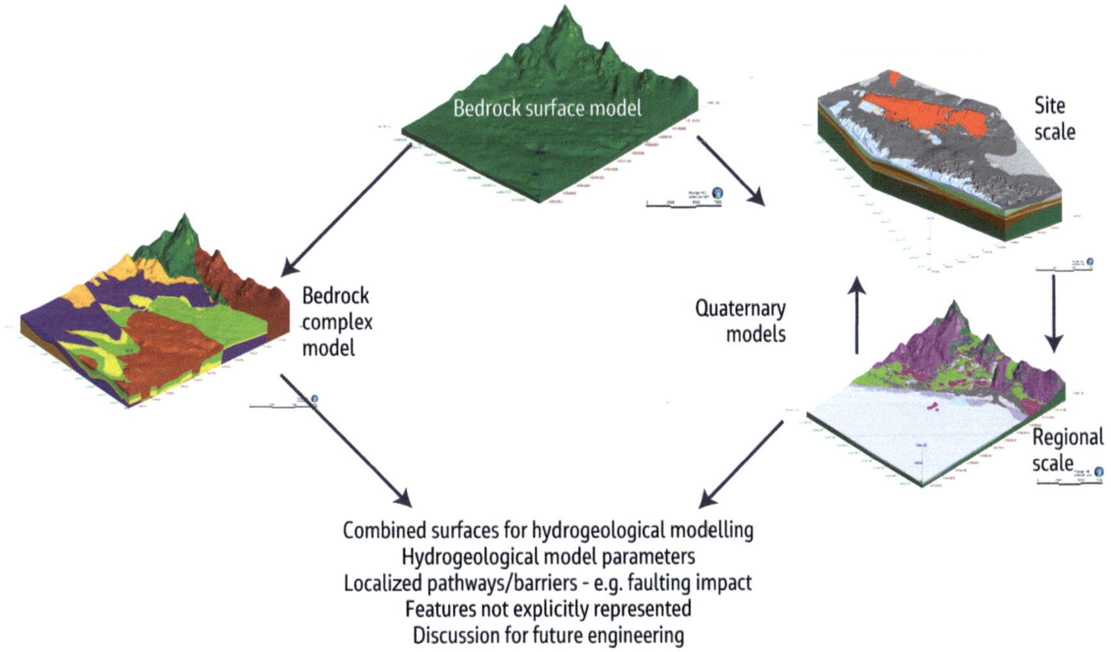

Fig. 4. Relationships of the four component models to produce a merged output that could support hydrogeological modelling for the ESC at LLWR, and additional considerations beyond the 3D geological models that support the LLWR from the developed geological understanding.

penetration through the sequence and collection methods that yield good recovery/high-resolution data in the deposits of the site – typically sonic boreholes and CPTs (Coleman *et al.* 2021). It was not uncommon to find that older boreholes were not described to a consistent standard, or to BS5930. The quality assessment considered the quality of logging descriptions relative to BS5930, and also additional sedimentological descriptions sufficient to support Quaternary deposit interpretations. This resulted in quality assessment relative to their potential use in the model, rather than outright quality compared to a standard.

The records that offer the highest quality data to the model are integrated first, meaning early models are founded on the data with the least descriptive and interpretational uncertainty. Lower quality, and less certain records are sequentially added to reduce data gaps, with comparisons drawn to any available nearby higher-quality records helping to reduce uncertainties and ambiguities with data. This method, as opposed to treating all data equally, is key to ensuring lower quality and higher uncertainty data do not exert a disproportionate control over the model given its relative confidence. Throughout the modelling process information is yielded from all available data through iteration.

Intrusive data were then interpreted within an initial stratigraphic framework. The stratigraphic framework required conceptualization to develop, utilizing an understanding of geological processes in the region, event stratigraphy, landsystem concepts, sequence stratigraphy and more, and is the foundation of the subsequent modelling. The stratigraphic framework is driven by the needs of the model to support hydrogeological modelling. Heterogeneity remains within the stratigraphic units, and they may not be directly applicable for other purposes (e.g. geotechnical modelling). At LLWR, given the history of investigations, stratigraphy has been developed and continually updated, and this existing

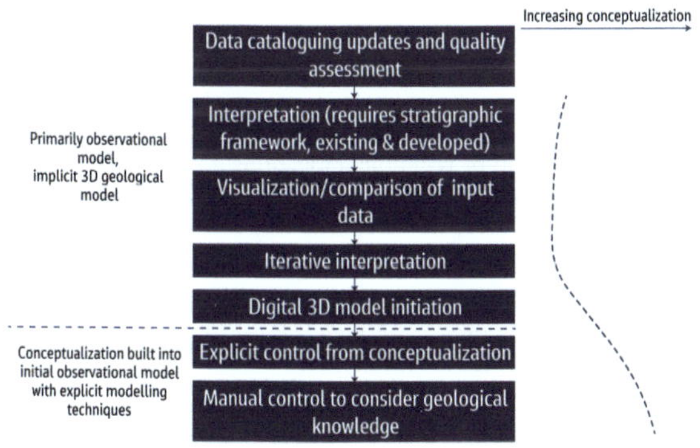

Fig. 5. Overview of the process to produce 3D digital geological models at LLWR. This process is applied to each of the component models, with the stages varying dependent on the model scale and input data; conceptualization shown is most relevant to the site-scale modelling.

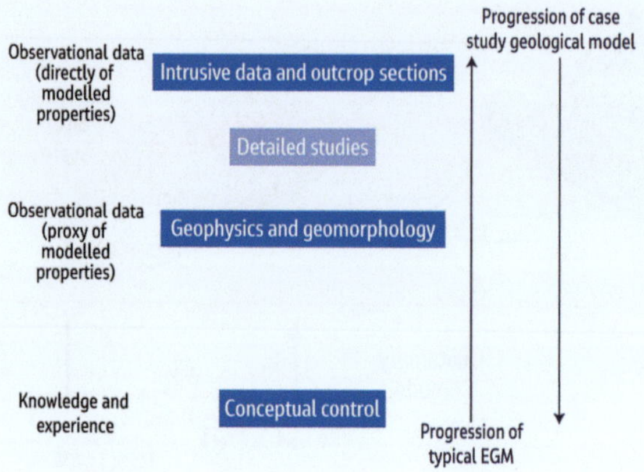

Fig. 6. Relative timing of the application of different data types to the 3D geological model in sequence, for LLWR in comparison to a typical EGM.

stratigraphy formed a base. This is the equivalent of an early conceptual model to support the observational model development in this case. Throughout interpretations, however, this framework was open to challenge and change. New data, new understanding, and competing points of view aid the development of the framework iteratively throughout the model development process. This iterative development is true of the framework and the data interpretations.

Input data (see Table 1) were added sequentially to the model relative to how directly they observed the geological information they provide (Fig. 6). Models were initially developed with intrusive data only, with later addition of geomorphological and geophysical interpretations. It should be noted intrusive data were utilized to support the interpretation of non-intrusive datasets, particularly to support depth control which tends to be less well resolved in geophysical datasets. Section interpretations offer key data on relationship of units; however, as they are limited in extent, they were included as small components of the overall model. The balance of these different data types varies dependent on the model, location and scale. Integration of such diverse data inevitably leads to the need to iterate for conflicts and reconcile the interpretations of different datasets. Reconciliation of observational input data took place prior to adding conceptual control. This means the initial models were observational data driven, without the need for significant conceptual input.

Leapfrog Works provided a single platform to view all input data in 3D, bringing together data from multiple reference sources and interpreted by different geoscience specialists. This visualization, including the opportunity to 'slice' through to aid checking and comparison, was a valuable process and supported validation of iterations. Table 3 provides information on the data types, their interpretation and integration into Leapfrog Works. Integration and reconciliation of these variable datasets earlier in the model development process proved invaluable and more efficient to the overall model build process.

The early role of the conceptual model, not produced in 3D digital form, is only seen in the stratigraphic framework. This is not resultant from a lack of consideration to the conceptualization, but recognition that providing the stratigraphic framework can be produced to support consistent interpretation of data in a way that can be represented in a model, then sites with this volume of observational data availability can benefit from an initially observational driven 3D digital model.

The ability to do this from the outset is dependent on the data availability. To illustrate this, for the initial updated

Table 3. Data types, interpretation software and format for integration into Leapfrog Works

Data type	Interpretation software & method	Format to apply to 3D digital geological model in Leapfrog Works
Intrusive data	Logs tabulated in Microsoft Excel 'Master Stratigraphy' spreadsheet with assigned stratigraphy.	CSV imports for XY (collars – the positions of the top of boreholes) locations and borehole log intervals with stratigraphic assignments.
Reflection 2D seismic	IHS Kingdom. Data underwent load, mis-tie correction, tie to well data, interpretation, and depth conversion.	X, Y, Z point data from depth-converted line interpretations in CSV format.
ERT, refraction, and reflection seismic where images were only available	Leapfrog Works. Polylines were used to interpret.	Polylines transferred from interpretation project to main modelling Leapfrog project as .lfpl format.
HVSR data	CSV point locations visualized in ArcGIS.	X, Y, Z point data in CSV format.
Geomorphology	ArcGIS. Shapefiles produced to map geomorphological features.	Shapefiles visualized in Leapfrog via geodatabase link. Polylines produced from shapefiles to be applied to the model.
Sections (e.g. cliff section)	ArcGIS for point data on section observations, section diagrams produced as image files.	Section images imported as cross-sections to Leapfrog. Control points produced based on sections to apply to relevant units.

geological models for the LLWR ESC in 2022, the regional Quaternary area suffered from very little data offshore and in some areas onshore away from the LLWR site. To support modelling, conceptual cross-sections, utilizing observational data where available, were drawn, providing direct input data from a conceptual model into the resultant 3D digital model for some areas. This was vital given the lack of data in this particular model at the time and recognizes that when data are limited a conceptual model is a vital starting point.

Once the 3D digital model founded on observational data reconciled the range of input data, control could be applied based on conceptual understanding and to manage any model outputs that were not consistent with geological understanding. Conceptual understanding developed and iteratively reviewed throughout the process is captured in interim reports, sketches, cross-sections and maps which support this process. The 3D digital model is a singular representation of the wider geological model/knowledge framework that represents all the understanding and is therefore presented alongside a report detailing the development of the digital output as well as a detailed description of the conceptual understanding. These items are essential and serve to communicate knowledge beyond the digital representation.

The modelling process applied at the LLWR founded a developing observational model on the most direct observations and highest quality data first and foremost. This is different from a typical EGM where useful, high-quality observational data are available at project completion but is not available at the start. Typical methods rely on a preliminary conceptual model and more proxy observational data to develop initial EGMs (Fig. 6). This approach is likely to become a more common practice as sites both in the UK and globally age, an increased volume of good quality historical data is utilized to develop or update EGMs.

The process reported here moves the focus of the application of conceptual ideas. In typical EGM development the conceptual model is a significant base model on which to develop future investigations and transition into observational as data are collected. In this case study the use of conceptualization, or conceptual ideas, while vital to underpin the start of geological model development through the stratigraphic framework, is reduced at the outset but then increases in use through the process as the limitations of observational data are reached.

There are both limitations and benefits in the change in focus of the application of conceptual ideas. It is, of course, vital to have an early understanding of the conditions expected, to support design of the ongoing work, whether that is modelling development or new on-site characterization. This should not be overlooked and a strong conceptual model, albeit not built in a 3D digital environment, is still required early in the model development process. In this example this is within stratigraphic framework development, provided by existing conceptualization for LLWR by previous works. Addition of control for conceptualization after interpretation of observational data has the benefit of reducing the impact of a potentially incorrect or inadequate conceptual framework, which can introduce errors (Baynes *et al.* 2021), and instead allows the developing conceptual model to be supported by visualization of observational data. The resultant model is therefore data driven and utilizes the most certain and highest quality data available as a foundation.

This is an overview, but as mentioned the process is slightly different for each component model at LLWR based on the input data available, the complexity and the scale relative to one another.

Examples of model output and control applied

Bedrock model

The complex bedrock model produced for the LLWR includes regional faulting and seven regional bedrock units, with the level of detail included driven by the requirements of hydrogeological modelling. Units represented are the Borrowdale Volcanic Group, Lake District Ordovician Felsic Plutonic Suite, St Bees Sandstone (Chester Formation), Wilmslow Sandstone Formation, Helsby Sandstone Formation split into upper and lower components, and the Mercia Mudstone Group (undifferentiated). Boreholes with wireline geophysical logs allow division of the Sherwood Sandstone Group and support subdivision onshore; these are primarily based on Nirex borehole investigations. Offshore, seismic reflection data, tied to offshore legacy wells, are the foundation of the model. Seismic interpretation (horizon and fault mapping) was carried out in IHS Kingdom software and imported to Leapfrog Works as depth-converted XYZ point data to be applied to the unit boundaries, or to fault boundaries where interpreted. In data gaps, control has been applied to produce a geologically reasonable model considering variation in unit thicknesses and structural setting. Many of the controls for this model are therefore relative to the seismic interpretations and sensitive to depth-conversion uncertainty.

Where direct observational data were not available, in some areas, existing maps were used to support the model, such as BGS Nirex maps (BGS 1993). Boundaries from existing georeferenced maps were traced as polylines in Leapfrog Works at the elevation of the modelled bedrock surface and applied to the corresponding unit boundary to ensure consistency with the maps in the absence of other data. Manual control points were also used at depth to maintain consistent thicknesses of units at depth away from observational data or mapping; this is shown in Figure 7 with an example south of the Seascale Gosforth Fault Zone.

Site-scale Quaternary deposits model

The Quaternary site model allows the modelling of more complex Quaternary deposit features that could not be resolved within the regional scale, such as narrow channel deposits. The model is limited to the area of highest density

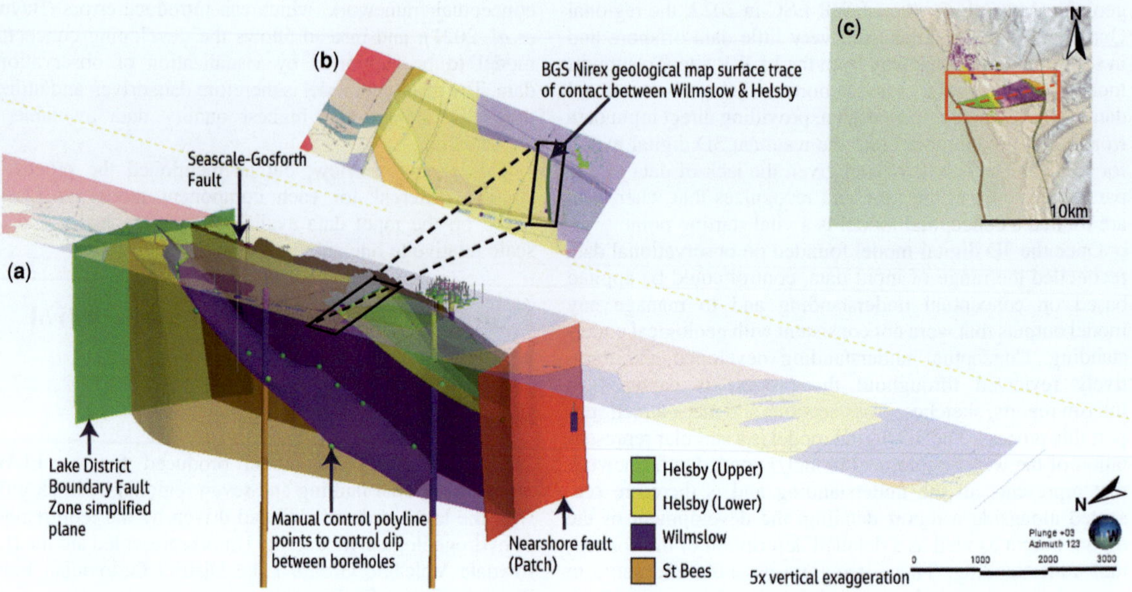

Fig. 7. Example of control applied to the bedrock complex model where observational data are limited. BGS mapping (BGS 1993) has been used to support control of the contact between the Wilmslow and Helsby sandstone units (shown in (**a**), highlighted in BGS map excerpt in (**b**)), as well as the Wilmslow to St Bees Sandstone further inland, with polyline points at depth controlling the dip of the surfaces between boreholes where necessary. Key applies to model imagery (a) and differs from BGS mapping (b); location is shown in (**c**). Scale in metres.

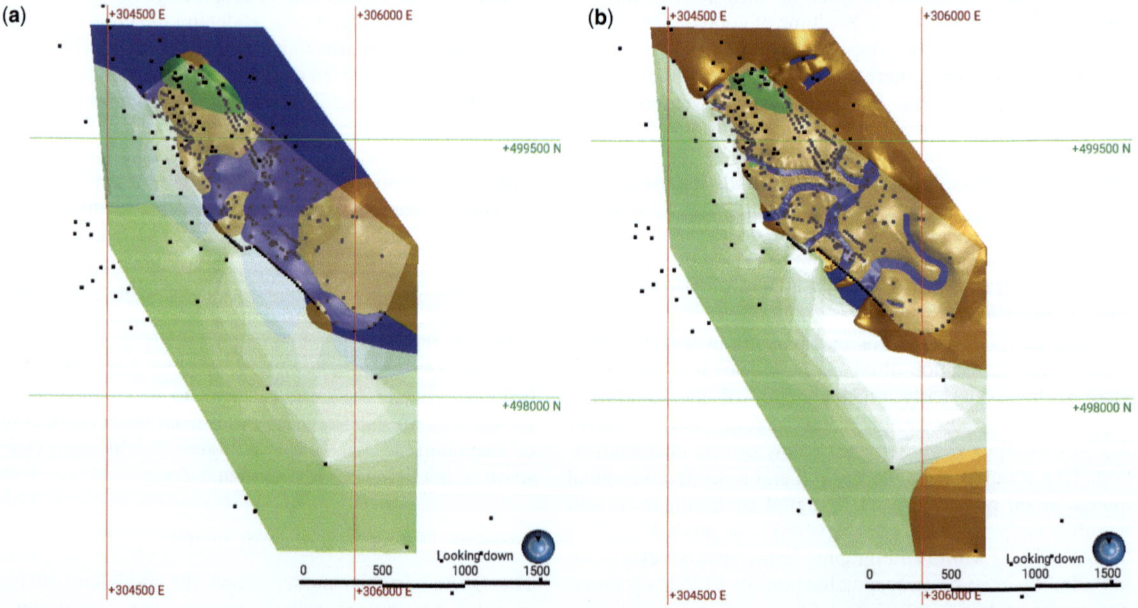

Fig. 8. Refinement of unit using conceptual control in the site-scale Quaternary deposits model. (**a**) Borehole-driven model without geophysics input data and conceptual control. (**b**) Geophysics input data and conceptual control applied. Scale in metres.

intrusive data surrounding the site and is therefore highly data driven, primarily by boreholes and CPT data. These data series were iteratively interpreted within the developing stratigraphic framework to produce a model whereby a predictive outcome could be sought with the integration of lower quality or newly collected data. The model is also supported by geophysics data including ERT, seismic refraction and reflection, particularly for channelized units.

Direct application of conceptual ideas is required to support the refinement of the model to be geologically realistic, particularly for channelized units that only appear in some boreholes and cross some 2D geophysics lines, but require interpretation in the gaps. Conceptualization in the site-scale model is supported by understanding gained from the regional-scale model, particularly interpretations of ice margins and the event-based stratigraphy.

An example of control applied is shown in Figure 8. A post-Last Glacial Maximum channelized unit does appear to be broadly channelized in the modelling software when driven only by borehole data, prior to any manual control, but the channels are wide, they do not align to geophysics interpretations, or to modern analogues. Following the application of control, the model is refined to be more aligned to the conceptualization as a meandering channelized system.

Regional-scale Quaternary deposits model

The regional-scale Quaternary deposits model provides a lower-resolution continuation of stratigraphic surfaces within the site-scale model and stretches beyond the boundaries of the existing regional hydrogeological model. The regional model also allowed a better understanding of the influence of larger-scale regional processes that may affect the LLWR site geology, such as the effects of glaciation.

The regional model utilizes the same underlying data as the site scale, plus boreholes beyond the site-scale extent, regional geomorphological mapping (including work supported by Sinead Birks, Imperial College London) and cliff

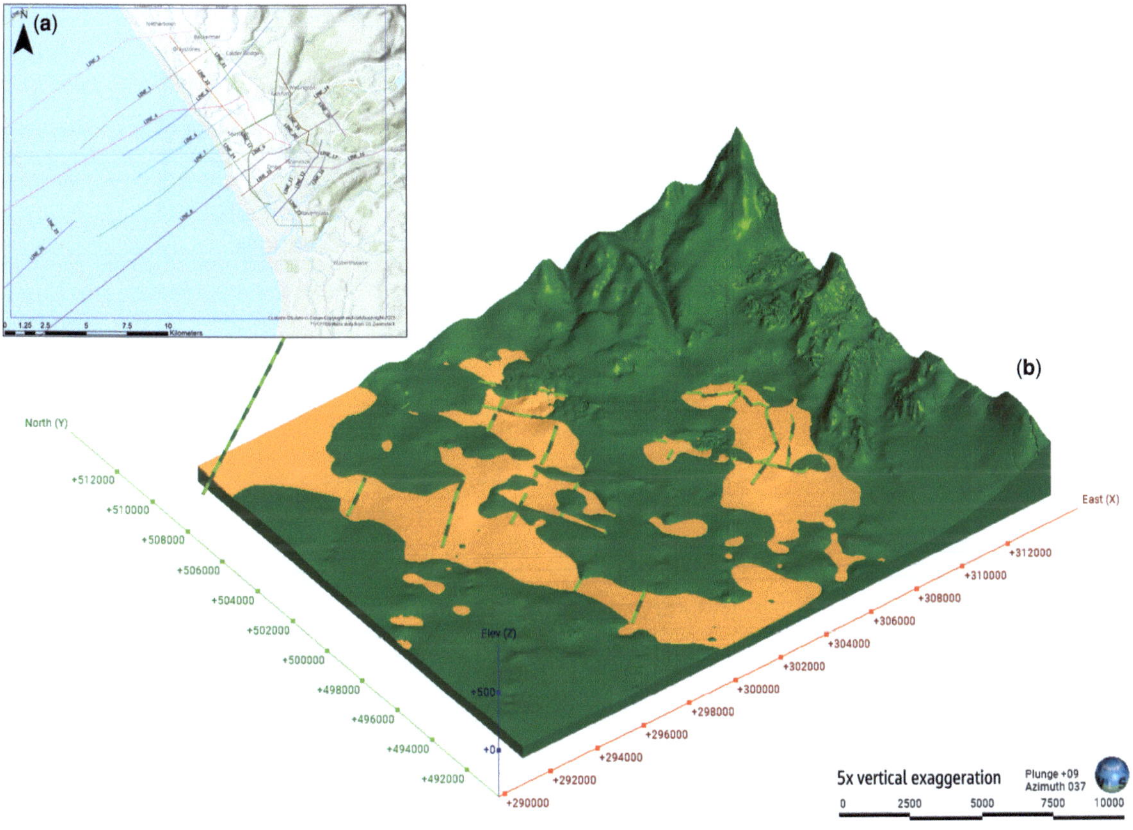

Fig. 9. Conceptual cross-section input to the regional Quaternary deposits model. Green volume shows modelled bedrock, separately controlled. Orange is the unit controlled by conceptual cross-sections in this example, and dashed green lines are the interpreted top of this unit from each cross-section, applied to the model surface as control. Scale in metres.

section observations including a recent logging of Seascale cliffs to inform conceptualization.

Boreholes regionally are sparse compared to the site scale, geomorphological data only reflect the near-surface deposits, and cliff sections are isolated and relatively shallow compared to a Quaternary sequence that is over 50 m thick in some boreholes at the LLWR site, and even thicker regionally. Offshore geophysics data available for the initial model in 2022 were limited and low quality. Given the scale of the model relative to the data coverage, the regional model required conceptualization to support an initial model. Conceptual cross-sections were created, using available data where possible, supported by geological theory and an understanding of regional processes. Sections were then digitized and used as control into Leapfrog as polylines for each modelled surface expected. This allowed the production of a 3D digital model that reflected the conceptual model being developed and was required given the data density available. An example of the cross-section control is shown in Figure 9.

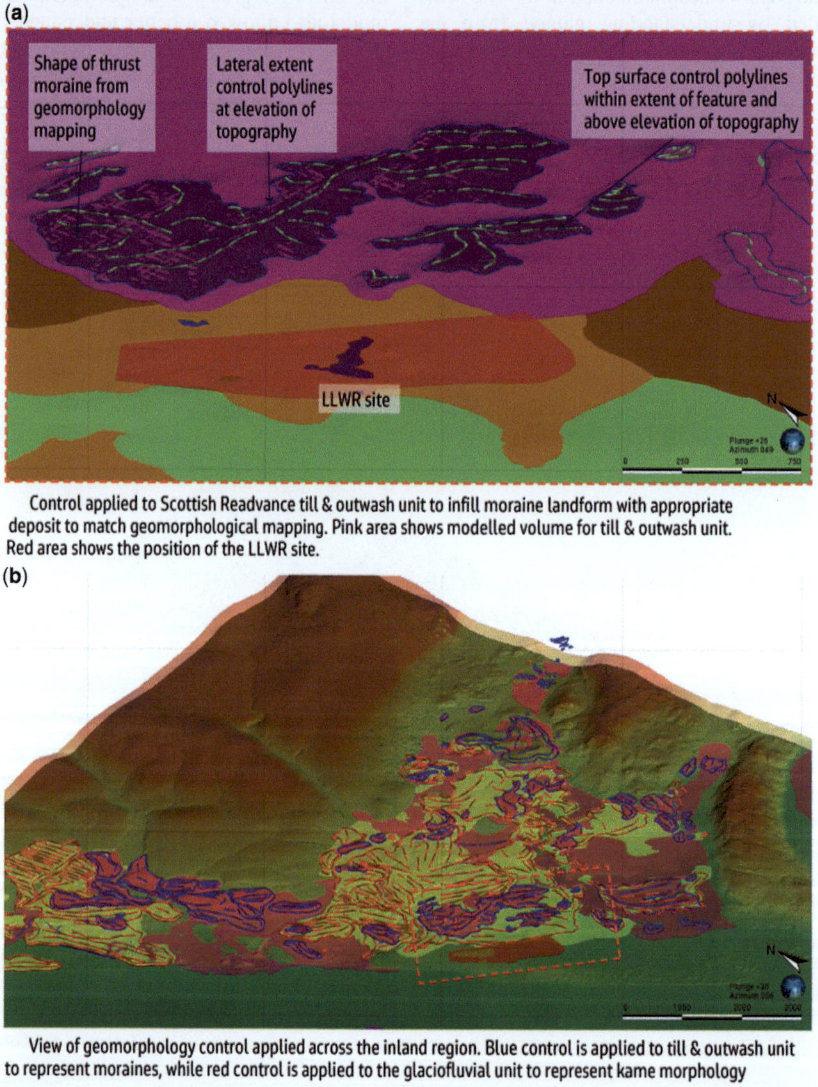

Fig. 10. Geomorphology control in the regional-scale Quaternary deposits model. (**a**) Specific control to a landform north of the LLWR site interpreted as a thrust moraine complex. (**b**) Control applied across the region with red dashes indicating the location of (a). Scale in metres.

The geological model updated with validation and quality control outputs and additional data, which at the time of writing is in progress, has the benefit of additional seismic reflection data offshore collected in 2015 but previously unavailable. This provides higher-resolution imagery of the Quaternary sequence, has been made available with depth conversion and has boreholes within the survey area to tie to seismic. The addition of this dataset offshore has allowed conceptual cross-sections to be replaced by specific manual control where required throughout the regional model. This further shows how the requirement for input data from a conceptual model to the 3D digital model is dependent on the data availability, with the regional-scale Quaternary deposits model developing in line with more typical examples which become increasingly observational as more data become available, in contrast to other areas of the model that have much higher data density.

For the near-surface Quaternary deposits, the geomorphological mapping has been applied directly to the model through mapped outlines of landforms being assigned to the unit the landform is interpreted to correlate to, and then additional control added to ensure the modelled volume of the stratigraphic unit encompasses the landform. Figure 10 shows an example where control is applied based on geomorphology mapping, to represent a moraine feature within the Scottish Readvance Till and Outwash unit. The Scottish Readvance is the most recent glacial event in the stratigraphic sequence, dated 19.2–18.2 ka (Chiverrell *et al.* 2018) and is the most likely cause of this geomorphology. The same process is also shown for kame and outwash geomorphology.

Conclusions

The conceptual model, and development of conceptualization is, and always will, underpin the development of geological models. C25 guidance offers support in producing geological models for engineering geology applicable to a range of end uses. The model build process varies significantly depending on the data availability as well as the end use of the model; the LLWR offers an example of geological model development for a nuclear site with a large volume of variable quality legacy data. The LLWR case study is an example of producing a data-driven, observational model where data allow, supported and developed with conceptualization, with integral consideration of data quality and the use of proxy observational data. The case study highlights the range of data, and conceptual control, that can support 3D geological modelling where the end use supports a complex 3D digital model.

Into the future, more sites will have an increasing volume of legacy data available to consider for geological modelling, providing opportunities to develop geological models that are increasingly data driven at an earlier stage in the process. Geoscientists will need to carefully consider the data quality, including data collection methods, to optimize the use of legacy data. This case study highlights that even with sufficient data availability to develop an observational model, the interpretation of geoscience data within a stratigraphic framework is underpinned by conceptual decisions to produce that framework, and geological models require refinement with geological knowledge, beyond the available data, even in data-dense areas. This case study should be borne in mind with the advancement of data modelling and automated (AI) based modelling approaches. The fundamentals of geological understanding and modification of models to suit, is required even in areas of high data density due to data quality and data gap issues present in all datasets.

Acknowledgements Thank you to the LLWR for allowing the model development process to be shared at the conference and within this Special Publication. We would like to thank the geoscience team at Amentum & Jacobs who supported the geological modelling for the LLWR presented, particularly Madeline Tovey and James Hamnett who were heavily involved in the implementation of this process. Thank you to Sinead Birks for discussions on the regional Quaternary landforms and processes, and Hannah Woollard for discussion on the influence of geological variables on the hydrogeological modelling for LLWR.

Competing interests The authors declare the following financial interests/personal relationships which may be considered as potential competing interests: AP, RT, MH, NJ and CC are employed by Amentum, who have provided consultancy to Nuclear Waste Services, including scientific investigations leading to the results presented here. JS is employed by Nuclear Waste Services.

Author contributions **AP**: methodology (equal), writing – original draft (lead), writing – review & editing (equal); **RT**: writing – original draft (supporting), writing – review & editing (equal); **MH**: methodology (supporting), writing – review & editing (equal); **NJ**: methodology (supporting), writing – review & editing (equal); **JS**: writing – review & editing (equal); **CGC**: methodology (equal), supervision (lead), writing – review & editing (supporting).

Funding This research received no specific grant from any funding agency in the public, commercial, or not-for-profit sectors.

Data availability The data that support the findings of this study are available from LLWR (Nuclear Waste Services) upon reasonable request, but restrictions may apply to the use of these data.

References

Akhurst, M.C., Chadwick, R.A. *et al.* 1997. *Geology of the west Cumbria District.* Memoir of the British Geological Survey, Sheets 28, 37 and 47 (England and Wales), British Geological Survey, https://webapps.bgs.ac.uk/Memoirs/docs/B01506.html

Ambrose, K., Hough, E., Smith, N.J.P. and Warrington, G. 2014. *Lithostratigraphy of the Sherwood Sandstone Group of England, Wales and South-West Scotland.* British Geological Survey

Research Report, **RR/14/01**, https://nora.nerc.ac.uk/id/eprint/507530/1/RR14001.pdf

Baynes, F.J. and Parry, S. 2022. *Guidelines for the Development and Application of Engineering Geological Models on Projects*. International Association for Engineering Geology and the Environment (IAEG) Commission 25 Publication No. 1, v1.0. https://iaeg.info/wp-content/uploads/2022/12/IAEG-C25-EGM-Guidelines-v1.0-14-December-2022.pdf

Baynes, F.J., Parry, S. and Novotny, J. 2021. Engineering geological models, projects and geotechnical risk. *Quaternary Journal of Engineering Geology and Hydrogeology*, **54**, qjegh2020-080, https://doi.org/10.1144/qjegh2020-080

British Geological Survey 1993. *Sellafield Data Integration Study Part 2: District*. BGS Technical Report, **WA/93/10C**.

Chaplow, R. 1996. The geology and hydrogeology of Sellafield: an overview. *Quarterly Journal of Engineering Geology and Hydrogeology*, **29**, S1–S12, https://doi.org/10.1144/GSL.QJEGH.1996.029.P1.01

Chaplow, R. 2001. *Sellafield Geological Investigations: Abandonment of Boreholes*. Nirex Report, **N/032**.

Chiverrell, R.C., Smedley, R.K. *et al.* 2018. Ice margin oscillations during deglaciation of the northern Irish Sea Basin. *Journal of Quaternary Science*, **33**, 739–762, https://doi.org/10.1002/jqs.3057

Coleman, C.G., Grimoldi, E., Woollard, H., Holton, D. and Shevelan, J. 2021. Developing 3D geological and hydrogeological models for the Low Level Waste Repository site, west Cumbria, UK. *Quarterly Journal of Engineering Geology and Hydrogeology*, **54**, qjegh2020-026, https://doi.org/10.1144/qjegh2020-026

Environment Agency 2015. *Decision document: application by LLW Repository Ltd under the Environmental Permitting (England and Wales) Regulations 2010 to carry on radioactive substances activities at: Low Level Waste Repository, Old Shore Road, Drigg, Holmrook, Cumbria, CA19 1XH*. Environment Agency, https://assets.publishing.service.gov.uk/media/5a75c59440f0b6488c78ed6f/LLWR_decision_document.pdf

Fookes, P.G. 1997. Geology for engineers: the geological model, prediction and performance. *Quarterly Journal of Engineering Geology and Hydrogeology*, **30**, 293–424, https://doi.org/10.1144/GSL.QJEG.1997.030.P4.02

Fookes, P.G., Baynes, F.J. and Hutchinson, J.N. 2000. Total geological history: a model approach to the anticipation, observation and understanding of site conditions. *Invited Paper, Proceedings of GeoEng2000, an International Conference on Geotechnical and Geological Engineering*, 19–24 November, Melbourne, Technomic, Pennsylvania, 370–460.

Jacobs 2023. ConnectFlow Technical Summary Version 13.0, https://www.connectflow.com/resources/docs/conflow_technical.pdf

Leapfrog Works, Version 2022.1.0. Seequent, The Bentley Subsurface Company, https://www.seequent.com/products-solutions/leapfrog-works/

LLW Repository Ltd 2011. *The 2011 Environmental Safety Case – Main Report*. **LLWR/ESC/R(11)10016**, https://webarchive.nationalarchives.gov.uk/ukgwa/20170412104646mp_/http://llwrsite.com/wp-content/uploads/2016/03/L1-LLWR-ESC-R-11-10016-The-2011-Environmental-Safety-Case-Main-Report--MASTER-April-2011.pdf

Merritt, J.W. and Auton, C.A. 1997. *The Quaternary Lithostratigraphy of the Sellafield District*. Nirex Science Report, **SA/97/045**.

Merritt, J.W. and Auton, C.A. 2000. An outline of the lithostratigraphy and depositional history of Quaternary deposits in the Sellafield district, west Cumbria. *Proceedings of the Yorkshire Geological Society*, **53**, 129–154, https://doi.org/10.1144/pygs.53.2.129

Michie, U., Hunter, J. and Towler, G. 2010. *LLWR ESC: The Geology of the LLWR Site and Surrounding Region*. Quintessa Report, **QRS-1443YR1**, https://webarchive.nationalarchives.gov.uk/ukgwa/20170412104851mp_/http://llwrsite.com/wp-content/uploads/2016/03/2.-QRS-1443Y-R1-Geology-Report-Final-Version-2-MASTER-05-10-10.pdf

Nirex 1995. *Sellafield Geological and Hydrogeological Investigations: Factual Report – Compilation of Maps and Drawings, 2 Volumes*. Nirex Report, **SA/95/002**.

Nirex 1997. *Sellafield Geological and Hydrogeological Investigations: The Geological Structure of the Sellafield Site*. Nirex Report, **SA/97/007**.

Parry, S., Baynes, F. J. *et al.* 2014. Engineering geological models: an introduction: IAEG Commission 25. *Bulletin of Engineering Geology and the Environment*, **73**, 689–706, https://doi.org/10.1007/s10064-014-0576-x

Smith, N. 2011. *3D Geological Modelling to Support 2011 ESC*. **NNL (10)11217 Issue 4**, https://webarchive.nationalarchives.gov.uk/ukgwa/20170412104854mp_/http://llwrsite.com/wp-content/uploads/2016/03/3.-11217-3D-Geological-Modelling-Issue-4-MASTER-11-04-11.pdf

Smith, N.T., Merritt, J.W. and Phillips, E.R. 2023. High-resolution 3D geological modelling of heterogeneity in poorly exposed glacial deposits using sedimentary and glaciotectonic architectural element analysis: a case example from Sellafield in west Cumbria, UK. *Quarterly Journal of Engineering Geology and Hydrogeology*, **56**, qjegh2022-022, https://doi.org/10.1144/qjegh2022-022

Understanding the geological model in a chalk environment: an example from the High Speed Two (HS2) railway

Charalampos Saroglou[1]* and Georgios Katsigiannis[2]

[1]Eiffage Keir Ferrovial BAM (EKFB) JV, Exchange House, 450 Midsummer Boulevard, Milton Keynes MK9 2EA, UK
[2]Ferrovial Construction UK, 3rd Floor Building, 5 Chiswick Business Park, London W4 5YF, UK

CS, 0000-0002-5253-7972; GK, 0000-0002-4889-8660
*Correspondence: harry.saroglou@ekfb.com

Abstract: Chalk is known to have a potential risk for natural cavities, thus developing an accurate geological model and documenting the location and nature of these geohazards is important to infrastructure projects. Building the geological model requires investigation of the geomorphology, engineering geology, geological history and behaviour of geological materials (soil and rock mass). The identification of dissolution features is based on geomorphological mapping, LiDAR surveys, intrusive ground investigation and surface geophysics, with the aim of understanding the geological environment and the associated risk. Peter Fookes' geological model approach is illustrated with a case study from High Speed Two (HS2) railway line in the North Chilterns Area section. This section includes cuttings, embankments and viaducts that will be designed and constructed in chalk formations in an area where the risk of dissolution features is moderate to high. These features appear either as infilled features near the base of cuttings and embankments, or as void or infilled features, that are present immediately below the toe of foundation piles or along their shaft. The exposed ground conditions during construction of the cutting are presented and compared with the predicted geological model. In brief, the model could not predict accurately the shape of the surface between the chalk and the clay infill material, and the excavation revealed a geometry for the infilled dissolution features in the chalk with irregular patterns that would have not been possible to be predict accurately.

Unforeseen ground conditions are occurring on construction projects and other schemes in chalk, often because of a lack of understanding of the ground model and the potential hazards in chalk (Lord *et al.* 2002). Fookes (1997) highlighted the importance of preparing geological models to help us better understand our project sites, including those in chalk environments, as a way of predicting the behaviour of the ground and effectively identifying and then mitigating potential problems during the design.

Developing an engineering ground model in a chalk environment is very important, as any potential adverse ground conditions and hazards related to infilled dissolution features and voids can be assessed and addressed with appropriate design and construction mitigation measures. The effectiveness of a ground model in chalk relies on understanding the stratigraphy, geomorphology, tectonics and geological processes, and proposing a conceptual ground model – based on the desk study, field observations and the results of site investigations – that needs to be designed in such a way that it can confirm the conceptual model and answer questions raised by it. The models should be reviewed continuously and improved as the investigation evolves, and it is essential to check and refine the initial model so that it becomes site specific and resolves most of the uncertainties related to any adverse ground conditions in the chalk. The study used to demonstrate the importance of a geological model in chalk is a cutting from the High Speed Two (HS2) railway line. This case provided the opportunity to validate a borehole-based ground investigation (GI) and ground model through the excavation of the cutting, offering the rare opportunity to evaluate the accuracy, performance and value of the ground model.

Geological model

Fookes, in his Glossop Lecture, stated that when preparing the geological model:

> it should be a realistic goal to be able to make a very close approximation to the actual site conditions, particularly if you know what you are looking for and what questions to ask
>
> Fookes (1997, p. 295).

Fookes also stated that a moderately close approximation of the actual geological conditions can be made from the desk study supplemented by ground and laboratory investigation, but in most sites the actual geological conditions would not

From: Griffiths, J. S., Giles, D. P., Davis, J. A. and Dunlop, J. J. (eds) 2025. *The Peter Fookes Engineering Geological Legacy in Geomodels, Geomaterials and Geomorphology*. Geological Society, London, Engineering Geology Special Publications, **30**, 93–109,
First published online 25 July 2025, https://doi.org/10.1144/egsp30-2024-12

be revealed until the construction is completed: that is, when the excavation is dug. In this context, Fookes called this approximation the 'Geological Model', which is defined as a representation of the geology of a particular location. The form of the model can vary widely and may include written descriptions, two-dimensional sections or plans, block diagrams, or may emphasize some particular aspect such as groundwater or geomorphological processes, rock structure, and so on.

Fookes (1997) presented a geological model of a periglacial environment developed on some of the rock types in southern Britain, in which different periglacial features would predominate. Common periglacial effects in the unglaciated parts of southern Britain and various features in chalk were explored in this model (Fig. 1). Waltham and Fookes (2003) developed an engineering classification of karst that defined various complexities of ground conditions by the geohazards that they create – mainly caves, sinkholes and rockhead relief.

Chalk and dissolution features

Dissolution features are a common geohazard in karst terrain and are formed through the chemical weathering of soluble rocks, such as limestone and chalk, and can be affected by changes to the groundwater levels across a region. The range of natural cavity types formed in limestone and chalk are presented in detail by Edmonds (2020), and these may occur in a variety of forms, as illustrated in Figure 2. Both shallow and deep dissolution features may be present in the chalk. Surface depressions are very common throughout the chalk of southern and eastern England, especially where there is a thin cover of Paleogene or superficial deposits, particularly clay with flints (Mortimore 2018).

The following provides a summary of each of the dissolution feature types shown in Figure 2, as described in CIRIA Report C574 (Lord *et al.* 2002).

- Dissolution sinkhole: forms by the dissolution of the surface of the chalk where it is not overlain by superficial deposits.
- Subsidence sinkhole: forms by the settlement of superficial deposits into a zone of dissolution in the underlying chalk.
- Collapse sinkhole: forms by the collapse of the roof of a cave at depth in the chalk.
- Dissolution pipe: a cone or pipe-like cavity in a vertical section, typically partially or completely infilled with overlying deposits that have subsided into the cavity created by dissolution of the soluble chalk host rock.
- Swallow holes: a surface feature where a void in the chalk continuously or intermittently 'swallows' wholly or partially a surface stream – that is, the point at which the stream disappears from the surface to flow underground.

Dissolution features may become unstable due to: (a) collapse of superficial deposits into a void, creating a zone of loose ground; (b) creation of a roofed cavity either in the superficial deposits or the chalk; and (c) collapse of a discontinuity in the chalk that has been enlarged by dissolution.

The geological model and site investigation in chalk

According to Lord *et al.* (2002), there are three stages in the engineering assessment of a chalk site:

(1) Understanding the broad geoframeworks of stratigraphy, sedimentology, tectonics and geomorphology through a desk study.
(2) Proposing a conceptual ground model based on the desk study, field observations and expert advice.
(3) Developing an engineering ground model based on the above and the outcomes of the site investigation designed to determine the unknowns raised by the conceptual ground model.

The second and third stages are largely capturing Peter Fookes' Geological Model approach, which aims to build a ground model in order to understand the geological environment and the geological and geomorphological history of the site, including the stratigraphy, structure, and past and present geomorphological processes and climatic conditions (Fookes *et al.* 2000). The conceptual ground model should be used to identify areas with significant potential for problems related to dissolution features and extensive deposits of clay with flints. The site investigation in chalk should be based on desk studies, walkover surveys, geomorphological mapping, LiDAR surveys, geophysical surveys and intrusive GI methods (i.e. rotary core boreholes, dynamic probing, etc.). Surface depressions can be identified from digital terrain models and LiDAR, topographical maps, or by referring to geological maps. These features can be natural or have an anthropological origin, such as pits dug for agricultural purposes, or clay extraction; or depressions caused by collapsed mine workings (Edmonds 2020). Geophysical surveys can be very useful in determining whether a surface depression is a dissolution feature, an old chalk or clay pit, or a combination of these (Jeffrey *et al.* 2020).

Fookes and Hawkins (1988) studied the Humber Lane cutting of the Kingsteignton Bypass in Devon, a large road cut mainly in Devonian limestone that had major karst features, and used this and similar cases to develop a five-fold classification of karstic features. Figure 3a presents the ground conditions anticipated in the cutting at tender, which did not correspond to any particular karst class but indicated a possible doline structure (depression in the rockhead with deep superficial deposits), and Figure 3b presents the actual conditions found during construction that corresponded to karst class V. It was Fookes' point that this situation had the potential to be identified prior to tender, by developing the geological model. The authors believe that this case study demonstrates that predicting the actual ground conditions and geometry of the karst/dissolution features in a limestone or chalk environment can be challenging and perhaps not

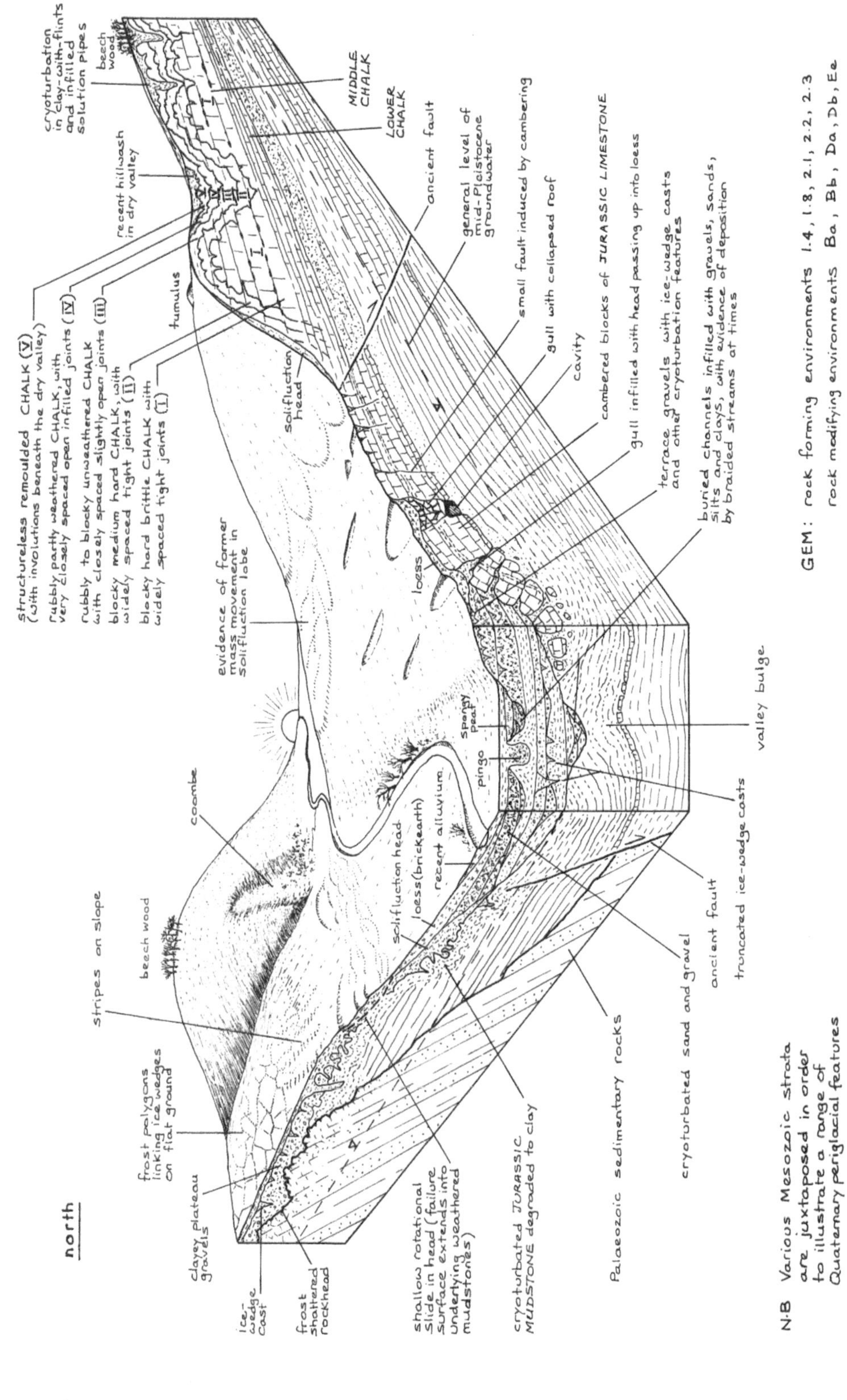

Fig. 1. Common periglacial effects in unglaciated parts of southern Britain. Source: from Fookes (1997).

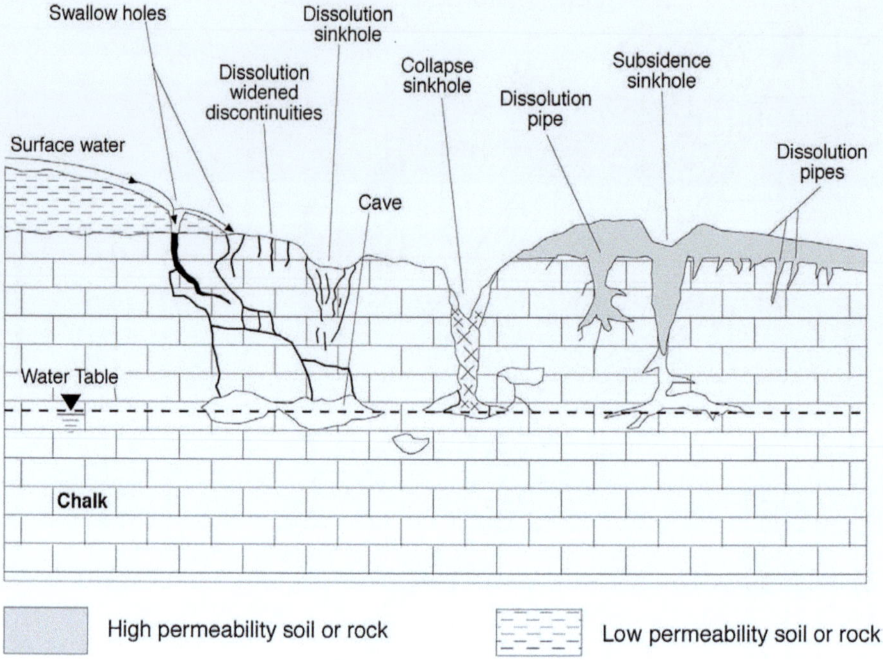

Fig. 2. Schematic cross-section of common dissolution feature types. Source: reproduced from Lord *et al.* (2002).

even possible. However, if the conceptual ground model had been more detailed, it might have been able to predict that the actual shape of the surface between the limestone and infill material was not smooth and continuous. This could have been predicted if the model was improved as the GI evolved and refined so that it became site specific.

Case study from the HS2 railway

The HS2 project is a high-speed railway line currently under construction in the UK. Phase 1 of HS2 runs between London and Birmingham across the English Midlands. This case study is part of the HS2 Main Works Civils Contract C23, awarded to Eiffage Keir Ferrovial Bam (EKFB) JV in partnership with the design JV of Arcadis Setec COWI (ASC). The HS2 route and C23 section are shown in Figure 4. The North Chilterns Area (NCA) section covers a length of *c.* 11 km; of which *c.* 8 km is formed of chalk belonging to the White and Grey Chalk Subgroup. The stratigraphic units of chalk found in the NCA section include the Lewes Nodular Chalk, Holywell Nodular Chalk, New Pit Chalk, Zig Zag Chalk and West Melbury Marly Chalk formations. The Southern Province Chalk stratigraphy is presented in Figure 5 (reproduced from Lord *et al.* 2002). The subject of the present paper is the first 500 m of the South Heath cutting (between chainage (Ch.) 47 + 250 and 47 + 750 m), which has been now excavated. The cutting has a length of 3.1 km and maximum height of 21.8 m, and is located between Ch. 47 + 320 and Ch. 50 + 420 m within Section C2 (within the NCA) immediately north of the Chiltern Tunnel.

The cutting is present within an area where karstic features are prevalent and the risk of dissolution features along the cutting during the scheme design stage was rated as moderate to high (Fig. 6). The potential dissolution feature risk assessment involved utilizing elements of Edmonds (2001) methodology whilst also considering site-specific information from the geological long sections, LiDAR data, water features, known dissolution features and indicators from the GI data. The risk identification indicators that were used to undertake the detailed assessment included scoring of the following risk components: susceptible topography, dissolution feature extent, cover deposit, contributing factors (water features, fault zones) and ground conditions from the GI data. The total score produced from the scoring of the components was categorized into three risk levels: low (0–10), moderate (11–19) and high (>20).

Geomorphological assessment

Moore *et al.* (2022) conducted a detailed engineering geomorphological assessment using LiDAR to identify geohazards on Phase 1 and Phase 2a of HS2. In that study, a total

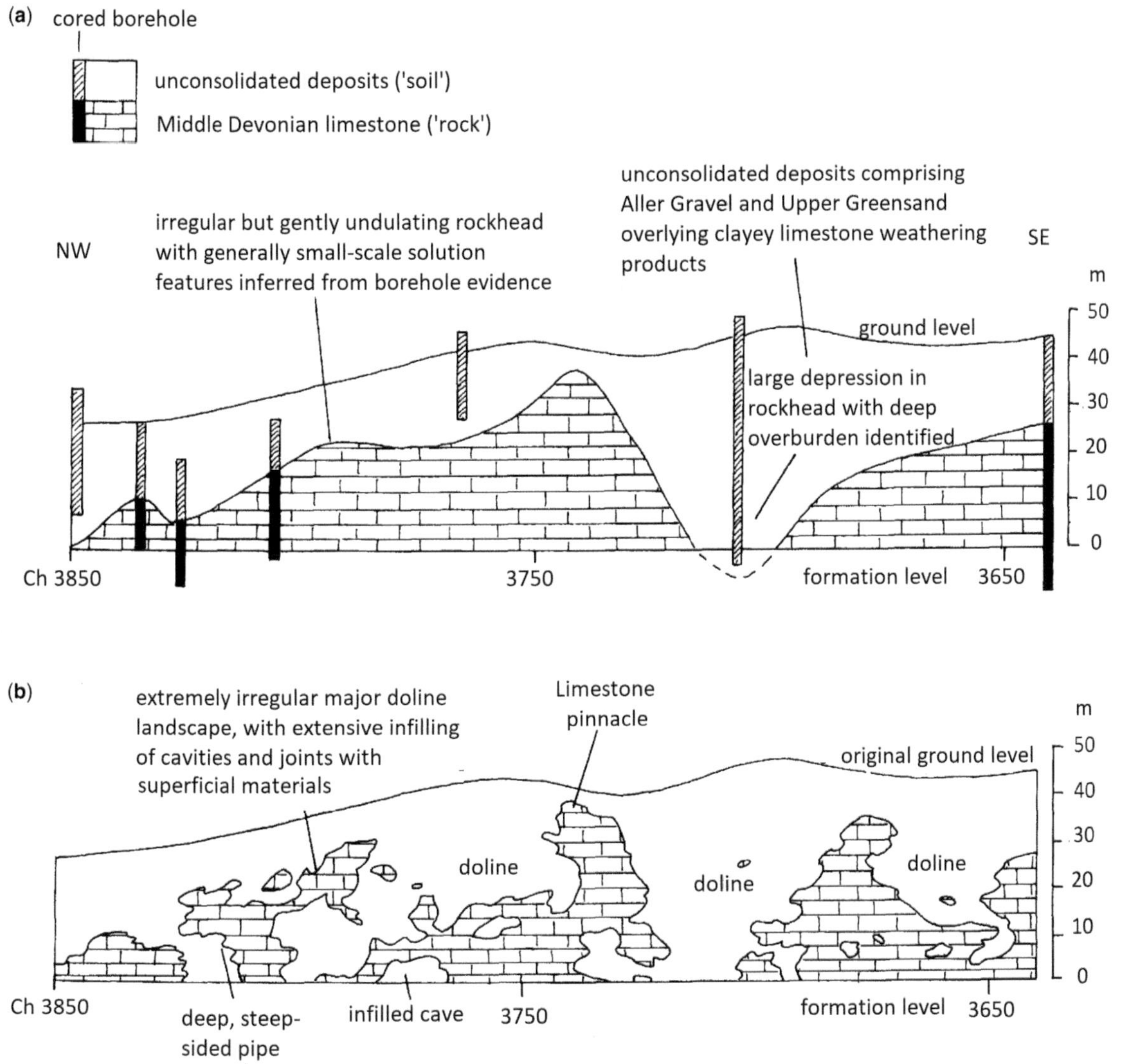

Fig. 3. (**a**) Conditions anticipated at tender. (**b**) Conditions found during construction. Source: after drawings in Fookes and Hawkins (1988).

of 245 hollows were recorded on the surface of the chalk formations along the Phase 1 route corridor and their relief could be clearly discerned in the LiDAR data. The geomorphological assessment, based on using composite 1 m DTM 2020 hill shade LiDAR survey data (available from the Environment Agency), shows the potential dissolution features in the broader area along the alignment. The geomorphology of the chalk belongs in domain 2, which is the dry valley with *in situ* degraded chalk buried by head and hill wash.

A review of the terrain unit map presented in the HS2 Phase 1 Engineering Geomorphology Assessment (HS2 Ltd 2017) shows the site to be located within an incised chalk hill terrain and the site to have 62 identified solution features.

This section of the route is located on a ridge line above the village of Great Missenden to the SW. The cutting is located on the uppermost slope of a broad ridge that gently rises *c.* 50–60 m to the north above a dry valley to the west to a maximum height of 196 m Ordnance Datum (OD) at Ch 48 + 300 m. Dry valleys have developed west of the alignment and several depressions were detected on the ground surface in the broader area, indicating the presence of infilled dissolution features, as illustrated in Figure 7.

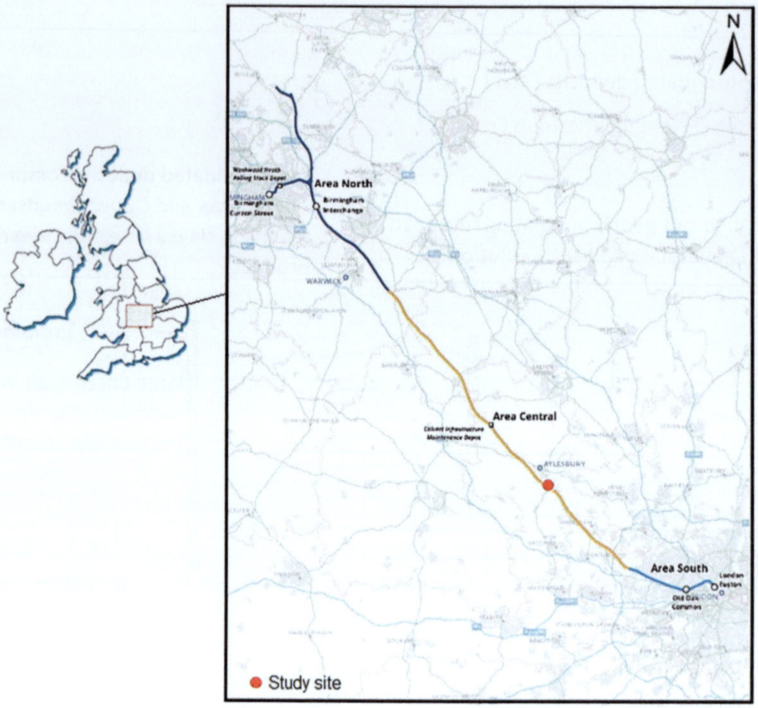

Fig. 4. HS2 Phase 1 route and C23 contract (Area Central).

Geological model: ground investigation

Two phases of GI works were undertaken for the HS2 Project in the cutting, namely the HS2 main GI and the EKFB supplementary GI. This included a suite of boreholes, cone penetration tests (CPTs) and trial pits, and as they were part of the detailed design phase they were not specifically targeted to investigate potential dissolution features. It was determined that these investigations should be undertaken as part of the construction programme to expedite the design works. However, as part of the detailed design investigation works, a preliminary programme of surface geophysics was undertaken. This comprised microgravity, electromagnetic and electrical resistivity surveys. The purpose of these works was to inform a preliminary qualitative assessment of the possible extent of dissolution features, in order that targeted investigation works during the construction phase could be undertaken (Pryce *et al.* 2024).

The plan of the cutting with GI locations for the study area is presented in Figure 8.

The geology of the site predominantly comprises 2.6–29.9 m of soft to stiff Clay with Flints (CWF) over the Lewes Nodular Chalk Formation (LECH). The areas of deeper superficial materials were considered to be associated with the infill of chalk solution features. The LECH is recorded along most of the asset as having a high variation in weathering depth. The New Pit Chalk Formation (NPCH) underlies the LECH at certain locations, which has been interpreted as possibly associated with inferred geological fault locations. The cutting will be excavated primarily in the CWF and the LECH. Preferential dissolution of the LECH has occurred to depth at a number of locations, which may have occurred at possible inferred geological fault locations.

The geological section developed during the detailed design is illustrated in Figure 9. Based on the borehole data, within the first 500 m of the cutting, a continuous karst chalk profile was inferred, while the presence of a dissolution feature was anticipated between Ch. 47 + 550 and 47 + 650 m due to a significant thickness of the CWF encountered by the site investigation in borehole ML047-CR004 to 24.50 m below ground level (bgl). The base of the CWF (depth to chalk) was not proven in borehole ML047-CR001. Chalk in adjacent exploratory hole positions ML047-CR014 and ML047-CR005 was encountered at depths of 9.60 and 5.80 m bgl, respectively. Juxtaposition between the CWF and LECH chalk units is considered to be evidence of the potential fault at Ch. 47 + 725 m. Downthrow of the fault is shown to the SE of the fault. The depth of the chalk on the downthrow of the fault is uncertain.

The location and extents of dissolution features shown on the geological section were uncertain, while the detailed design specified additional GI prior to construction to

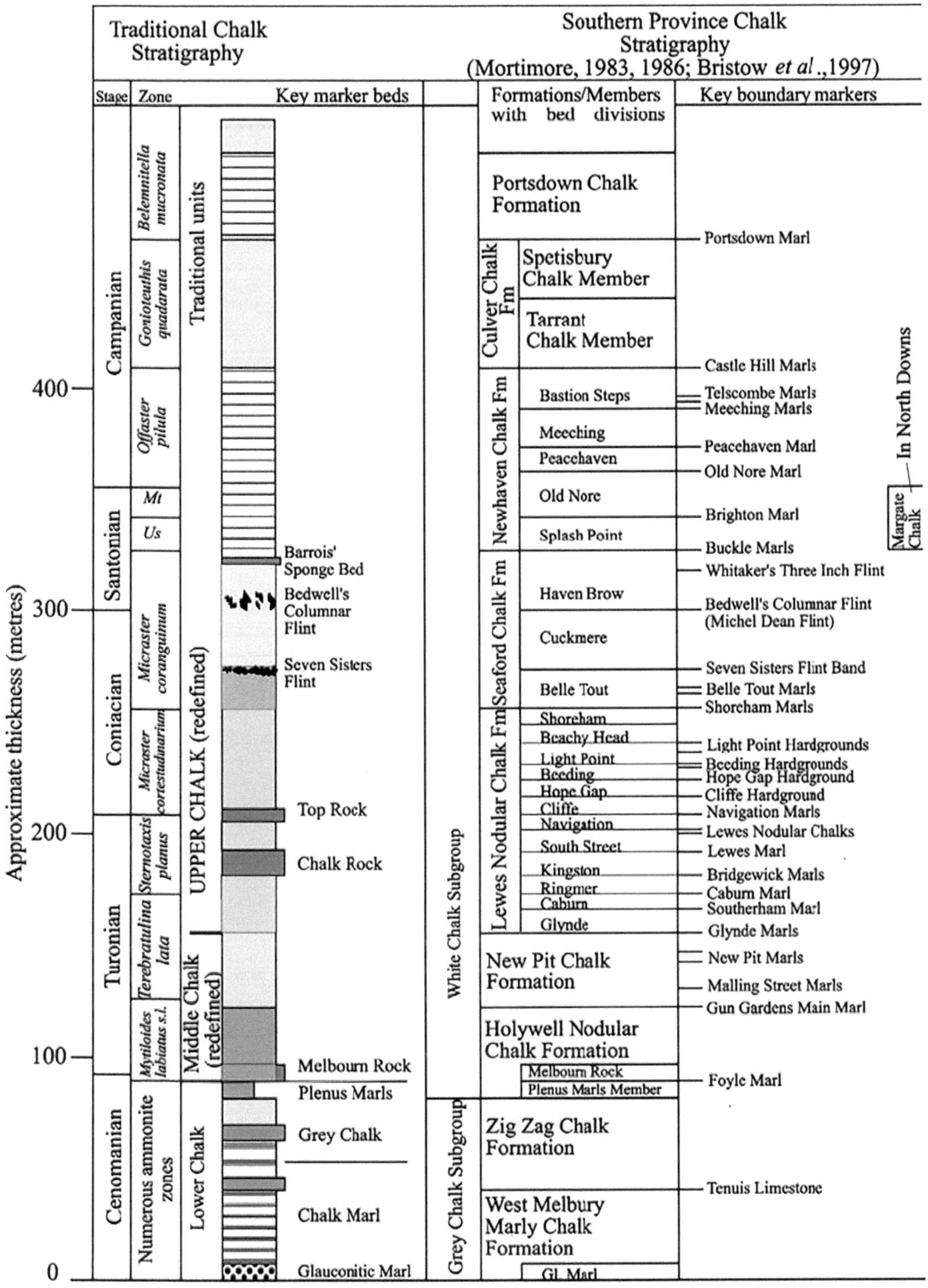

Fig. 5. Chalk stratigraphic units in the Southern Province. Source: reproduced from Lord *et al.* (2002).

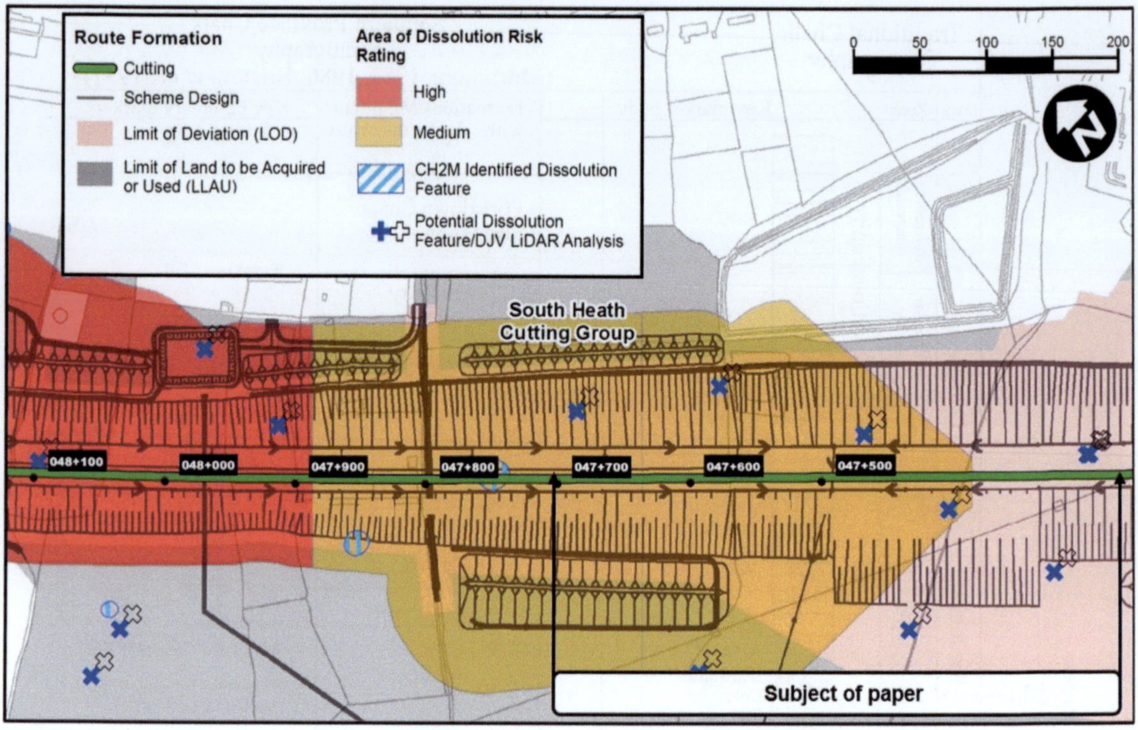

Fig. 6. Dissolution feature risk assessment of the cutting area. Potential dissolution features are indicated.

identify the extents and geometry of the potential features consisting of geological inspections and mapping, surface geophysics covering the full extent of the formation level, and additional intrusive investigation (boreholes and/or probing) if necessary. It should be noted that the assessment of the location and geometry of any dissolution feature requires a combined interpretation of all the available information.

Chalk was generally recovered as a structureless weathered chalk, predominantly a silty/clayey Gravel (CIRIA Grade Dc, representing clast-dominated strata) and varying to include a slightly sandy gravelly Silt/Clay (CIRIA Grade Dm, representing a matrix-dominated strata) (Fig. 10). The gravel fraction was principally chalk, although some localized flints were recorded. An example from the LECH structureless chalk is shown in Figure 8. Where identified, the underlying structured chalk was generally recorded as having closed discontinuity apertures (Grade A classification), with some open or infilled apertures less than 3 mm in size (Grade B classification). The presence of karstic features was identified in a few boreholes near the culvert structure, which intercepts the cutting at Ch. 47 + 550 m (see Fig. 8). These features have been interpreted based on different factors, including disturbed ground, assumed zones of core loss (AZCL) and weakened or softened layers of deposits.

No faults were mapped on the solid geology British Geological Survey map (BGS 1990) within the cutting area. However, a review of recorded ground conditions identified evidence of potential faulting. The orientation of the faults is uncertain and their inferred location in the long section is approximated. The potential fault in the section presented here at Ch. 47 + 725 m is evidenced due to the juxtaposition between the CWF and the LECH.

Geophysical investigation

An initial programme of geophysical investigation was undertaken across the cutting and for the preliminary identification of subsurface anomalies (i.e. possible dissolution features) that could then be targeted by intrusive investigation for the purpose of ground truthing. The following survey techniques were undertaken as part of the geophysical investigation:

- electromagnetic conductivity mapping;
- electrical resistivity tomography; and
- microgravity mapping.

The first two techniques aimed to identify relatively shallow anomalies. The microgravity mapping targeted deeper anomalies and was considered as the best technique available for

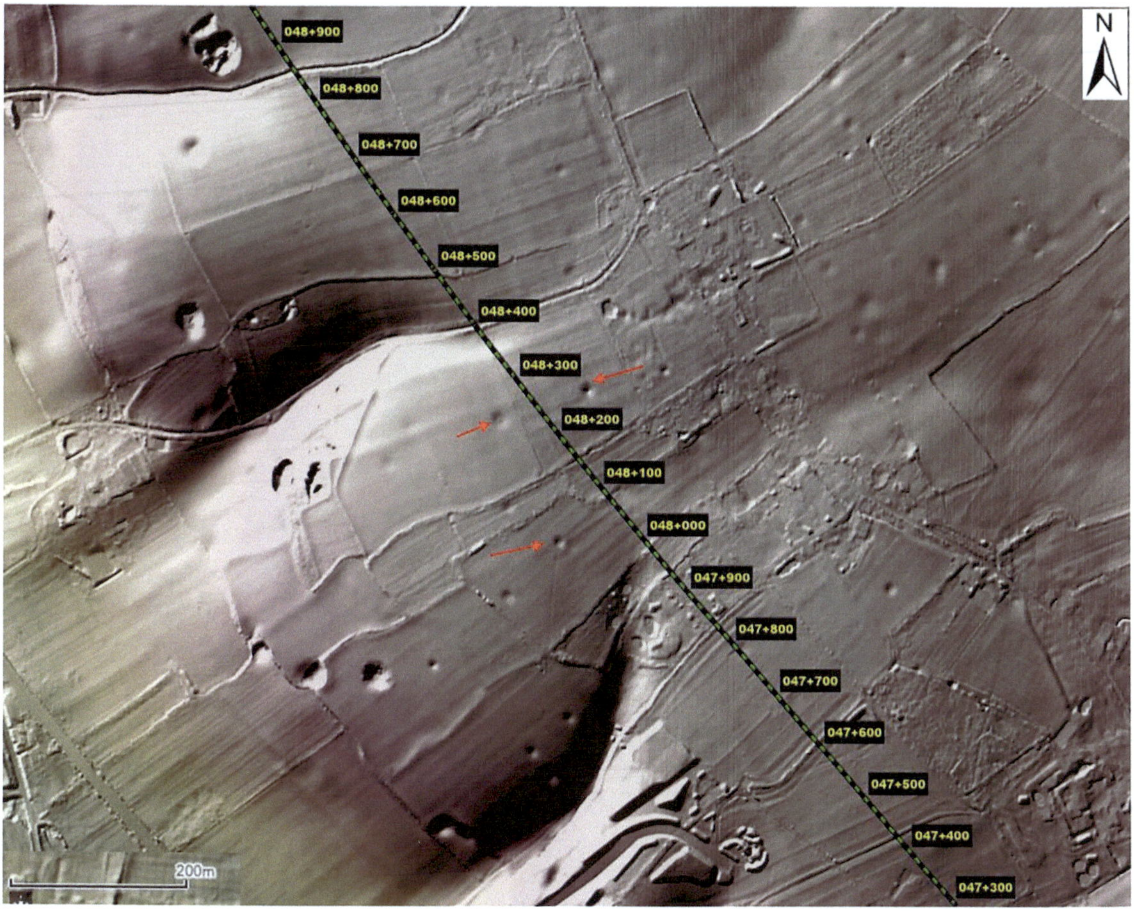

Fig. 7. Geomorphological assessment using the LiDAR survey. Arrows indicate signs of dissolution in the chalk.

the investigation of features at the study site. The findings from the resistivity survey performed in the section of the cutting between Ch. 47 + 480 and 47 + 850 m are discussed. The location of the survey is shown in Figure 8 and the resistivity section is illustrated in Figure 11. The available borehole data indicate that this area is underlain by a variable thickness of superficial deposits (CWF), typically *c.* 5–15 m thick, overlying the chalk. Borehole ML047-CR435 indicates superficial deposits to a depth of *c.* 14 m (178 m OD), which is consistent with the data from the resistivity survey. The presence of a pipeline has resulted in anomalous readings in the resistivity section at around Ch. 190 m, centred on the service and fanning out with depth (see Fig. 11). One microgravity survey in the area produced variations in residual Bouguer readings, which confirmed the variable thickness of the superficial deposits. The results of the resistivity and microgravity surveys, as well as the available borehole data, indicate a variable thickness of CWF over this area, indicating that infilled dissolution features or erosional channels are present.

Encountered ground conditions

During construction of the cutting, the interface between the CWF and the underlying chalk proved to be irregular with a significant number of dissolution features identified on the slopes and at the base of the excavation, as illustrated in Figure 12. The encountered ground conditions were generally more complex in terms of the geometry of the chalk dissolution when compared to the predicted smooth surface between the chalk and the clay infill material that was suggested by the geological section. The geological model during the design stage predicted the presence of dissolution features between Ch. 47 + 550 and 47 + 650 m due to significant thicknesses of CWF encountered by the site investigation.

An example of the recording of as-revealed ground conditions on the western slope using digital tools to map the extent of the dissolution features and the interface of the CWF with the underlying chalk is presented in Figure 13 (Barker *et al.*

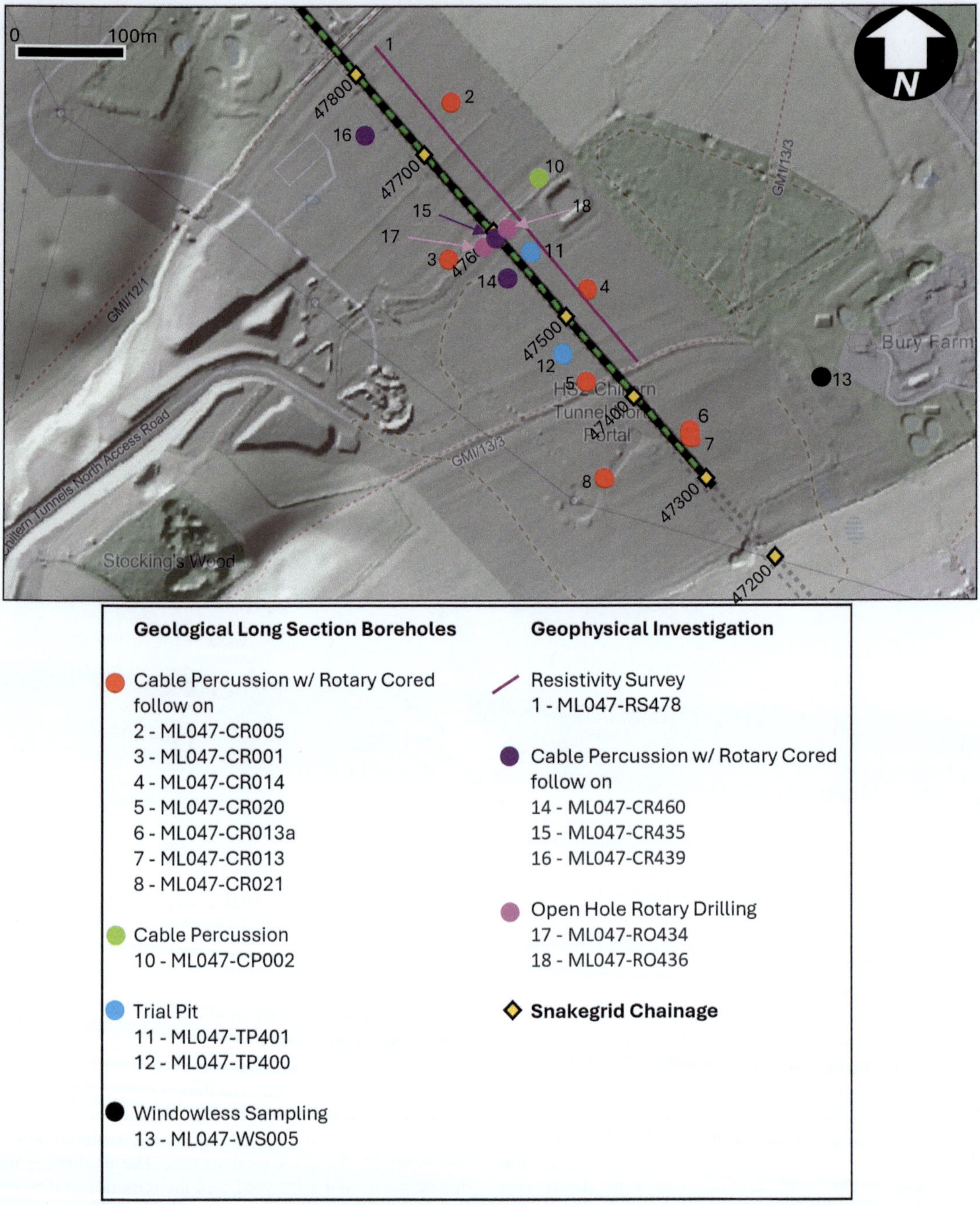

Fig. 8. Plan of the cutting section from Ch. 47 + 250 to 47 + 750 m with ground investigation (GI) locations.

2023). A total of 41 features were mapped and investigated further at the formation level and slopes of the cutting. The dimensions of the encountered features at the base of the cutting were generally less than 5 × 5 m. The features at the formation level were easily identified, investigated and treated where necessary.

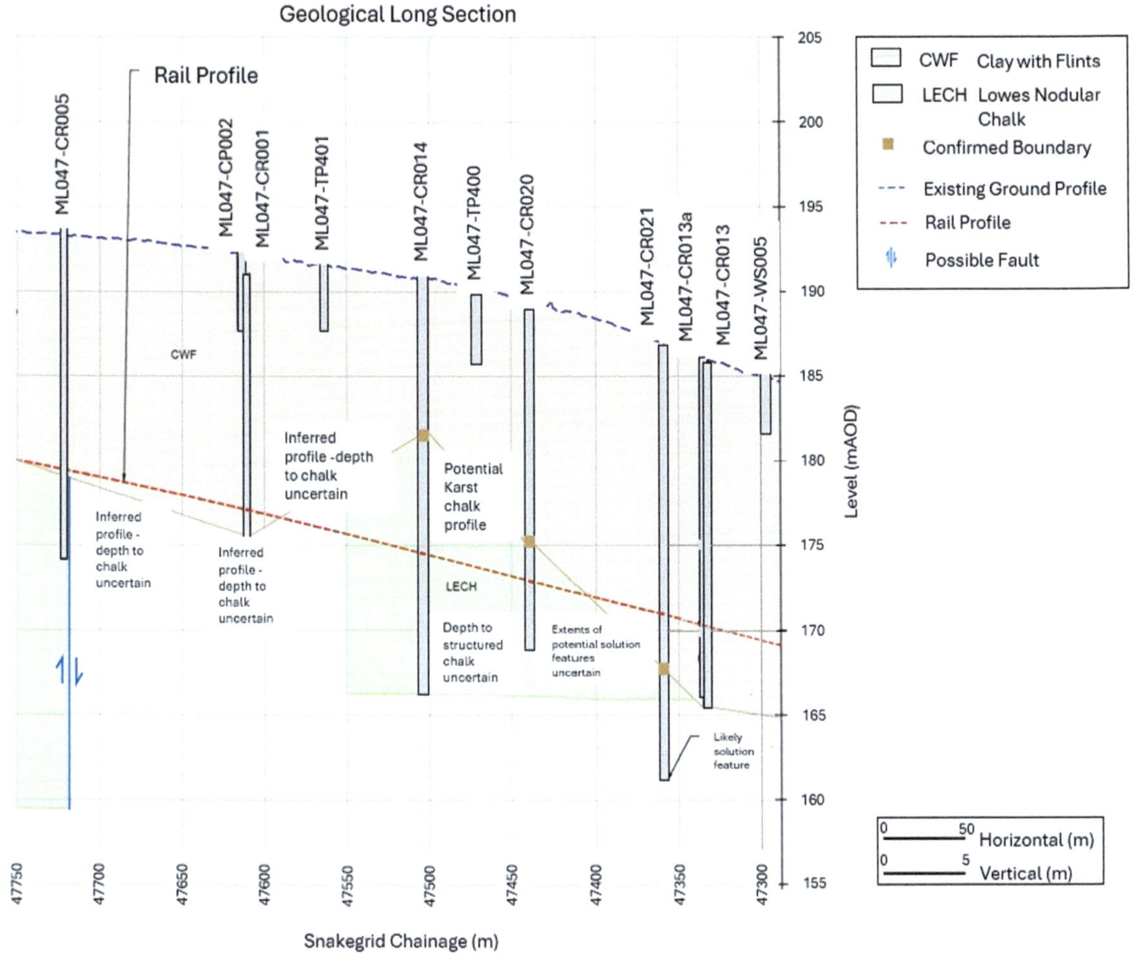

Fig. 9. Geological section based on the design ground investigation report. CWF, Clay with Flints (shown in a brown colour); LECH: Lewes Nodular Chalk Formation. m AOD, metres above Ordnance Datum.

Dissolution feature risk mitigation

In terms of risk mitigation, the design specification requires inspection and geological mapping of the exposed ground during construction, identification and investigation of features, compilation of a dissolution feature register, and two-phase investigation by geophysical survey and intrusive ground investigation (either dynamic probing or rotary core drilling). The risk of dissolution feature is then categorized based on its diameter. Features are treated as per standard earthworks details, and robust solutions for earthworks foundation treatment and slope stabilization are followed. The extent of treatment is assessed on a case-by-case basis for the different scenarios identified, and mitigation options include excavation and replacement of the near-surface dissolution features with temporary works to ensure a safe working environment, geogrid bridging, and drill and grout of open voids.

The features that were exposed at the formation level were further investigated to determine the depth and extent of the infill and the possible presence of voids below the cutting base. A total of 230 cone penetration probes were performed at 4 m centres, as illustrated in Figure 14. The depth of the cone penetration probing was determined based on the size of the features. The specified termination depth below ground level was 1.5 times the diameter. The maximum depth of penetration into the infill material or structureless chalk was up to 10 m, while the maximum depth of infill material was 4 m bgl. The probing proved that there was no presence of voids. The features at the base of the cutting were mitigated on a case-by-case basis by removing soft, cohesive soil, and backfilling with compacted material (either granular or chalk fill).

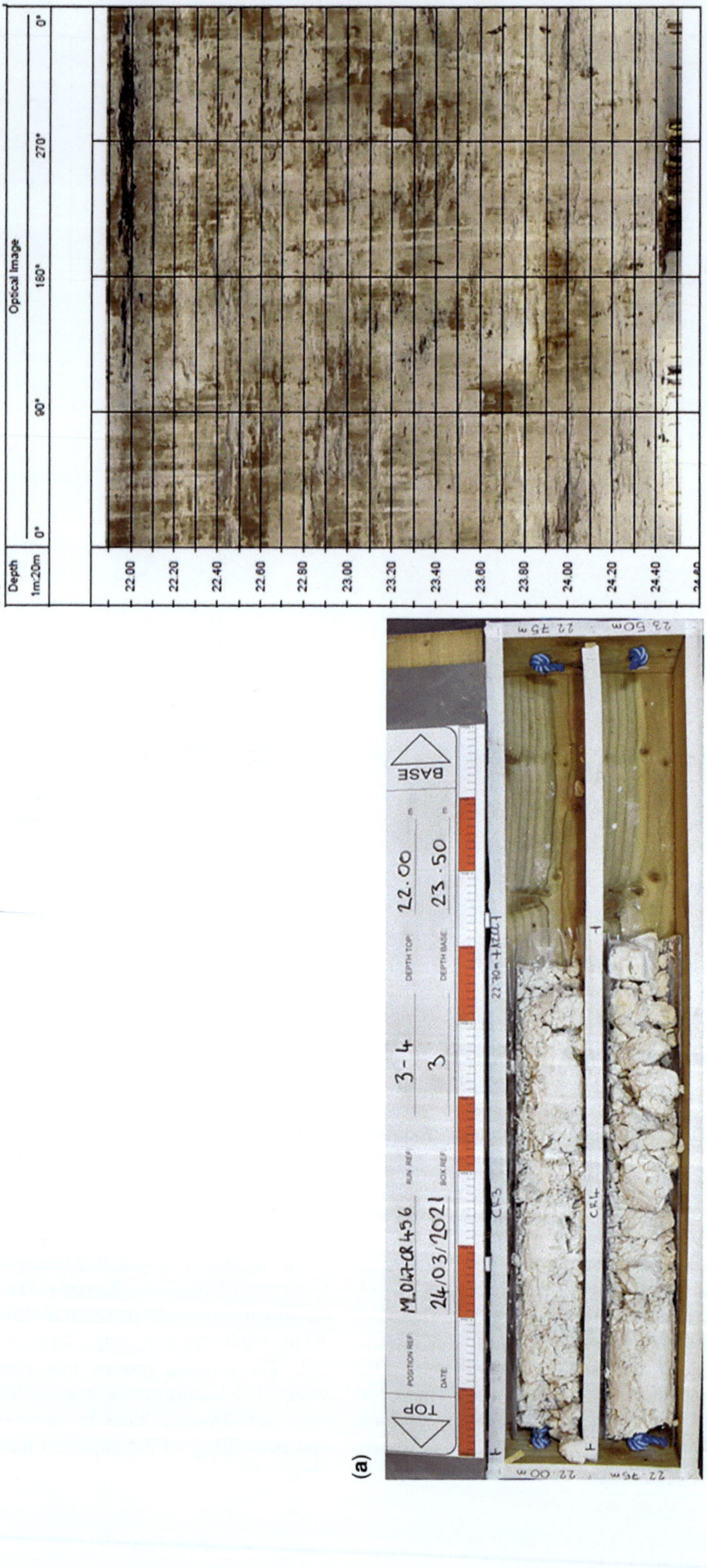

Fig. 10. Structureless chalk as encountered **(a)** in borehole ML047-CR456 and **(b)** in wireline geophysical log ML047-CR435.

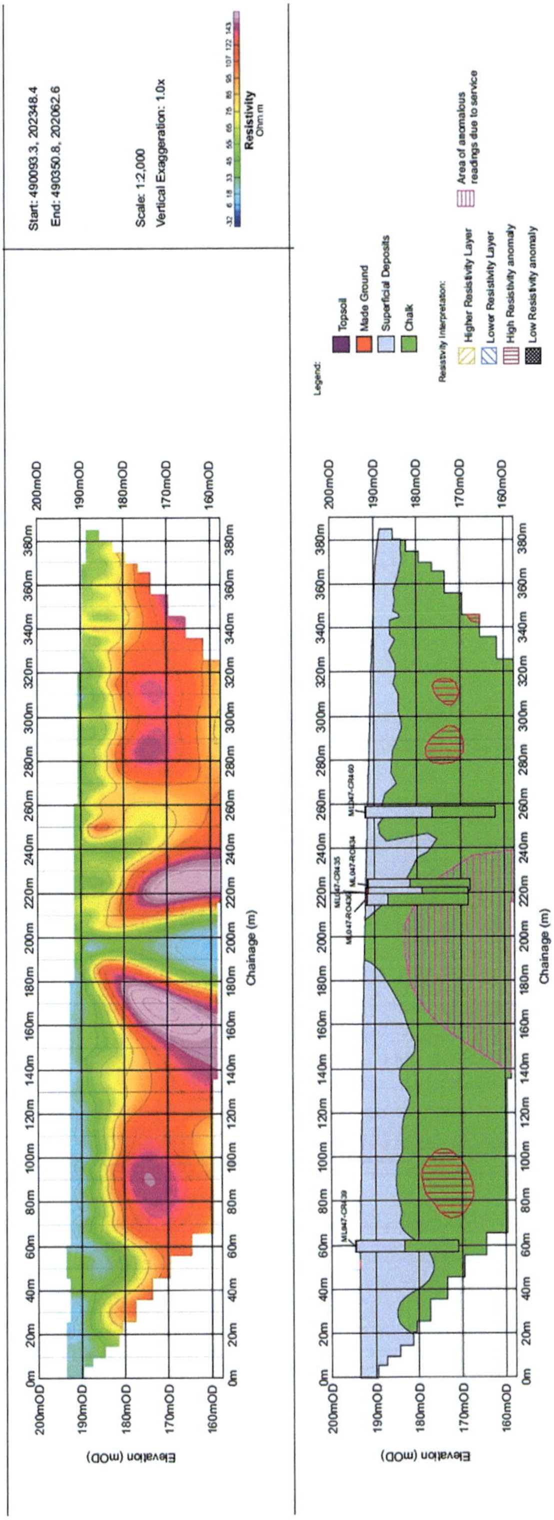

Fig. 11. Electrical resistivity section in the study area. Source: from HS2 Ltd (2022).

Fig. 12. View of the cutting looking towards the eastern slope, showing the presence of infilled dissolution features in the chalk. The cutting is excavated at the formation level. The photograph was taken in July 2022.

Regarding the features encountered on the lower part of the slopes, trial excavations were performed to investigate the infill material of the features (Fig. 15a). The dissolution features terminated at a maximum depth of 3 m. The sandy infilling material encountered in the features was found to be competent, with a very low risk of slipping and thus causing a slope failure. It was decided to remove the infill material at the surface when the slopes were capped to mitigate the risk of it being washed out during heavy rainfall. A standard method of treatment for a feature in the cutting slope is presented in Figure 15b. This consists of removal of the shallow infill, benching of the remaining infill material and capping with compacted granular fill, providing lined/sealed drainage at the crest.

Discussion

The actual ground conditions encountered during the excavation in a karst environment, both as presented in the case studied by Fookes and Hawkins (1988) (Fig. 3b) and in the example from HS2 (Fig. 12), suggest that the geometry and distribution of dissolution features in chalk can exhibit irregular patterns that may have not been visible or extensively detected in the GI. The key finding from the case study is that the conceptual model, which is based on an assessment of the geomorphology and the GI data available during the design stage, may have not been able to predict accurately the shape of the surface between the chalk and the clay infill material. This is a reminder that the geological ground model needs to be updated and refined during construction with targeted GI to determine the anticipated complexity of the ground.

Peter Fookes' pioneering work in understanding the geological processes and building a Geological Model is more than relevant and useful today. His work demonstrates that GI should be tailored to identify the key aspects/problems suggested by the ground model. In chalk environments, a geomorphological risk assessment using LiDAR and other technologies (i.e. surface geophysics, including microgravity, electrical resistivity and dynamic probing or drilling of boreholes in potential locations of features) can assist significantly in delineating and interpreting the level of risk of dissolution features.

As demonstrated in this paper, a geological model can provide an assessment of the potential geometry of dissolution features in a chalk environment, keeping in mind that it is very difficult to determine their exact location and geometry. It is therefore important that a robust recording and investigation procedure is in place during the construction stage to reduce the level of risk to as low as reasonably possible. This is when construction activities expose large areas of the subsurface, hence enabling the identification and

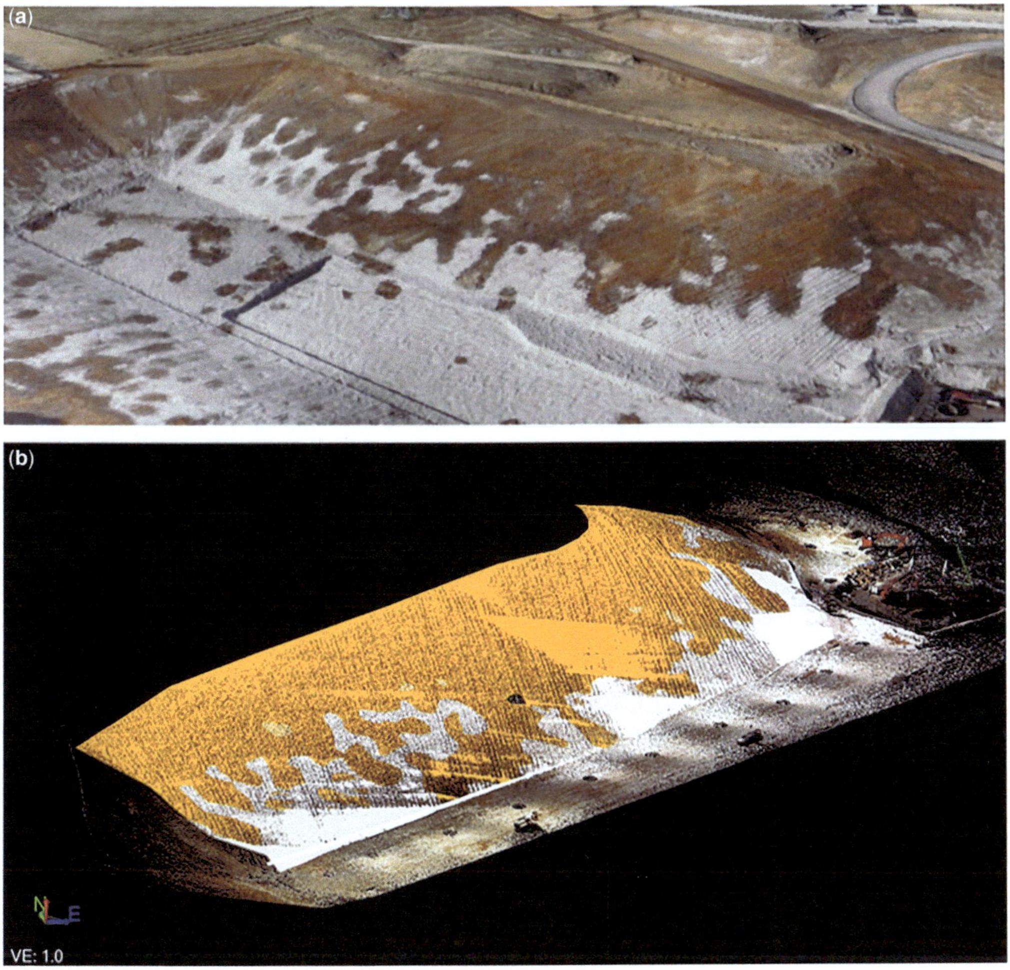

Fig. 13. (**a**) Western slope of the cutting. The photograph was taken in July 2022. (**b**) Digital mapping of the extent of dissolution features. Clay with Flints (CWF) infill material is shown in a brown colour. VE, vertical exaggeration. Source: Barker *et al.* (2023).

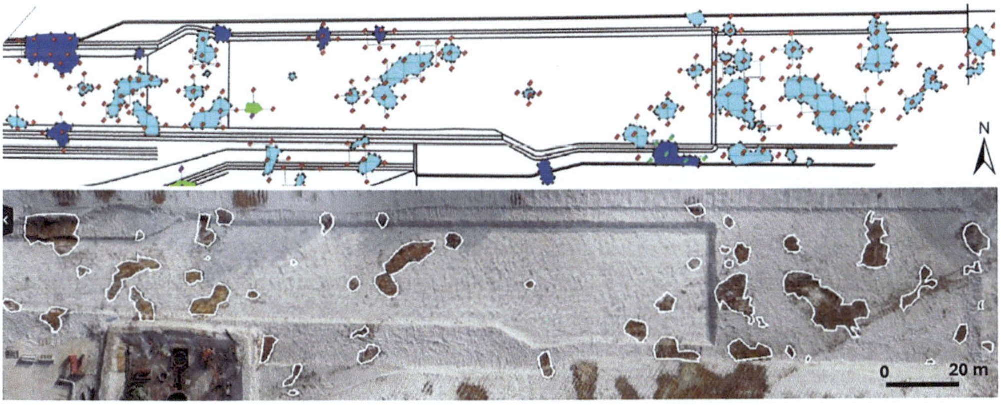

Fig. 14. Mapping of investigated dissolution features (in blue) and probing locations at 4 m centres.

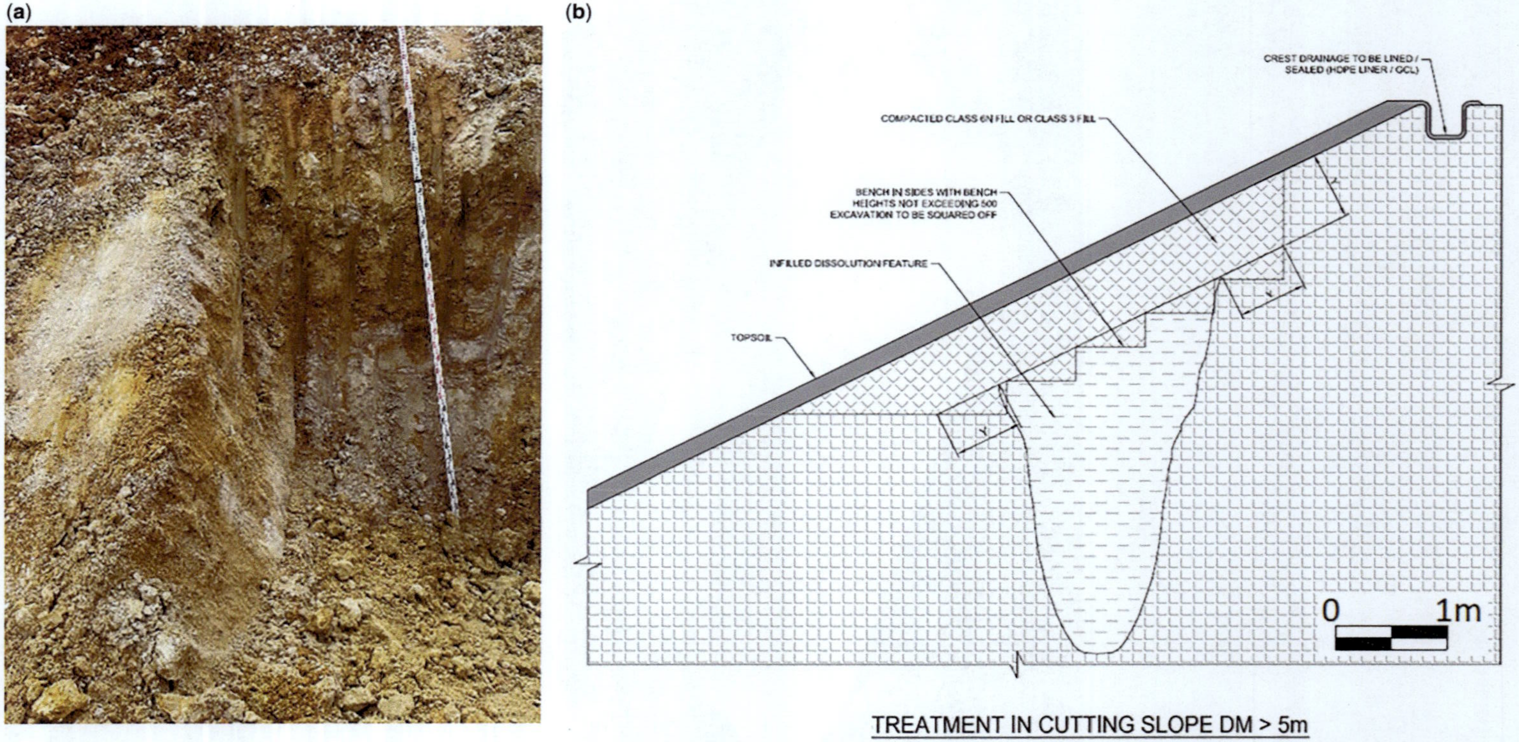

Fig. 15. (**a**) Trial excavation in dissolution infill material on the slopes. (**b**) Standard treatment of features in the cutting slope.

assessment of risk from dissolution features. It is also highlighted that a robust process must be followed, as in the example presented from HS2, to identify and effectively mitigate the risk from dissolution features during construction.

Acknowledgements The authors would like to thank G. Barker, EKFB Senior Geologist, for assisting in the preparation of figures. The authors acknowledge ASC DJV for providing the extract from the geological section. The authors also acknowledge James Dudfield, Senior Project Engineer in HS2, for reviewing the paper, and are grateful for the support of EKFB JV and HS2 Ltd in the preparation of this paper.

Competing interests The authors declare that they have no known competing financial interests or personal relationships that could have appeared to influence the work reported in this paper.

Author contributions **CS**: conceptualization (lead), formal analysis (lead); **GK**: conceptualization (lead), data curation (equal).

Funding This research received no specific grant from any funding agency in the public, commercial, or not-for-profit sectors.

Data availability The data that support the findings of this study are available from HS2 Ltd but restrictions apply to the availability of these data, which were used under licence for the current study, and so are not publicly available. Data are, however, available from the authors upon reasonable request and with permission of HS2 Ltd.

References

Barker, G., Hemmings, G. and Dinan, T. 2023. *Construction Stage Geo-Data Collection, Reporting and Visualization – Digital Workflow*. HS2 Learning Legacy, Birmingham, UK, https://learninglegacy.hs2.org.uk/document/construction-stage-geo-data-collection-reporting-and-visualisation-digital-workflow/

BGS 1990. *Aylesbury. England and Wales Sheet 238. Drift Geology. 1:50 000*. British Geological Survey (BGS), Keyworth, Nottingham, UK.

Bristow, C.R., Mortimore, R.N. and Wood, C.J. 1997. Lithostratigraphy for mapping the Chalk of southern England. *Proceedings of the Geologists' Association*, **109**, 293–315, https://doi.org/10.1016/S0016-7878(97)80014-4.

Edmonds, C.N. 2001. Predicting natural cavities in chalk. *Geological Society, London, Engineering Geology Special Publications*, **18**, 29–38, https://doi.org/10.1144/GSL.ENG.2001.018.01.05

Edmonds, C.N. 2020. Dissolution – carbonates. *Geological Society, London, Engineering Geology Special Publications*, **29**, 389–401, https://doi.org/10.1144/EGSP29.15

Fookes, P.G. 1997. Geology for engineers: the geological model, prediction and performance. *Quarterly Journal of Engineering Geology*, **30**, 293–424, https://doi.org/10.1144/GSL.QJEG.1997.030.P4.02

Fookes, P.G. and Hawkins, A.B. 1988. Limestone weathering: its engineering significance and a proposed classification scheme. *Quarterly Journal of Engineering Geology*, **21**, 7–31, https://doi.org/10.1144/GSL.QJEG.1988.021.01.02

Fookes, P.G., Baynes, F. and Hutchinson, J.N. 2000. Total geological history: a model approach to the anticipation, observation and understanding of site conditions. *In*: *Proceedings of the International Conference on Geotechnical and Geological Engineering, Melbourne, Australia, Volume 1*. Technomic Publishing, Lancaster, PA, 370–460.

HS2 Ltd 2017. *Phase 1 Engineering Geomorphology Assessment Final Report*. High Speed Two (HS2) Ltd, Birmingham, UK.

HS2 Ltd 2022. *1MC06 – Stage One C2 – MWCC – North Portal of Chiltern Tunnels to Brackley. Interpretation of Electromagnetic Conductivity, Electrical Resistivity Tomography and Microgravity Geophysical Data*. High Speed Two (HS2) Ltd, Birmingham, UK.

Jeffrey, Z.E., Penn, S., Giles, D.P. and Hastewell, L. 2020. Identification, investigation and classification of surface depressions and chalk dissolution features using integrated LiDAR and geophysical methods. *Quarterly Journal of Engineering Geology and Hydrogeology*, **53**, 620–644, https://doi.org/10.1144/qjegh2019-098

Lord, J.A., Clayton, C.R.I. and Mortimore, R.N. 2002. *Engineering in Chalk*. CIRIA Report C574. Construction Industry Research and Information Association, London.

Moore, R., Fish, P., Trinder, S., Czarnomski, C., Dabson, O. and Fitzgerald, R. 2022. Engineering geomorphology of HS2: management of geohazards. *Quarterly Journal of Engineering Geology and Hydrogeology*, **55**, https://doi.org/10.1144/qjegh2021-122

Mortimore, R.N. 1983. The stratigraphy and sedimentation of the Turonian–Campanian in the Southern Province of England. *Zitteliana*, **10**, 27–41.

Mortimore, R.N. 1986. Stratigraphy of the Upper Cretaceous White Chalk of Sussex. *Proceedings of the Geologists' Association*, **97**, 97–139, https://doi.org/10.1016/S0016-7878(86)80065-7

Mortimore, R. 2018. Developing ground models for chalk engineering: links across the disciplines. Presented at the *Engineering in Chalk 2018 Conference*, 17–18 September 2018, London, UK.

Pryce, S., Black, M., Saroglou, H., Katsigiannis, G. and Muniyasami, S. 2024. Ground investigation methods used in high-speed railway for mitigation of dissolution features risk in chalk. *In*: *Proceedings of the XVIII ECSMGE 2024*. CRC Press, London, 1291–1296, https://doi.org/10.1201/9781003431749-235

Waltham, A.C. and Fookes, P.G. 2003. Engineering classification of karst ground conditions. *Quarterly Journal of Engineering Geology and Hydrogeology*, **36**, 101–118, https://doi.org/10.1144/1470-9236/2002-33

Desk studies, geological models and avoiding construction disputes – your records will be used against you

John Davis

Geotechnical Consulting Group, 150 Brompton Road, London SW3 1HX, UK
0000-0002-9932-2055
Correspondence: j.a.davis@gcg.co.uk

Abstract: This paper contains the author's personal views on how and why construction problems and disputes related to unforeseen ground conditions occur. The 'how and why' relate to: (i) a common contaminated land focus in the initial project desk studies that is then not updated as projects develop; (ii) a focus on known factual data; and (iii) a lack of focus on the conceptual data that inform what construction hazards might be present in gaps between observed data points.

Potential technical and commercial mitigations for these shortcomings are suggested. These include structuring geotechnical deliverables around the concepts set out in the IAEG Commission 25 on engineering ground models: increased training of ground engineering practitioners on the contractual context of their work and on increasing the self-awareness of what the engineering geological input into projects should be and the competencies required to produce that input.

Context

The abbreviated title of this paper – 'your records will be used against you' – is a quote by Dr Mike de Freitas, delivered to the author's MSc class in 1988. Another quote with a similar theme comes from Dr Barry New's January 2017 Harding Memorial Lecture. That evening Dr New told the British Tunnelling Society audience that 'managing engineers' needed to 'value their signatures'.

The overarching theme of this paper is that many ground-based construction disputes could be avoided if more care was taken over a project's initial conceptual characterization of the ground. In the author's experience and opinion 'unforeseen ground' in a contractual sense is much more common than it ought to be. Professor Peter Fookes in his 1997 Glossop Lecture stated 'the nature of the ground is rarely unforeseeable, but is often unforeseen'.

This paper presents a personal opinion based on the author's UK-based experiences of working forensically on ground-related construction problems, disputes or insurance claims related to the built environment. As such, these opinions are biased towards those occasions where disputes have arisen. These problems typically involve construction difficulties or damage to property that arose from unexpected encounters with the ground.

'Disputes' are not necessarily always 'legal' in a formal expert witness sense. An example of a formal expert witness role is set out in the 'UK Civil Procedure Rules' (Ministry of Justice 2021). Formal disputes generally involve: a formal written agreement (a contract or insurance policy); more than one party; and at least one party who asserts it has suffered a loss. The author also has experience of supporting contractors and insurers with unforeseen ground issues that were not part of the formal legal processes.

A common characteristic of working on construction problems or disputes is that they usually involve looking back forensically in time to review the content of pre-contract or pre-construction documentation and comparing these with a review of the records of an 'adverse' encounter with the ground during construction. In a dispute, this is usually done in the context of the wording of a 'test' that is set out in a contract or an insurance policy.

This paper focuses on disputes or problems that involve 'unforeseen ground' rather than disputes that might involve some perceived failure in workmanship or design. The author is an experienced engineering geologist, he is not a lawyer, nor is he an expert on contracts.

The author's view is that the importance of appropriate desk studies and the associated conceptual ground models they produce in relation to construction is not always properly recognized. Note the abbreviated title of this paper in this context. This personal view is, of course, biased by the author's regular exposure to ground-related construction problems and disputes. Happily, most projects avoid disputes, despite sometimes being based on inadequate ground models.

The author's experience is that there are several, often overlapping, factors that commonly underpin unforeseen ground condition disputes. These are:

- Desk studies that are overly focused on contaminated land and planning consent issues, and conversely lack a focus on hazards related to design and construction.

From: Griffiths, J. S., Giles, D. P., Davis, J. A. and Dunlop, J. J. (eds) 2025. *The Peter Fookes Engineering Geological Legacy in Geomodels, Geomaterials and Geomorphology*. Geological Society, London, Engineering Geology Special Publications, **30**, 111–121,
First published online 6 May 2025, https://doi.org/10.1144/egsp30-2025-2

- Desk studies that do not set out conceptual models that specifically relate to the *ground engineering* required by the project.
- Desk study, ground investigation and interpretative reports that focus on design and do not adequately address potential construction or ground engineering-related difficulties.
- Inadequate geological insight (by all parties to a contract). 'Geological insight' is used in this paper as shorthand for Peter Fookes, Fred Baynes and John Hutchinson's 'total geological history' concept (Fookes 2001). This is discussed further below.
  Design processes set out in Standards force a cautious and factored approach to the adoption of design parameters. This is to build sufficient resilience into the design such that it can cope with some adverse variation in the encountered ground conditions. Encounters with unforeseen ground that adversely impact the actual construction often do not have the benefit of similar built-in resilience.
- A lack of recognition by consultants and contractors of the inherent limitations and sources of error in ground investigation techniques and equipment.
- Desk study authors and checkers with a lack of relevant broad-based experience and training. This partly relates to the increasing separation and specialization of 'geo-environmental'/contaminated land engineers when compared to specialized engineering geologists who have a more of a geotechnical focus. This separation tends to arise from an initial desk study focus on matters relating to managing the contaminated land planning application process.
- Desk studies and interpretative reports that sometimes fail to properly consider or interpret basic freely available desk study information.
- Inappropriate use or presentation of images of ground models.
- Failure to use and express conceptual engineering ground models (EGMs).

This paper is aimed at consultants and contractors. Some of the factors above could potentially arise from an employer's unwillingness to provide sufficient funds or time for the whole geotechnical investigation and reporting process.

External cost and time constraints should never prevent reports or correspondence stating what the desk study authors believe should be said or done. Stating these views in reports and correspondence could be considered to be a form of 'reasonable skill and care', and would be likely to be helpful in any later dispute setting. Such statements would also be consistent with professional Codes of Practice.

Disputes can involve significant additional unplanned costs for all sides (regardless of who 'wins'), reputational losses, significant increases in insurance premiums and, unfortunately, sometimes they also result in the winding up of companies. It is worth noting that once a dispute is formally triggered, the defending side tends to cede control of the following processes to its insurers.

Definitions

- Desk study: the initial gathering together and reporting of all the relevant available information on the ground on and around the project location.
- Ground model: any visual or text-based representation of ground properties used to assist design and construction. Several types of ground model are described in this report', these include engineering geological models (EGMs) as defined by Baynes (2024) and discussed below.
- Conceptual site model (CSM): a specialized ground model used in the assessment of contaminated land (see BSI 2020*b* and its precursors).

Ground models

Many unforeseen ground-based construction disputes could easily be avoided by use of appropriately experienced people to critically appraise all the available ground data to produce project specific EGMs. These EGMs should address both the nature of the ground and the hazards it poses to the design and to the proposed construction techniques. It is worth noting that conventionally: Risk = Hazard × Consequence.

At the present time for most construction projects in the UK, there are (or should be) a minimum of two types or styles of ground model that arise out of desk study data, each has a different focus.

One is the long-established CSM approach to the assessment of contaminated land and the related planning consent issues. The others are similar to the EGMs described in the IAEG's important 2022 report (revised in 2024) *Guidelines for the Development and Application Engineering Geological Models* (Baynes 2024), referred to from now on as 'C25'.

In both cases, the initial tool for developing these models involves the data gathered in the desk study.

C25 defines several types of EGM. The basic EGM models relevant to this paper – the conceptual EGM and the observational EGM – are defined in C25, as detailed below.

Conceptual EGM

'A model based mainly on engineering geological concepts and interpretations and the knowledge that certain engineering geological conditions and processes are likely to have certain engineering characteristics.' In essence, this is a view or an idea of 'what might be found' beneath a site, it is broadly analogous to a desk study stage and contains interpretations that do not necessarily have a precisely defined location.

Observational EGM

'A model based mainly on engineering and geological observations and measurements that are constrained in space by 3D data (xyz) or in space and time by 4D data (xyz plus time)'. An observational model is a model based on observed 'facts' that critically have a defined location. A borehole log would be an example of an observed set of facts. Knowledge that the ground at a site is thought to increase in strength with depth would not be a fact.

Desk studies and contaminated land

Commonly, initial project desk studies focus on planning issues related to contaminated land. In these a typically standardized 'recipe book' structure is frequently used for all scales of project in the UK. At its most basic level, this involves purchase of a digital desk study data pack, which is appended to a report, that factually summarizes the data pack contents. At its simplest such a report tends to make generalized points about the ground and the project.

If done correctly, this of itself is not necessarily a problem. At very early stages, project proposals can be vague and subject to change.

However, problems can arise in a later dispute where this initial, often broadly based, general advice is incorporated into a contract. If this initial advice is not subsequently updated or qualified as observed data are acquired, then it could be used to, perhaps, unfairly support arguments in a dispute. This is an example of 'your records being used against you'. These issues arise where desk studies and the models that arise from them are regarded as 'done' once they are complete. This is often the case where the main purpose of the desk study is to support a contaminated land planning process.

Desk study reports should include unambiguous statements about the purpose of the report and make it clear that the report's conclusions may require revision later in a project's life. This is particularly important if an organization has just been employed to provide pre-planning contamination or pre-purchase due diligence desk studies.

Desk study reporting using a C25 EGM framework would simplify these issues by providing a focus on what is not known.

In early contamination-focused desk study reports there is often little critical appraisal of the relevance of the gathered facts to the ground engineering that the project construction requires. In the author's view, this lack of critical appraisal appears to be related to:

(i) a focus on what is 'known' (an observational EGM) rather than the design and construction-related hazards that could arise from what is not known but is plausible. In other words, they lack a conceptual EGM, a 'what might be there' thought process with respect to design and construction; and
(ii) a lack of experience on the part of the desk study author/checker/approver that can miss the significance of 'what might be there' in ground engineering design and construction terms.

In contrast, the contaminated land and planning aspects of potential unknowns ('what might be there') are often dealt with efficiently and comprehensively via the use of CSMs.

Digital desk study data packs are excellent products, but by themselves are *not* a desk study, and will probably never contain all the information that a comprehensive desk study requires.

The publication of Part 2A of the Environmental Protection Act 1990 (UK Government 1990) embedded pollution-focused desk studies and CSMs very early into a UK project's life. As a result, local authority planning consent processes required desk studies containing site-specific conceptual site models and contaminated land- risk assessments.

Prior to this, desk studies, if they were done at all, focused on establishing ground investigation scopes. As a young geotechnical engineer working for the PSA (a now defunct Government Agency) in the late 1980s, the author recalls routinely carrying out desk studies to scope ground investigations, but only later writing them up as part of a single subsequent ground investigation report.

Later published contaminated land guidance, the most recent examples of which are *Contaminated Land Risk Assessment. A Guide to Good Practice* (Construction Industry Research and Information Association Report C552: CIRIA 2001) and Investigation of Potentially Contaminated Sites – Code of Practice (BS 10175:2011 + A2:2017: BSI 2017), have a great deal to say on the subject of conceptual site models for contamination risk assessment.

In the author's opinion, this early planning requirement to produce pollution-focused conceptual models based on desk studies alongside the development of geo-environmental contamination specialists and cheap digital desk study data packs seems to have led to a widely adopted desk study approach where potential ground engineering issues get less and lower quality attention than the hazard minimization requires.

These low-cost early project stage planning requirements seem to have firmly embedded the CSM source pathway receptor model as the principal initial desk study end point.

At these early pre-ground investigation planning stages, the project proposals are often loosely formed and subject to change. As such, any geotechnical advice contained within these desk studies tends to be necessarily vague and so is potentially inappropriate if the project proposals change or the ground is different from that indicated by the desk study.

In a dispute setting, these early-stage desk studies views on engineering matters can be fertile ground for experts seeking arguments on which to base their views, especially if 'what might be there' conceptual EGMs are not described or updated.

A safer approach to reporting desk study information (discussed later) would be to entirely separate the 'geo' (engineering) elements from the 'environmental' (planning) elements into two desk study reports, each of which include text that clearly sets out the basis of the reports and clearly limits the scope the reports. Other suggestions for mitigating the impact of necessarily vague early project stage information in contract documents are given below.

Desk studies and ground engineering

From a ground engineering perspective, the initial purpose of the desk study is to gather together all the relevant available information on the ground on and around the project location, to establish design data gaps and to scope appropriate investigations to fill those gaps. This desk study information should also populate the design and construction risk registers required by Standards. Ideally, in a desk study this process would follow the conceptual and observational EGM approaches set out in C25.

The author's experience is that it is unusual for desk study authors with a more geotechnical background to opine on contaminated land issues, or for contaminated land specialists to comment on engineering matters. However, the commonly used, but often inexperienced, early career hybrid 'geo-environmental' engineer can find themselves in situations where their ground engineering experience is insufficient to safely answer the questions they are being asked by their clients.

This appears to be principally a failure of supervision and checking by more senior colleagues. In the author's opinion, this problem is particularly acute in smaller geo-environmental contractor/consultancies and in smaller civil and structural practices that lack access to specialist geotechnical resources. In both these situations, appropriate checking experience may not exist.

What is often missing in both cases is the view of an engineering geologist with some construction experience or knowledge, who is able to focus on the 'what might be there' conceptual EGM unknowns. This absence tends to produce generalized and formulaic 'recipe' book approaches to engineering advice. When this is combined with insufficient geological insight then problems can arise.

In some cases, the quality of advice provided by engineering geologists is also poor

Some recent simple and avoidable UK examples of common errors the author has come across are given below. These errors were made by engineering geologists, geo-environmental engineers, geotechnical engineers and civil/structural engineers:

- Reliance on just two layers ('bedrock geology' and 'superficial deposits') in the BGS Geoindex online mapping tool. The author has experienced several instances of consultants working in well-known landslide areas not consulting the 'mass movement' layer and completely missing the presence of landslides at their sites.
- Failing to consult the actual geological maps, most of which are freely available online at all scales down to 1:10 000.
- An inability to interpret geological maps and sections. This often involves civil or structural engineers, but also geologists of all types. Examples include misunderstanding dip directions of coal seams and failing to recognize the likely presence of faulting when the geological map indicates none is present. There is often a focus on reporting the formal strata names rather than explanation of what the materials actually are and, importantly, how they were formed.
- A lack of awareness of the hazards related to Quaternary processes, particularly slope processes (e.g. solifluction and cambering/valley bulge) and the hazards (e.g. obstructions) associated with erosion surfaces within glacial and post-glacial materials.
- A lack of recognition of the potential design and construction problems relating to soils and rocks containing high pyrite concentrations, even when the threats to buried concrete are recognized and addressed.
- Poorly scoped ground investigations and inappropriate scheduling of instrumentation, and *in situ* and laboratory testing.
- A lack of explanation where physical constraints prevent appropriate investigations from taking place.
- A lack of critical assessment of basic desk study materials.
- A lack of insight into the interplay between the ground and ground engineering techniques.
- Failings in the interpretation of topographical maps.
- A lack of awareness and use of freely available detailed digital time-stamped topographical and other data (e.g. LiDAR data and BRITICE GIS layers). See the list of selected data sources given in Appendix A.

These problems are not restricted to smaller companies. The author's experience is that large projects with big geotechnical teams can fall into a similar 'group think' traps, even where extensive observational EGM data exist. Here there can sometimes be an overreliance on gathered facts, and the 'what might be there' conceptual element is forgotten. Sophisticated coloured-in 3D digital block models and their difficulties in communicating uncertainty are not helpful in this regard because they tend to obscure the data gaps between observed data points. These gaps are often where contractual 'unforeseen ground' is encountered.

Geological models and geotechnical standards

Stressing the importance of desk studies has been long been a feature of guidance and Standards. The current version of the UK Code of Practice for Ground Investigations (BS 5930:2015 + A1:2020: BSI 2020*a*) contains excellent advice on desk studies and on the development of ground models.

The 1999 version of BS 5930 was weaker on ground models, but is still fundamentally sound with regard to the content and purpose of desk studies.

Eurocode 7: Geotechnical Design, Part 2: Ground Investigation and Testing (BS EN 1997-2:2007: BSI 2007) describes the purpose and content of a desk study but in a less detailed and structured way than BS 5930. It does not mention conceptual ground models.

Eurocode 7: Geotechnical Design Part 1 (current version) (BS_EN_1997-1:2004 + A1:2013: BSI 2013) does not contain the phrases 'desk study' or 'ground model'. It does, however, set out requirements for the gathering of 'preliminary data' and it links these data to the nature of 'design investigations'. It also mentions, but does not stress, the importance of considering construction hazards.

New versions of Eurocode 7 were published in September 2024. These are briefly discussed below in the section on minimizing the risks of disputes in relation to ground models.

Unsurprisingly, the focus of the Eurocode 7 documents (and Network Rail or Highway Authority Standards) is design, and the selection of cautious characteristic design input values and their subsequent factoring for use in calculations. In C25 terms, the end point of this process is a production of a simplified and cautious analytical ground model.

Construction encounters with the ground that lead to disputes rarely arise because an analytical model is too cautious. These disputes tend to arise because of an unexpected encounter involving the ground, or an unexpected interaction between ground and ground engineering plant or techniques. It follows that an analytical model is unlikely to be useful when assessing the possibility of construction difficulty.

These Eurocode Standards deal with design, so this minimal coverage of desk studies, ground models and 'what might be there' is, perhaps, understandable. However, these are the Standards that designers are likely to refer to most, and so, perhaps, the light touch coverage of desk studies and ground models has led to an underappreciation of the power and importance of desk studies and the conclusions that they could reach.

Once identified, the 'what might be there' conceptual EGM hazards should be captured and communicated via the Eurocode 7 requirement to produce design and construction risk registers. The author finds that if these registers exist, they are often generic, bland and lack references to conceptual EGMs.

Note that in a formal legalistic dispute scenario, arguments based on non-compliance with Standards are often deployed. Also note that in general for a structure to be Eurocode compliant, all the relevant codes must be complied with.

A comparison with CSMs here is useful. Through the planning process a CSM produces a risk assessment and typically mitigations are built into the planning conditions with requirements for written validation. The risk assessment element forces a consideration of 'what might be there'.

Why do avoidable unforeseen ground disputes occur?

Given that the desk study/ground model concept is well established and covered by Standards and C25, why do avoidable disputes occur?

The author's view is that where unforeseen ground disputes arise some of the factors below often apply to documents produced by project consultants:

- Early-stage desk studies for planning or for pre-land purchase due diligence are insufficiently engineering focused. Sometimes these desk studies are not updated and do not state that they need to be updated. This can lead to optimistic misconceptions in the minds of clients and developers. The financial and emotional impact of a later surprise can be an important factor in decisions to instigate dispute proceedings.
- Desk studies and conceptual ground models that lack appropriate geological and engineering geological insight. This is often because of inexperience or a lack of knowledge on the part of the desk study authors and checkers. The 'what might be there' aspect of conceptual ground models requires geological and engineering geological knowledge. This insight point is often a difficulty for civil engineers.
- A related insight point is the presentation of 'digital twin' models of the ground, including 'coloured-in' geological sections where these are presented without provision of commentary on the inherent uncertainty within those models.
- A lack of knowledge of the ground engineering techniques that a project may or will require, and in particular how the techniques interact with and change the ground. This often appears to be a largely experience-based issue. This point is often a difficulty for geologists with an environmental/contamination-focused background.
- Poor scoping of ground investigations in relation to design and construction. This point is also often a difficulty for geologists with an environmental/contamination-focused background

These failings often lead to the provision of inadequate or inappropriate ground investigation data, and in this way lead to a potential for a project to be surprised by the ground.

In addition, construction contractors and subcontractors often lack the ability to identify ground risk. This is not a criticism, they necessarily (and contractually) largely rely on the tender documentation on which they base their price and programme.

As noted above, 'geological insight' is used here as a shorthand for Peter Fookes, Fred Baynes and John Hutchinson's 'total geological history' concept set out in an article in the March 2001 edition of *Ground Engineering* magazine (Fookes 2001) and freely available online.

The Fookes (2001) article points out that the 'Ground conditions at any site are a product of its total geological and geomorphological history'.

It follows that assessing the ground conditions at a site requires knowledge of all of the geological and geomorphological history of the site. This is much more than referencing a formal strata name in a desk study and referencing topographical maps. The conclusions of this important article are repeated below; note the emphases on the Quaternary and on experience:

- There must be a specific and determined endeavour to understand the engineering geology and geomorphology environment of the site and to incorporate that understanding into the project design.

- Around the world it is often the problems related to the Quaternary geology and geomorphology events that dominate.
- The engineering geologist must be competent, well trained, experienced and a good geologist if they are to be either the lead engineering geologist in the team, or the only engineering geologist in the team.
- The engineering geologist must have a good knowledge of geomorphology and on occasion will need to work with a geomorphologist. Appropriate experience is most important in the professional development of qualified geologists and geomorphologists.
- Each site evaluation, however small, should ideally have at least one engineering geologist involved in the work. This does not always occur in practice.
- The preliminary stage is when the engineering geologist probably can have the most significant influence on the project by indicating potential hazards and their consequence on the economy of design, matters of construction and expected performance of the works.

The ability to provide the necessary geological insight is largely exclusive to geologists and in this context largely exclusive to engineering geologists.

Foreseeability, contracts and EGMs

The foreseeability of ground conditions during construction varies according to the:

- contract definition of unforeseen;
- the quality and quantity of ground investigation data, and the recognition of the limitations of ground investigation techniques;
- the quality of the EGMs; and
- the knowledge, experience and background of the observer.

What might be foreseeable to an experienced engineering geologist may not be foreseeable to a civil engineering-trained geotechnical engineer, a specialist subcontractor, a general contractor or an employer.

It is important to note that in a dispute the contract provisions and their interpretation take precedence over any other views. This interplay between the ability of a party to foresee and a separate contractual foreseeability test can be complex.

Unforeseen ground can and does exist in a purely geological or technical sense (no ground model is perfect). Unforeseen ground disputes often revolve around the extent to which an organization's view of the ground was 'reasonable'.

An example of this is provided by the NEC construction contract that defines unforeseeable in this way (the emphasis here has been added by this author and does not appear in the original):

> [T]he (encountered adverse 'unforeseen') physical conditions are those which, 'an experienced Contractor would have judged at the Contract Date *to have such a small chance of occurring that it would have been unreasonable to have allowed for them*'.

Unforeseeability clauses often also contain a 'publicly available' information text. The NEC form of contract provides a good example (the emphasis here has been added by this author and does not appear in the original):

> In judging the physical conditions for the purpose of assessing a compensation event, the Contractor is assumed to have taken into account: the Site Information, *publicly available information referred to in the Site Information*, information obtainable from a visual inspection of the Site, and other information which an experienced Contractor could reasonably be expected to have or to obtain.

Such clauses are particularly important with respect to desk studies and digital desk study data packs because these documents can reference, but not incorporate, sources of material that could indicate conceptual 'what might be there' hazards.

The author's experience is that it is not necessarily sufficient for desk study authors to rely on employers or tenderers reading and understanding 'the small print' in their desk study appendices. If the relevant publicly available information indicates a ground hazard, the desk study authors should set out the potential hazard in their own text in their documents. This is an extremely important point. Courts will tend to expect specialists to explain the potential hazards rather than assume that the non-specialist understands.

Claims based on the impact of unforeseen ground conditions that are aimed at consultants or ground investigation contractors usually frame the claim around an allegation of negligence in relation to how their duties under their contract are carried out: for example, use or not of 'reasonable skill and care'.

This could include not identifying a clearly foreseeable aspect: for instance, missing the reference to a Geoindex mass movement layer, as noted in the point noted earlier in this paper.

Claims based on the impact of unforeseen ground conditions that are instigated by contractors are often directed towards employers and are typically based on the contractual 'unforeseen ground' tests.

In ground that is commonly thought to be well understood, like London Clay, it is not unusual for some complacency to arise in all parties regardless of their experience. This is a 'geological insight' point. For example, 'it's just London Clay' has been a dangerous presumption for many projects.

The contractual 'not facts' problem

This can be a difficult problem. Contract definitions of 'unforeseen' usually rely on tests that refer to a set of contract 'facts' provided for tendering (e.g. desk study and ground investigation reports).

By their nature, early desk studies can be largely based on conceptual EGMs: that is, interpretations of 'what might be there'. These by definition are often not 'facts'.

As well as facts about a site, digital desk study data packs usually contain risk-based hazard assessments for a site: for example, the BGS hazard ratings for soluble rocks. These are also not facts, they are assessments based on the data supplier's available knowledge at and around a site. The author has experience of disputes where the ground conditions were substantially more onerous than hazard assessments suggested.

In addition, all geological maps are interpretations rather than facts. These 'not facts' and their inherent uncertainty should be appraised and the appraisal recorded in any reports.

In practice, all of these products would tend to be treated as facts in a dispute setting. This could be unhelpful to a party in a dispute if the uncertainty inherent in these 'not facts' was not explicitly addressed. So, it is important for conceptual and observational EGMs to critically reappraise these 'not facts' in reports as site-specific data become available.

It follows from this that incorporating desk studies and interpretative reports into contracts as contract documents that can be relied on can create unintended consequences in dispute settings if there is no obvious documented reappraisal of interpreted desk study information. There is usually very little a desk study author can do about this beyond discussing the uncertainty, having a tightly defined contract scope, and applying 'reasonable skill and care' in the delivery of that scope.

How to minimize the risk of disputes

There is no perfect solution for eliminating the possibility of unforeseen ground and its interaction with development proposals and ground engineering techniques. However, technical and contractual improvements are always likely to be possible. No EGM or other ground model can be relied upon to be entirely correct.

Training and experience

There are several aspects to this which can only briefly be mentioned here.

(1) Over the last few years there has been a reduction in the availability of UK post-graduate engineering geology-focused training opportunities, particularly at the Master's level. This reduction is ongoing. It is partly driven by a lack of interest from graduate geologists and partly driven by university funding difficulties. This is a serious strategic issue for the UK construction industry.

(2) The Geological Society of London (GSL) Code of Conduct (and the Institution of Civil Engineers (ICE), etc. equivalents) sometimes appear to be forgotten, and geotechnical and geo-environmental engineers sometimes appear to work well beyond their experience and competence in relation to EGMs and to ground engineering techniques.

(3) This is particularly the case with respect to Quaternary materials and processes, and also to geochemical processes.

(4) A lack of recognition of the constraints of common ground investigation techniques, inappropriate investigation scoping, and inappropriate scheduling of *in situ* and laboratory testing are common features of smaller-scale disputes. This brings to mind Professor Littlejohn's 1991 well-known quote 'You pay for a site investigation whether you have one or not'.

Mitigation for (2), (3) and (4) can only come from the practitioners and their employers through identifying and training weaknesses (CPD) and via procurement (e.g. via 'buying in expertise' where required). These mitigations are unlikely to be cost-free. Employers need to consider the potential cost of not protecting their professional indemnity insurance in this regard.

In the author's opinion, there appears to be a long-standing practice of employers asking their least experienced staff (engineers and geologists) to produce desk studies. This is welcome, but in some cases there are serious failures in checking and approval by more senior staff. The earlier examples of simple errors of omission are proof that these checking failures exist. The Dr Barry New 'value your signature' and Dr Mike de Freitas 'your records will be used against you' quotes made earlier in this paper are important here. The author would add 'value your professional indemnity insurance' to this list.

Improved practitioner knowledge of the contractual context of geotechnical deliverables and the associated insurance arrangements would be of great benefit for all levels of experience.

It is likely that project construction contracts would not be available when desk studies and geotechnical reports are being written. So what is required is a generalized awareness of how the information in geotechnical reports is used during tendering, construction and potentially during a dispute.

An obvious specific contractual example here is the understanding the importance of *publicly available information*, and recognizing the (C25) distinctions between factual data (observed) and interpreted data (partly or wholly conceptual).

Ideas for reorganizing deliverables to better match contractual practice, C25 and existing and new revisions of Eurocode 7 are given below.

One of the insuperable challenges for the broader construction industry is the lack of geological insight inherent in most civil engineering-based geotechnical engineers and contractors. This is not a criticism; 'geological insight' is not their role. It is the role of the engineering geologist to provide that insight, and in the author's experience simple and avoidable errors are being made. Most unforeseeable ground is foreseeable.

Reporting

The following ideas are suggestions for changing conventional reporting practice (i.e. desk study, factual report and interpretive report templates). These ideas are compatible with Eurocode 7 reporting requirements:

(1) Stating the purpose of each geotechnical report and risk register in C25 EGM terms. An explanation of the meanings of those EGMs should be included in each report.

(2) Split desk studies into three reports: a 'factual' data report, and separate desk study interpretative reports on contamination and on construction/design issues (CSMs and EGMs).

The conclusions of the EGM report must contain a risk-based assessment of the hazards that plausibly 'might be there' for design and for the *likely construction methods*. The EGM report should use the conceptual/observational/analytical terminology set out in C25 to structure the report. The CSM and EGM reports should be revised every time new data are procured or development plans are updated.

If report templates were structured around the C25 EGMs, then that ought to lead to a greater awareness of the distinction between facts and uncertainty.

At each report revision required by further data acquisition or project change, there should be a statement of the remaining potential design and construction hazards generated by the conceptual 'what might be there' gaps in the ground data. This is effectively mirroring the Eurocode 7 requirement to produce geotechnical risk registers. The author's experience is that these Eurocode 7 risk registers are often unhelpfully generic and insufficiently detailed to be of much use to tenderers. National Highways document CD622 'Management of Geotechnical Risk' is a good example of a process that specifically deals with this point.

Repeatedly producing and revising these registers using a C25 structure ought to produce a greater focus on uncertainty in relation to construction and its potential consequences. Adverse unforeseen ground tends to be encountered during construction. During construction, unlike during design, there are no factors of safety or cautious characteristic values to account for uncertainty.

This three-part desk study structure has another significant advantage. Splitting factual and interpretive reports is potentially very helpful for contract compliers. It provides an opportunity to avoid interpretations, particularly early unamended desk study interpretations, being treated as facts in a dispute.

There are contractual methods of providing tenderers with the technical benefit of interpretations that avoid contractors being able to 'rely' on them as if they were facts in a contractual sense. An example of such a technique is appending interpretative reports to the instructions for tenderers. This provides sight of the reports, but once a contract is signed the reports do not become 'contract documents'.

This approach is potentially useful because contractual tests for unforeseen ground are often based on a comparison of 'before and after' facts. Tenderers need to understand what are facts and what are interpretations when pricing and programming projects. Interpretations are inherently uncertain, and tenderers require sight of them to be able use their experience to price that uncertainty.

(3) The new revisions of Eurocode 7 (BSI 2024) include a new requirement to explicitly set out a 'ground model' as a deliverable. Unfortunately, this requirement is largely set out in observational EGM terms and is, unsurprisingly, design focused. The author's view is that this new ground model deliverable should be C25 based (to address the shortcomings in the new Standard approach) and should explicitly address uncertainty as it relates to construction.

Other current draft Eurocode proposals (see Nyambayo 2024) require enhanced consideration of ground risk, geotechnical complexity and 'consequence classes'. These appear to be promising developments.

(4) Any graphical representation of ground models (e.g. cross-sections and digital 3D models) should always be accompanied by a written critical appraisal of the data contained within the model, and any graphical representation should distinguish explicitly between factual and interpreted content. C25 contains extensive advice on the graphical presentation of ground models.

Identifying engineering geologists

Engineering geology and geo-environmental engineering are increasingly separate specialized disciplines, and rightly so. Here, the author's experience leads him to presume that geo-environmental engineers are practitioners who specialize in contaminated land, and have limited experience of engineering geology and ground engineering, and vice versa. It is noteworthy that this engineering/contamination separation between the two disciplines is also embedded in the application processes for Chartered Geologist status.

Identifying competency in engineering geology is not straightforward. Seeking Chartered Geologists who are listed on the ICE/GSL/Institute of Materials, Minerals & Mining (IoM3)/Energy Institute (EI) Register of Ground Engineering Professionals (RoGEP) is one simple and effective method. Unfortunately, it is clear that engineering geologists are significantly under-represented on this register compared to civil engineers when the membership of the GSL Engineering Group is compared to the smaller membership of the British Geotechnical Association. Employers are encouraged to promote RoGEP registration to their engineering geology staff to redress this imbalance.

The Register of Specialists in Land Condition (SiLC) contains a surprisingly low proportion of Chartered Geologists. This means that identifying geologists who are competent contaminated land specialists is also difficult.

Geotechnical baseline reports

The main role of a geotechnical baseline report (GBR) is to remove commercial/contractual ground uncertainty from a tendering process by stating what the ground conditions *will be* in the conceptual 'what might be there' spaces between factual observed data points. The links to C25 EGM concepts in the current UK GBR guidance (CIRIA 2023) are deliberate. These non-factual ground conditions are set out in 'baseline statements'. Tenderers are assumed to have allowed for encountering the baselined conditions in their tenders.

If, during construction, the contractor encounters a ground condition more adverse than a baseline and suffers a loss, then the contractor is entitled to compensation for the extra over cost/time created by the exceedance of the baseline. This can make claims processing very efficient.

Baseline statements can take many forms:

- they can use geotechnical parameters directly measured during construction ('the strength of X will be between Y and Z in locations A–B');
- or they can use indirectly measured parameters during construction via instrumentation on ground engineering plant;
- they can baseline geological conditions (e.g. strata boundaries);
- they can baseline ground or groundwater behaviour relative to specific ground engineering techniques; or
- they can presume the use of specific construction techniques, whilst allowing the contractor to adopt alternative techniques at their own risk.

The baselines must be measurable during construction; for many ground engineering techniques this can be challenging as the ground is often not visible or accessible during construction, and many ground engineering techniques also alter the properties of the ground or any spoil produced by the ground engineering technique. This is why indirect measurements of the ground are mentioned above.

GBRs arose in North American tunnelling practice in the late twentieth century and were first used in a significant way in the UK on the Crossrail project in London in the 2010s. There are differences in the approaches to GBRs between these two locations. Currently, only one standard form of contract (The FIDIC 'Emerald Book' for tunnelling works: FIDIC 2019) includes GBRs. All other standard forms require amendment to incorporate GBRs.

GBRs are effectively a form of commercial EGM, a point noted in CIRIA Report C807 (CIRIA 2023). The guidance on drafting GBRs in C807 is based on the EGM concepts set out in a 2014 version of C25. The 'total geological history' points made above and the ability to clearly distinguish between conceptual and observed EGMs are critical for competent drafting of GBRs.

GBRs are able to present a skewed view of what might be considered to be a reasonable and realistic view of the ground. This allows employers to tailor the GBR to match their appetite for sharing ground risk in a construction contract. For this reason, ground data set out in GBRs should never be used to support design. GBRs are purely commercial documents. They are not technical factual or interpretative reports.

Conclusions

Within the wider construction industry engineering geology is the discipline that is most able to identify and communicate ground hazards, as it is the discipline that is most able to generate plausible conceptual EGMs based on 'total geological history'.

Unfortunately, in UK disputes, it is often found that the 'total geological history' concept has been poorly applied, despite the nature of the ground being potentially foreseeable at the very early stages of projects. The author's experiences of formal dispute processes and projects where ground-related problems arise during construction suggest that many contractual unforeseen conditions were at best hinted at by the available information and at worst were entirely foreseeable. In these cases, something clearly went wrong, and the ground conditions were mischaracterized. Such failures cost the construction and insurance industries a great deal of time and money.

In the UK, the most basic desk study data are almost instantly available in digital data packs, and this is common to almost all projects. There also are many decades of published design and construction guidance, many case studies around common problems, and Standards covering most aspects of construction in the ground. UK examples would include CIRIA reports, the ICE's *Manual of Geotechnical Engineering*, and journals, special publications and conference proceedings published by the ICE and the GSL. The problems that arise do not necessarily arise from a lack of available information.

Most projects are successful and avoid unforeseen ground issues. Whilst accepting that truly unforeseeable ground can and does sometimes exist, it is clear that many failures to recognize the foreseeable are human failures.

In the author's view, mitigations for this situation need to mostly come from within the industry and they require changes in behaviour. Some simple steps would be:

- Changes in report templates that force consideration of remaining uncertainty. This means a greater emphasis on reporting conceptual EGMs in a C25 framework. This in turn requires engagement from practitioners around training, awareness of lessons already learned and a view of 'total geological history' that goes beyond a site boundary. Changes in Eurocodes may eventually help here.
- Increasing emphasis on reporting construction risk alongside the existing emphasis on design hazards imposed by Standards.
- More focus on institutional codes of conduct and the related links to self-awareness of competency.
- Addressing the difficulties in identifying competency in engineering geology. At the time of writing, applications

to become a Chartered Geologist (CGeol) are being submitted in large numbers; most applications are from engineering geologists and contaminated land specialists. Similar growth ought to be reflected in subsequent applications to join specialist registers like RoGEP or SiLC. This does not appear to be occurring.

Other matters that are harder for the industry to mitigate include the shrinking supply of engineering geology university courses and the unwillingness of geology graduates to take them.

The perennial (and often true) complaint that clients are unwilling to pay more for geotechnical services is a factor here, but it is not necessarily a direct cause for the issues discussed. Better identification and reporting of potential ground hazards would limit the client's ability to recover costs through disputes.

This paper is a personal view based on exposure to construction-based encounters with 'unforeseen' ground. The author anticipates and welcomes debate and disagreement. It is through debate and disagreement that improvements will be made.

Acknowledgements The author acknowledges the contribution of Dr Tom Morgan for his assistance in creating the data that are set out in Appendix A. The author also thanks the reviewers for their insightful comments.

Competing interests The author declares that he has no known competing financial interests or personal relationships that could have appeared to influence the work reported in this paper.

Author contributions **JD**: conceptualization (lead), writing – original draft (lead), writing – review & editing (lead).

Funding This research received no specific grant from any funding agency in the public, commercial, or not-for-profit sectors.

Data availability Data sharing is not applicable to this article as no datasets were generated or analysed during the current study.

Appendix A: Data sources

Selected basic UK onshore desk study data sources

This is not a comprehensive list. Note that there some licensing and copyright constraints related to reproduction of images from these sources:

- BGS Maps portal: viewer for all geological maps and sections (historical and current). It now includes 1:10 000 and 1:10 560 maps where available: https://webapps.bgs.ac.uk/data/MapsPortal/
- BGS Onshore Geoindex: primary source for BGS borehole logs and many other datasets. Always also check the actual maps using the link above: https://mapapps2.bgs.ac.uk/geoindex/home.html
- NLS Historical Map Viewer: historical UK OS map viewer. There are also basic LiDAR and satellite images: https://maps.nls.uk/
- EGMS European Ground Motion Service: low-resolution pan-European InSAR coverage that provides indications of wide-area vertical ground movement: https://egms.land.copernicus.eu/
- Local Authority Planning portals: Search on the Internet for ''your site town' planning portal'.
- Coal Authority data viewer: on the 'Planning' dropdown list (top right) select 'Coal Mining Data': https://mapapps2.bgs.ac.uk/coalauthority/home.html
- UK Onshore Geophysical Library: map-based repository for onshore and nearshore well logs, seismic profiles, regional interpretations, and reports: https://ukogl.org.uk/
- 2020 national seismic hazard model for the UK: maps and literature summarizing key seismic hazard and ground motion modelling for the UK: https://link.springer.com/article/10.1007/s10518-021-01281-z
- Historic Towns Trust: contains maps, atlases and articles for major historical British towns and cities: https://www.historictownstrust.uk/
- BGS TextViewer: digitized versions of all current BGS memoirs and many other BGS publications, with searchable text: https://webapps.bgs.ac.uk/Memoirs/
- British History Online: repository of historical literature sources; can be very useful for pre-OS development history: https://www.british-history.ac.uk/
- BGS Seismic Hazard in the UK: database containing earthquake catalogues, characterization models for seismic sources and ground motion, and geospatial data: https://www.earthquakes.bgs.ac.uk/hazard/UKhazard.html
- Ordnance Survey Datum Correction: use to correct for OS datum changes on old OS maps and borehole logs: https://www.ordnancesurvey.co.uk/gps/legacy-control-information/liverpool-to-newlyn
- BRITICE Online Viewer: interactive map of glacial features in the British and Irish landscape from the last glacial maximum: https://shefuni.maps.arcgis.com/apps/webappviewer/index.html?id=fd78b03a74bb477c906c5d4e0ba9abaf

LiDAR and other map-based data sources

- England: https://environment.data.gov.uk/survey
- Scotland: https://remotesensingdata.gov.scot/
- Wales: https://nrw.maps.arcgis.com/apps/webappviewer/index.html?id=3f8007ca0b8847948ca47de0a27319e2
- GSL Engineering Geology Special Publications: https://www.lyellcollection.org/book/egsp. In particular:

Engineering Geology and Geomorphology of Glaciated and Periglaciated Terrains – Engineering Group Working Party Report. Geological Society, London, Engineering Geology Special Publications, **28**, 2017, https://doi.org/10.1144/EGSP28-000

Geological Hazards in the UK: Their Occurrence, Monitoring and Mitigation Engineering Group Working Party Report. Geological Society, London, Engineering Geology Special Publications, **29**, 2020, https://doi.org/10.1144/EGSP29-000

References

Baynes, F. 2024. *Guidelines for the Development and Application of Engineering Geological Models on Projects. Version 2.0.* International Association for Engineering Geology and the Environment (IAEG) Commission 25 Publication, **1**, https://iaeg.info/wp-content/uploads/2024/12/IAEG-C25-EGM-Guidelines-v2.0-9-August-2024.pdf

BSI 2007. *BS EN 1997-2:2007 Eurocode 7: Geotechnical Design, Part 2: Ground Investigation and Testing*. British Standards Institution (BSI), London.

BSI 2013. *BS EN 1997-1:2004 + A1:2013 Eurocode 7: Geotechnical Design Part 1*. British Standards Institution (BSI), London.

BSI 2017. *BS 10175:2011 + A2:2017 Investigation of Potentially Contaminated Sites – Code of Practice*. British Standards Institution (BSI), London.

BSI 2020*a*. *BS 5930:2015 + A1:2020 Code of Practice for Ground Investigations*. British Standards Institution (BSI), London.

BSI 2020*b*. *BS EN ISO 21365:2020 Soil Quality: Conceptual Site Models for Potentially Contaminated Sites*. British Standards Institution (BSI), London.

BSI 2024. *BS EN 1997-2:2024 Eurocode 7. Geotechnical Design – Ground Properties*. British Standards Institution (BSI), London.

CIRIA 2001. *Contaminated Land Risk Assessment. A Guide to Good Practice*. CIRIA Report C552. Construction Industry Research and Information Association (CIRIA), London.

CIRIA 2023. *Geotechnical Baseline Reports: A Good Practice Guide*. CIRIA Report C807. Construction Industry Research and Information Association (CIRIA), London.

FIDIC 2019. *FIDIC Conditions of Contract for Underground Works (2019 Emerald Book)*. International Federation of Consulting Engineers (FIDIC), Geneva, Switzerland.

Fookes, P. 2001. Total geological history: a model approach to understanding site problems. *Ground Engineering Magazine*, March, 42–47, https://cdn.ca.emap.com/wp-content/uploads/sites/13/2001/03/2001-03_Pages_42-47.pdf

Ministry of Justice 2021. *Civil Procedure Rules, Part 35, Experts and Assessors*. Ministry of Justice, London, https://www.justice.gov.uk/courts/procedure-rules/civil/rules/part35

Nyambayo, V. 2024. EC7 Next Gen: Georisk is intrinsic to reliability of structures – what aspects must we consider when determining the risk categorisation of a structure? *AGS Magazine*, September, 22–27, https://www.ags.org.uk/magazine/ags-magazine-september-2024/

UK Government 1990. *Environmental Protection Act 1990: Part 2A*. HMSO, London.

The value of biostratigraphy in site investigation and its application to the development of 'ground models' (*sensu* Peter Fookes)

Malcolm B. Hart

School of Geography, Earth & Environmental Sciences, University of Plymouth, Drake Circus, Plymouth PL4 8AA, UK
0000-0002-5649-8448
Correspondence: M.Hart@plymouth.ac.uk

Abstract: Micropalaeontology is a core component of biostratigraphy and, as a subject, is constantly in use for the solution of stratigraphical and palaeoenvironmental problems. The application to the field of engineering geology, and specifically site investigation, is probably less well appreciated despite its role in the construction of two major infrastructure projects in SE England: namely the Channel Tunnel and the Thames Barrier.

During his long career in engineering geology Peter Fookes was noted for his passion for 'ground models' of an area being investigated for development (see, for example, Fookes 1997; Fookes *et al.* 2001, 2015; Davison *et al.* 2003; Charman 2008; Griffiths 2014; Norbury 2020). He was concerned that the geology of an area was comprehensively understood, and he would often suggest investigations that many thought rather 'left field': a good example being the investigations for the site of the Thames Barrier (see later).

Biostratigraphy

'Biostratigraphy is the branch of stratigraphy that focuses on the identification and organization of strata based on their fossil content' (Lucas 2021, p. 9). There are many definitions in current use, although most are almost identical. The subject was developed in the early nineteenth century, most notably by William Smith (in the United Kingdom) and Alcide d'Orbigny (in France). Smith was more associated with the 'superposition' of strata (and their fossil content), while d'Orbigny – who was more of a palaeontologist – concentrated on zonation and the development of 'stages': see Hart (2023) for a recent resumé. William Smith is, of course, best known for his application of stratigraphy to applied geology and (rightly) lauded for that linkage. Alcide d'Orbigny was also a micropalaeontologist and, without using the term, began the sub-discipline of microbiostratigraphy which is where the fossils being applied to the geological succession fall into the category of microfossils (foraminifera, ostracods, calcareous nannofossils, diatoms, radiolarians, holothurian sclerites, silicoflagellates, calcispheres, dinoflagellate cysts, spores and pollen, etc.). The first documented use of microfossils in applied geology is thought to be a water well in Vienna (in 1877), the sediments from which were dated as Middle Miocene using foraminifera (Jones 1956). The first significant use of microfossils in hydrocarbons exploration was by Jósef Grzybowski (1869–1922), although – as documented by Czarniecki (1993) – he was not the first palaeontologist to study microfossils in what is now Poland. In the years between 1832 and 1843, Wojciech Zbożewski (a secondary school teacher from Krzemieniec) described fossil foraminifera from the Wołyn region. Józef Bolesław Grzybowski was born in Kraków on the 17 March 1869 and grew up in difficult circumstances. He entered the Jagiellonian University (Kraków) in 1887 studying Law, graduating in 1891. In 1886 the university had appointed the first Chair in Geology (Władysław Szajnocha) who had studied at the University of Vienna. The young Grzybowski was appointed as his first geological assistant. In 1894 he completed his formal training in Geological Sciences and, in the same year, published his first micropalaeontological monograph. Czarniecki (1993) carefully describes Grzybowski's training and development including the publication of 60 papers on microfossils between 1894 and 1921. In 1897, Gryzbowski published a paper on 'Microscopical investigation of borehole muds of oil field. L. The Potok-Krosno Belt. Il General Remarks'. As Grzybowski explained:

> For a given geological stage, would it not be possible to use foraminifera to distinguish individual layers i.e. in an individual well or in the field, and in so doing provide more precise information for the petroleum industry? (Czarniecki 1993, p. 5).

In his work, Grzybowski described 1000 sediment samples from 40 wells; the description of, the mainly agglutinated

From: Griffiths, J. S., Giles, D. P., Davis, J. A. and Dunlop, J. J. (eds) 2025. *The Peter Fookes Engineering Geological Legacy in Geomodels, Geomaterials and Geomorphology*. Geological Society, London, Engineering Geology Special Publications, **30**, 123–142,
First published online 21 March 2025, https://doi.org/10.1144/egsp30-2024-3

foraminifera, in ~4000 samples. Czarniecki (1993) and Kaminski *et al.* (1993) go on to describe Grzybowski's life-long contribution to micropalaeontology and petroleum geology.

That is, essentially, the beginning of a long tradition of microfossils being used in hydrocarbon exploration which was enhanced by the establishment of the Cushman Laboratory for Foraminiferal Research in Sharon (Massachusetts) in 1923 by Joseph Augustine Cushman (Huber *et al.* 2013). While Cushman may have heard of Gryzbowski, it is unlikely that he ever read many of his works as they were written in an old, rather archaic, style of nineteenth-century Polish. Many regard this application to the hydrocarbons industry as the ***only*** industrial application of the subject. That view is incorrect, as there is a ~65-year-history of micropalaeontology being applied to engineering geology.

Peter Fookes and biostratigraphy

In the 1960s, the engineering geologists were located on the 2nd floor (east wing) of the Royal School of Mines building in Imperial College (London). At that time, palaeontology and stratigraphy occupied most of the north side, which included a small museum and display area. Adjacent to the central staircase, and opposite the entrance to the lift, there were three small rooms dedicated to micropalaeontology. One was a laboratory, one was a 'kitchen' for processing samples, while the other room housed postgraduate students with, in a small office, the head of the group, David James Carter DFC (Fig. 1). Behind David, in a wall of metal cabinets, were processed samples from the cores drilled during the Channel Tunnel Site Investigation. This work, which was related to the construction of the Channel Tunnel, remains one of the most comprehensive – and successful – micropalaeontological investigations in support of engineering geology anywhere in the world (Harris *et al.* 1996).

Following the Second World War, oil and gas resources were deemed a priority for many governments and the resulting exploration created a surge of interest in micropalaeontology including, in the United Kingdom, the stratigraphy of the Chalk Group (Hart and Bailey 2013). A landmark publication was that of Williams-Mitchell (1948) who documented the foraminifera of the chalk recovered in the Portsdown Borehole, which had been drilled by the D'Arcy Exploration Company Ltd in 1936. Williams-Mitchell did not record the number of samples examined in this investigation of the 1070 ft (326 m) of chalk recovered in the core. The ranges of the foraminifera were recorded in terms of the macrofossil stratigraphy. We do know that the correlation with the macrofaunal zonation was supported by the collection, and investigation, of surface samples from Norfolk, Lincolnshire, Surrey, Sussex, Hampshire and Devon. In his later description of the Dean Hill Anticline in Wiltshire, Williams-Mitchell (1956, p. 222) states that 'foraminifera, though not specifically studied, proved helpful in indicating horizons at localities where macrofossils were not found'. This, presumably, used the data from his earlier paper in which all the foraminifera had been recorded in terms of the macrofossil zonation.

At this time, there was a vibrant research ethos in the micropalaeontologists housed in the laboratories on the 2nd floor of the Royal School of Mines building in Imperial College. This included Martin Norvick (Paleogene of the Hampshire Basin), Graham Williams (London Clay Formation) and MBH (Gault Clay Formation and the chalk of southern England). At the time there were also students on an MSc programme, many of whom went on to other research positions or industry. All these projects focused on microbiostratigraphy as well as other related subjects such as palaeogeography and palaeoecology. It was with this group that Peter Fookes began a dialogue that would last for many years and lead to a number of projects that worked on the interface between micropalaeontology and engineering geology. Many of these began the 'dialogue' around the idea of ground models and how information from one discipline could help the other. It is important, however, to acknowledge that there are three levels of involvement:

Fig. 1. David James Carter DFC (1922–2013) at home in Hawkhurst (Kent) in November 2012. Source: image provided by the late Prof. John Murray.

(1) projects where the microfossil information was used to develop, or enhance, the information about an area and its stratigraphical setting contribute to the 'ground model' development;
(2) investigations of microfossils within sediments and how their presence may change the physical (and engineering) properties of the lithological units that may be encountered on a site (Tanaka and Locat 1999; Weimer and Kopf 2016); and
(3) projects that went well beyond this level of involvement and where the micropalaeontological expertise was used during the construction phase of a project so as to ensure the most effective route to construction.

Perhaps the best examples of (3) are in the process of directional drilling in the oil industry (see Bailey *et al.* 2009) and the construction of the Channel Tunnel (see below).

The micropalaeontology of the Channel Tunnel

This preliminary work on the biostratigraphy of the chalk using foraminifera (see Williams-Mitchell 1948) just predated the establishment of the Channel Tunnel Study Group (Bruckshaw *et al.* 1961). The post-war political situation in Europe, and the formation of the European Economic Community (EEC) in 1957, re-kindled an interest in building the tunnel, a project first proposed in 1751 by Nicholas Desmaret. In 1802, Albert Mathieu-Favier had suggested that a tunnel be built between England and France although, in 1803, Tessier du Mottray suggested that a submerged tube might be a better alternative to a tunnel. The Napoleonic Wars and the worsening political situation in Europe terminated any talk of such a major construction but, in 1855, the idea was resurrected by Aimé Thomé de Gamond, who began a full geological survey of the area, including the collection of sea-bed samples. De Gamond was certainly dedicated to the task, weighting his belt with stones and sinking rapidly to the sea floor where he collected a sample, released the belt and floated back to his boat. His application to this survey was truly 'heroic'. The building of a tunnel began in 1878 with a French Company beginning excavations near Sangatte. In 1880, Sir Edward Watkin sank the first shaft at Abbot's Cliff, near Dover, and tunnelled ~800 m. In 1881 he floated the Submarine Continental Railway Company, and another pilot tunnel (of some 800 m) was constructed below Shakespeare Cliff. While some of the French construction methods were rather dangerous (explosives followed by hand-digging), Sir Edward turned to more advanced technology and a compressed-air tunnelling machine designed by Colonel Frederick Beaumont. This attempt was abandoned in 1883, having drilled about 2100 yards (2000 m) of tunnel in the clay-rich 'Chalk Marl' of the Lower to Middle Cenomanian chalk. In the 1950s the Channel Tunnel Study Group was formed and this first investigation for the post-war railway tunnel involved micropalaeontology and David J. Carter (of Imperial College, London) was given the task of generating a detailed zonation of the lower part of the Chalk Group and Gault Clay Formation for the region between Dover/Folkestone and Sangatte/Calais. This was both the shortest distance across the English Channel and, fortuitously, coincided with an outcrop of the clay-rich Lower to Middle Cenomanian which had been excavated during the 1881–83 construction. The award of this 'contract' to David Carter created a rift with Tom Barnard (then at University College, London, and also an expert on chalk foraminifera); see Hart and Bailey (2013).

David Carter, with his experience in industry, used commercial rather than academic methodologies and rapidly produced a workable zonation (Hart and Bailey 2013) that was published in Bruckshaw *et al.* (1961): see Figure 2. The Channel Tunnel Study Group established a full-scale site investigation of the English Channel using 100+ boreholes drilled along the projected route of the tunnel. Thousands of chalk samples were collected from these cores and investigated for clay mineralogy, calcimetry and foraminifera (Carter 1966). The micropalaeontological work was undertaken by David Carter in a laboratory established for this purpose in Dover Castle, with students from Imperial College (including MBH) helping during vacations and at weekends. This work was done in collaboration with BRGM, the French Geological Survey, based in Orléans. The research led to the production of a series of massive reports, but these were never published as they were only to be used in the construction of the two rail tunnels and the service tunnel between Folkestone and Sangatte. Work on the construction of the tunnel began in 1973 but was terminated shortly afterwards, beginning again in the 1980s. At that time, Colin Harris (ex-Plymouth PhD student with industrial experience in the Far East) undertook site-based micropalaeontological work in Dover aided, occasionally, by MBH. The successful completion of the tunnel allowed publication of the stratigraphy, micropalaeontology and engineering work (Harris *et al.* 1996). The similarity of the foraminiferal zonation used in the construction phase (Harris *et al.* 1996, fig. 26.2) is striking, even though many of the microfossil names have been changed over time.

It was recognized, quite early in the investigations, that the boundary between the Gault Clay Formation, 'Glauconitic Marl' and the Chalk Group would be important and David Carter, with MBH, investigated the area of East Wear Bay (Folkestone) for the complexities (and misunderstandings) of this part of the Cretaceous succession (Fig. 3). The history of these excavations has only recently been described by Hart and Fox (2020) who have explained the out-of-place parts of the succession, and the problems caused by the misinterpretation of 'Bed XII' of the Gault Clay Formation (De Rance 1868; Price 1874*a*, *b*, 1876, 1879; Chapman 1891–98; Jukes-Browne and Hill 1900). The changes in the foraminifera within this part of the succession are quite subtle (Figs 4, 5 & 6), but this research in the 1960s proved important as the design for the eventual tunnel progressed.

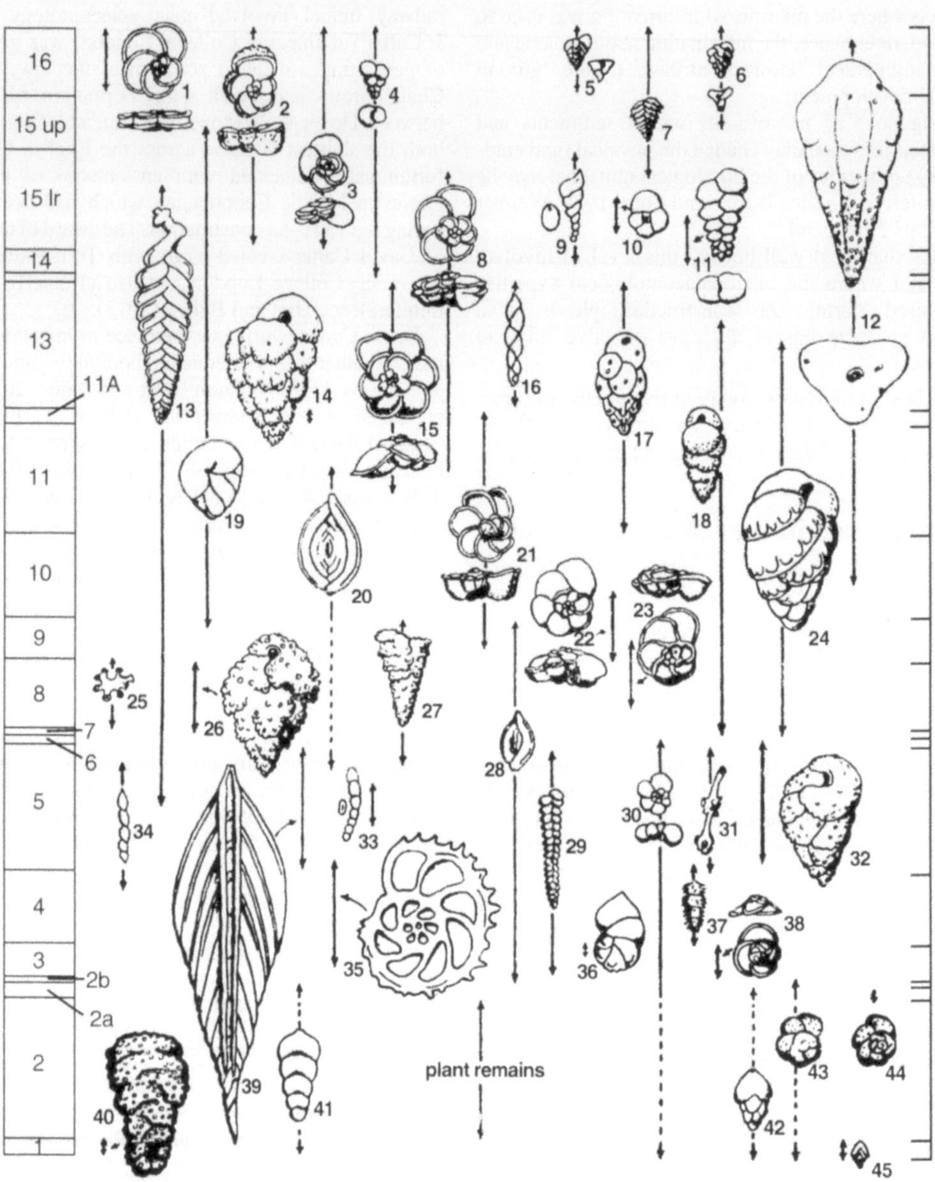

Fig. 2. The foraminiferal zonation of the Albian–Turonian, which was generated by D.J. Carter as part of the preliminary investigations for the Channel Tunnel Site Investigation. Zones 3–6 represent the Gault Clay Formation. Some key taxa are identified with both the original names (applied by D.J. Carter), updated where possible, but others remain indeterminate. (1) *Marginotruncana pseudolinneiana*; (2) *Helvetoglobotruncana helvetica*; (3) *Dicarinella imbricata*; (4) *Heterohelix globulosa*; (8) *Dicarinella hagni*; (10) *Lingulogavelinella globosa* (*Biorostella* sp., C.T.P.); (12) *Tritaxia pyramidata* and *T. macfadyeni*; (13) *Spiroplectinata annectens*; (14) *Flourensina mariae* (*Flourensina crenata* C.T.P.); (15) *Rotalipora cushmani*; (16) *Pleurostomella* sp. 26 C.T.P.; (17) *Plectina cenomana*; (18) *Plectina mariae*; (19) *Pernerina* sp. C.T.P., now *Arenobulimina depressa*; (20) *Spiroloculina papyracea*; (21) *Thalmanninella reicheli* and *T. greenhornensis*; (22) *Praeglobotruncana* sp. cf. *P. delrioensis*; (23) *Thalmanninella globotruncanoides*; (24) *Arenobulimina advena*; (25) spicules of hexactinellid sponges; (26) *Flourensina intermedia*; (27) *Marssonella ozawai*; (28) *Quinqueloculina antiqua*; (29) *Dorothia filiformis*; (30) *Muricohedbergella* spp.; (31) *Nodobacularia nodulosa*; (32) *Arenobulimina chapmani*; (33) *Dentalina* sp. 38 C.T.P.; (35) *Epistomina spinulifera*; (37) *Sagrina asperula*; (38) *Conorboides lamplughi*; (39) *Citharinella pinnaeformis*; (40) *Ammobaculoides* sp. C.T.P.; (43) *Hoeglundina chapmani*; (44) *Hoeglundina carpenteri*. Source: initially published in Bruckshaw *et al.* (1961) and is reproduced with the permission of the Institution of Civil Engineers.

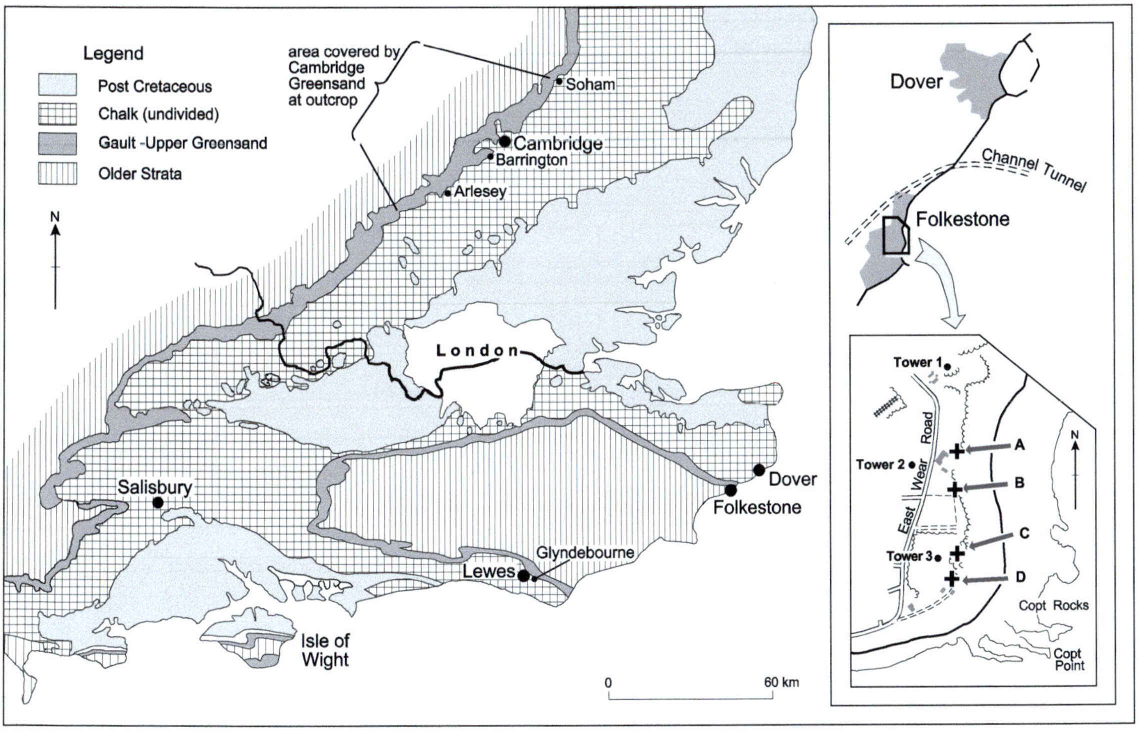

Fig. 3. Locality map for the Albian–Cenomanian boundary successions in southeastern England. The location of the M11 near Barrington (Cambridge) is indicated. The inset map shows the location of the Copt Point (Folkestone) succession, Martello Towers 1–3 and the excavated sections A–D, investigated by David Carter and Malcolm Hart. Source: initially published in Hart (1993, fig. 7.13) and updated (with permission of Springer Nature, Heidelberg) in Hart and Fox (2020), and reproduced with permission of the Geological Society of London.

At two locations in mid-Channel the 'running tunnels' had to come together and a 'crossover' created where trains could be switched from one tunnel to the other for maintenance or in an emergency. That required the safety tunnel to be relocated as shown in Figure 7. All of these changes in the tunnel alignment meant crossing the Albian/Cenomanian boundary (several times) and some additional cores were drilled (Fig. 8) in the vicinity of the UK crossover. The fieldwork done in the 1960s, and the micropalaeontological results, proved invaluable in the understanding of the crossover stratigraphy, where an additional cycle of sedimentation was detected between the Gault Clay Formation and the Glauconitic Marl at the base of the chalk succession. This detailed work on one small part of the larger project is a classic case of problem-solving within the context of chalk correlation across the complete route of the tunnel using microfossils.

This project demonstrated the significance of micropalaeontology in site investigation, especially in a lithology – such as the chalk – that can appear almost uniform when extracted from a drill core. This work clearly demonstrated that the micropalaeontologists had something to offer engineering geologists and a strong bond developed between the two extremities of the 2nd floor in the RSM building of Imperial College. Over time this turned into a 'coffee club' and the various parties (DJC, MBH, along with Graham Williams, Martin Norvick and others) would meet with Peter Fookes, John Knill (occasionally), Bruce Denness, Mike Saunders, Al Burnett, and others, to discuss engineering projects that required the stratigraphical knowledge and – in some cases – the expertise of the micropalaeontologists.

Landslides

The landslides determined by David Carter and MBH in East Wear Bay and described subsequently by Hart and Fox (2020), are only a part of the major landslide complex of Folkestone Warren. This landslide, which became known as 'The Great Fall', occurred in December 1915 and, after the initial train derailments, the tracks moved *c.* 50 m towards the sea with an estimated 1.5 million cubic metres of chalk falling into the sea following a period of heavy rainfall. The Folkestone–Dover railway line remained closed until 11 August 1919. There had been instability on this railway line since its opening in 1844, and it remains unstable even

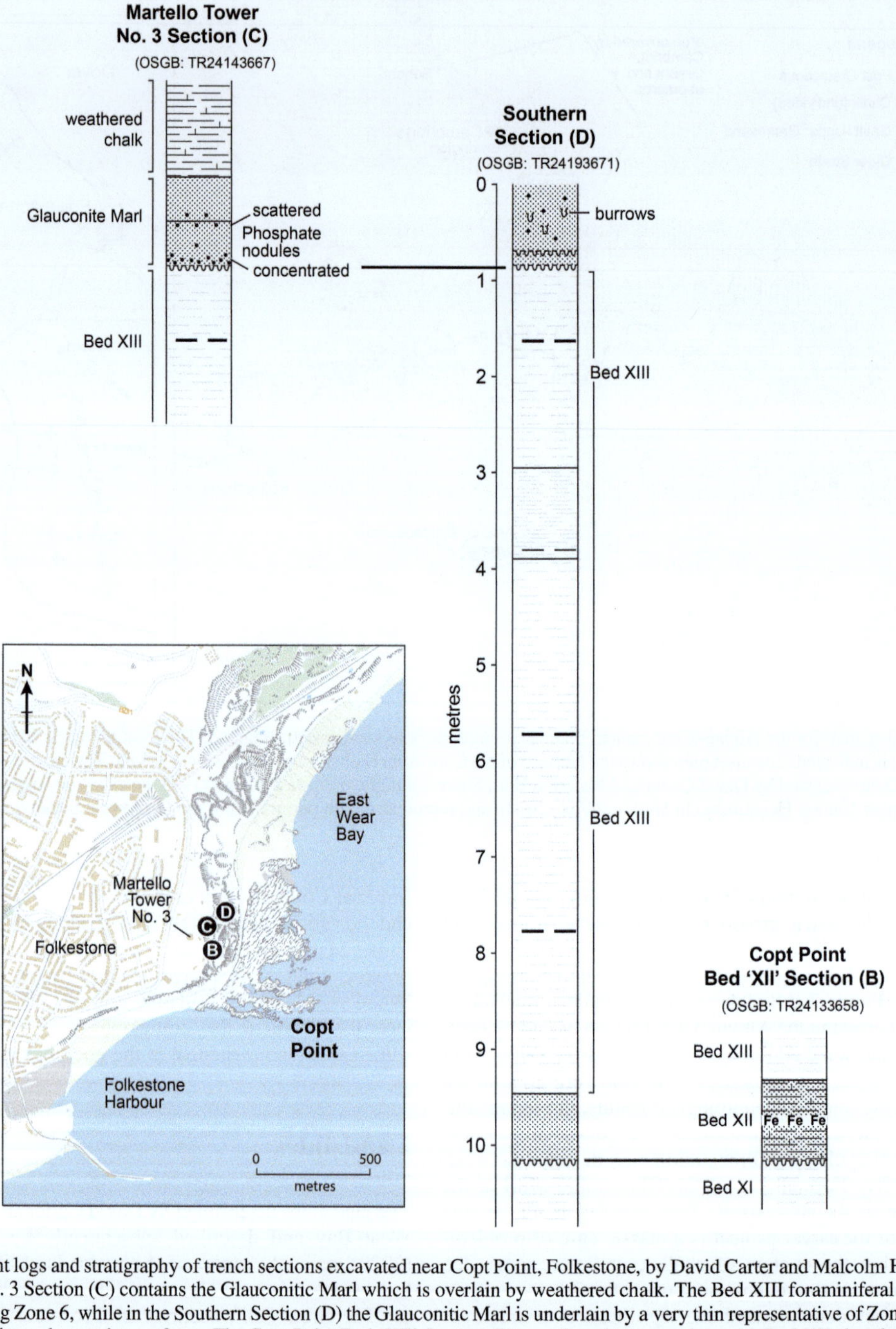

Fig. 4. Sediment logs and stratigraphy of trench sections excavated near Copt Point, Folkestone, by David Carter and Malcolm Hart. The Martello Tower No. 3 Section (C) contains the Glauconitic Marl which is overlain by weathered chalk. The Bed XIII foraminiferal assemblage is that representing Zone 6, while in the Southern Section (D) the Glauconitic Marl is underlain by a very thin representative of Zone 6a, bounded above – and below – by erosion surfaces. The Copt Point Bed 'XII' Section (B) was thought to correlate with Bed XII elsewhere in the Copt Point area, but it is now known that the Bed XI, XII and Bed XIII attributions are incorrect and that this represents a slipped block of the Glauconitic Marl. It was the suggestion that this was Bed XII (on macrofaunal evidence) that caused the incorrect representation of the succession reported in Hart (1973*a*, *b*). The log should be moved up the figure and the correlation line attached to the base of the Glauconitic Marl, with the Bed XI/XII/XIII attributions changed. The complex stratigraphy of the East Wear Bay successions is described by De Rance (1868), Price (1874*a*, *b*, 1876, 1879), Chapman (1891–1898), Jukes-Browne and Hill (1900) and Hart and Fox (2020). Source: reproduced from Hart and Fox (2020), with permission of the Geological Society of London.

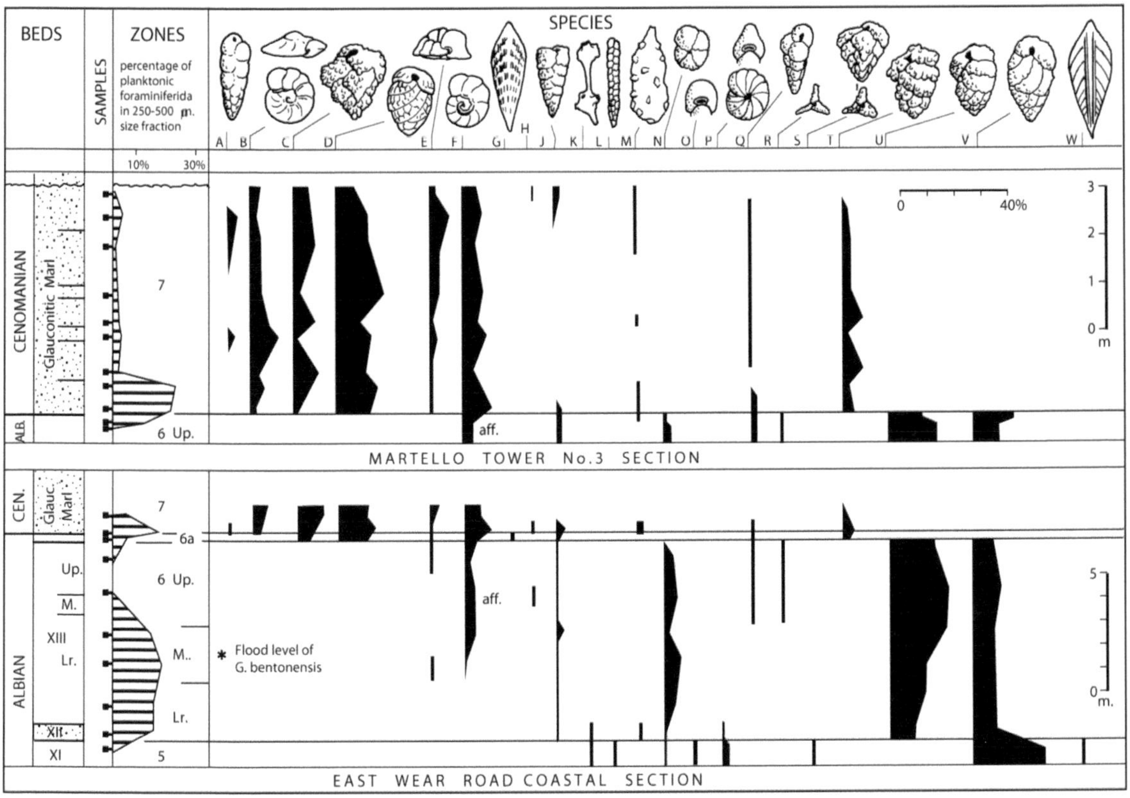

Fig. 5. Changes in foraminiferal assemblage across the Albian/Cenomanian boundary and the distribution of some key species of in the successions near Copt Point (see Fig. 3). A, *Plectina mariae* (Franke); B, *Berthelina* sp. cf. *G. berthelini* (Keller); C, *Flourensina intermedia* Ten Dam; D, *Arenobulimina advena* (Cushman); E, *Lingulogavelinella jarzevae* (Vasilenko); F, *Berthelina cenomanica* (Brotzen); G, *Citharinella laffitei* Marie; H, *Cribratina cylindracea* (Chapman); J, *Marssonela ozawai* Cushman; K, *Nodobacularia nodosa* (Chapman); L, *Dorothia filiformis* (Berthelin); M, '*Haplostiche sherborniana* Chapman'; N, *Haplophragmoides chapmani* Morozova; O, *Cribrostomoides nonioninoides* (Reuss) forma *rotunda* Magniez-Jannin; P, *Cribrostomoides nonioninoides* (Reuss) forma *angulosa* Magniez-Jannin; Q, *Arenobulimina frankei* Cushman; R, S, *Tritaxia* sp. cf. *T. singularis* Magniez-Jannin (including two forms with straight (R) and convex (S) sides); T, *Arenobulimina sabulosa* (Chapman), variety showing crenulated chamber margins; U, *Arenobulimina sabulosa* (Chapman), typical form; V, *Arenobulimina chapmani* Cushman; W, *Citharinella pinnaeformis* (Chapman). Source: from Hart and Fox (2020) courtesy of the Geological Society of London.

today with speed restrictions occasionally imposed at times of significant rainfall. A new sea wall was opened in September 2016 in the latest attempt to stabilize the railway. There are numerous historical and geological accounts of the landslide available in, for example, Toms (1946), Muir Wood (1955), Viner-Brady (1955), Hutchinson (1969), Hutchinson *et al.* (1980) and Warren *et al.* (2000). Sadly – but predictably – the Folkestone Warren landslide remains active and further investigations (augmented by a drilling campaign) are on-going (in 2024).

In the 1960s there was also a growing interest in the landslides (in Jurassic strata) of the Dorset Coast with Brunsden *et al.* (1955), Brunsden and Jones (1972, 1976), Chandler and Brunsden (1995) and Brunsden and Chandler (1996) suggesting that there was a probable link between rainfall totals, stratigraphy and landslide activity (Fig. 9a, b). After the work in East Wear Bay, and the continuing interest in Folkestone Warren, Peter Fookes began to wonder if micropalaeontology could identify the stratigraphical location of landslide 'surfaces', with Bruce Denness and MBH being despatched to Folkestone and a series of quarries in SE England to look at instability within the Gault Clay Formation. As an Isle of Wight 'native', Bruce Denness also wanted to study the landslide problems on the island and, eventually, the island became a large part of his PhD thesis (Denness 1969). At the time the most active landslides were between St Catherine's lighthouse (near Niton) and the tourist destination of Blackgang Chine. The geology coincided exactly with the stratigraphical research of MBH and the work was done with some enthusiasm, spurred on by Peter Fookes who –

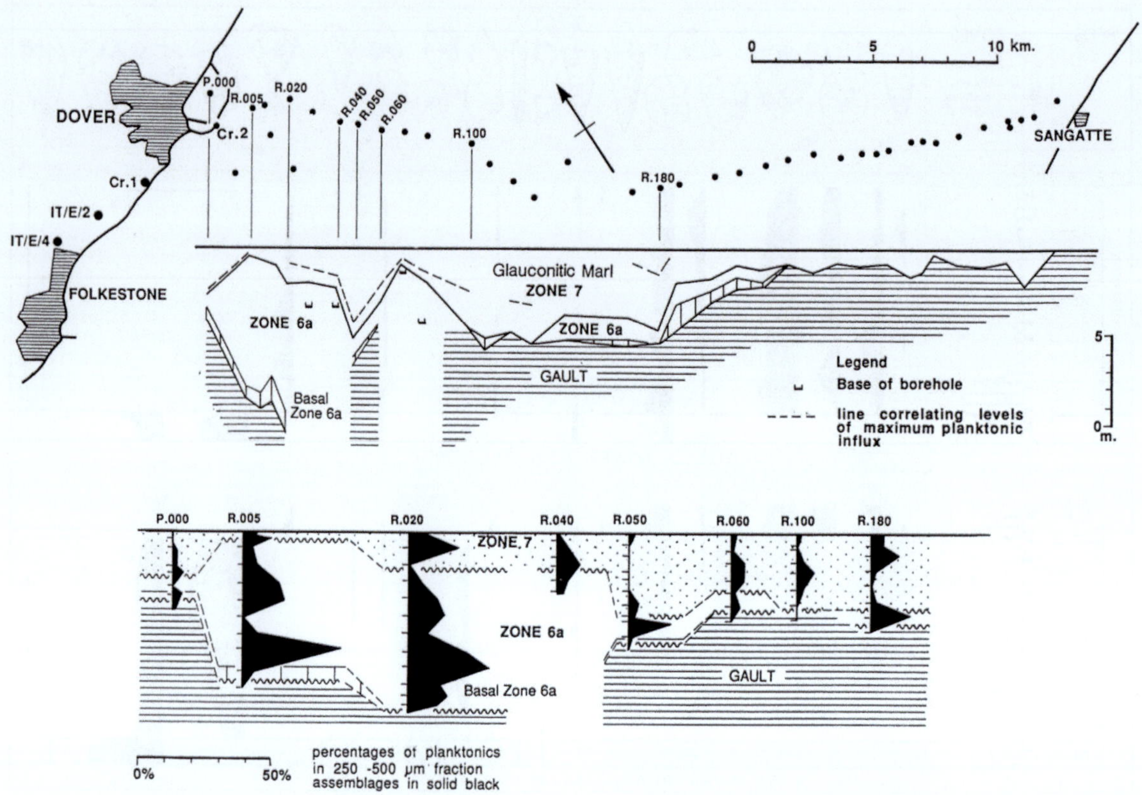

Fig. 6. Correlation of the Albian/Cenomanian boundary along the route of the Channel Tunnel. The top of the 'Glauconitic Marl' has been used as a datum simply as a convenient line across the profile. The marine boreholes were initially numbered with an 'R' prefix, while onshore boreholes were given different numbers. The lower part of the diagram shows how levels of abundance of *Hedbergella* spp. can be used in the correlation of Zone 6a and Zone 7. It should be noted that Zone 6a was not identified in the onshore successions and was only detected in the analysis of the offshore boreholes. Source: based on Hart (1993, fig. 7.6) and reproduced with permission of Springer Nature (Heidelberg).

at the time – was clearly developing his interest in 'ground models' as a way of understanding site investigation work.

As was discovered quite early in this work, the Gault Clay Formation on the Isle of Wight is much more silt-rich than the stiff, over-consolidated mudstones in Kent and much of East Sussex. This made the mudstones significantly more permeable and some of the marker foraminifera (mainly the epistominids), being constructed of aragonite, were lost during burial, compaction and modern weathering. This thwarted attempts at detailed correlation with, for example, Folkestone and the Channel Tunnel. Nevertheless, it was possible to construct a detailed map and cross-sections of the landslide (Hart 1993, fig. 7.17) using lithological features alone. Figure 10 (upper) shows the cliff at the landward edge of the landslipped area, as well as the tell-tale signs of tilted blocks showing the levels on which there had been rotation of relatively intact blocks of the Cretaceous succession. While the St Catherines to Blackgang landslide is still active, attention has moved across to the area inland of Ventnor (Moore *et al.* 2010; Jenkins *et al.* 2011; Carey *et al.* 2015; Moore 2020) and the area near Bonchurch, which is often known as the 'Landslip' in tourist literature. This area also remains active (Palmer *et al.* 2007) and could be investigated using microfossils in order to determine the horizons on which the Gault Clay is sliding seawards. These potential dislocation surfaces are recorded as sudden changes in plastic and liquid limits (Palmer *et al.* 2007, fig. 7) within what is described as the 'middle layer' within the recorded succession of the Gault Clay Formation.

The Lower Jurassic succession on the Dorset Coast also contains foraminifera, but the assemblage is quite limited (Clémence and Hart 2013) and, in any case the detailed description of the lithological markers by Lang and Spath (1926, and references therein) seemed adequate: see Cope (2016), for a summary. This means that, in areas such as Black Ven (between Lyme Regis and Charmouth) and on Stonebarrow (east of Charmouth) there is little need for micropalaeontological investigations to construct a viable stratigraphy. In the 1970s there was considerable interest in these West Dorset landslides with Bruce Denness (then

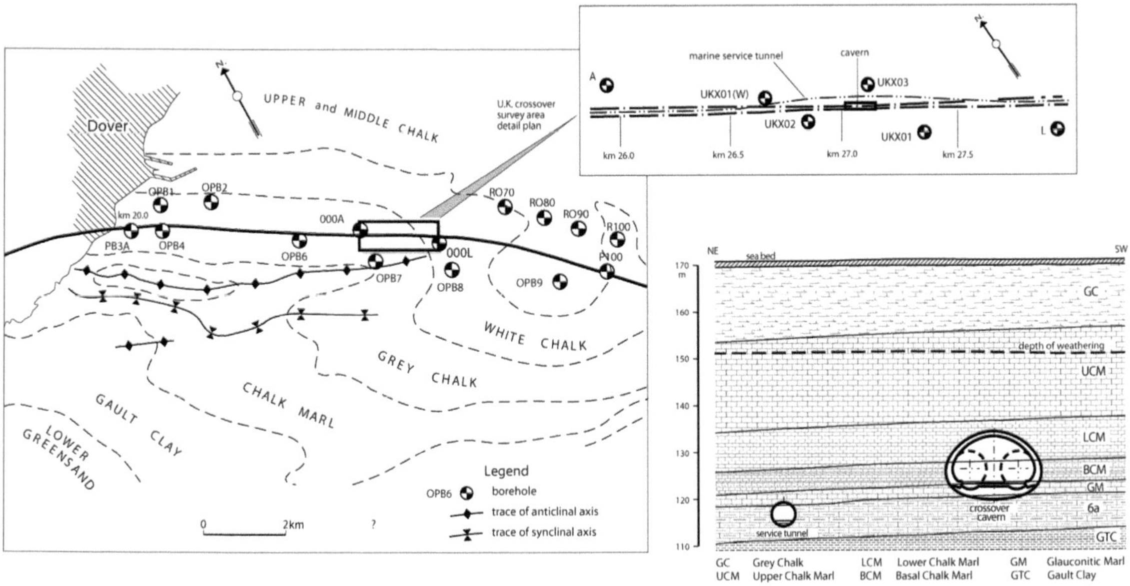

Fig. 7. The UK 'crossover' location to the east of Dover, which is approximately one-third of the way across the English Channel. The two running tunnels are located in the 'crossover cavern' while the Service Tunnel is displaced to the NE. It is located in Zone 6a of the micropalaeontological zonation, immediately below the Glauconitic Marl and just above the Gault Clay Formation. The 'R' and 'P' boreholes were part of the 1960s investigation while other borehole sites were drilled in advance of the construction phase in order to confirm the details of the stratigraphy. Source: diagram produced by John Abraham, former geology cartographer, University of Plymouth.

based at the Institute of Geological Sciences) applying his knowledge of landslide mechanisms in the Cretaceous to those of the Jurassic (Denness 1972; Conway 1974; Denness *et al.* 1975). Landslides still dominate geological work on the Dorset Coast with the area around Lyme Regis, and the stabilization works in recent years being described by Gallois and Davis (2001), and the area of the Great Bindon Landslide being documented by Edmonds (2022) using modern techniques such as LiDAR imagery and drones for aerial photography.

Thames Barrier Site Investigation

In the hiatus between submission of the major Channel Tunnel results and construction beginning in 1973, another project arose as a direct result of a series of major reports (Waverley 1953; Bondi 1967) on the potential flooding of London. In 1953 much of Essex, and parts of the Greater London area, were inundated by a major flooding event caused by a storm surge between 29 January and 1 February 1953 (Heaps 1977; Hart 2000) in which 307 lives were lost and *c.* 24 000 homes destroyed. After the event there were a number of major investigations on why this flooding occurred and how another event of the same magnitude could be avoided. At the time of these deliberations post-glacial isostatic re-adjustments were taken into consideration although modern concerns over climate change and sea-level rise, as a consequence of global warming, were not on the agenda in the 1970s.

The Site Investigation was conducted between December 1971 and May 1972 by Foundation Engineering Limited under the guidance and direction of Rendel Palmer and Tritton Ltd, the Consulting Engineers to the Department of Public Health Engineers of the Greater London Council. Foundation Engineering were advised by the Consulting Engineers – prompted by Peter Fookes – that the site investigation should include a micropalaeontological study of the underlying bedrock, which was largely chalk, overlain by a thin succession of Thanet Sand (Paleocene) and alluvium. David Carter was engaged in the study, aided at times by MBH, who both visited the site to examine selected cores, extract the samples required for the micropalaeontological work and subsequently assist with the microfossil determinations. The final report was submitted within the overall report from Foundation Engineering Ltd that went to the Consulting Engineers and the Greater London Council in May 1972. A paper derived from the micropalaeontological work was presented at the Regional Meeting of the Engineering Group of the Geological Society held in Exeter University 13–16 September 1976 (Anonymous 1977) and published in the conference volume by Carter and Hart (1977).

Peter Fookes visited the site with David Carter and MBH, and on a number of other occasions. He suggested that, while

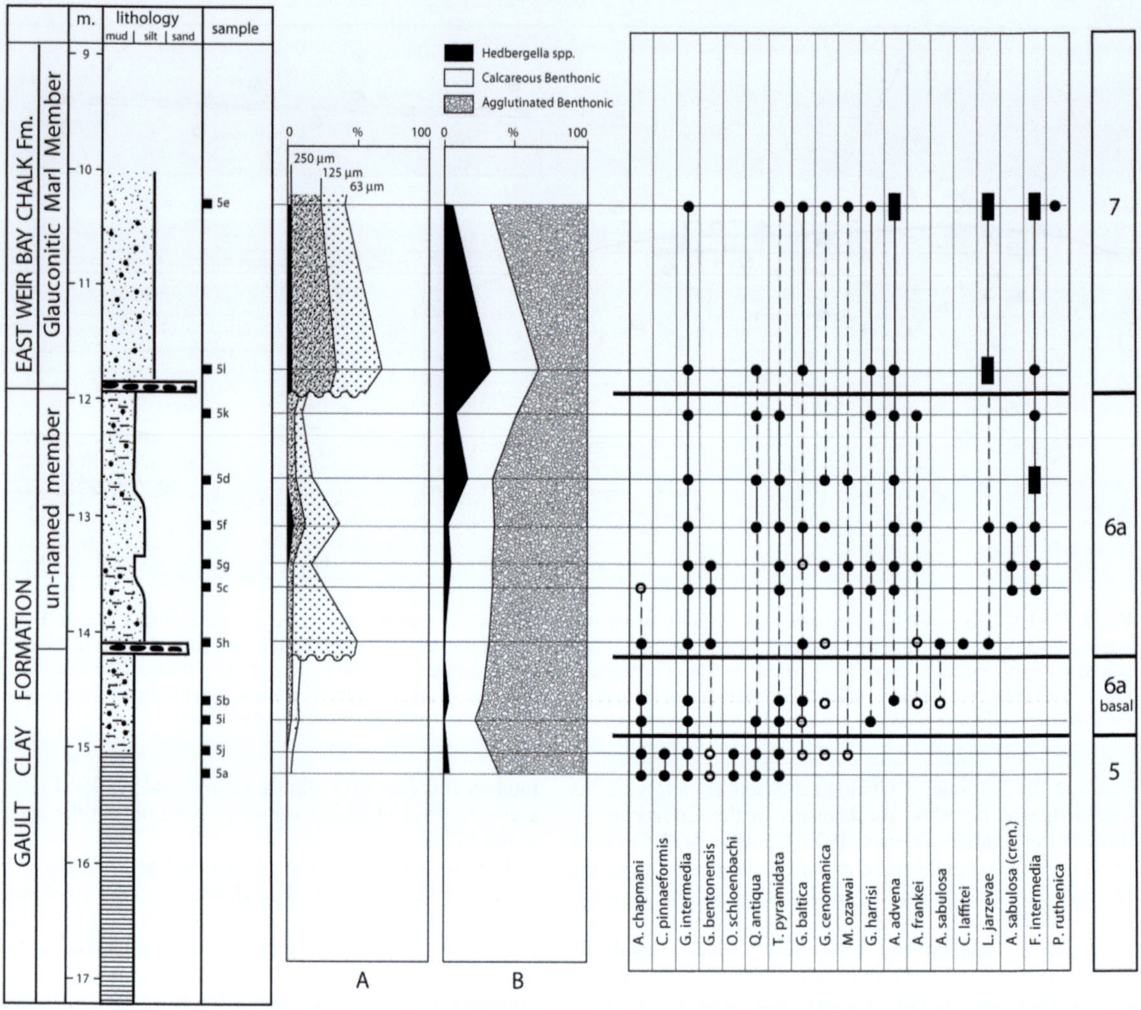

Fig. 8. One of the additional boreholes associated with the UK 'crossover'. In this example, Zone 6a is identified, as is a basal Zone 6a unit. This basal unit was quite locally developed, especially in the area of the 'crossover'. Zone 6a is un-named, lying between the Glauconitic Marl and the Gault Clay Formation *sensu stricto*. Only the stratigraphically diagnostic foraminifera are plotted as the total assemblage at this level can total *c.* 100 species. Source: unpublished figure drawn by John Abraham, former geology cartographer, University of Plymouth.

we were fully engaged with the micropalaeontological research, we should be circumspect in what we told the engineers lest they consider our work as being too research oriented. For example, although we identified all the species very precisely (Carter and Hart 1977, fig. 3), we did not openly discuss the differences between, for example, *Stensiöina exsculpta exsculpta* and *Stensiöina exsculpta gracilis* despite having to make use of these taxa in developing the microbiostratigraphy of the site. Instead, we 'hid' this level of detail by using the red, blue, orange, violet and yellow zones and that is how they appeared in Carter and Hart (1977, fig. 4). Further work on the samples from the site investigation was included in the PhD research of Haydon Bailey (Bailey 1978) as there was a considerable amount of microbiostratigraphy to do following the tight schedules of the site investigation.

The study showed the presence of several small north–south faults crossing the River Thames in the vicinity of the planned barrier, as well as some very small anticlines and synclines (Fig. 11). These were regarded as inconsequential to the construction of the barrier and, as we know, the Thames Barrier still stands, operates effectively, and has protected London from flooding as it was designed so to do (Fig. 12). The fact that it is nearing the end of its design life (or even past that date) at a time of climate change and rising sea levels should be a concern for all those with a responsibility for the

(a)

(b)

Fig. 9. Black Ven landslide from the sea (**a**) and from the cliff-top (**b**). Source: images courtesy of Richard Edmonds.

London area (Fig. 13). The climate situation is now much different to that which existed in the 1970s when the concern was the possible onset of the next ice age. With the development of an understanding of stable isotopes and the Marine Isotope Stages (Shackleton 1969, 2000, 2006; Lisiecki and Raymo 2007; Railsback *et al.* 2015) in the 1970s (and subsequently) we now know the glaciation chronology of the Pleistocene and, in any case, that has been completely overtaken by modern climate and sea-level change concerns (Gehrels 2009).

(a)

(b)

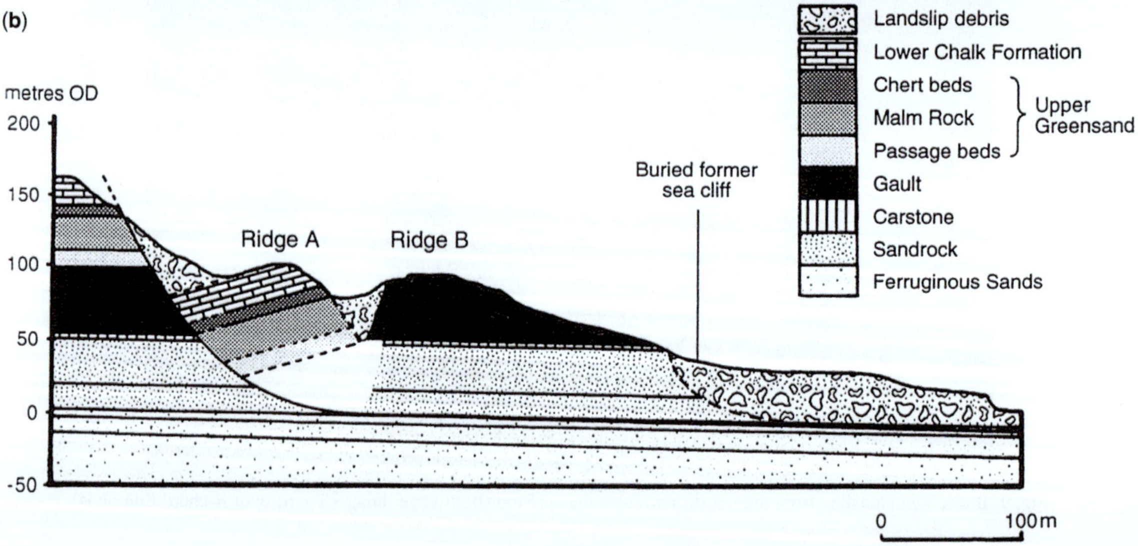

Fig. 10. Niton to Blackgang Chine area of the Isle of Wight, showing the vertical cliff at the landward side of the landslide and the rotated blocks of chalk and 'Chert Beds' showing the typical dip into the cliff, which demonstrates the rotation. Highly weathered Gault Clay Formation forms the foreground. The Gault Clay Formation sits above a Lower Greensand succession of 'Carstone', 'Sandrock' and 'Ferruginous Sands' which forms the sea cliffs, thereby stopping the accumulation of a 'toe' that might help stabilize the landslide (temporarily). Source: image courtesy of the author and the cross-section was drawn by John Abraham (former geology cartographer, University of Plymouth).

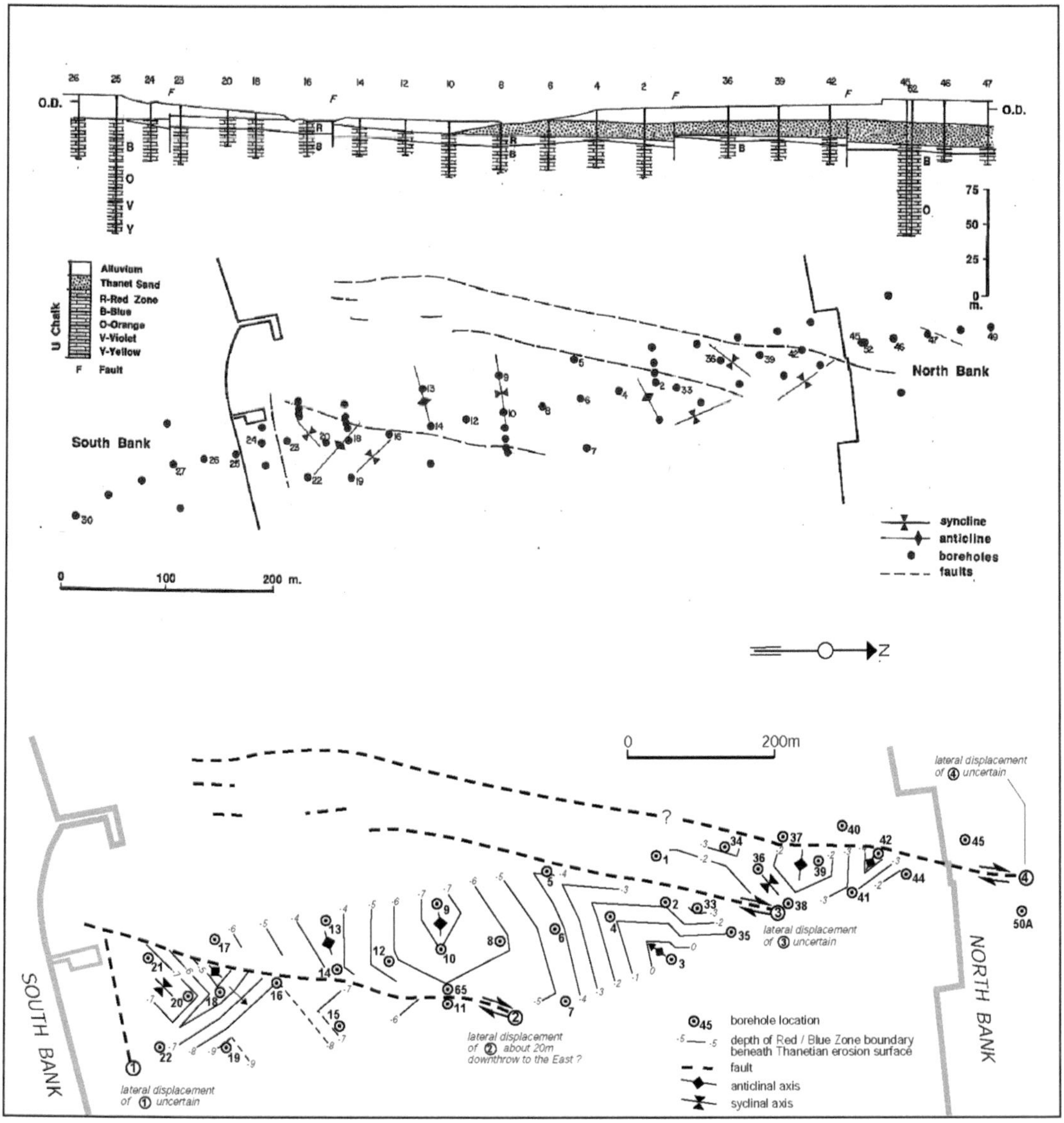

Fig. 11. Location of the boreholes used in the site investigation for the Thames Barrier. Boreholes 25 (south) and 52 (north) were located on the banks of the river, while most of the other boreholes were collected from a barge. The line of the barrier is seen in the line of boreholes crossing the river, whilst the piers are marked with lines of boreholes at 90° to the cross-river sites. The 'colour' zones are shown as are, in the lower part of the diagram, the location of the faults and very minor anticlines/synclines detected from the micropalaeontological analysis. It must be remembered that these are very small amplitude features which is why they posed no threat to the structure of the barrier. Source: modified from Carter and Hart (1977) courtesy of the Geological Society of London.

At the Regional Group meeting in 1976, David Carter – who presented the paper – was asked by one of those present about the costs of the micropalaeontological study relative to the overall project. Peter Fookes, with his customary aplomb answered, on behalf of the authors, that the microfossil work cost less than 1% of the site investigation budget and, that while the work had not really shown anything of significance that would impact the construction project, the Consulting Engineers knew that they could sleep peacefully at night knowing that the stratigraphers had not described anything that should be a problem. This approach was very much part of Peter Fookes' thought processes in the development

Fig. 12. Thames Barrier with the distinctive appearance of the piers in the river. Source: image courtesy of the Environment Agency.

of his view of 'ground models': getting confirmation of the stratigraphy and the nature of the succession involved in an area under investigation.

Some other site investigations involving microfossils

With micropalaeontology now part of the 'vocabulary' of the engineering geologists at Imperial College (especially Peter Fookes), both David Carter and MBH were regularly quizzed about problem-solving in site investigation work. One of the first problems was a road cutting on the A2 road in Kent (near Canterbury) where the excavations had collapsed. One end of this relatively short cutting looked, and behaved, like normal *in situ* chalk, but the other end of the cutting appeared to lack any stratigraphical organization. There was a suspicion that this chalk might be soliflucted and behaving like 'putty chalk' (Bundy 2013) and I can recall being asked whether this could be detected from any microfossils present. On processing a few samples it was clear that the potentially

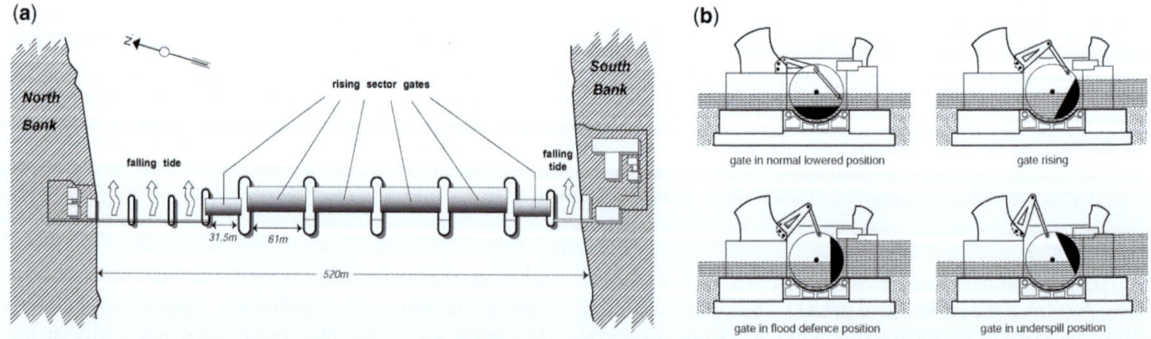

Fig. 13. Plan of the Thames Barrier showing the four large gates and the two smaller gates. The operation of the barrier is shown in the lower part of the diagram. The gate in the underspill condition (tilted 20° beyond the vertical) gives extra height; this operation is explained in Hart (2000, p. 452). Source: adapted from a diagram provided by the Environment Agency and reproduced with their permission.

soliflucted chalk had the same microfossils as the *in situ* chalk and there were no foraminifera that could detect transported material: all of the microfossils were local. There were, however, a few small, terrestrial snails and some rather unusual glassy 'beads' which were shown to be siliceous and best determined as phytoliths. These are the siliceous granules found in the leaves of modern grasses and certainly not something that would have been in the chalk sea when the chalk was forming during the Cretaceous. Modern grasses are a Cenozoic phenomenon and their presence indicated modern re-working of the chalk being analysed. This small-scale investigation of a road cutting predated another project that – in terms of cost – was significantly more important. This was the construction of the M11 near Cambridge (Fig. 14).

The initial site investigation information suggested that the planned road was cutting through glacial or periglacial gravels, (boulder) clays and sands. At first, MBH declined the invitation to visit the site, not being an expert in the glacial sediments of East Anglia. Persuaded to visit the site, it was quickly evident that this was an *in situ* Cretaceous succession and that the clays were best regarded as the Gault Clay Formation and the so-called gravels were, almost certainly, the Cambridge Greensand: see Figure 3. During construction, water ingress had become so severe that construction was terminated, and 5 trial pits dug into the profile of the road (Fig. 15). These quite clearly showed the Gault Clay–Cambridge Greensand–Chalk succession and it appeared that, having excavated through the boundary interval, water in the chalk succession was filling the road cutting which had cut down into the upper part of the Gault Clay succession: a total aquiclude. Someone, during the site visit, quite clearly stated that 'there was no chalk in the area', despite – to the SW – the visible chimney at the Barrington Cement Quarry (Fig. 14) generating clouds of white smoke as the chalk from that huge quarry was being crushed and dried as part of the cement-making process. While some micropalaeontological work was done on a few samples, it was only necessary to confirm the stratigraphy that was clearly visible in the field. These data were compared to the succession at Arlesey described by Hart (1973*a*) and other locations previously studied by MBH as a part of his on-going research into Cretaceous micropalaeontology.

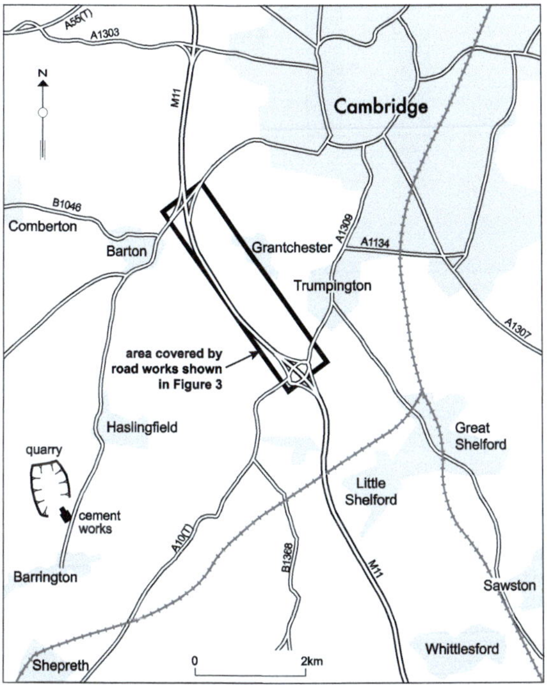

Fig. 14. Sketch map of the area to the SW of Cambridge that shows the location of the completed M.11 motorway. The area shown in Figure 15, near Grantchester, was that investigated during the site visit and where trial pits were excavated. The chalk quarry at Barrington is indicated as that is where, prior to recent flooding, the Cambridge Greensand and the uppermost part of the Gault Formation was exposed. Source: from Hart and Fox (2020) with the permission of the Geological Society of London.

Other investigations

Near Pitstone (Hertfordshire) the Lower and Middle chalk have long been quarried for the manufacture of cement. The process (in the 1970s) involved the mixing of more clay-rich chalk (from the Lower Chalk) with chalk from higher stratigraphical levels (uppermost Lower Chalk and Middle Chalk). The foraminifera from the succession were investigated (Hart 1993, fig. 7.20) and linked to the calcimetry of the chalk which contributed to a 'ground model' of the quarried area. This was helpful in planning the quarrying operations such that the correct mix of chalk was fed into the processing of the cement. The quarries are now closed, and the plant removed: the former quarries now being converted into Nature Reserves with limited geology being visible. The original quarry 3 (Hart 1993, fig. 7.19), in the clay-rich lower part of the Lower Chalk is now a lake (Fig. 16), while part of quarry 2 has had the Plenus Marls succession identified for conservation (Fig. 17).

Another interesting application of biostratigraphy and site investigation is the construction of an electricity cable from France to Jersey (Consolaro *et al.* 2014). In this case, the significant number of boreholes drilled in the shallow marine area simply recorded the presence of sediment in which to bury the cable. The limited micropalaeontological investigation essentially confirmed that these sediments were – as expected – of Holocene age, although some of the cores appear to record, in depth, the presence of either Eocene sediments or more modern sediments in which there are re-worked Eocene microfossils. As the cable was to be buried within the Holocene sediments no further microfossil work was part of the site investigations,

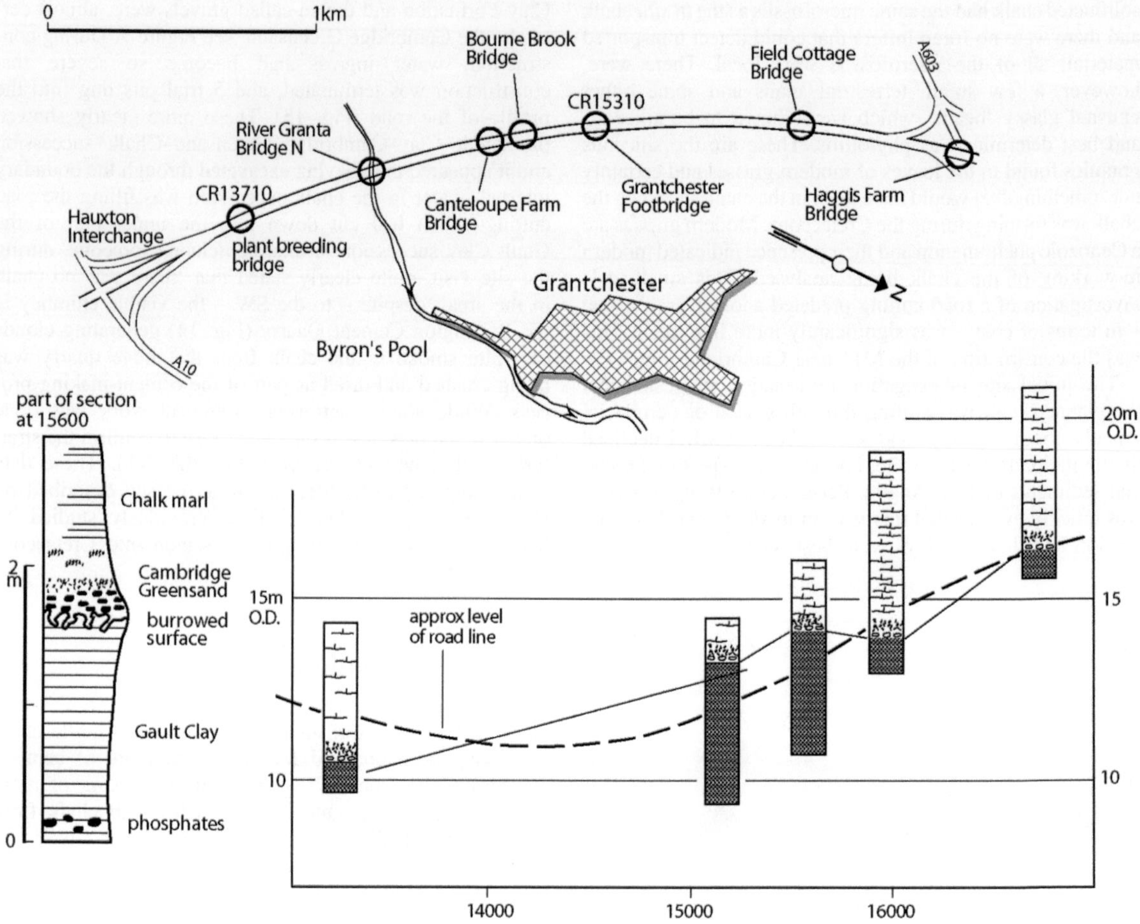

Fig. 15. Road profile near Grantchester showing the regional dip of the chalk, the stratigraphical position of the Cambridge Greensand and the underlying Gault Clay Formation. During the site visit, more samples than were needed for the interpretation of the engineering geology were taken and used in subsequent research. At the time, this was the best exposed succession of the Cambridge Greensand (see Hart and Fox 2020) seen in many years. All the samples collected from the trial pits contained typical foraminiferal assemblages with nothing unexpected. Source: from Hart and Fox (2020) with the permission of the Geological Society of London.

though there are on-going research interests in the area and its record of sea-level change (Massey *et al.* 2008; Gehrels 2009).

Contribution to 'ground models'

Hart (1993), in a paper on the role of micropalaeontology in engineering and economic geology, outlined the ways in which a study of microfossils could impact on site investigation:

(1) where detailed stratigraphical knowledge is required;
(2) where subsurface/borehole material is being employed and, as a result, sample size may be limited;
(3) where it is necessary to identify the presence of faults and/or repeated successions;
(4) where stratigraphical knowledge is required in an area of complex facies relationships; and
(5) where there is a need to identify the presence or absence of reworked or allochthonous material.

Hart (1993, pp. 226–228, fig. 7.3) went on to explain that fitting any given site into a scheme of international correlation is not the purpose of site investigation and, very often, the area involved is limited: e.g. the width of the River Thames in the case of the Thames Barrier (Fig. 11). Microfossil 'patchiness' can be an issue in such small areas and, quite often, solving such problems requires graphical techniques. There are now powerful Geographical

Fig. 16. Pitstone quarry 3 (Hart 1993, fig. 7.19) which is now a lake as it infills the clay-rich part of the Lower Chalk succession. Source: image courtesy of Dr Haydon Bailey of the Hertfordshire Geological Society.

Information System (GIS) methods of doing such work but, in the past, it was often a case of pouring over large charts of hand-drawn graphical data. I still have vivid memories of crawling over the Channel Tunnel correlations with David Carter in Dover Castle to the never-ending strains of '*What's new pussy cat*'; those of a certain vintage will remember that song!

Of course, the determination of a chalk sample as being of – for example – Cenomanian age brings with it the concept of the UK being located at 40–45° N of the Equator, in a warm

Fig. 17. Part of Pitstone quarry 2 (Hart 1993, fig. 7.9) showing a conserved part of the Plenus Marl Member of the chalk succession (Hart 1993, fig. 7.20). Source: image courtesy of Dr Haydon Bailey of the Hertfordshire Geological Society.

climate with exceptionally high sea levels (Davies and Simmons 2023; Scotese *et al.* 2024) and, as in the case of the M2 road in Kent, not containing evidence of terrestrial fossils and grasses. Almost all microfossil groups can bring palaeoenvironmental interpretations into the debate on the nature of the subsurface succession and, in his famous 3D 'models' (Fookes 1997, figs 13, 14, 19, 20, 21, 28–33, etc.) this is precisely what Peter Fookes was trying to achieve.

Fookes (1997, p. 348), in that major paper on 'ground models', makes the following statement:

> It is also my contention that when the main ground investigation is complete, i.e. use of boreholes, trenches, pits, geophysics and whatever other methods are deemed appropriate, then a *minimum of*, say, 95% of the geology should be known, i.e. if it is correct to say knowledge of the geology equates with identification of the potential problems, then 95% of potential problems should be capable of being identified.

Taking one's mind back to 1971–72, when the site investigation for the Thames Barrier was undertaken, one can now see where Peter Fookes was coming from when he asked for a thorough microbiostratigraphical appraisal of the site and why he thought that the engineers could sleep peacefully, knowing that many of the ***potential*** geological problems had been resolved (and largely dismissed). It was the same at the site of the M11 near Cambridge where a geological site visit, a few well-selected trench sections, and microfossil confirmation of the age of the succession, almost completely solved the cause of the flooding – or at least explained the problem. Knowing the problem meant that the engineers could take remedial action (even if expensive) and upgrade the drainage of the planned motorway.

Sadly, we often fail to learn from our endeavours and, after completion of the Channel Tunnel, when many of the tunnelling engineers went across to Denmark to build the tunnel under the Great Belt, a full micropalaeontological and stratigraphical survey was not undertaken and there was a flooding event that a stratigraphical survey could, most surely, have predicted. The costs of such work are often minimal (1% of the site investigation costs for the Thames Barrier) and only *c.* 0.0025% of the cost of the whole project (covering the site visits, the additional cores and the microfossil work).

In his major work on the development and application of 'ground models', Peter Fookes was, most surely, remembering back to these earlier projects where his thirst for the geological information never changed throughout his career. He also had an uncanny ability to determine those who could help solve his perceived problems, whether it be in the field of geophysics, clay mineralogy or micropalaeontology.

Acknowledgements The author wishes to thank all those who were involved in the projects described in this paper and from whom guidance and advice was forthcoming. The author also records his thanks to the University of Plymouth who, despite 'retirement' in 2010, is still provided with facilities to undertake research. Many of the figures presented here were created by John Abraham, cartographic technician in geology at the University of Plymouth before his retirement. Mr James Quinn (cartographic technician, University of Plymouth) assisted in accessing some of the figures from the university digital archives and – where necessary – making improvements.

Competing interests There are no competing interests as I am sole author for this paper.

Author contributions **MBH**: conceptualization (lead), writing – original draft (lead), writing – review & editing (lead).

Funding This work was not supported by any research funding.

Data availability There are no issues over data availability as no new data (other than cited material) are used in the paper.

References

Anonymous 1977. Introductory Statement. *Quarterly Journal of Engineering Geology*, **10**, 175–176, https://doi.org/10.1144/GSL.QJEG.1977.010.03.01

Bailey, H.W. 1978. *A foraminiferal biostratigraphy of the Lower Senonian of Southern England*. PhD thesis, CNAA/Plymouth Polytechnic (now University of Plymouth).

Bailey, H.W., Hart, M.B. and Swiecicki, A. 2009. Evolutionary lineages of benthic foraminifera in the Chalk Seas of N.W. Europe. *SEPM Special Publication*, **93**, 233–249.

Bondi, H. 1967. *London Flood Barrier*. Unpublished Report to the Ministry of Housing and Local Government.

Bruckshaw, J.M., Goguel, J., Harding, H.J.B. and Malcor, R. 1961. The work of the Channel Tunnel Study Group 1958–1960. *Proceedings of the Institution of Civil Engineers*, **18**, 149–178, https://doi.org/10.1680/iicep.1961.11531

Brunsden, D. and Chandler, J.H. 1996. Development of an episodic landform change model based upon the Black Ven mudslide, 1946–1995. *In*: Anderson, M.J. and Brookes, J. (eds) *Advances in Hillslope Processes*, Vol. **2**. J. Wiley and Sons, Chichester, 869–896.

Brunsden, D. and Jones, D.K.C. 1972. The morphology of degraded landslide slopes in South West Dorset. *Quarterly Journal of Engineering Geology*, **5**, 205–222, https://doi.org/10.1144/GSL.QJEG.1972.005.03.01

Brunsden, D. and Jones, D.K.C. 1976. The evolution of landslide slopes in Dorset. *Philosophical Transactions of the Royal Society, Series A*, **283**, 605–631, https://doi.org/10.1098/rsta.1976.0097

Brunsden, D., Ibsen, M.-L., Lee, M. and Moore, R. 1955. The validity of temporal archive records for geomorphological processes. *Questions Geographicae, Special Issue*, **4**, 1–13.

Bundy, S.P.S. 2013. *Geotechnical properties of chalk putties*. PhD thesis, University of Portsmouth. https://doi.org/10.1007/s10346-014-0538-1

Carey, J.M., Moore, R. and Petley, D.N. 2015. Patterns of movement in the Ventnor landslide complex, Isle of Wight, southern England. *Landslides*, **12**, 1107–1118, https://doi.org/10.1007/s10346-014-0538-1

Carter, D.J. 1966. Channel Tunnel Study Group Site Investigation: Appendix 3 (Micropalaeontology Report), pp. MI-1 to MA-14; Appendix 4 (Figures).

Carter, D.J. and Hart, M.B. 1977. Micropalaeontological investigations for the site of the Thames Barrier, London. *Quarterly Journal of Engineering Geology*, **10**, 321–338, https://doi.org/10.1144/GSL.QJEG.1977.010.03.11

Chandler, J.M. and Brunsden, D. 1995. Steady State behaviour of the Black Ven mudslide: the application of archival analytical photogrammetry to studies of landform change. *Earth Surface Processes and Landforms*, **20**, 255–275, https://doi.org/10.1002/esp.3290200307

Chapman, F. 1891–98. The foraminifera of the Gault of Folkestone. *Journal of the Royal Microscopical Society, London*, (Part 1) 1891, 565–575; (Part 2) 1892, 319–330; (Part 3) 1892, 749–758; (Part 4) 1893, 579–595; (Part 5) 1893, 153–163; (Part 6) 1894, 421–427; (Part 7) 1894, 645–654; (Part 8) 1895, 1–14; (Part 9) 1896, 581–591; (Part 10) 1898, 1–49.

Charman, J. 2008. Peter George Fookes: father of modern British engineering geology. *Quarterly Journal of Engineering Geology*, **41**, 201–216, https://doi.org/10.1144/1470-9236/07-061

Clémence, M.-E. and Hart, M.B. 2013. Proliferation of Oberhauserellidae during the recovery following the Late Triassic extinctions: Paleoecological implications. *Journal of Paleontology*, **87**, 1004–1015, https://doi.org/10.1666/13-021

Consolaro, C., Hart, M., Chambers, P., Evans, G., Nichols, R., Smart, C. and Whittaker, J. 2014. The Holocene separation of Jersey from France: the microfossil evidence. *Geophysical Research Abstracts*, **16**, EGU2014-4149.

Conway, B.W. 1974. *The Black Ven Landslip*. Institute of Geological Sciences, Report No. 74/3.

Cope, J.C.W. 2016. *Geology of the Dorset Coast*, 2nd edn. Geologists' Association Guide No. 22, London.

Czarniecki, S. 1993. Grzybowski and his school: the beginnings of applied micropalaeontology in Poland at the turn of the 19th and 20th centuries. *In*: Kaminski, M.A., Geroch, S. and Kaminski, D.G. (eds) *The Origins of Applied Micropalaeontology: The School of Józef Grzybowski*. Grzybowski Foundation Special Publication, no. 1, Alden Press, Oxford, 1–15.

Davies, A. and Simmons, M.D. 2023. Placing constraints on the nature of short-term eustatic curves. *Basin Research*, **36**, https://doi.org/10.1111/bre.12832

Davison, L., Fookes, P. and Baynes, F. 2003. Total geological history: a web-based modelling approach to the anticipation, observation and understanding of site conditions. *In*: Rosenbaum, M.S. and Turner, A.K. (eds) *New paradigms in subsurface prediction*. Lecture Notes in Earth Sciences, **99**, Springer, Berlin, https://doi.org/10.1007/3-540-48019-6_21

Denness, B. 1969. *Fissures and related studies in selected Cretaceous rocks of S. E. England*. PhD thesis, University of London.

Denness, B. 1972. *The reservoir principle of mass movement*. Institute of Geological Sciences, Report No. 72/7.

Denness, B., Conway, B.W., McCann, D.M. and Grainger, P. 1975. Investigation of a coastal landslip at Charmouth, Dorset. *Quarterly Journal of Engineering Geology*, **8**, 119–140, https://doi.org/10.1144/GSL.QJEG.1975.008.02.03

De Rance, C.E. 1868. On the Albian or Gault of Folkestone. *Geological Magazine*, **5**, 163–171, https://doi.org/10.1017/S001675680020767X

Edmonds, R. 2022. A new interpretation of the Great Bindon Landslide. *Geoscience in South-West England*, **15**, 43–55.

Fookes, P.G. 1997. Geology for engineers: the geological model, prediction and performance. *Quarterly Journal of Engineering Geology*, **30**, 293–424, https://doi.org/10.1144/GSL.QJEG.1997.030.P4.02

Fookes, P., Baynes, F. and Hutchinson, J. 2001. Total geological history: a model approach to understanding site conditions. *Ground Engineering*, 42–47.

Fookes, P., Pettifer, G. and Waltham, A. 2015. *Geomodels in Engineering Geology: An Introduction*. Whittles, Caithness

Gallois, R.W. and Davis, G.M. 2001. Saving Lyme Regis from the sea: recent geological investigations at Lyme Regis, Dorset. *Geoscience in South-West England*, **10**, 183–189.

Gehrels, R. 2009. Sea-level changes since the Last Glacial Maximum: an appraisal of the IPCC Fourth Assessment Report. *Journal of Quaternary Science*, **25**, 26–38, https://doi.org/10.1002/jqs.1273

Griffiths, J.S. 2014. Feet on the ground: engineering geology past, present and future. The 14th Glossop Lecture. *Quarterly Journal of Engineering Geology and Hydrogeology*, **47**, 116–143, https://doi.org/10.1144/qjegh2013-087

Harris, C.S., Hart, M.B., Varley, P.M. and Warren, C.D. (eds) 1996. *Engineering Geology of the Channel Tunnel*. Thomas Telford, London.

Hart, M.B. 1973*a*. Foraminiferal evidence for the age of the Cambridge Greensand. *Proceedings of the Geologists' Association*, **84**, 65–82, https://doi.org/10.1016/S0016-7878(73)80007-0

Hart, M.B. 1973*b*. A correlation of the macrofaunal and microfaunal zonations of the Gault Clay in southeast England. *In*: Casey, R. and Rawson, P.F. (eds) *The Boreal Lower Cretaceous*. Geological Journal, Special Issue, **5**, 267–288.

Hart, M.B. 1993. Engineering and economic geology. *In*: Jenkins, D.G. (ed.) *Applied Micropalaeontology*. Kluwer, Dordrecht, 225–262.

Hart, M.B. 2000. Construction of the Thames Barrier: An application of micropalaeontology to the solution of an environmental problem. *In*: Martin, R.E. (ed.) *Environmental Micropaleontology*. Topics in Geobiology, **15**. Kluwer/Plenum, New York, 429–459.

Hart, M.B. 2023. Prologue: the founding of the Cretaceous System. *Geological Society, London, Special Publications*, **544**, https://doi.org/10.1144/SP544-2023-60

Hart, M.B. and Bailey, H.W. 2013. Key figures from the history of research on the Foraminifera of the Chalk Group in the UK. *In*: Bowden, A.J., Gregory, F.J. and Henderson, A.S. (eds) *Landmarks in Foraminiferal Micropalaeontology: History and Development*. The Micropalaeontological Society, Special Publications, London, 85–102.

Hart, M.B. and Fox, L.R. 2020. Micropalaeontology and stratigraphical setting of the Cambridge Greensand. *Geological Society, London, Special Publications*, **498**, 147–163, https://doi.org/10.1144/SP498-2018-144

Heaps, N.S. 1977. Development of storm surge models at Bidston. Internal Report No. 51, Institute of Oceanography.

Huber, B.T., Buzas, M.A., Drew, A. and Jett, J.A. 2013. History of foraminiferal collections at the Smithsonian Institution. *In*: Bowden, A.J., Gregory, F.J. and Henderson, A.S. (eds) *Landmarks in Foraminiferal Micropalaeontology: History and development*. The Micropalaeontological Society, Special Publications, London, 71–83.

Hutchinson, J.N. 1969. A reconsideration of the coastal landslides at Folkestone Warren, Kent. *Geotechnique*, **19**, 6–38, https://doi.org/10.1680/geot.1969.19.1.6

Hutchinson, J.N., Bromhead, E.N. and Lupini, J.F. 1980. Additional observations on Folkestone Warren landslides. *Quarterly Journal of Engineering Geology*, **13**, 1–31, https://doi.org/10.1144/GSL.QJEG.1980.013.01.01

Jenkins, G.O., Foster, C. and Hopson, P.M. 2011. Geology as a control on landslides on the Isle of Wight: an overview. *Proceedings of the Geologists' Association*, **122**, 906–922, https://doi.org/10.1016/j.pgeola.2011.09.003

Jones, D.J. 1956. *Introduction to Microfossils*. Harper & Brothers, New York.

Jukes-Browne, A.J. and Hill, W. 1900. *The Cretaceous Rocks of Britain. Volume 1, The Gault and Upper Greensand of England*. Memoirs of the Geological Survey, Great Britain.

Kaminski, M.A., Geroch, S. and Kaminski, D.G. (eds) 1993. *The Origins of Applied Micropalaeontology: The School of Józef Grzybowski*. Grzybowski Foundation Special Publication, no. 1, Alden Press, Oxford.

Lang, W.D. and Spath, L.F. 1926. The Black Marl of Black Ven and Stonebarrow, in the Lias of the Dorset coast. *Quarterly Journal of the Geological Society, London*, **82**, 144–187, https://doi.org/10.1144/GSL.JGS.1926.082.01-04.11

Lisiecki, L.E. and Raymo, M.E. 2007. Plio-Pleistocene climate evolution: trends and transitions in glacial cycle dynamics. *Quaternary Science Reviews*, **26**, 56–69, https://doi.org/10.1016/j.quascirev.2006.09.005

Lucas, S.G. 2021. Biostratigraphy. *In*: Alderton, D. and Elias, S.A. (Editors-in-Chief), *Encyclopedia of Geology*, Vol. **3**, 2nd edn, 89–95, https://doi.org/10.1016/B978-0-08-102908-4.00076-X

Massey, A.C., Gehrels, W.R., Charman, D.J., Milne, G.A., Peltier, R., Lambeck, K. and Selby, K.A. 2008. Relative sea-level change and postglacial isostatic adjustment along the coast of South Devon, United Kingdom. *Journal of Quaternary Science*, **23**, 415–433, https://doi.org/10.1002/jqs.1149

Moore, R. 2020. The Isle of Wight Undercliff. *In*: Goudie, A. and Migón, P. (eds) *Landscapes and Landforms of England and Wales*. World Geomorphological Landscapes, Springer, 145–167, https://doi.org/10.1007/978-3-030-38957-4_8

Moore, R., Carey, J.M. and McInnes, R.G. 2010. Landslide behaviour and climate change: predictable consequences for the Ventnor Undercliff, Isle of Wight. *Quarterly Journal of Engineering geology and Hydrogeology*, **43**, 447–460, https://doi.org/10.1144/1470-9236/08-086

Muir Wood, A.M. 1955. Folkestone Warren landslips: investigations 1948–50. *Proceedings Institution Civil Engineers, Railway Paper*, **56**, 410–428, https://doi.org/10.1680/ipeds.1955.11707

Norbury, D. 2020. Ground models; a brief overview. *Quarterly Journal of Engineering Geology and Hydrogeology*, **54**, https://doi.org/10.1144/qjegh2020-018

Palmer, M.J., Moore, R. and McInness, R.G. 2007. Reactivation of an ancient landslip, Bonchurch, Isle of Wight: Event history, mechanisms, causes, climate change and landslip potential. *In*: McInness, R., Jakeways, J., Fairbank, H. and Mathie, E. (eds) *Landslides and Climate Change: Challenges and Solutions*. Proceedings of the International Conference on Landslides and Climate Change, Ventnor, Isle of Wight, UK, 21–24 May 2007, Taylor & Francis, London, 355–364.

Price, F.G.H. 1874*a*. On the Gault of Folkestone. *Quarterly Journal of the Geological Society, London*, **30**, 342–366, https://doi.org/10.1144/GSL.JGS.1874.030.01-04.42

Price, F.G.H. 1874*b*. On the Lower Greensand and Gault of Folkestone. *Proceedings of the Geologists' Association*, **4**, 135–150, https://doi.org/10.1016/S0016-7878(75)80080-0

Price, F.G.H. 1876. Excursion to Sandgate and Folkestone. *Proceedings of the Geologists' Association*, **4**, 554–556, https://doi.org/10.1016/S0016-7878(76)80067-3

Price, F.G.H. 1879. *The Gault*. Taylor & Francis, London.

Railsback, L.B., Gibbard, P.L., Head, M.J., Voarintsoa, N.R.G. and Toucanne, S. 2015. An optimized scheme of lettered marine isotope substages for the last 1.0 million years, and the climatological nature of isotope stages and substages. *Quaternary Science Reviews*, **111**, 94–106, https://doi.org/10.1016/j.quascirev.2015.01.012

Scotese, C.R., Verard, C., Burgener, L., Elling, R.P. and Kocsis, A. 2024. The Cretaceous world: plate tectonics, paleogeography, and paleoclimate. *Geological Society, London, Special Publications*, **544**, https://doi.org/10.1144/SP544-2024-28

Shackleton, N.J. 1969. The last interglacial in the marine and terrestrial record. *Proceedings of the Royal Society*, **B174**, 135–154.

Shackleton, N.J. 2000. The 100,000-year ice-age cycle identified and found to lag temperature, carbon dioxide and orbital eccentricity. *Science (New York, NY)*, **289**, 1897–1902, https://doi.org/10.1126/science.289.5486.1897

Shackleton, N.J. 2006. Formal Quaternary stratigraphy – What do we expect and need? *Quaternary Science Reviews*, **25**, 3458–3462, https://doi.org/10.1016/j.quascirev.2006.11.008

Tanaka, H. and Locat, J. 1999. A microstructural investigation of Osaka Bay clay: the impact of microfossils on its mechanical behaviour. *Canadian Geotechnical Journal*, **36**, 493–508, https://doi.org/10.1139/t99-009

Toms, A.H. 1946. Folkestone Warren landslips: research carried out in 1939 by the Southern Railway. *Proceedings Institution Civil Engineers, Railway Paper*, **19**, 3–25, https://doi.org/10.1680/idivp.1946.13109

Viner-Brady, N.E.V. 1955. Folkestone Warren landslips: remedial measures, 1948–1954. *Proceedings of the Institution of Civil Engineers, Railway Paper*, **57**, 429–441, https://doi.org/10.1680/ipeds.1955.11708

Warren, C.D., Palmer, M.J., Bromhead, E., Dixon, N. and Ibsen, M.L. 2000. Observations on the nature of landslipped strata, Folkestone Warren, United Kingdom. *Proceedings of the 8th International Symposium on Landslides*, Cardiff, 26–30 June, Vol. 3, 1551–1556.

Waverley Report 1953. *Flood and Tempest*. Waverley Committee Report, The National Archives, Kew, 91235/2/14/3.

Weimer, G. and Kopf, A. 2016. Influence of diatom microfossils on sediment shear strength and slope stability. *Geochemistry, Geophysics, Geosystems*, **18**, 333–345, https://doi.org/10.1002/2016GC006568

Williams-Mitchell, E. 1948. The zonal value of foraminifera in the Chalk of England. *Proceedings of the Geologists' Association*, **59**, 91–112, https://doi.org/10.1016/S0016-7878(48)80014-3

Williams-Mitchell, E. 1956. The stratigraphy and structure of the Chalk of the Dean Hill Anticline, Wiltshire. *Proceedings of the Geologists' Association*, **67**, 221–227, https://doi.org/10.1016/S0016-7878(56)80023-0

Implications for ground engineering through deterioration of landforms and geological materials caused by atmospheric reactions involving sulfide and sulfate minerals

Mourice A. Czerewko[1]* and John C. Cripps[2]

[1]AECOM, Royal Court, Basil Close, Chesterfield S41 7SL, UK
[2]School of Mechanical, Aerospace and Civil Engineering, University of Sheffield, Sir Frederick Mappin Building, Mappin Street, Sheffield S1 3JD, UK
MAC, 0000-0003-0061-1234; JCC, 0000-0002-7676-1469
*Correspondence: mourice.czerewko@aecom.com

Abstract: The implications of sulfide and sulfate minerals on landforms and foundations vary significantly between temperate, arid and equatorial zones. Sulfur minerals, especially pyrite and gypsum, underlie significant engineering problems due to volumetric and mineralogical changes that occur following exposure to surface weathering environments. Whilst pyrite is stable under the anoxic reducing conditions in which it forms, exposure to oxygenated, damp environmental conditions result in rapid oxidation that produces chemically aggressive acidic, sulfate-rich solutions. If carbonates or cementitious material are present, potentially expansive selenite–gypsum is likely to be formed. In dry arid environments, these processes will occur more slowly, but their effects are often observed as dramatic slope failures. Dry environments are also conducive to the preservation of sulfate minerals such as gypsum and anhydrite in the ground. However, these minerals are liable to undergo dissolution and volume changes when disturbed by construction that may also create conditions that are chemically aggressive for concrete and other materials.

The possibility of long-term impacts of weathering and changes brought about by the engineering works must be considered in ground investigation and geomaterials assessments, accounted for in the ground model, and addressed in the proposed design and method of construction.

The intention of this paper is to outline some of the implications of sulfur minerals in geomaterials, provide guidance on the identification of potential problems in construction and landform evaluations, and suggest ways of avoiding difficulties.

Although a wide range of factors encompassing physical, chemical and biological processes may bring about chemically aggressive ground conditions, a high proportion of such problems arise from the presence of sulfur compounds and sulfate ions in groundwater (Hawkins and Pinches 1987; Pye and Miller 1990; Reid *et al.* 2005). This gives rise to attack on concrete and steel, heave of *in situ* rock masses and fill, and deterioration of the geotechnical properties of ground materials. The source of the sulfates can either be dissolution of existing sulfate minerals or oxidation of sulfide minerals. Sulfur compounds can rapidly transform from one form to another, particularly when disturbed by construction. In the presence of atmospheric oxygen and water, pyrite and other iron sulfide minerals, such as marcasite and pyrrhotite, become oxidized, producing acidic sulfate-rich solutions and insoluble orange-brown hydrous iron oxide, which precipitates out of solution with deleterious impacts on the ecological and chemical status of receiving water bodies. Therefore, the nature and quantity of sulfur compounds in ground strata and construction materials need to be determined and the risks assessed to avoid detrimental reactions occurring during or after construction.

The sulfide 'pyrite' and the sulfate 'gypsum' are by far the most encountered types of sulfur minerals in geological materials (Czerewko and Cripps 2003; Czerewko *et al.* 2016). The problems that pyrite creates are associated with the production of sulfuric acid, and the potentially expansive formation of the crystalline form of gypsum termed 'selenite'. The consequential removal of pyrite and carbonate cements from the rock results in degradation of material and causes instability of natural and man-made slopes. However, gypsum and other sulfate minerals may occur as a primary constituent of geological deposits (see Fig. 1), the dissolution of which will generate sulfate ions in solution and cause damaging expansive reactions with concrete. The removal of sulfates from the ground will result in subsidence and weakening of materials, with adverse consequences for the affected ground. Barytes, celestine and organic sulfur are examples of sulfur minerals that tend not to be troublesome in engineering situations, although this depends on the environmental conditions.

Since the 1930s it has been well known by engineers in the UK that soluble sulfates produce chemically aggressive ground conditions for buried concrete structures (Cripps *et al.* 2019). However, awareness of the role played by pyrite

From: Griffiths, J. S., Giles, D. P., Davis, J. A. and Dunlop, J. J. (eds) 2025. *The Peter Fookes Engineering Geological Legacy in Geomodels, Geomaterials and Geomorphology*. Geological Society, London, Engineering Geology Special Publications, **30**, 143–177,
First published online 2 July 2025, https://doi.org/10.1144/egsp30-2024-22

Fig. 1. Examples of gypsum/selenite from diverse sources: (**a**) laminated gypsum of the Lisan Formation, Saudi Arabia; the scale is 300 mm; (**b**) lenses of the alabaster form of siliceous gypsum with secondary veins of satin spar in Triassic-age, Blue Anchor Formation, near Watchet, north Somerset, UK; (**c**) selenite crystals formed as a consequence of pyrite oxidation at a newly exposed surface of Ampthill Clay near Aylesbury, UK; and (**d**) stratified selenite resulting from oxidation of pyrite in Precambrian schists in Mauritanian west Sahara.

and other iron sulfide minerals was seemingly limited to the research community and not appreciated by engineers until much later. The Building Research Establishment (BRE) first drew general attention to pyrite as a source of sulfuric acid and sulfate ions in BRE (1968), but no testing procedures or limits to sulfide content were specified at this time. However, by the early 1980s several prominent cases of construction failures, including failure of the Carsington Dam in Derbyshire and heave during construction of the M40 motorway near Banbury (Snedker and Temporal 1990; Cripps *et al.* 2019), were attributed to pyrite oxidation processes. These and other problems and failures occurred because, although the ground was tested for the presence of sulfates in accordance with the practices then in place (BRE 1991), at the time of construction these values were low. However, the construction process and subsequent weathering reactions resulted in pyrite oxidation that gave rise to the creation of damaging chemically aggressive ground conditions.

The aim of this paper is to provide a clear overview of the main sulfur minerals that are found in the ground, and the implications they pose to ground engineering and landform processes due to the development of aggressive conditions caused by rapid weathering often initiated by construction activities. The discussion considers an assessment of geological material and presents guidance for management of the geotechnical risk associated with sulfide and sulfate minerals.

Occurrence of sulfur species associated with aggressive ground

Sulfur occurs in geological materials of all ages and origins, and in a variety of forms (King 1999). It is also one of the most reactive of elements and readily combines with others to form metallic sulfides. Sulfate minerals are the stable form of sulfur in oxygen-rich atmospheric conditions; and sulfide minerals, which form under anaerobic conditions typically devoid of free oxygen, occur widely in geological materials, meaning that sulfur is frequently encountered in ground engineering.

Pyrite and other iron sulfide minerals

Pyrite, marcasite and pyrrhotite are by far the most common and widely occurring sulfide minerals, thus presenting the greatest concerns for ground engineering. Pyrite is the most prevalent of these minerals, whereas marcasite and pyrrhotite, although highly reactive under atmospheric conditions, occur less frequently. While several other metallic sulfide minerals, including the lead sulfide, galena, the zinc sulfide, sphalerite, and the copper iron mineral, chalcopyrite, may be occasionally found in rock material, they have not been credited with any significant effects for ground engineering to date (Bromley and Petifer 1997).

Pyrite occurs in a visible 'macro' form as brass-yellow, hard brittle cubic, octahedral and pyritohedral crystals (examples are shown in Fig. 2a–c), or as 'microscopic' minute dark-coloured crystals or in its framboidal form (Fig. 2d), often as microscopic framboidal agglomerations or 'polyframboidal' clusters (Raiswell and Berner 1985; Ricard 2015). These latter forms are too small to discern with the naked eye or a hand lens, and are difficult to identify using conventional petrographical techniques, particularly when present in an argillaceous host. Therefore, reliance on more expensive and sophisticated techniques such as scanning-electron microscopy (SEM) coupled with advanced elemental analytical techniques are necessary for positive mineralogical and chemical confirmation of microscopic forms. Usually in ground engineering investigations, the presence of sulfide minerals is inferred by chemical analysis rather than them being identified by petrographical means (see BSI 2021).

Pyrite occurs widely in rocks of all types and ages but is a significant constituent in the UK of marine sedimentary

Fig. 2. Pyrite is encountered in a diverse range of forms and sizes: (**a**) macroscopic striated cubic forms in mineral deposits from Peru; (**b**) nodular form of pyrite in Mid-Jurassic mudrock from near Port Mulgrave, UK; (**c**) macroscopic pyrite crystals in veins from Shap Granite, UK; and (**d**) in a microscopic framboidal form in Eocene London Clay.

formations, as seen in Figure 3. These include grey mudrocks and stiff clays, as well as impure limestones, siltstones and sandstones of various ages. It is also present as an accessory mineral in many igneous and metamorphic deposits and associated ores, in which it may be distributed through the host rock or concentrated in hydrothermally deposited mineral veins or dense masses of well-formed crystals (Deer *et al.* 1992; Bowles *et al.* 2011; Hawkins 2014). Although these non-sedimentary lithologies tend to be less frequently encountered during civil engineering works in the UK, they

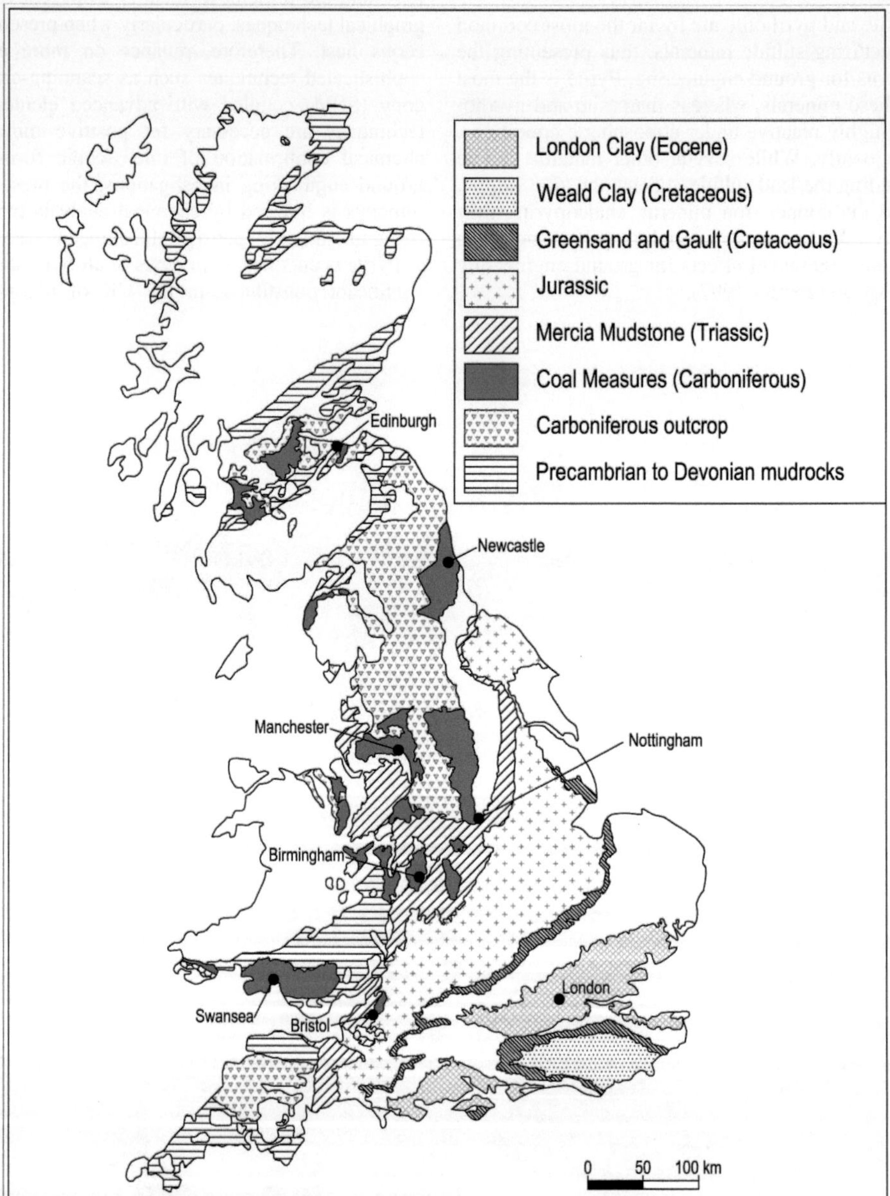

Fig. 3. The distribution of predominantly mudrock bedrock formations in the UK, all of which, except for Mercia Mudstone, contain significant pyrite. Mercia Mudstone contains sulfate deposits, including gypsum. Major routeways, industrial developments and urban centres are situated on these formations.

constitute a major source of construction aggregates, raising the potential for pyrite-bearing materials to be introduced as construction materials.

The morphology and grain size of pyrite has a significant influence on its susceptibility to oxidation under atmospheric conditions (Arrouvel and Eon 2019). All forms of pyrite possess the same internal arrangement of iron and sulfur atoms but the temperature, pressure, acidity, partial gas pressures and relative concentrations of the solutes in the liquid phase during formation affect the morphology of the crystals produced (Murowchick and Barnes 1987; Ricard 2015; Arrouvel and Eon 2019). Well-crystallized pyrite is typically found in rocks that have been subjected to moderate–high temperatures and pressures. In these situations, pyrite has a densely packed structure and relatively small specific surface area, and the crystals tend to be relatively stable in weathering environments (Lowson 1982). The cubic and pyritohedral forms of pyrite are relatively stable; whereas Arrouvel and Eon (2019) report that the octahedral form is less stable. Conversely, microscopic forms of pyrite are more reactive due to their relatively large surface area. But the substitution of trace elements into the lattice of such pyrite can affect the oxidation rate; cobalt and nickel are believed to stabilize the structure, whereas arsenic can result in faster oxidation (Kwong 1993). Studies on metalliferous deposits suggest that surface striations often seen in cubic forms of pyrite that occur in metamorphic and igneous lithologies are an expression of greater integrity of the internal structural form and greater chemical stability (see Fig. 4a) compared with their smooth-faced counterparts found in sedimentary deposits (Arrouvel and Eon 2019) (see Fig. 4b–d).

The microscopic framboidal form of pyrite is the most troublesome type in ground engineering works. It occurs widely in dark-coloured clays, mudrocks, argillaceous limestones, metamudrocks, low-grade metamorphic counterparts and low-temperature hydrothermal deposits. It has also been found in certain alluvial sand and gravel deposits (Reid *et al.* 2005). Such pyrite forms under oxygen-deficient anaerobic

Fig. 4. Examples of typical 'macro' forms of pyrite crystals: (**a**) striated cubic form from Peru; (**b**) smooth-faced octahedral crystals in Carboniferous limestone aggregate from the SW Midlands of England; (**c**) smooth-faced, partly tarnished cubic crystals in Lower Jurassic Charmouth Mudstone from Dorset, UK; and (**d**) smooth-faced, partly tarnished pyritohedral crystals in Late Triassic Westbury Mudstone from near Lilstock, north Somerset, UK.

Fig. 5. Example of framboidal pyrite: (**a**) discrete well-ordered pyrite framboids in London Clay showing some disorder of microcrystals due to construction disturbance; and (**b**) ordered polyframboidal agglomerations of pyrite in Jurassic Lower Lias Clay.

conditions that occur in many marine and some terrestrial environments such as swamps, in which sulfur becomes added to the iron-bearing sediments through bacterial action. Framboids, as seen in Figure 5, are discrete or grouped raspberry-like spherules typically 2–80 µm in size comprising agglomerations of pyrite grains typically <0.3–2 µm in size (Raiswell and Berner 1985; Ricard 2015). The framboidal structure results in spherules with a large surface area in proportion to their volume and the clusters are only weakly bonded. The large collective surface area renders them highly susceptible to oxidation.

Pyrite oxidation may proceed 'abiotically' by purely chemical means following exposure of pyrite to atmospheric conditions; however, the process may be accelerated by several orders of magnitude by the bacterial action of species such as *Acidothiobacillus* sp., which are ubiquitous in the surface environment (Temple and Delchamps 1953; Silverman 1967; Sasaki *et al.* 1998). Under dry conditions that occur during dry spells of weather in temperate zones and in arid climates, bacterially enhanced oxidation of pyrite is generally ruled out. Although the relevant bacteria tend to be most active in warmer conditions in the range of 30–35°C (Nixon 1978), below a relative humidity (RH) of 95% they become largely inactive (Fenlon and Petrera 2019), so they also require humidity or moisture to be present. Therefore, current understanding recognizes that the bacteria implicated in pyrite oxidation are active under wet or intermittently wet conditions typical of wet

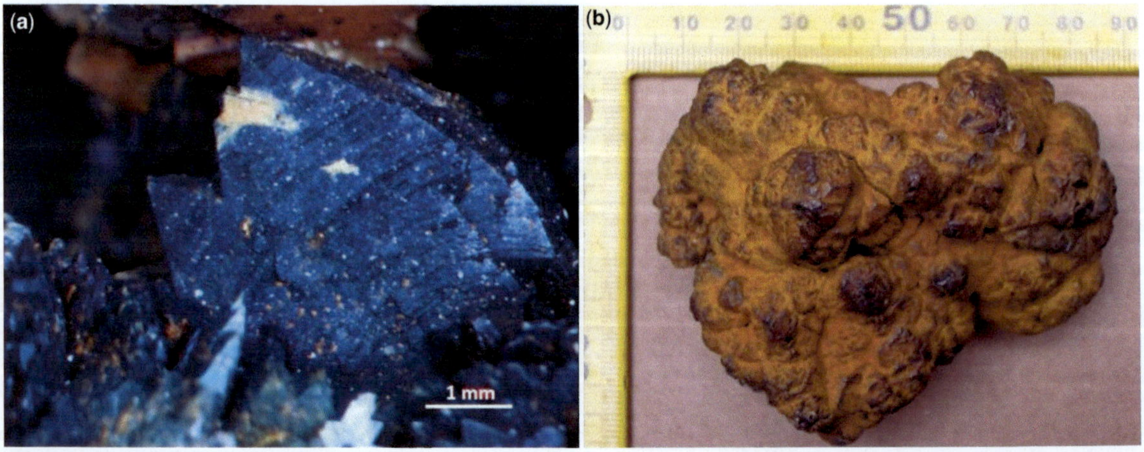

Fig. 6. Examples of marcasite: (**a**) cockscomb or 'spear-shaped' crystals present in Carboniferous limestone; (**b**) the nodular form from chalk deposits on the Isle of Wight with crude-stacked, broad spear-shaped crystals and a brown oxidized surface.

temperate, subtropical and tropical climates, but not in low humidity or dry air conditions.

Marcasite, which is sometimes referred to as white iron pyrite, is a dimorph of pyrite that forms brittle silvery-gold-coloured 'spear-shaped' crystals when fresh (Fig. 6a), but which very rapidly oxidizes to brown on exposure (see Fig. 6b). Like pyrite, it may also occur in nodular forms. In the UK, it is typically found in argillaceous and carbonate deposits of Carboniferous and Cretaceous ages, whilst in the USA it occurs in coal deposits of Carboniferous age. It may also occur as a late-formed mineral in low-temperature mineral veins. Although Wiese *et al.* (1987) document a number of environmental problems with marcasite-bearing coals in the USA, other cases are not forthcoming from the literature. Marcasite is uncommon but, when found, it is often associated with pyrite, in which case it undergoes rapid oxidation. This means that even if present, the problem tends to be attributed to pyrite (Dougherty and Barsotti 1972).

Pyrrhotite ($Fe_{1-x}S$) is another iron sulfide mineral that has rarely been reported in UK construction works, although it has been occasionally reported in well-indurated deposits in other countries. It occurs as bronze-brown masses or pseudo-hexagonal tabular crystals (Fig. 7a) but more often it is finely disseminated in rocks, making it difficult to identify, as seen in Figure 7b.

Pyrrhotite tends to be absent from sediments such as clays and mudrocks, which Horng and Roberts (2006) ascribe to it being formed very slowly and at temperatures above 180°C. It is more common in basic igneous rocks, pegmatites, hydrothermal veins and associated metamorphic rocks of Precambrian and Paleozoic age that may be exploited as construction aggregates (see Fig. 7c, d). According to Craig and Vokes (1993), loss of sulfur during deep burial and metamorphism

Fig. 7. Examples of pyrrhotite in aggregate; (**a**) stacks of hexagonal pyrrhotite crystals from the Carboniferous Great Limestone, Weardale, UK; (**b**) finely disseminated pyrrhotite with pyrite in metamorphic Hegaf Formation, Saudi Arabia; (**c**) cored samples showing a mass of platy bronze-coloured pyrrhotite crystals in schist (on the left) compared with fine multifaceted pyrite crystals (on the right) in Precambrian rocks from western Mauritania; and (**d**) pseudo-hexagonal tabular pyrrhotite crystals in Carboniferous dolomitic limestone, Derbyshire, UK.

results in its conversion from pyrite, so it may be present in metamudrocks and indurated impure limestones (see Fig. 7d).

It has been shown that the morphology of pyrite is a very important factor in the rate of its oxidation (Arrouvel and Eon 2019). For example, Grattan-Bellew and Eden (1975) record that fine-grained pyrite is more disposed to bacterial action. Pyrrhotite is more reactive than marcasite, which is more reactive than pyrite, and framboidal pyrite is much more reactive than massive pyrite (Pugh *et al.* 1981).

Microscopic forms of pyrite, particularly the framboidal form, are often referred to as the 'reactive' form of pyrite, whereas macroscopic crystalline forms may be referred to as the 'non-reactive' form (BRE 2005; Reid *et al.* 2005). However, these terms may be misleading given that, over engineering timescales, oxidation of 'non-reactive' pyrite may well occur, albeit at a slow rate. This may not be considered significant where construction is concerned (BRE 2005; Reid *et al.* 2005; Hawkins 2014). However, physical damage to crystals, low crystal symmetry and surface defects (Belzile *et al.* 2004), or the catalysing presence of pyrrhotite (Hagelia *et al.* 2001; Hawkins and St John 2014), may also render macroscopic forms more reactive (see also Czerewko and Whitworth 2025).

Many cases of degradation of concrete related to the presence of sulfide minerals have been reported by Bromley and Petifer (1997) in SW England. Referred to as the 'Mundic' issue, these are attributed to the oxidation of pyrite and pyrrhotite in mine waste used widely for the production of cementitious building blocks. However, pyrrhotite-related problems have been documented more extensively from outside of the UK, with examples from the USA, particularly in Connecticut and Massachusetts, where widespread failure of foundations occurred due to pyrrhotite in locally sourced construction aggregates (Mauk *et al.* 2020). Penner *et al.* (1973) note that pyrrhotite caused swelling of rock aggregates in large areas of Canada. Moum and Rosenqvist (1959), Hagelia *et al.* (2001) and the Nordic Concrete Federation (2018) describe sulfate attack resulting from oxidation of pyrrhotite in Cambrian shales in Norway, Sweden and Finland. While Chinchón-Paya *et al.* (2012) reported problems of concrete degradation caused by oxidation of pyrrhotite present in aggregates sourced from the Pyrenees in Spain.

Occurrence of gypsum and other sulfate minerals

Sulfates, of which gypsum is the most common form found at or near the ground surface, tend to originate as ancient or modern evaporite deposits from the dehydration of saline water bodies, as can be seen in Figure 1a. They are also deposited by precipitation from circulatory solutions, or by deposition as exhalative deposits from volcanic gasses or heated waters. Secondary forms of sulfate minerals, particularly the selenite form of gypsum, also evolve as naturally occurring salts of sulfuric acid through oxidation of pre-existing sulfide minerals such as pyrite (Fig. 1d).

Primary sulfate minerals, including gypsum and anhydrite, occur extensively in the UK within evaporite deposits of Permian and Triassic age that underlie parts of the Midlands and NW of England (see Figs 3 & 8a). They are also present in minor amounts in some marine sedimentary clay deposits of Mesozoic and Cenozoic age in central and southern England. Secondary gypsum derived by the dissolution of primary gypsum by percolating groundwaters may occur at depth through reprecipitation in voids and pores (Jarzyna *et al.* 2024). Secondary gypsum derived from pyrite

Fig. 8. (**a**) Primary gypsum as thin beds of 'satin spar' in Mercia Mudstone, from south Nottinghamshire, UK. (**b**) Secondary form of gypsum as disc-shaped selenite crystals at the desiccated weathering front in Oxford Clay from near Calvert, Oxfordshire, UK.

oxidation, as seen in Figure 8b, also occurs at shallow depths of between 2 and 8 m within the ground profile in areas where pyrite is present in the bedrock (see Cripps and Edwards 1997). Other sulfate minerals may also be present, including the highly soluble forms of hydrous magnesium sulfate, 'epsomite', and hydrous sodium sulfate, 'mirabilite', as well as the relatively insoluble recalcitrant strontium sulfate, 'celestine', and barium sulfate, 'barytes'.

In the UK and other temperate climatic areas, significant quantities of gypsum may be encountered in weathered Carboniferous Coal Measures deposits and the weathering profiles developed on ancient Mesozoic- and Cenozoic-age marine clay deposits, as well as in some recent sand and alluvial deposits. These strata, which contain pyrite, are of significance as they form large areas of outcrop on which many British urban centres are located, as seen in Figure 3. The distribution of sulfate minerals can vary widely within the weathered zone, with the top few metres having negligible sulfate contents due to rain leaching, but elevated levels may be present at the base of the root zone at *c.* 2–3 m depth and at the base of the weathered zone (see Fig. 8b), although this may be affected by past groundwater levels (Cripps and Edwards 1997; Reid *et al.* 2024). It may take many thousands of years for weathering to progress a few metres into these clay deposits (Chandler 1972), so where the weathered horizons have been removed or groundwater levels changed, the distribution may not have become established. Such variations can make it difficult to characterize the ground accurately in terms of its sulfur content (see St. John 2020). However, in areas with a mature weathering profile, it is likely that most soluble sulfates will have been leached out of the near-surface zone where most terrestrial civil engineering works such as routeways and residential developments are carried out. But they may be present below the leached zone and encountered in excavations for substructures, foundations, shafts and tunnels, where they can cause serious problems to civil engineering works, including degradation of construction materials, ground movements and subsidence.

Ritsema and Groenenberg (1993) explain that in soils the presence of gypsum formed through the oxidation of pyrite tends to be seasonal, in that gypsum readily dissolves during the wet season and tends to persist during the dry season. Seasonal variations have also been recorded during earthworks in the UK, with development of discrete gypsum crystals occurring in the dry periods following heavy rainfall events soon after initial rapid efflorescent development (Czerewko and Cross 2015). Although gypsum has relatively low solubility (Deer *et al.* 1992), in mobile water conditions it is slowly but continuously dissolved. The resulting sulfate-rich waters will be chemically aggressive to any concrete and cementitious material they encounter. However, large discrete selenite crystals tend not to be so readily redissolved or removed by groundwater unless an acidic environment develops (Hawkins and Pinches 1987). This transportation of reaction products is a significant factor for ground heave involving selenite–gypsum formation, as seen in Figure 9a, as expansion observed in rock masses or engineered fills tends to be caused by growth of discrete crystals, and not by systematic filling of void space (see Fig. 9b).

Sulfate in arid environments

Gypsum and anhydrite also occur widely in arid environments where they can be problematic for construction (Fookes *et al.* 2013). Most authorities regard arid environments as places where the annual evapotranspiration exceeds rainfall, resulting in increased salinity of groundwater and surface water, such that precipitation of soluble minerals occurs (Bell 2000). Fookes (1978) explains that

Fig. 9. Mam Tor landslip in Derbyshire, UK: (**a**) damage to the former A-trunk road due to a landslip in which oxidation of pyrite-bearing Carboniferous-age Bowland Shales is occurring; and (**b**) expansion in Bowland Shale caused by selenite growth in discontinuities.

accumulations of gypsum are usually found in centrally draining deserts, including salt playas or salinas or in coastal sabkhas. Its presence, and that of other soluble minerals, leads to chemical conditions that are aggressive to construction materials, the expansive growth of minerals, and the hydration and dehydration of minerals that result in volume changes and cause problems with locally sourced aggregates through contamination (Taylor and Cripps 1984). Not all arid areas are hot but rainfall may be sparse or seasonal. In some areas, including Arabia and North Africa, variations in sea level over time have resulted in accumulations of alternating beds of calcarenites, carbonate sandstones, gypsum, sands and cemented sands (Fookes *et al.* 2013). The effects of wind erosion, capillary action and evaporation have produced extensive sabkha deposits in certain areas, particularly along shorelines and inlets that contain hypersaline deposits, including gypsum and anhydrite. Further inland these and other soluble minerals are likely to be the products of major climatic changes experienced through the Quaternary where gypsum was deposited through evaporation of saline palaeolakes (Fookes 1978; Fookes and Lee 2018).

In arid environments, silts and other sediments that accumulate in ephemeral lakes and floodplains become cemented or partially cemented by soluble minerals. The absence of water infiltration from the surface provides conditions in which soluble salts precipitate in the intergranular pore space, resulting in cementation. This may lead to a cemented gypcrust at the surface, or the development of cemented nodules or layers within the soil profile (see Fig. 10a). Soil profiles may possess cemented crusts of gypcrete or calcrete, or subsurface layers where upward movement of water by capillary action and evaporation causes cementation (see Fig. 10b). Capillary action may raise water by between 2

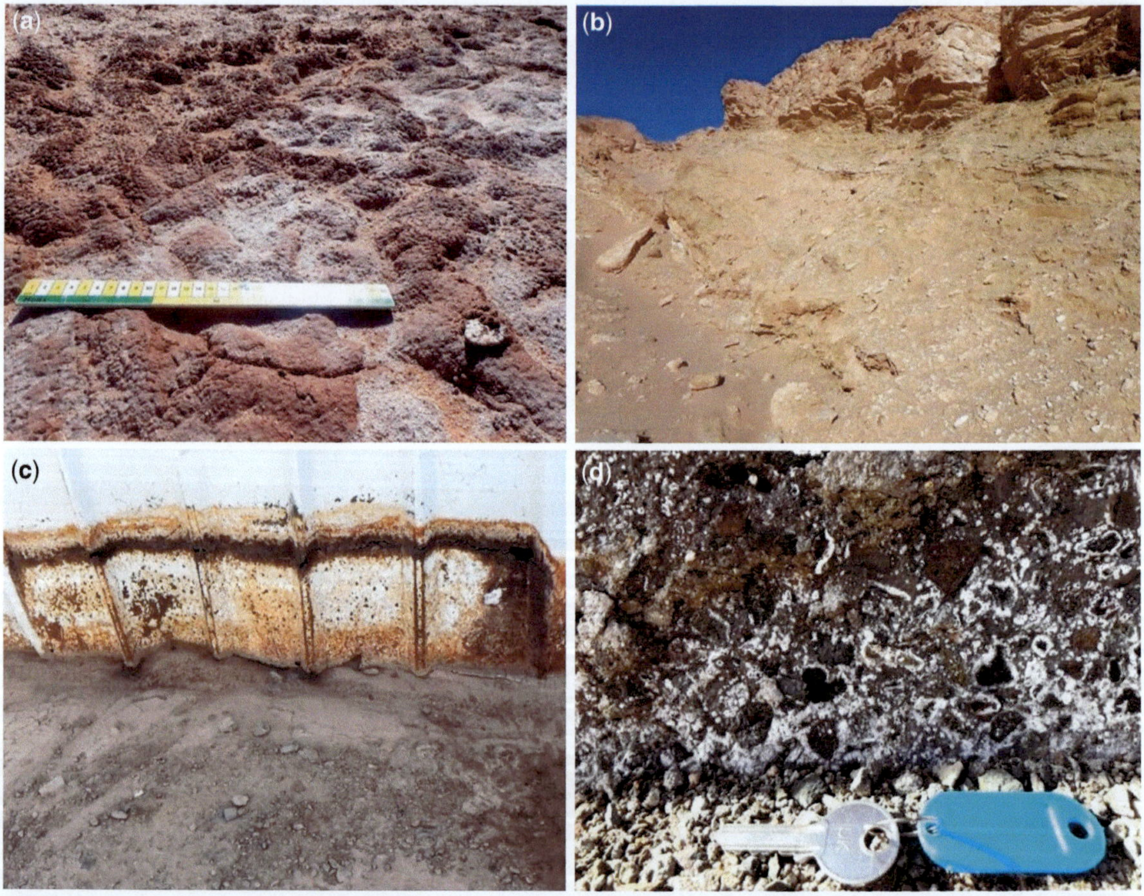

Fig. 10. Examples of conditions formed in dry arid environments: (**a**) sabkha surface nodular crust with a concentration of gypsum and halite salts, Gulf of Aquaba; (**b**) rapid gullying and slope deterioration in gypsiferous weak sedimentary rock, Lisan Formation, NW Saudi Arabia; (**c**) corrosion of steel in contact with windblown, halite salt-enriched ground, Libyan Sahara; and (**d**) physical deterioration of ground-bearing concrete due to crystallization of sulfate and chloride minerals mobilized through capillary suction, Duba, Saudi Arabia.

and 10 m from the water table to the site of precipitation, depending on the smallest voids in the soils (Bell 2000; Barnes 2010).

Inland and coastal sabkhas, which are extensive areas of relatively flat terrain, pose challenging conditions for engineering as strong cemented layers and duricrusts are liable to be interbedded with weaker, compressible layers. Salt, including sulfates and chlorides, formed in such areas will create chemically aggressive ground conditions, and the movement of sulfates and other salt particles by wind action results in the transfer of potentially aggressive chemical conditions over great distances (see Fig. 10c). The deposits are usually normally or lightly overconsolidated with low strength and high compressibility. Due to changes in the position of coastlines, and changes to the sedimentation, groundwater conditions and diagenetic processes, the deposits are usually vertically and laterally variable. In coastal sabkhas in hot arid areas, gypsum is usually precipitated in the zone between the high and low water levels, and anhydrite can be precipitated in the zone reached only during storms or exceptionally high tides (Fookes *et al.* 1985). Due to the dominance of physical weathering processes over chemical ones, granular soils tend to dominate in contemporary soil profiles with particles consisting of rock and mineral fragments, and sometimes grains comprising clusters of cemented particles. Soils are liable to be in an unsaturated condition such that the relief of porewater suctions when they are wetted results in a dramatic reduction in strength and the collapse of voids. In coastal areas there is a risk of an accumulation of sulfate ions in concrete structures due to drying and wetting cycles or capillary suction that can lead to rapid progressive damage, as seen in Figure 10d

Impacts and the key causes of problems

Environmental controls of pyrite oxidation

The process of pyrite oxidation is a complex phenomenon, and the actual reaction and rates tend not to correspond exactly with any conceptualized process. Although the oxidation process and formation of secondary minerals (e.g. selenite) are frequently represented by a series of chemical equations, such as those in Figure 11, these are a simplistic representation of thc processes involved such that they can cause misunderstanding and confusion (Nordstrom 1982). Therefore, the following section focuses on aspects of pyrite oxidation that are liable to be affected by environmental conditions which occur because of engineering and environmental construction works. Readers who require further details, a more detailed review and a source of further references may wish to refer to Czerewko and Cripps (2023).

In its simplest context, the oxidation of pyrite is initiated by iron atoms through interaction with both oxygen and water. The ensuing reaction is electrochemical, involving the loss of an electron from the ferrous iron in pyrite, producing ferric iron. The oxidizer is the electron acceptor, which can be ferric iron or oxygen dissolved in liquid water or water vapour. Jerz

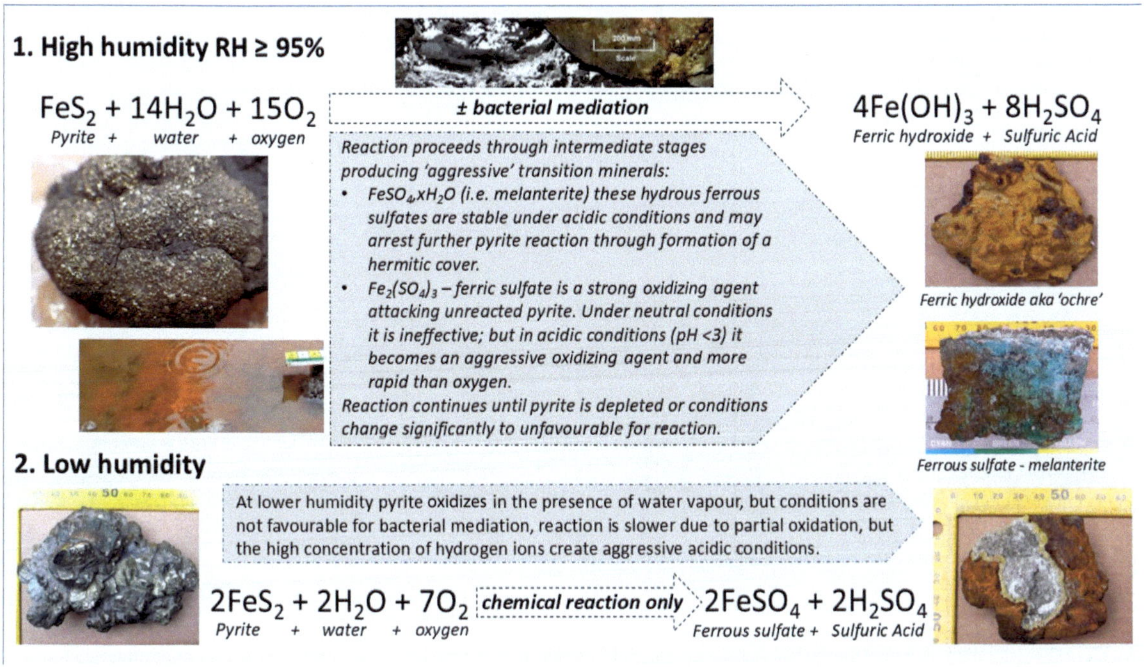

Fig. 11. Simplified visual schematic representation of the pyrite oxidation reaction consequential to the built environment.

and Rimstidt (2004) have shown that over short timescales the oxidation rates are higher in air than in water. Oxidation of reduced iron sulfide minerals occurs rapidly with conversion of the sulfide sulfur atom through exchange of an electron or hydroxyl ion to produce a sulfate ion, which being soluble in water forms sulfuric acid and is removed in solution. The While the pyritic iron precipitates as the insoluble reddish-brown iron hydroxide. Therefore, the natural weathering of pyrite requires both water and oxygen to be present (Silverman 1967; Edwards *et al.* 1998), both of which are readily available in all commonly occurring construction environments.

Howie (1977) explains that the oxidation processes in aqueous and non-aqueous environments are fundamentally different. In ground engineering situations in the UK, the weather conditions and groundwater regime are liable to vary widely during the construction period, such that the impact of pyrite changes as construction proceeds and depends on the season. Exposure to air and water during construction will result in pyrite oxidation processes. Temporary or permanent lowering of the water table is also liable to be responsible for oxidation. Consequently, although any adverse effects of pyrite oxidation may occur during construction, they may not come about until many years after construction has been completed. Not anticipating and accommodating these processes in engineering designs and construction methods has the potential for serious consequences, including the need for changes to the design or construction method, or complex and expensive remedial works during or after construction. In extreme cases there has been a failure of an engineered structure to perform its designed function, such as failure of the Carsington reservoir dam described by Reid and Cripps (2019). The key aspects pertinent for ground engineering are briefly considered in the following section, but if further details are required, they may be found in the vast scientific literature available on the subject (Lowson 1982; Pye and Miller 1990; Hawkins and Pinches 1997; Rimstidt and Vaughan 2003; Vaughan 2006; Hawkins 2014; Ricard 2015; Nordstrom 2020; Tacker 2020).

Experience with curation of pyrite-bearing samples in museums reported by Hodgkinson and Martin (2004) suggests that to prevent deterioration they should be stored at a RH of below 60%; unless they are highly organic, in which case storage should be at a RH of below 30%. These factors are pertinent to ground engineering in the UK as carbonaceous and pyrite-bearing mudrocks and clay soil bedrock feature prominently in construction environments, often in conditions where RH is above 30% or free water is present. Deterioration of stored core material is not uncommon and is occasionally observed where microbial-mediated oxidation occurs under controlled low RH storage conditions, as seen in Figure 12a. Such changes are important as they result in cores being unrepresentative of the *in situ* ground conditions.

In dry conditions, such as that of semi-arid and arid climates, and periodic hot dry spells that occur in temperate climates, such as in the UK, water may be available intermittently during rainstorms or dew when the onset of rapid reaction may then be initiated (see Fig. 12b). However, during dry periods, purely abiotic partial oxidation tends to occur such that pyrite survives with an oxidized veneer, as seen in Figure 13a; while in temperate settings, efflorescent minerals and visible crystals may develop in dry periods, as seen in earthwork stockpiles (see Fig. 13b). Hawkins (2014) suggests that surface coatings may inhibit further oxidation. Their formation in crystal fractures or within disaggregated framboidal forms, however, may induce the break-up of mineral and rock fragments, sometimes in an

Fig. 12. (**a**) Deterioration of pyrite-bearing Blue Lias core material from north Somerset, UK under controlled storage conditions maintained at a relative humidity (RH) of *c.* 40–50% and temperature of 7–10°C. (**b**) Rapid deterioration of the Ancholme Clay Formation with selenite development following a short heavy rain event, despite it having previously been stable for several days under dry sunny conditions.

Fig. 13. (**a**) Pyrite cubes under a veneer of iron oxide remain stable in exposure for many years under dry arid conditions in Saudi Arabia. (**b**) Selenite crystal up to 120 mm in length formed in less than 5 weeks in a stockpile of Blue Lias Clay in the East Midlands of England. (**c**) Rapid oxidation of pyrite-bearing strata under humid conditions in Gabon, with deterioration of a pyritic amphibolite dyke leaving an orange-brown ferric oxide residue.

explosive manner, and expose fresh surfaces that are susceptible to further reaction, which accelerates reactions. Furthermore, efflorescent minerals are capable of capturing water from the air, thus facilitating the transportation of sulfates in solution even in the absence of other sources of liquid water.

Enhanced rates of pyrite oxidation are a feature of hot wet climates typical of tropical regions but may occur elsewhere in wet construction conditions. In tropical regions, chemical weathering of the primary minerals embodied in geological material is more intense and occurs to greater depth than elsewhere, often extending to many tens of metres below the ground surface, as seen in Figure 13c. Pyrite oxidizes rapidly to 'ferric' iron sulfate species that convert in wet conditions to hydrated iron oxides (Chandra and Gerson 2010). As hydrated iron oxide has low solubility, it remains *in situ* and becomes integrated into red lateritic soils that overlie saprolitic deposits as a residue-weathered pyrite-bearing bedrock (Fig. 13c). Therefore, pyrite tends to be less frequently encountered in tropical environments due to the presence of deep weathering profiles but will be present at depth in fresh unweathered bedrock or in the unweathered parts of corestones. The impacts of pyrite becoming weathered in these situations needs to be evaluated.

As under aqueous conditions, the reaction of pyrite is largely electrochemical, the rate of oxidation is principally controlled by the solution Eh, although the solution must contain dissolved oxygen for the reaction to persist. Therefore, pyrite can remain unchanged when in contact with static unaerated water, but the introduction of air through agitation or groundwater flow associated with engineering works is liable to result in oxidation. Other factors affecting the reaction rate include the solution pH, oxidant type and concentration, pyrite grain size, and the surface area and form, as well as the temperature and pressure conditions (Czerewko and Cripps 2023). In damp-humid conditions where water vapour or liquid water are abundant, typical of tropical and many subterranean tunnelling and mining environments, oxidation will be initiated as a chemical reaction that through bacterial catalysis will proceed at a greatly increased rate (Hawkins and Pinches 1997; Sasaki *et al.* 1998; Chandra and Gerson 2010; Hawkins 2014).

As the oxidation of pyrite proceeds under aqueous conditions sulfuric acid is produced; this is extremely corrosive to both geological bedrock and engineering materials such as aggregates, concrete and steel. If calcium carbonate (calcite) is present as disseminated cement, or as veins, shell or limestone fragments, reaction with the sulfuric acid will produce selenite–gypsum as discrete crystals; in more indurated rock-like mudrocks, crystallization generally occurs along bedding planes and discontinuities, and this involves expansion (see Fig. 14a). While in less well-indurated soil-like marine compaction mudrocks and clay deposits in which pyrite occurs as a fine-grained, disseminated component, oxidation and subsequent reactions involve calcite, which is likely to be present in a fine-grained disseminated form or as shell fragments, and usually result in randomly distributed selenite–gypsum crystals being formed within the weak rock-soil mass, as seen in Figure 14b. In the absence of calcium carbonate, reactions with potassium and iron minerals and feldspar in the deposit may result in the formation of the acidic sulfate, jarosite.

The replacement of pyrite and calcite by selenite involves a volume increase (Czerewko and Cripps 2023) and may cause heave of the rock mass or filled ground. In wet conditions, the oxidation products may be transferred in solution by water flow or capillary migration, so that crystallization of selenite occurs remotely from the pyrite in low-stress zones. Typically, these selenite crystals form as discrete crystals at the fabric and structural interfaces such as joints or bedding surfaces in rock masses (see Fig. 15a), or at embankment, foundation and subgrade interfaces where nucleated growth may produce large multiple crystal forms (see Fig. 15c), or at construction boundaries such as between ground-bearing concrete structures and fill (see Fig. 15b), causing heave (Nixon 1978; Hawkins and Pinches 1987; Wilson 1987; Hoover and Lehmann 2009). Discrete selenite crystals expand the space in which they are situated as less surface energy is required to enlarge an existing crystal than to create a new one. Therefore, there is a tendency for existing crystals to grow even if they are constrained by the size of the space in which they are situated. This process may increase the size of pores without them being completely filled, so the extent of expansion can be much greater than the volume of the precipitated minerals or commensurate with the original pyrite and calcite from which the solutions were derived. The pressure crystals are capable of exerting depends on the concentration of ions in solution, the temperature and the molecular volume of the compound concerned, but they are sufficient to cause uplift of lightly loaded structures, including the concrete floor slabs of two- or three-storey dwelling houses supported on pyrite-bearing ground or fill (see Czerewko and Cripps 2022).

Fig. 14. (**a**) Heave in 'rock'-like mudrock resulting from precipitation of well-formed selenite along joints (Jurassic Blue Lias Mudrock). (**b**) Heave in 'soil'-like mudrocks with randomly distributed selenite crystals within the mass (Jurassic Oxford Clay).

These chemical alterations are frequently accompanied by rapid deterioration in the engineering properties of geomaterials, including softening and increased water content. Furthermore, the significant volume changes involved in the removal or alteration of pyrite, dissolution of calcite or other cements, and the hydration and growth of sulfate minerals produce ground heave that may cause damage to structures and earthworks, and may also initiate slope failure (Cripps *et al.* 1993). The dissolution of selenite crystals also constitutes a source of sulfate-rich solutions that create chemically aggressive conditions for construction materials.

Implications of pyrite oxidation in ground engineering

While there are many instances of successful engineering construction in, on and out of soils and rocks containing significant amounts of pyrite, as is clear from the forgoing text, problems with pyritic soil and rock extend globally and across many disciplines. It has been reported by Penner *et al.* (1973) that quantities of pyrite even as small as 0.1% by weight have been sufficient to cause problematic heave of building foundations in Canada. Ground conditions can rapidly deteriorate, as oxidation of disturbed or exposed pyrite-bearing material can be swift following entry of oxygenated water during construction. Because of this, geotechnical parameters determined during a ground investigation may not accurately reflect the properties of the materials at the time of construction or service life of the structure. Furthermore, depending on the groundwater conditions, the

Fig. 15. (**a**) Selenite crystals forming as a discrete 'tooth'-like arrangement prising apart rock joints in Carboniferous Bowland Shale. (**b**) Well-formed selenite crystals growing at the interface between rock fill sourced from the Lower Carboniferous Malahide Formation near Dublin and a ground-bearing concrete slab. (**c**) Large cluster of multiple selenite crystals in Jurassic Lower Lias Clay from Cotgrave, south of Nottingham, UK.

products of pyrite oxidation processes may be carried away from the reaction site, thereby compromising the exposed ground, and causing softening and an increase in water content (see Czerewko *et al.* 2011). In low-permeability crystalline or cemented pyrite-bearing rock, oxidation may be slow, but once it has commenced it has the potential to give rise to troublesome acidity, deterioration of geotechnical properties and volume changes.

Significant weathering processes can occur during the course of civil engineering construction works, particularly when materials are exposed in excavations for periods of time or excavated and placed as fill with exposure to percolating rainwater and air. Reid *et al.* (2024) explain that a free-draining embankment fill is one of the most oxidizing environments in nature. Aggressive acidic conditions can develop in these situations, leading to a greatly increased solubility of sulfate and the potential for chemical attack of construction materials. The oxidation of pyrite can also result in changes to samples used to determine key geotechnical properties that are the basis of geotechnical designs, which can result in designs that are more conservative than they need to be.

Many cases are reported where undisturbed pyritic materials underlie foundations in which pyrite oxidation processes cause mechanical damage to structures through expansive mineral formation (Penner and Eden 1972; Hawkins and Pinches 1987; Hoover *et al.* 2004). Similar mechanical damage to structures has been attributed to the expansion of pyrite-bearing fill material (Nixon 1978; Carr *et al.* 2006; Maher *et al.* 2011). Although inundation with air-free water would arrest the reaction, this would not be a feasible option for problem prevention in the construction environment. In any case, exclusion of water may not stop the reactions once they have started. It is also accepted that a high content of calcium carbonate will neutralize the acidic environment, rendering conditions unfavourable for microbial mediation and slowing down the rate of pyrite oxidation (Murad and Rojik 2003). However, studies of cave environments by D'Angeli *et al.* (2018) indicate that by generating acidic biofilms, microbes can create microcosms in which acidic conditions persist and the same could occur in construction settings. But in the authors' experience, even in impure pyritic limestone, once started a reaction can continue rapidly until all available pyrite is consumed (see Fig. 16).

The amount of ground movement caused by selenite formation can be quite catastrophic to lightly loaded structures such as highways. In recent years, serious distress due to foundation uplift has been observed in residential developments involving buildings up to three stories high in Japan (Yamanaka *et al.* 2002), the USA (Anderson 2014), Canada (Maher 2014) and Ireland (Hawkins 2012, 2014). Czerewko

Fig. 16. Rapid oxidation of pyrite hosted in limestone, with the generation of local acidic water before neutralization with the formation of granular selenite in Lower Carboniferous (Dinantian) limestone from Derbyshire, UK.

and Cross (2015) describe cases of damage to highways in the UK, but similar problems occur in other countries where pyrite-bearing ground is encountered. In these situations, in well-indurated and cementation mudrocks, sulfate minerals grow on well-developed bedding planes and fissility, which results in expansion of the bedrock or aggregate fragments. The sulfate ions in acidic solution generated through the sulfide oxidation process will react with any carbonate minerals present to produce the selenite form of gypsum that may be precipitated through nucleated growth at a site remote from the source minerals due to groundwater movement. Hawkins and Pinches (1987) describe that the growth of stratified bladed selenite crystals in discontinuities is a dominant factor in vertical heave of structures, as can be seen in Figure 14a. In the less well-indurated deposits, it would appear that 0.5–8% $CaCO_3$ provides optimum conditions for rapid pyrite oxidation and expansion. However, higher concentrations of calcium carbonate may be tolerated in situations where a throughflow of water exists, such as in stockpiles or spoil heaps (Fig. 13b), as an acidic environment is maintained and the buffering reaction contact time is reduced by a high water-flow rate. Not only does a higher calcium carbonate content result in the acid being neutralized, but its cementing action makes a stronger rock in which the formation of selenite crystals is less likely to cause widening of fractures; however, reactions do occur where the rock becomes fragmented or processed for use as aggregate, as seen in Figure 16

The amount of expansion bought about through the formation of new sulfate minerals depends on the volume gain of the new products (Hawkins 2014). For example, the formation of jarosite ($KFe_3(SO_4)_2(OH)_6$) involves a 115% volume increase compared with the original volume occupied by pyrite. A 536% volume increase occurs when melanterite ($FeSO_4{\cdot}7H_2O$) is precipitated, and the subsequent formation of gypsum resulting from the reaction of oxidation products of pyrite and calcite produces a 103% volume increase (Taylor 1988). Cases of heave resulting from pyrite oxidation have estimated resultant heave pressures ranging widely from 14 to 500 kPa (Taylor and Cripps 1984) and forces as high as 5 MPa within fragments of rock (Strogen 2021) with the development of differential ground heave. The complexity of the chemical reactions and variations in the ground conditions result in a large range of end products and effects.

Although the mechanism of pyrite oxidation and the types of deleterious reactions that may occur have been discussed, the implications to construction depend on the morphology of the pyrite and are controlled by site-specific conditions related to the type and properties of the ground host and construction materials present, and the current and future site environment (Czerewko and Cripps 2023). Air and water are required in appropriate quantities to maintain the reaction and remove reaction products from the system. In static saturated conditions, such as below the water table, the reaction is usually very slow and may be arrested, whereas mobile water transfers ions in solution that expose fresh surfaces to reaction, such that reactions continue to occur. In roadway pavements and housing construction, the heave process can be driven kinetically through temperature contrasts that develop in response to heat. Moisture and dissolved oxidation products may be drawn up by capillary rise generated through the heat from the dwelling or sun-heated road surface, resulting in the precipitation of expansive reaction products beneath the structure. The resulting heave and structural damage (see Fig. 17) may be manifested sometime after construction.

Damage to structural concrete and buried steel

The deleterious effect of sulfates on concretes and any contained steel reinforcement has long been appreciated by engineers (Fookes and Collis 1975; Hawkins and Pinches 1987). Buried and ground-bearing concrete, cementitious materials, and steel are vulnerable to acid and sulfate attack caused by sulfide oxidation and the presence of sulfate minerals in the ground. The source of sulfate is generally external, including soluble sulfate minerals such as gypsum, or sulfate generated by pyrite oxidation. Acidic conditions result in corrosion of concrete and steel, and mobile sulfate ions react with concrete constituents with destructive consequences. An internal sulfate source may also be present (see Fig. 18), as reported by Shayan (1988) who noted that pyrite or pyrrhotite-bearing aggregates were used for the manufacture of concrete. In hot arid regions where high concentrations of soluble salts occur locally in the groundwater, sulfates may also be introduced during concrete production by the water added to the mix. Fookes and Lee (2019) also allude to Middle East concrete being affected by aggregate contaminated with evaporite salts and chemical impurities, especially chlorides and sulfates. They also mention that pyrite can be a problem if it is present in aggregate, although this is not common in the Middle East.

When acidic conditions develop, damage may occur through volumetric loss of the carbonate, calcium hydroxide and calcium hydrate in the cement binder (Hagelia *et al.* 2001), as seen in Figure 19, caused by dissolution. But the most frequent damage occurs through penetrative chemical attack by mobile sulfate ions on the cement binder in concrete, which results in the expansive formation of new sulfate minerals in pores and cracks. Such problems can be avoided by designing the concrete to resist the chemically aggressive conditions and/or by the provision of appropriate protection measures (Penner *et al.* 1973; BRE 2005).

Fig. 17. Structural damage to ground-bearing structures caused by ground heave: (**a**) and (**a′**) damage to an industrial floor impairing use, caused by the precipitation of sulfates from the Coal Measures fill by capillary rise; (**b**) and (**b′**) heave of pre-split hydraulically bound road pavement caused by selenite development at the construction interface; and (**c**) and (**c′**) heave of domestic a ground-bearing floor slab seen through a vinyl flooring overlay caused by thermally driven selenite development at the foundation interface from burnt colliery shale fill.

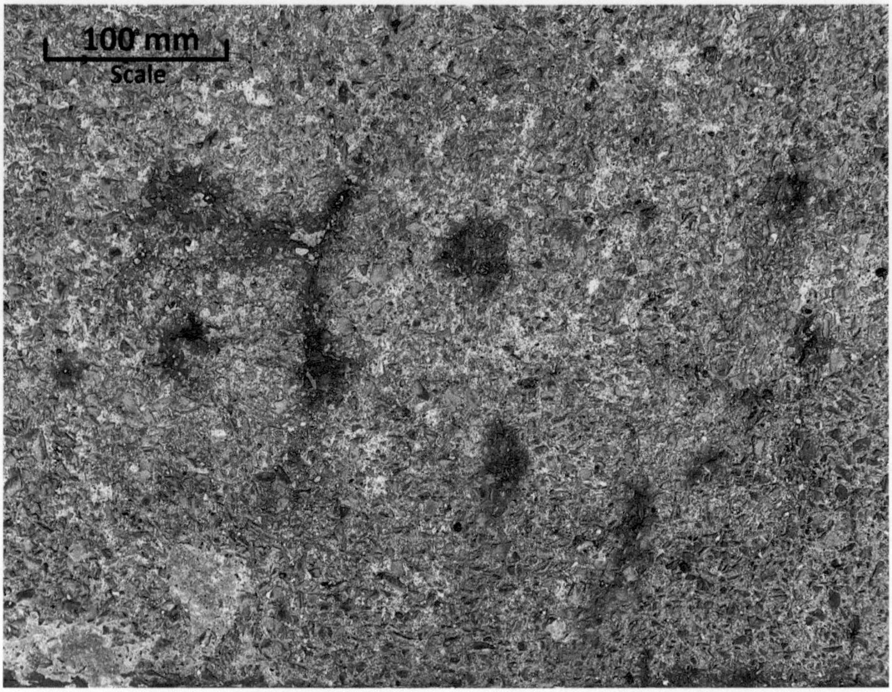

Fig. 18. Deterioration of pyritic limestone aggregate used in concrete manufacture.

The BRE (2005) report explains that there are two recognized forms of sulfate attack on concrete. The first type, which has long been recognized, is referred to as the 'conventional form of sulfate attack'. This involves the conversion of calcium hydroxide and hydrated calcium aluminate in the cement paste into expansive gypsum and ettringite (a hydrous calcium aluminium sulfate mineral), which causes a physical break-up of the cement or concrete (see Fig. 20a, b).

The second form of sulfate attack is referred to as the 'thaumasite form of sulfate attack' and causes the cement binder to be converted into thaumasite, which is a weak hydrated calcium silicate, carbonate and sulfate mineral that forms a residual soft paste (see Fig. 20c, d). Although known about since the early 1960s, it did not come into prominence until the late 1990s as identification techniques improved (Crammond 2003). Thaumasite formation entails the presence of sulfate anions in mobile water, which react with hydrated calcium silicate present in the cement and calcite present in the ground or in concrete aggregates. The process is more prevalent where temperatures are below 15°C and requires pH conditions greater than 10.5, which are typical of the alkali nature of cements and concretes (Crammond 2003).

Dissolved sulfate ions in mobile groundwater are also a prime factor in chemical attack on engineered ground, particularly where lime or cement stabilization techniques have been used (Longworth 2004). Ground improvement involving the addition of lime and or cement to soils results in a reduced water content, increased strength and a reduction in the capacity of certain clay minerals to swell. The process produces alkali soil conditions (pH $\geq$12) but, as pyrite is unstable where pH >10, rapid oxidation occurs (Thomas *et al.* 1989; Britpave 2019). This results in the formation of expansive minerals, including gypsum, ettringite and thaumasite, in the treated soil (Higgins *et al.* 2002), which may result in the heave of structures.

Impact of pyrite oxidation in UK construction and environmental projects

Despite well-documented instances of problems in ground engineering due to pyrite, the deleterious reactions caused by the aggressive ground conditions and mineralogical changes associated with pyrite and sulfate minerals often come as a surprise to UK ground engineering professionals. Reid *et al.* (2024) attributed this to the fact that most engineers lack a background in geochemistry and assume that chemical reactions only happen in laboratories or industrial processes. Soils and rocks are assumed to be inert and will not undergo structural or chemical change other than changes in moisture content through and beyond the construction period. As Figure 3 shows, many UK geological deposits contain sulfur minerals for which this assumption is erroneous, and the same problems recur despite the basic principles being known and appropriate supporting guidance being available (Hawkins 2014; Cripps *et al.* 2019).

Fig. 19. Examples of corrosive effects of pyrite-generated acidity: (**a**) dissolution of cementitious blockwork caused by acidic mine water; (**b**) dissolution of a cementitious binder in a concrete floor slab seen on the left side of the image exposing the internal aggregate due to the flow of acidic groundwater from pyrite oxidation, while the intact concrete on the right was not affected by the flow; (**c**) dissolution of mortar in a brick-lined tunnel caused by acidic groundwater from pyrite oxidation; and (**d**) deterioration of a concrete culvert lining caused by pyrite oxidation-generated acidic water.

Aggressive ground conditions arising from the presence of pyrite and gypsum have been responsible for widespread and often serious impacts on ground engineering in the UK and Ireland. Although many cases are not reported due to confidentiality or legal issues, a number of instances have been cited by authors, and serve as useful reference and guidance for practitioners, including Pye and Miller (1990), Hawkins (2012), Czerewko *et al.* (2016) and Cripps *et al.* (2019). The types of deleterious categories can be summarized as follows:

- Rapid degradation of earthwork fill containing pyritic mudrocks of Namurian age, degradation of limestone drainage layers and fatal production of carbon dioxide are reported in Pye and Miller (1990) and Reid and Cripps (2019).
- Impacts of pyrite oxidation on groundwater composition at the major slope failure at Mam Tor in Derbyshire, UK (see Fig. 21a), are described in Vear and Curtis (1981), and effects of these on the residual strength of mudstone were investigated by Steward and Cripps (1983).
- Corrosion of buried steel culvert linings on the A564 Hatton–Hilton–Foston road in Derbyshire, UK, as seen in Figure 21b, caused by the oxidation of pyritic backfill is reported in TRL Report 447 (Reid *et al.* 2005) and corrosion of stainless-steel reinforcement in spent oil-shale backfill to a retaining wall is described by Winter *et al.* (2002).
- Enning and Garrelfs (2014) review the corrosion of buried iron gas pipelines by sulfate-reducing bacteria in an anoxic, sulfate-containing soil.
- Acid attack on buried structural concrete or concrete in contact with the ground, including foundations, culverts, piles and tunnel linings (see Fig. 19c, d), is reviewed by Eglinton (1975). Bracegirdle *et al.* (1996) describe corrosion of cast iron linings in London Underground tunnels by acid released through oxidation of pyrite present in a

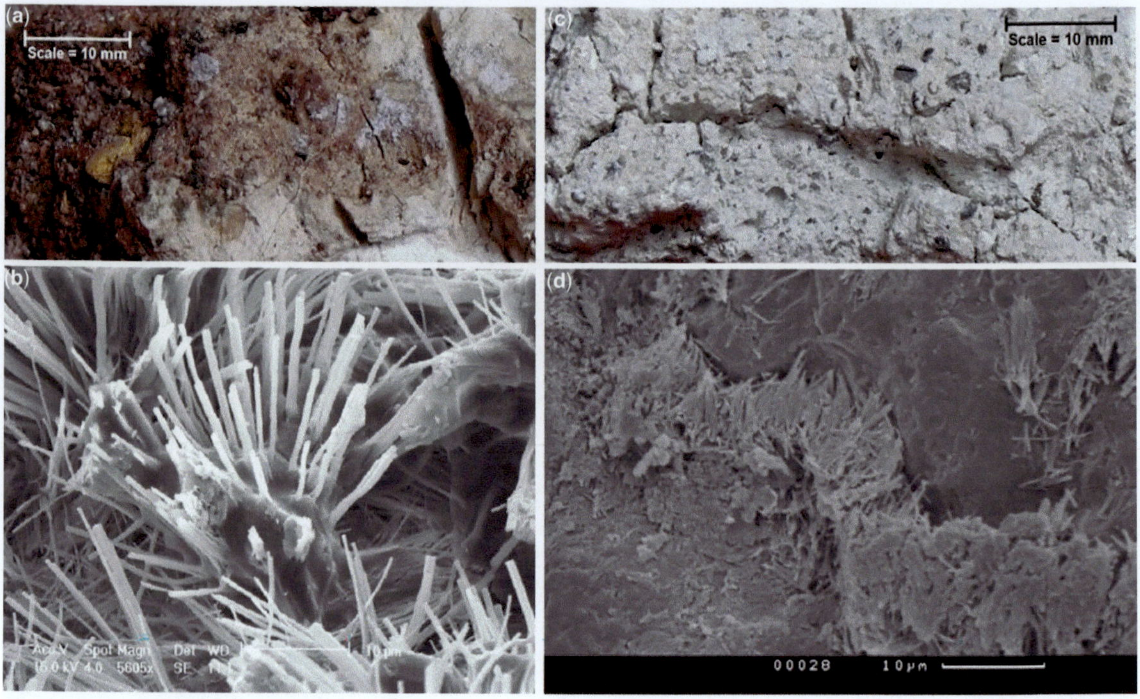

Fig. 20. Forms of sulfate attack affecting cement and concrete: (**a**) ettringite-damaged concrete; (**b**) acicular ettringite crystals; (**c**) example of the thaumasite form of sulfate attack on concrete; and (**d**) matted development of thaumasite affecting the cementitious mortar binding aggregate grains.

sandy horizon in what are likely to be Paleocene–Eocene deposits comprising the sandy Harwich and basal London Clay formations (Ellison *et al.* 2004).

- Extensive research literature and overviews on sulfate attack on buried structural concrete are given by the Thaumasite Expert Group (DETR 1999), in Crammond (2003) and in BRE Special Digest 1 (BRE 2005).
- Ettringite-form of sulfate attack that caused expansion and warping of floor slabs laid on sulfate-bearing ground and fill, and lime/cement-treated pyritic fill is described by Longworth (2004, 2008).
- Thaumasite-form of sulfate attack that caused significant deterioration of buried concrete structures is reported by Crammond (2003), who explains in the abstract of the paper (p. 809) that the thaumasite form of sulfate attack 'completely destroys the binding ability of concrete by transforming it into a mush' (see Fig. 21c). Rapid disintegration and expansion of crushed concrete used as fill material has also been reported by Watts and Charles (2015) who identified this as a new category of attack in the UK.
- Heave of roadways due to a volume increase of subgrades and earthwork fill containing pyritic mudrocks are reported by Czerewko *et al.* (2011) involving Jurassic Scunthorpe Mudstone deposits, and by Czerewko and Cross (2015) involving Oxford Clay and Ancholme Group mudstone deposits, also of Jurassic age.
- Heave problems caused by ground stabilization for highway construction and foundations involving pyrite-bearing Jurassic stiff clays treated with lime and cement are reported by Thomas *et al.* (1989), Snedker and Temporal (1990) and Longworth (2004).
- Expansion of construction fills containing pyrite-bearing mudstone due to oxidation leading to heave of overlying floor slabs are described by Nixon (1978), Hawkins and Pinches (1987) and Wilson (1987). In recent years, this hazard has affected thousands of residential and some municipal buildings in the Dublin area of Ireland (Tuohy *et al.* 2012; Sutton *et al.* 2013; Hawkins and Stevens 2014; Czerewko and Cripps 2015; McCabe *et al.* 2015).

Effects of sulfates in arid environments

Sulfates are among many soluble minerals capable of causing problematic volume changes and chemical attack of construction materials that occur in arid environments. Hydration and dehydration, and dissolution and precipitation of anhydrite and gypsum often lead to heave and subsidence. As groundwater levels and composition are important controlling issues, changes in these are liable to impact on engineering. The dissolution of gypsum resulting from leakage of pipes or surface-water ingress leading to its removal in solution is the cause of many cases of ground subsidence.

Fig. 21. Examples of catastrophic damage to construction attributed to pyrite oxidation: (**a**) damage to the A625 road at Mam Tor, Derbyshire, UK; (**b**) rapid corrosion of buried steel culverts due to the presence of pyritic fill; and (**c**) deterioration with loss of strength and integrity of concrete due to formation of thaumasite.

The main concerns with buried and surface concrete in arid environments involve the ingress of salts, particularly gypsum and halite, which leads to both a physical deterioration through crystallization and a chemical attack of the cement matrix and damage to steel reinforcement bars (Fookes and Lee 2019). All of these processes also require the presence of moisture. The rate of chemical attack of concrete is dependent on the concentration of aggressive ions and the ease at which they are replenished at the surface of the concrete or interface of cementitious material. There is also a risk of an accumulation of sulfate ions in the concrete due to drying and wetting cycles, or capillary suction typically in near-coastal environments and inland playas, which can lead to rapid progressive damage, as seen in Figure 10d.

Rapid degradation through solution, hydration and dehydration is likely to occur where these sulfate deposits are exposed in natural and man-made slopes and rock cuttings, resulting in slope failures, as seen in Figure 10b. Engineering works that change the groundwater levels, or chemistry, or exposure of soils and evaporite-bearing rocks to erosion must be carefully planned and executed to avoid these problems.

Assessment of pyritic and gypsiferous ground conditions for construction

Understanding if and how much pyrite, gypsum or other sulfur minerals are present in the ground, or in construction materials is fundamental to avoiding problems in engineering and construction works. The possibility of the processes discussed in the previous sections arising from the presence of sulfur species should be recognized and considered at the outset of ground engineering projects, in particular when planning ground investigation or selecting construction materials such as aggregates.

Consequently, assessment of ground and aggregate materials for their potential aggressiveness in engineering situations sensibly follows a systematic staged approach. Once the requirements for the intended work have been determined, the geological and geotechnical ground assessment starts by performing preliminary studies to assess the nature of the ground and the likelihood of pyrite or sulfates being present, and the potential for deleterious conditions to be created during and after construction. The investigations also need to provide values of relevant geotechnical parameters on which the engineering design and method of construction are based with due regard for the management of risk and uncertainties (see Fig. 22).

An initial review of the geological setting may confirm the presence of known sulfide- and sulfate-bearing strata, and further assessments of available archival investigation and research records may give some indication of potential severity. Only then can appropriate field assessments and intrusive ground investigations be designed to target specific aspects of the work and select the most appropriate methodologies for ground truthing and material assessment, and otherwise determine what unknowns may occur. This requires a clear understanding of the intended engineering design and construction methodology options, as well as an awareness of the implications of potentially aggressive materials. A clear understanding is adopted of what can be achieved through a focused mineralogical assessment and chemical testing programme in line with the proposed design and recommendations for construction sequencing through a contemplated study of both, which may need to be refined, with considerations for mitigation and requirements for long-term monitoring inferred.

A well-planned assessment process should include a review of ground conditions and implications for the

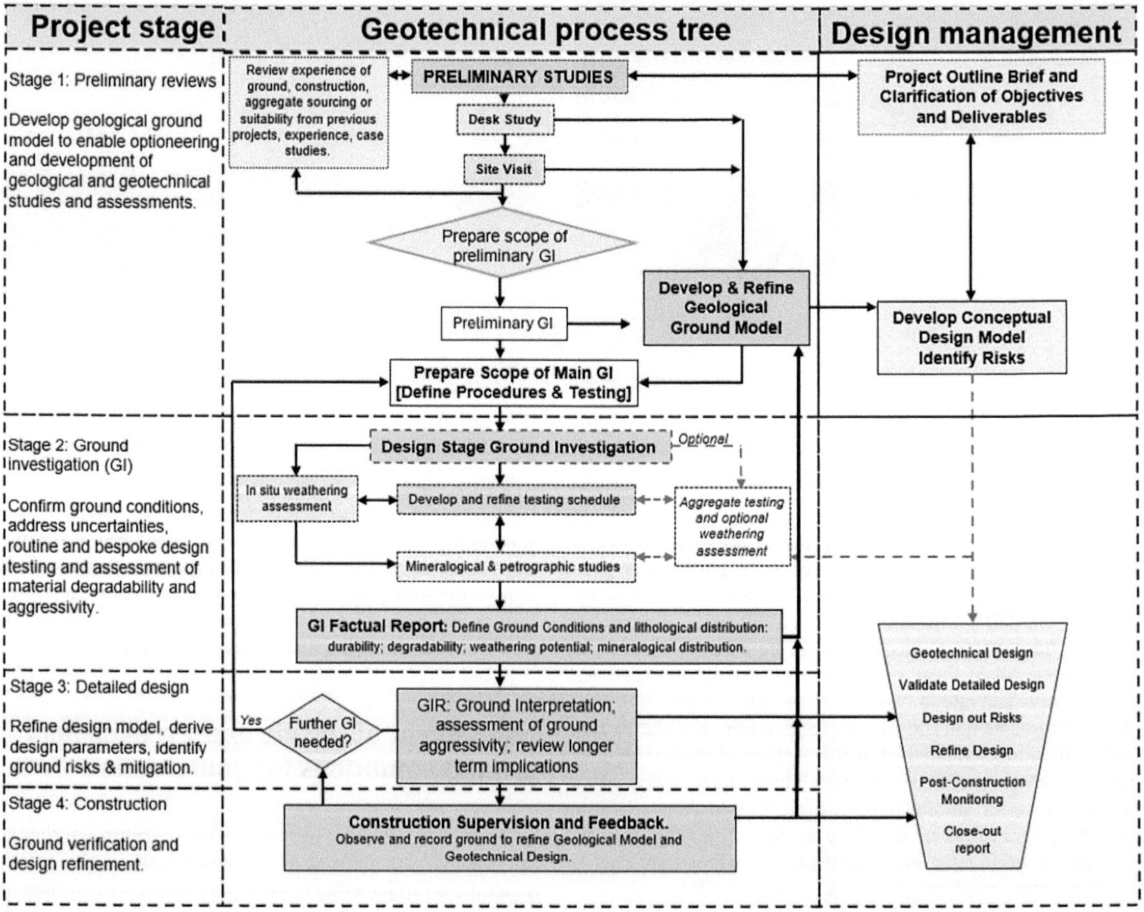

Fig. 22. Decision chart for the investigation and management of risk from potentially aggressive ground. GIR, Ground Investigation Report.

proposed development and provide advice to the designer at all stages of investigation and review. The issues should be recognized during the preliminary studies and addressed through ground investigation and testing, as outlined in Figure 22. Depending on the nature and scale of the scheme, the ground investigation may need to be carried out in several stages as a more refined understanding of the ground conditions develops. The investigation and assessment need to be managed in a systematic sequential manner and a working geological ground model developed to guide the design and construction processes as staunchly advocated in the first Glossop Lecture delivered by Peter Fookes in 1996 (Fookes 1997). An effective ground model continues to evolve as additional relevant data are collected and a better understanding of the ground conditions develops. Design standards and guidance such as Eurocode 7 (BSI 2007) and BRE Special Digest 1 (BRE 2005) rely on the derivation of characteristic values for design purposes using meaned datasets with rejection of data extremes. However, this approach is inappropriate as all credible data should be considered including hotspots and anomalous zones present at the site. Although such features may occur only in localized areas, their implication to the intended development needs to be considered or further investigated so that any risks they pose are accordingly managed.

The fieldwork and ground investigation should fundamentally ensure that a good visual description of the ground to the full extent and depth of ground affected by the construction forms the basis for any assessment of potentially aggressive ground conditions. It may be necessary for the conditions outside the immediate construction site boundaries to be considered, as local concentrations of detrimental minerals such as pyrite, pyrrhotite and gypsum may have an impact on the design and construction method. The presence of potentially problematic conditions may be implied by the colour of the host lithologies, colour changes caused by weathering or groundwater seepages, effects on vegetation, and other features. Despite laboratory testing giving precise values for the concentrations of chemical constituents of tested lithologies at specific locations, they will not usually be fully representative of a site.

The occurrence of pyrite is often associated with presence of organic carbon in aquatic sediments, and as little as 0.5% organic carbon produces a grey colour in mudstones, siltstones and sandstones. During the natural process of weathering such grey deposits undergo a colour change to brown with patchy or mottled yellow and orange discoloration caused by the oxidation of reduced iron minerals such as pyrite, as seen in Figure 23a in a continuous sequence of samples recovered from a ground boring. The colour changes are influenced by the hydration states of the iron oxide and hydroxide products of pyrite oxidation. With the presence of calcite, discoloration is often accompanied by precipitation of granular saccharoidal selenite crystals, which are often mistaken in material descriptions for layers of sand because of yellowish-brown surface staining by iron hydroxide (see Fig. 23b). These simple colour differences and chemical precipitates are indicative of chemical changes, and form a visual basis of assessment of the ground material.

In pyritic strata, weathering occurs slowly under normal conditions of soil formation (Reid *et al.* 2024) and shows subtle changes that are best observed in large supported or benched trial excavations, such as seen in Figure 24a. In the UK the upper 1–2 m of natural strata often contain very low concentrations of sulfates and sulfides such as pyrite, as they have been oxidized and leached out by rainfall over preceding millennia. The leached material, including the residual sulfates, often accumulates within the zone of tree roots or former groundwater levels as ferruginous selenite nodules (see Figs 23c & 24d) or as discrete concentrated clusters of selenite crystals (see Figs 23d & 24e) that pose a ground chemistry hazard for engineering structures, including culverts, tunnels and foundations. At greater depths, and below the permanent groundwater table, pyrite tends to be the dominant form of sulfur mineral (see Fig. 24f), except in evaporite sequences.

The transition from sulfate to sulfide corresponds to the depth of surface weathering, typically consisting of brownish-coloured materials often with yellow and orange mottling in the weathered zone grading often quite rapidly down into dark grey unweathered deposits at depth. Such changes are observed from continuous borehole sampling profiles, such as that seen in Figure 23a recovered from stiff (overconsolidated) clays and mudstone formations, which are widespread in the UK (Czerewko and Cripps 2024). Complex profiles with large variations in sulfate and sulfide concentrations may develop over short vertical distances due to the influence of previous groundwater levels or where the distribution of pyrite is affected by lithological variation (e.g. in interbedded sequences containing mudstones, limestones and sandstones: St. John 2020), or associated with changes in vegetation or land use. These characteristics may be missed by the conventional spot-sampling strategies employed in conventional ground assessment, which involve restricted borehole investigation procedures with material testing based on set depth intervals or changes in strata (St. John 2020). A more indicative understanding of the variability in ground conditions can be obtained from a detailed examination of the ground in carefully managed excavations, which reveal a more representative volume of ground, as seen in Figure 24. The sampling and testing of materials should be targeted to elucidate the chemical changes apparent from these investigations. Continuous sample recovery to greater depths may also be considered using cable percussive boring, continuous dynamic sampling with recovery of samples in clear plastic protective liners or high-quality rotary coring methods, but avoiding the use of water flush where possible as this may cause dissolution of sulfate minerals or accelerate the oxidation of pyrite (Czerewko and Cripps 2024).

The selection of material for laboratory testing, mineralogical assessment and weathering trials should focus on the construction zone, but also appraise the underlying and adjacent

Fig. 23. (**a**) Colour change from orange-brown to grey in a weathered pyritic London Clay profile. (**b**) Ferric oxide-stained saccharoidal selenite resulting from pyrite oxidation in Lower Lias Clay, often mistakenly described as sand. (**c**) Nodules of selenite crystals in ferric iron cement precipitating along former groundwater interface horizons in weathered Oxford Clay. (**d**) Nucleated selenite 'star' recovered from weathered London Clay.

lithologies, which may be affected by construction activities in terms of stress relief, disturbance and impact on the groundwater regime. The exploratory investigations will depend on the size of a project and the available budget, and will need to address the complexity of the geology. No fixed rules can be given for the depth or spacing of exploratory boreholes and trial excavations, which will depend on the nature and scale of each scheme, the specific circumstances of each site, including the complexity of the ground conditions, and the amount and quality of existing information. Detail guidance for investigations is provided in standards including BS 5930 (BSI 2020) and Eurocode 7 (BSI 2007), but these and other guidance documents should be used for advice only and not taken as a specification for investigation. Ideally, the designer and investigation team should be experienced ground practitioners who must use their professional judgement for determining the necessary depth and spacing of exploratory holes and the exploratory techniques to be used, so that sufficient robust information is obtained for adequate design and management of geotechnical risk for the proposed development. These people need to have the training and experience commensurate to this task.

The ground investigation should ensure that a suitable mass of material is recovered using the most appropriate

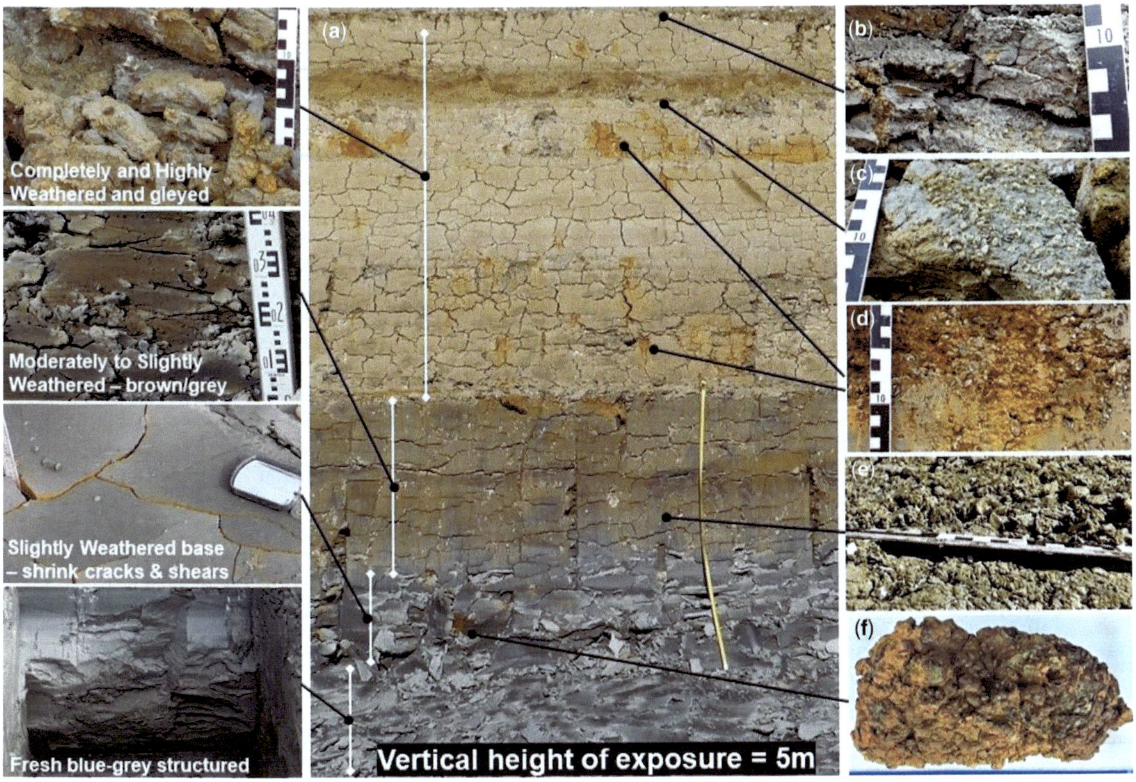

Fig. 24. Detailed ground assessment for the design of earthworks through a weathered Oxford Clay profile: (**a**) the weathering profile can be seen clearly in the mass excavation; the weathered state is shown on the left of the profile, and details of the nature and distribution of features observed from close examination are shown on the right; (**b**) concentrated saccharoidal selenite along root zones at shallow depth; (**c**) concentration of small euhedral selenite crystals in an area of a gleyed, formerly waterlogged, horizon; (**d**) nodules of concentrated selenite crystals in ferruginous cement at former groundwater interfaces; (**e**) the base of a weathered zone where the interface is marked by low-angled shears with abundant large selenite crystals overlying dark-coloured unweathered mudrock; and (**f**) partly oxidized nodular pyrite within slightly weathered clay.

method for the soil and rock type likely to be encountered. The samples should be recovered over the full ground profile. All samples intended for chemical characterization and mineralogical assessment should be retained and stored in a suitable manner to prevent deterioration, ensuring that the integrity of the material is maintained for testing and the process managed using a quality audit system (see Reid *et al.* 2005; BSI 2021). Samples of groundwater should also be collected for testing when encountered and stored in the appropriate manner (BRE 2005; Reid *et al.* 2005). The selection of material for testing should consider variations of ground conditions identified by visual description, and samples should be tested to define strata and lithotypes and to determine values of sulfur species. Where concentrations vary widely, a representative dataset may need to consider a minimum of between five and nine results for each stratum and aggregate source. But where complex ground conditions occur and where budgets permit, such as on larger projects, 10 or more results for each separate lithology and/or stratum identified are desirable. Guidance on the number of samples to be taken and the calculation of characteristic values for comparison with limiting values are given in guidance such as TRL Report 447 (Reid *et al.* 2005) and BRE Special Digest 1 (BRE 2005).

Where the presence of pyrite is suspected this needs to be confirmed by chemical testing, ensuring that appropriate chemical procedures are applied. The test procedures are outside the scope of this paper, but readers should be aware that some procedures specified in BSI (2012) for example, for the determination of sulfide and sulfate species are unreliable and unfit for the purpose – full details are given in BSI (2021) and Czerewko *et al.* (2016). Chemical testing alone cannot evaluate the nature or exact phases of the constituent, or predict if they are potentially reactive. Investigation to confirm the presence, nature and potential reactivity of microscopic pyrite forms may require X-ray diffraction (XRD) analysis and scanning-electron microscopy (SEM) accompanied by energy-dispersive X-ray (EDX) techniques for positive mineralogical and chemical confirmation. For an assessment of more indurated and crystalline material for potential aggregates applications, however, specialist petrographical techniques need to be utilized. As transmitted light microscopy may not be best suited for the study of sulfide minerals such as pyrite or pyrrhotite, reflected light microscopy may be needed to investigate these minerals.

In most civil engineering schemes, assessment of the ground conditions and determination of the foundation suitability focuses on the determination of immediate ground-bearing characteristics in terms of strength and stiffness. These data underpin the design of foundations that aim to provide adequate support that limit immediate and long-term settlements of structures to within acceptable limits. The parameters for design are determined by carrying out laboratory-based and *in situ* testing. While laboratory tests can establish the sensitivity of materials to changes in moisture conditions from which the long-term bearing characteristics are derived, *in situ* tests will usually have been carried out on prepared ground in favourable weather conditions. The potential for deterioration of design parameters during the pre-construction, construction and post-construction periods tends to be overlooked, which can have significant consequences when pyritic or sulfate-rich materials are present. The potential for deterioration can be simply appraised using a combination of laboratory index tests, including Atterberg limits and the response to wetting and drying with water immersion in the static jar slake test, and the cyclic dynamic slake durability tests that assess the resilience of material (Czerewko and Cripps 2024). Determination of the sulfate content of the immersion solutions will indicate the involvement of chemical changes in the performance of materials. Observations of the growth of sulfates such as selenite during the drying of samples will assist this assessment (Czerewko and Cripps 2019). The damaging effects of pyrite oxidation where they occur, including acid production, can develop rapidly, often within the construction period. Therefore, the engineering properties of the ground may be altered, leachates produced and minerals precipitated (Cripps *et al.* 1993), all of which may result in the ground not performing in the anticipated manner. Knowledge of the effects and rates of weathering reaction offers guidance for engineering design, mitigation and long-term monitoring when sulfur-bearing materials are involved.

The environmental implications from sulfur minerals affected by the engineering works must also be considered as these pose short- and long-term risks that need to be managed. Risks include the generation of acidic high ionic strength solutions that may cause pollution of groundwater and surface waters. Such an assessment will underpin the design of appropriate mitigation controls and construction programming to avoid problems from occurring and the need for long-term monitoring (Czerewko *et al.* 2011). Knowledge of the mechanical behaviour of the host rock and the changes brought about during construction, particularly from exposure to the atmosphere and changes to the groundwater conditions, need to be considered. Simple field weathering trials, as described by Czerewko *et al.* (2011), coupled with examination of natural exposures in similar material around the area of construction will provide an indication of the likely weathering changes that may occur during construction. However, accelerated laboratory weathering tests that subject material to disaggregation such as that which they will experience during construction, and exposing materials to cycles of wetting and drying are useful means of providing insights for understanding possible changes to the materials at different stages of construction (see Fig. 25).

The adverse reactions observed following exposure during earthworks and construction activities have been seen to occur rapidly, often in a matter of weeks or months, so significant changes may occur in the properties of material during construction. However, the conditions aggressive to engineering material can be insidious and may not become apparent during construction, only to cause problems in some cases many years after construction is completed (Pye and Miller 1990; BRE 2005; Reid *et al.* 2005; Czerewko and Cross 2015). Therefore, to avoid or manage problems caused by gypsum dissolution or by pyrite oxidation, it is necessary to know not just that it is present, but in what form it is present, how reactive it is and its distribution, such as knowledge of which lithologies or locations contain these minerals, but also whether the engineering works or subsequent environment and groundwater conditions will cause it to become problematic (Czerewko *et al.* 2016).

Simple field-monitored weathering observations are invaluable for assessment of likely behaviour of material following exhumation and exposure to atmospheric conditions, as shown in Figure 26a–c and described by Czerewko *et al.* (2011). Such field procedures have no stringent test requirements, but they have the benefit of simplicity and lack of ambiguity, and so provide a credible basis of assessment that may be implemented at an early stage of a project and with observations continued for however long is considered necessary by the geopractitioner. This implies early involvement of suitably qualified and experienced personnel at the inception of projects, as an awareness of the likely behaviour of material allows appropriate planning for investigation and management of risk during construction operations. For example, in the case shown in Figure 26, rapid weathering characteristics determined through field trials were employed to condition the excavated calcareous pyritic mudrock by weathering the material in loose windrows that promoted the oxidation of pyrite and generation of selenite to produce acceptable earthworks fill material within a 3 week period, as seen in Figure 26d, e.

Evaluation of aggregate resources

To ensure that structures are designed and built so that they perform their intended purpose in a safe and economical manner, it is essential that all the materials that form the structure, including the ground supporting the foundations and aggregates used for structural fill or as integral constituents of engineering materials such as concrete and asphalt, behave in a

Fig. 25. Example of accelerated laboratory weathering on pyritic slate through which tunnelling using a TBM is to be constructed with test sample preparation using an abrasive cutting tool: (**a**) pyrite in slate; and (**b**) selenite stars and pH 4.6 conditions recorded after four cycles of accelerated wetting–drying of the sample in (a).

manner that can be predicted before construction begins. The potential for oxidation and the reaction of pyrite or other iron sulfide minerals that may be present in the hard rock aggregate, or the presence of gypsum and other sulfate minerals, may give rise to deleterious changes in the construction materials.

Crushed and sized aggregates are used for various fill applications, in asphalt-bound materials and concrete, and these changes are difficult to predict under the various exposure conditions and so present a huge challenge to ground engineering and geotechnical specialists. Therefore, a detailed characterization of material and consideration of the risks that may develop will allow for forward planning to manage the risk. In view of the widespread occurrence of pyrite in geological materials and the occasional presence of pyrrhotite, it is not practical to avoid the use of such materials, although some may not be suitable for certain applications. Therefore, there is a need to understand the potential changes that may be brought about in specific construction situations and predict the impacts of different engineering operations on pyrite so that mitigation measures for adverse impacts may be considered and incorporated into the engineering design and construction processes.

The assessment of the potential for oxidation of reduced-sulfur compounds in rock aggregates proposed for use in earthwork may be carried out though observation of exposed stockpiles if time is available or by using accelerated weathering tests. These involve subjecting material to simulated weathering cycles in an environmental chamber, as described by Reid and Avery (2009). These performance-related assessments offer a pragmatic evaluation of what reaction may occur and which often cannot be deduced through use of routine industry standard chemical-testing protocols. Although determinations of total sulfate and reduced sulfides will indicate if oxidation is likely to occur, the results do not evaluate what if any reactions are likely to occur given the individual environmental conditions. These approaches offer the ground professional sufficient justification for appropriate management of the works, and a selection of appropriate mitigation solutions.

Ground engineers undertaking an assessment of pyrite-bearing lithologies and aggregates should also be aware that the morphology of the pyrite present in the material will influence its stability following exposure to atmospheric conditions. But the coexistence of pyrrhotite alongside pyrite, even in minor concentrations, will catalyse oxidation of otherwise unreactive pyrite. Therefore, in materials containing both pyrrhotite and pyrite, the possibility of the rapid oxidation of all forms of pyrite needs to be carefully considered, depending upon the intended end use.

Impact mitigation

Mitigation measures adopted for aggressive ground conditions depend on the causes and character of the aggressivity, site conditions, and the nature of the development. The effective treatments include isolation of the structure, control of groundwater, compaction limiting the air and water content, provision of capillary barriers, impermeable barriers, flexible granular interfaces, impermeable construction interfaces, and ground treatment. These and other measures may be selected in isolation or in combination as most appropriate for the development.

Although the mechanism of pyrite oxidation, which is one of the main cases of ground aggressivity, has been discussed, the implications to construction and the success of the

Fig. 26. Example of field weathering trials to determine the rate of change to pyritic Blue Lias Mudstone: (**a**) day 1 of the weathering trial; (**b**) day 18, showing significant development of selenite, with chemical testing confirming a substantial loss of pyrite; (**c**) day 41, showing no further changes in the residual soil product recording loss of sulfate and sulfide; (**d**) earthwork management of the material through weathering in loose windrows; and (**e**) change in the 2 week stockpiled mudrock accompanied by selenite development, as seen in the inset image.

earthworks tend to be controlled by site-specific conditions largely related to the type and properties of the material present and the site environment. The site investigation may confirm the presence of significant quantities of pyrite, gypsum and calcite, but these values alone do not facilitate assessments of the reaction rate and significance to construction. Determination and evaluation in terms of the engineering operations, of the type and distribution of sulfur compounds in the ground, and the geological construction materials are of fundamental importance.

The consequences of pyrite reactions when unexpected may become a significant hit for the construction budget and progress. The oxidation of pyrite-bearing deposits during earthworks and construction has been observed to progress rapidly over a matter of weeks or months, producing conditions chemically aggressive to engineering materials. Therefore, to avoid or manage potential problems attributed to pyrite oxidation, it is necessary to know not just that it is present but also its distribution, its form and reactivity. Investigations and construction may overlook the potential for material deterioration but this can be determined at a site level through observation of changes following exposure and targeted chemical testing.

Pyrite-bearing materials can often be safely utilized in construction provided that appropriate measures are taken. Methods of avoiding problems due to the presence of pyrite include:

- Inclusion of granular layers for controlling the upward migration of sulfate-bearing solutions and to act as strain-absorbing units (see Fig. 27a).
- Knowledge of the mechanical behaviour of the host material and the changes brought about through exposure during construction may expedite management of the construction process by facilitating the reuse of a favourably weathered product in which, for example, pyrite had rapidly oxidized on exposure in stockpiles and selenite crystal had developed on reaction with calcite, thus mitigating for potential heave in earthworks, reuse of which would otherwise constitute an unsuitable material (see Fig. 27b).

Fig. 27. Various methods of mitigation used for aggressive ground: (**a**) granular barriers; (**b**) pre-weathering of pyritic subgrade; (**c**) isolation of structures with suspended floors above ground liable to undergo movements; (**d**) provision of drainage showing an array of raking drains formed into a slope below a highway; (**e**) stabilization and blending of potentially aggressive ground; and (**f**) encapsulation of aggressive ground using suitable bonding and filling agents.

- Avoiding the use of ground-supported floor slabs on pyritic fill or bedrock by constructing suspended floors (see Fig. 27c).
- Management of groundwater levels with provision of targeted drainage (see Fig. 27d) to reduce the possibility of providing air to pyrite, the transportation of reaction products and the dissolution of soluble minerals.
- Use of inert carbonate-free aggregates for drains and filters in situations where acidic or sulfate-rich groundwaters are liable to exist.
- Ground stabilization with appropriate binders determined through testing and performance trials or blending with inert material (see Fig. 27e).
- Encapsulation of aggressive material using cementitious and chemical binders with pozzolanic additives (see Fig. 27f).
- Provision of impermeable barriers and surface coatings to buried concrete or steel structures (see Fig. 28a).
- Inclusion of flexible synthetic construction interfaces as barriers in structures and membranes in tunnels (see Fig. 28b).
- Inclusion of sacrificial thickness for buried steel structures with protective backfill of inert structural aggregate (see Fig. 28c).
- Inclusion of sacrificial concrete to buried concrete structures (see Fig. 28d).
- Use of stainless steel as a replacement for mild steel for reinforcement and other applications of steel.

Ultimately, management of material avoiding costly offsite disposal may be achieved through informed investigation. Materials that have been satisfactorily utilized in previous projects, but which fail current specifications, are not necessarily to be discarded. Use is acceptable provided that their suitability and performance can be verified by monitoring of existing structures, comprehensive laboratory characterization and testing designed to simulate the effects of weathering, preferably accompanied by field trials.

Guidance is available on the acceptable amounts of pyrite in materials in contact with buried steel or concrete and other hydraulically bound materials. Where buried concrete structures are concerned, BRE (2005) provides comprehensive guidance on procedures to predict, prevent and mitigate against deterioration of concrete due to the presence of aggressive ground conditions. Various preventive and protective measures are possible, including modified mix designs, the use of bitumen coatings to concrete, the addition of a capillary break to prevent contact with aggressive solutions and the use of precast concrete products. TRL Report 447 (Reid *et al.* 2005) provides guidance on acceptable amounts of sulfide and sulfate minerals in fills to be placed with 500 mm of buried concrete and steel structures in UK highway schemes.

Mitigating for the effects of gypsum challenges in arid environments

In dry arid environments the risk of an accumulation of sulfate ions in buried concrete must be considered as accumulation is driven by capillary suction or cycles of wetting and drying due to the extreme climatic conditions.

For buried reinforced concrete in arid environments, the need arises for protection; but a balance is sought between protection from chlorides against the need for protection against sulfates. Where necessary, a concrete resistant to both is necessary, and designed to a high quality with low permeability, with surface protection, such as that seen in Figure 28a, to minimize water and salt ingress, and, where justifiable, the use of non-metallic or stainless-steel reinforcement.

In arid environments, surface crusts and cemented layers in soil profiles may appear to provide adequate support for foundations, but these properties may be destroyed by natural environmental changes or changes brought about by the engineering works. The presence of weaker layers within the soil/bedrock profile needs to be considered and the precipitation of minerals in the soil profile is liable to result in surface heave. Simply covering an area of ground may cause this, and roads are particularly vulnerable to damage due to the resulting changes in the patterns of soil moisture movement. Factors to be considered during design include the changes that may occur in the properties of materials cemented with soluble minerals. Incorporating impermeable barriers into the structure to reduce evaporation from the surface and the use of uniformly graded granular fill to arrest the capillary movement of moisture may prevent undesirable changes in the conditions (see Fig. 27a).

In arid regions, design is required to satisfy local design standards in which the development of ground engineering has become established on the acceptable levels of total salt (gypsum and halite) contents for concrete and other applications outlined in local technical documentation (see Concrete Society 2008). The design class limits for soils in such arid environments recognize that mixtures of soluble salts are usually more aggressive than pure solutions. Assessment procedures such as Saudi Building Code 304 (SBC 2007) requires that the 'total' water-soluble sulfate content of the ground material is determined (equivalent to Acid-Soluble Sulfate determination of UK practice), along with the 'water-soluble chloride' content of the material for classification purposes. Where water is introduced into the ground as either rainfall or leakage from pipes, etc., soluble minerals may be removed, leading to reduced strength, loss of material and metastable conditions. Excavations below the water table are liable to cause water of lower salinity to flow into the deposit, even if this is itself saline, and thus cause dissolution. So, the impacts of engineering on the groundwater and surface water conditions, both during and after the construction phase, must be

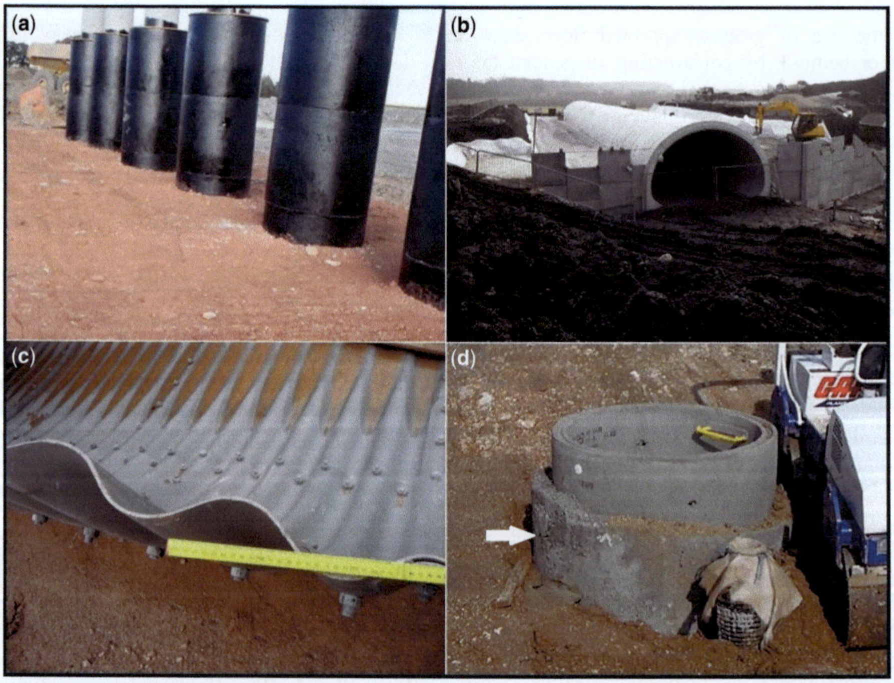

Fig. 28. Various mitigation methods used for buried structures: (**a**) bituminous 'tanking' barriers with a surround of inert granular structural fill; (**b**) impermeable construction barriers included to buried structures against pyritic fill and subgrade; (**c**) sacrificial surface cover to buried steel structures with a surround of inert granular structural fill; and (**d**) provision of sacrificial concrete (outer layer with a white arrow) to precast concrete structures.

carefully planned for: for example, appropriate ground investigation, assessment of ground materials and assessment of hydrology, including the possibility of unexpected drought, rainfall, pipe leakage or flooding event.

Conclusions

Sulfur mineral species, especially reduced sulfides (e.g. pyrite) and sulfates (e.g. gypsum), are widely distributed within British sediment, rock and fill materials, and are similarly found widely in international settings. As such, they are commonly encountered during engineering works. Although the sulfate content may be determined in routine ground investigations, sulfides are often ignored. In part, this may be due to a lack of appreciation of their importance, although adequate evaluation procedures have been available for many years. Sulfide minerals, in particular pyrite, rapidly oxidize once exposed to atmospheric conditions, such that they generate aggressive chemical conditions that cause the corrosion of concrete and steel, deterioration of ground materials, the expansive precipitation of minerals, and the pollution of groundwater and surface water. Deposits containing sulfates may result in chemically aggressive ground conditions, and dissolution of sulfates will result in subsidence and loss of strength of geological materials.

Failure to recognize the presence of sulfide and sulfate minerals in the ground and take these into account when designing engineering structures and construction methods is likely to cause costly and time-wasting changes in design and construction methods, during construction. In some cases, structural failure and/or post-construction remedial measures and premature maintenance will be required.

In assessment procedures, the aggressivity of ground materials and aggregates is based on the premise that all the reduced sulfur will be oxidized, which, in most cases, is a conservative or worst-case position. Although, it is known that coarser-grained crystalline pyrite is much less reactive than fine-grained and framboidal forms, the rate and extent of reaction depend on many site-specific factors. Thus, mineralogical and chemical testing alone will not provide sufficient information for the impact of pyrite to be accurately predicted in most civil engineering project situations. The permeability of the ground and the presence of buffering minerals are obviously important considerations but, for avoidance of problems due to sulfur minerals, cognisance needs also to be taken of the groundwater conditions and environmental changes to be caused by the construction processes and the design itself. This requires a more holistic approach beyond strict adherence to published guidance for ground investigations for construction typical in the industry, with understanding of the processes leading to problematic situations coupled with an appreciation of the effects of construction on the materials involved at a site-specific level.

It is therefore essential to consider the types and distribution of sulfides and sulfates in the ground and any construction materials that may affect buried steel, concrete and stabilized ground, and impact on foundations and slope formation. It is also worth remembering when assessing construction scenarios that sulfide minerals such as pyrite may provide no hazard to buried construction materials in the absence of oxygen and mobile water. But when ground disturbance is brought about by excavation and replacement, soil mixing or where present in aggregates that are used as fill to structures and construction above the water table, then sulfide minerals are liable to pose a hazard.

Supplementary material assessment and characterization, along with detailed site observations, may help to establish the susceptibility of materials to changes due to the presence of sulfur compounds, particularly in partially weathered materials in which the oxidation of pyrite may be ongoing. Changes such as material discoloration, softening and presence of selenite are indicators of sulfide oxidation, in particular. Simple procedures may be employed to support chemical and petrological studies, including field and accelerated laboratory weathering tests, compaction performance trials, and durability assessments. Chemical testing of the leachates derived during these performance tests should also be carried out as these may indicate mineralogical changes. The findings further inform the design of structures, engineering of landforms and the construction process, such that these enable formulation of the most effective construction sequence and longer-term mitigation methods.

Acknowledgements The authors would like to thank the reviewers for their dedication, patience and guidance with preparation of the paper.

Competing interests The authors declare that they have no known competing financial interests or personal relationships that could have appeared to influence the work reported in this paper.

Author contributions **MAC**: writing – original draft (lead); **JCC**: writing – review & editing (supporting).

Funding This research received no specific grant from any funding agency in the public, commercial, or not-for-profit sectors.

Data availability Data sharing is not applicable to this article as no datasets were generated or analysed during the current study.

References

Anderson, W.H. 2014. Pyrite oxidation and structural problems in the Chattanooga (Ohio) Shale, east central Kentucky. *In*: Hawkins, A.B. (ed.) *Implications of Pyrite Oxidation for Engineering Works*. Springer, Cham, Switzerland, 243–273.

Arrouvel, C. and Eon, J.-G. 2019. Understanding the surface and crystal growth of pyrite FeS2. *Materials Research*, **22**, e20171140, https://doi.org/10.1590/1980-5373-MR-2017-1140

Barnes, G. 2010. *Soil Mechanics: Principles and Practice*. 3rd edn. Palgrave Macmillan, London.

Bell, F.G. 2000. *Engineering Properties of Soils and Rocks*. 4th edn. Blackwell Science, Oxford, UK.

Belzile, N., Chen, Y.W., Cai, M.F. and Li, Y. 2004. A review of pyrrhotite oxidation. *Journal of Geochemical Exploration*, **84**, 65–76, https://doi.org/10.1016/j.gexplo.2004.03.003

Bowles, J.F.W., Howie, R.A., Vaughan, D.J. and Zussman, J. 2011. *Rock-Forming Minerals, Volume 5A: Non-Silicates: Oxides, Hydroxides and Sulphides*. 2nd edn. Geological Society, London.

Bracegirdle, A., Jefferis, S.A., Tedd, P., Crammond, N.J., Chudleigh, I. and Burgess, N. 1996. The investigation of acid generation within the Woolwich and Reading Beds at Old Street and its effect on tunnel linings. *In*: Mair, R.J. and Taylor, R.N. (eds) *Geotechnical Aspects of Underground Construction in Soft Ground*. A.A. Balkema, Rotterdam, The Netherlands, 653–658.

BRE 1968. *Concrete in Sulphate-Bearing Soils and Groundwaters*. BRE Digest 90. Building Research Establishment (BRE), Watford, UK.

BRE 1991. *Sulfate and Acid Resistance of Concrete in the Ground*. BRE Digest 363. Building Research Establishment (BRE), Watford, UK.

BRE 2005. *Concrete in Aggressive Ground: Assessing the Aggressive Chemical Environment*. BRE Special Digest 1. Building Research Establishment (BRE), Watford, UK.

Britpave 2019. *Stabilisation of Sulfate-Bearing Soils: Guidelines for Best Practice*. Report BP/51. Britpave, Camberley, UK, https://ukcsma.co.uk/wp-content/uploads/2016/08/BP51-Stabilisation-of-Sulfate-Bearing-Soils.pdf

Bromley, A. and Petifer, K. 1997. *Sulfide-Related Degradation of Concrete in Southwest England (The 'Mundic' Problem)*. BRE Report 325. Building Research Establishment (BRE), Watford, UK.

BSI 2007. *BS EN 1997-2:2007: Eurocode 7 – Geotechnical Design – Part 2: Ground Investigation and Testing*. British Standards Institution (BSI), London.

BSI 2012. *BS EN 1744-1:2009 + A1:2012. Tests for Chemical Properties of Aggregates – Part 1: Chemical Analysis*. British Standards Institution (BSI), London.

BSI 2020. *BS 5930:2015 + A1:2020. Code of Practice for Site Investigations*. British Standards Institution (BSI), London.

BSI 2021. *BS 1377-3:2018 + A1:2021. Methods of Test for Soils for Civil Engineering Purposes – Part 3: Chemical and Electrochemical Tests*. British Standards Institution (BSI), London.

Carr, P.G., Thesier, J.L. and Kimball, M.B. 2006. An 'invisible menace' – the impact of pyrite induced expansive forces on long-term building failure – a case study. *In*: Bosela, P.A. and Delatte, N.J. (eds) *Forensic Engineering – Proceedings of the 4th Congress*. American Society of Civil Engineers (ASCE), Reston, VA, 154–168, https://doi.org/10.1061/40853(217)14

Chandler, R.J. 1972. Lias clay: weathering processes and their effects on shear strength. *Geotechnique*, **22**, 403–431, https://doi.org/10.1680/geot.1972.22.3.403

Chandra, A.P. and Gerson, A.R. 2010. The mechanism of pyrite oxidation and leaching: fundamental perspective. *Surface Science Reports*, **65**, 293–315, https://doi.org/10.1016/j.surfrep.2010.08.003

Chinchón-Paya, S., Aguado, A. and Chinchón, S. 2012. A comparative investigation of the degradation of pyrite and pyrrhotite under simulated laboratory conditions. *Engineering Geology*, **127**, 75–80, https://doi.org/10.1016/j.enggeo.2011.12.003

Concrete Society 2008. *Guide to the Design of Concrete Structures in the Arabian Peninsula*. Concrete Society Special Publications, **CS163**. Concrete Society, Camberley, UK.

Craig, J.R. and Vokes, F.M. 1993. The metamorphism of pyrite and pyritic ores: an overview. *Mineralogical Magazine*, **57**, 3–18, https://doi.org/10.1180/minmag.1993.057.386.02

Crammond, N.J. 2003. The thaumasite form of sulfate attack in the UK. *Cement and Concrete Composites*, **25**, 809–818, https://doi.org/10.1016/S0958-9465(03)00106-9

Cripps, J.C. and Edwards, R.L. 1997. Some geotechnical problems associated with pyrite bearing rocks. *In*: Hawkins, A.B. (ed.) *Proceedings of International Conference on the Implications of Ground Chemistry/Microbiology for Construction*. A.A. Balkema, Rotterdam, The Netherlands, 77–87.

Cripps, J.C., Hawkins, A.B. and Reid, J.M. 1993. Engineering problems with pyritic mudrocks. *Geoscientist*, **3**, 16–19.

Cripps, J.C., Reid, J.M., Czerewko, M.A. and Longworth, I. 2019. Tackling problems in civil engineering caused by the presence of pyrite. *Quarterly Journal of Engineering Geology and Hydrogeology*, **53**, 481–500, https://doi.org/10.1144/qjegh2019-024

Czerewko, M.A. and Cripps, J.C. 2003. The development of a new testing protocol for sulphur compounds in structural backfills. *Quarterly Journal of Engineering Geology and Hydrogeology*, **36**, 133–142, https://doi.org/10.1144/1470-9236/2002-26

Czerewko, M.A. and Cripps, J.C. 2015. Investigation of destructive ground heave affecting new-build properties constructed using pyrite bearing fill in the Dublin area of Ireland. *In*: Winter, M.G., Smith, D.M., Eldred, P.J.L. and Toll, D.G. (eds) *Geotechnical Engineering for Infrastructure Development. Proceedings of the XVI European Conference on Soil Mechanics and Geotechnical Engineering*. ICE Publishing, London, 2361–2366.

Czerewko, M.A. and Cripps, J.C. 2019. The consequences of pyrite degradation during construction – UK Perspective. *In*: Shakoor, A. and Cato, K. (eds) *IAEG/AEG Annual Meeting Proceedings, 2018, San Francisco, California, Volume 2*. Springer Nature, Cham, Switzerland, 143–151.

Czerewko, M.A. and Cripps, J.C. 2022. Investigation of destructive ground heave attributed to pyritic fill affecting new-build properties in the Dublin area of Ireland. *Engineering Geology*, **299**, https://doi.org/10.1016/j.enggeo.2022.106534

Czerewko, M.A. and Cripps, J.C. 2023. Implications of sulfur mineralogy and consequences of pyrite oxidation for ground engineering. *Journal of the Geological Society, London*, **180**, https://doi.org/10.1144/jgs2022-101

Czerewko, M.A. and Cripps, J.C. 2024. Mudrocks, clays and pyrite implications. *In*: Brown, M., Burland, J., Chapman, T., Higgins, K., Skinner, H. and Toll, D. (eds) *ICE Manual of Geotechnical Engineering, Volume 1: Geotechnical Engineering Principles, Problematic Soils and Site Investigation*. 2nd edn. Emerald Publishing, Leeds, UK, 517–557.

Czerewko, M.A. and Cross, S.A. 2015. The benefits of a granular interface over pyritic subgrade. *Geotechnical Research*, **2**, 97–122, https://doi.org/10.1680/jgere.15.00003

Czerewko, M.A. and Whitworth, M. 2025. Geological and engineering characteristics of the pyritic Triassic Westbury Formation and assessment of weathering implications for construction and material reuse. *Quarterly Journal of Engineering Geology and Hydrogeology*, **58**, https://doi.org/10.1144/qjegh2024-188

Czerewko, M.A., Cross, S.A., Dumelow, P.G. and Saadvandi, A. 2011. Assessment of pyritic Lower Lias mudrocks for earthworks. *Geotechnical Engineering*, **164**, 24–46, https://doi.org/10.1680/geng.2011.164.2.59

Czerewko, M.A., Longworth, I., Reid, J.M. and Cripps, J.C. 2016. Standardised terminology and test methods for sulphur mineral phases for the assessment of construction materials and aggressive ground. *Quarterly Journal of Engineering Geology and Hydrogeology*, **49**, 245–265, https://doi.org/10.1144/qjegh2016-010

D'Angeli, I.M., Carbone, C., Nagostinis, M., Parise, M., Vattano, M., Madonia, G. and De Waele, J. 2018. New insights on secondary minerals from Italian sulfuric acid caves. *International Journal of Speleology*, **47**, 271–291, https://doi.org/10.5038/1827-806X.47.3.2175

Deer, W.A., Howie, R.A. and Zussman, J. 1992. *An Introduction to the Rock-Forming Minerals*. 2nd edn. Longman, Harlow, UK.

DETR 1999. *The Thaumasite Form of Sulfate Attack: Risks, Diagnosis, Remedial Works and Guidance on New Construction*. Report of the Thaumasite Expert Group. Department of the Environment, Transport and the Regions (DETR), London.

Dougherty, M.T. and Barsotti, N.J. 1972. Structural damage and potentially expansive sulfide minerals. *Bulletin of the Association of Engineering Geologists*, **9**, 105–125.

Edwards, K.J., Schrenk, M.O., Hamers, R. and Banfield, J.F. 1998. Microbial oxidation of pyrite: experiments using microorganisms from an extreme acidic environment. *American Mineralogist*, **83**, 1444–1453, https://doi.org/10.2138/am-1998-11-1233

Eglinton, M.S. 1975. *Review of Concrete Behaviour in Acidic Soils and Groundwater*. CIRIA Technical Note 69. Construction Industry Research and Information Association (CIRIA), London.

Ellison, R.A., Woods, M.A., Allen, D.J., Forster, A., Pharaoh, T.C. and King, C. 2004. *Geology of London: Special Memoir for 1:50 000 Geological Sheets 256 (North London), 257 (Romford), 270 (South London), and 271 (Dartford) (England and Wales)*. British Geological Survey, Keyworth, Nottingham, UK.

Enning, D. and Garrelfs, J. 2014. Corrosion of iron by sulfate-reducing bacteria: new views of an old problem. *Applied and Environmental Microbiology*, **80**, 1226–1236, https://doi.org/10.1128/AEM.02848-13

Fenlon, A. and Petrera, L. 2019. Pyrite oxidation: a history of treatments at the Natural History Museum, London. *Geological Curator*, **11**, 9–18, https://doi.org/10.55468/GC403

Fookes, P.G. 1978. Middle East – inherent ground problems. *Quarterly Journal of Engineering Geology*, **11**, 33–49, https://doi.org/10.1144/GSL.QJEG.1978.011.01.04

Fookes, P.G. 1997. Geology for engineers: the geological model, prediction and performance. *Quarterly Journal of Engineering Geology*, **30**, 293–424, https://doi.org/10.1144/GSL.QJEG.1997.030.P4.02

Fookes, P.G. and Collis, L. 1975. Problems in the Middle East. *Concrete*, **9**, 12–17.

Fookes, P.G. and Lee, E.M. 2018. The engineering geology of playas, salt playas and salinas. *Quarterly Journal of Engineering Geology and Hydrogeology*, **51**, 287–298, https://doi.org/10.1144/qjegh2017-084

Fookes, P.G. and Lee, E.M. 2019. The engineering geology of concrete in hot drylands. *Quarterly Journal of Engineering Geology and Hydrogeology*, **52**, 257–279, https://doi.org/10.1144/qjegh2018-185

Fookes, P.G., French, W.J. and Rice, S.M.M. 1985. The influence of ground and groundwater geochemistry on construction in the Middle East. *Quarterly Journal of Engineering Geology*, **18**, 101–128, https://doi.org/10.1144/GSL.QJEG.1985.018.02.01

Fookes, P.G., Hart, A.B. and Lee, E.M. 2013. Some near surface desert features of significance in engineering geology evaluations. *Quarterly Journal of Engineering Geology and Hydrogeology*, **46**, 259–266, https://doi.org/10.1144/qjegh2012-083

Grattan-Bellew, P.E. and Eden, W.J. 1975. Concrete deterioration and floor heave due to biogeochemical weathering of underlying shale. *Canadian Geotechnical Journal*, **19**, 108–111, https://doi.org/10.1139/t82-010

Hagelia, P., Sibbick, R.G., Crammond, N.J., Gronhaug, A. and Larsen, C.K. 2001. Thaumasite and subsequent secondary calcite deposition in sprayed concrete in contact with sulfate bearing Alum Shale, Oslo, Norway. Conference Paper presented at the 8th Euroseminar on Microscopy Applied to Building Materials, 4–7 September 2001, Athens, Greece.

Hawkins, A.B. 2012. Sulphate heave: a model to explain the rapid rise of ground-bearing floor slabs. *Bulletin of Engineering Geology and the Environment*, **71**, 113–117, https://doi.org/10.1007/s10064-011-0408-1

Hawkins, A.B. 2014. Engineering implications of the oxidation of pyrite: an overview, with particular reference to Ireland. *In*: Hawkins, A.B. (ed.) *Implications of Pyrite Oxidation for Engineering Works*. Springer, Cham, Switzerland, 1–98.

Hawkins, A.B. and Pinches, G.M. 1987. Cause and significance of heave at Llandough Hospital, Cardiff – a case history of ground floor heave due to gypsum growth. *Quarterly Journal of Engineering Geology*, **25**, 17–30, https://doi.org/10.1144/GSL.QJEG.1992.025.01.02

Hawkins, A.B. and Pinches, G.M. 1997. Understanding sulphate generated heave resulting from pyrite degradation. *In*: Hawkins, A.B. (ed.) *Ground Chemistry: Implications for Construction*. A.A. Balkema, Rotterdam, The Netherlands.

Hawkins, A.B. and Stevens, M. 2014. Problems with the use of pyritiferous fill at Ballymun youth facility, Dublin. *In*: Hawkins, A.B. (ed.) *Implications of Pyrite Oxidation for Engineering Works*. Springer, Cham, Switzerland, 133–166.

Hawkins, A.B. and St John, T.W. 2014. Iron sulphides and surface heating: further engineering considerations for the Dublin area. *In*: Hawkins, A.B. (ed.) *Implications of Pyrite Oxidation for Engineering Works*. Springer, Cham, Switzerland, 275–307.

Higgins, D.D., Thomas, B. and Kinuthia, J. 2002. Pyrite oxidation, expansion of stabilised clay and the effect of ggbs. *In*: Zoorob, S.E., Collop, A.C. and Brown, S.F. (eds) *Performance of Bituminous and Hydraulic Materials in Pavements*. Swets & Zeitlinger, Lisse, The Netherlands, 161–167.

Hodgkinson, E.S. and Martin, S. 2004. *Curation History and Mineralisation of Highly Degraded Pyrite Fossil Collection*. BGS Internal Report IR/04/037. British Geological Survey (BGS), Keyworth, Nottingham, UK.

Hoover, S.E. and Lehmann, D. 2009. The expansive effects of concentrated pyritic zones within the Devonian Marcellus Shale Formation of North America. *Quarterly Journal of Engineering Geology and Hydrogeology*, **42**, 157–164, https://doi.org/10.1144/1470-9236/07-086

Hoover, S.E., Wang, M.C. and Dempsey, B. 2004. Structural damage induced by pyritic shale. *In*: *International Conference on Case Histories in Geotechnical Engineering*, 2, https://scholarsmine.mst.edu/icchge/5icchge/session07/2

Horng, C.S. and Roberts, A.P. 2006. Authigenic or detrital origin of pyrrhotite in sediments?: Resolving a paleomagnetic conundrum. *Earth and Planetary Science Letters*, **241**, 750–762, https://doi.org/10.1016/j.epsl.2005.11.008

Howie, F.M.P. 1977. Pyrite and conservation. Part 1: historical aspects. *Newsletter of the Geological Curators' Group*, **1**, 457–465, https://doi.org/10.55468/GC1033

Jarzyna, A., Babel, M., Czajkowska, M. and Lugowski, D. 2024. Morphometry and morphology of the gypsum tumuli from the Sorbas karst reqion, SE Spain. *Geomorphology*, **465**, 109403, https://doi.org/10.1016/j.geomorph.2024.109403

Jerz, J.K. and Rimstidt, J.D. 2004. Pyrite oxidation in moist air. *Geochimica et Cosmochimica Acta*, **68**, 701–714, https://doi.org/10.1016/s0016-7037(03)00499-x

King, R.J. 1999. Minerals explained 27: Sulphur. *Geology Today*, **15**, 157–160, https://doi.org/10.1046/j.1365-2451.1999.1504009.x

Kwong, Y.T.J. 1993. *Prediction and Prevention of Acid Rock Drainage from a Geological and Mineralogical Perspective*. Mine Environment Neutral Drainage (MEND) Project 1.32.1, https://mend-nedem.org>uploads>1.32.1

Longworth, T.I. 2004. Assessment of sulfate-bearing ground for soil stabilisation for built development. *Ground Engineering*, **37**, 56–59 (also published in *Concrete*, **38**, 36–39).

Longworth, T.I. 2008. *Sulfate Damage to Concrete Floors on Sulfate-Bearing Hardcore: Identification and Remediation*. RIBA Publishing, London.

Lowson, R.T. 1982. Aqueous oxidation of pyrite by molecular oxygen. *Chemical Reviews*, **82**, 461–497, https://doi.org/10.1021/cr00051a001

Maher, M.L.J. 2014. The Canadian pyrite experience and comparisons with the Irish problems. *In*: Hawkins, A.B. (ed.) *Implications of Pyrite Oxidation for Engineering Works*. Springer, Cham, Switzerland, 207–241.

Maher, M.L.J., Azzie, B., Gray, C. and Hunt, J. 2011. A large-scale laboratory swell test to establish the susceptibility to expansion of crushed rock containing pyrite. Presented at the 2011 Pan-Am CGS Geotechnical Conference, 2–6 October 2011, Toronto, Canada.

Mauk, J.L., Crafford, T.C., Horton, J.D., San Juan, C.A. and Robinson, G.R., Jr, 2020. *Pyrrhotite Distribution in Conterminous United States*. United States Geological Survey Fact Sheet 2020-3017, https://doi.org/10.3133/fs20203017

McCabe, B.A., McKeon, E.P., Virbukiene, R.J., Mannion, P.J. and O'Connell, A.M. 2015. Pyritiferous mudstone–siltstone: expansion rate measurement and prediction. *Quarterly Journal of Engineering Geology and Hydrogeology*, **48**, 41–54, https://doi.org/10.1144/qjegh2013-067

Moum, J. and Rosenqvist, I.T. 1959. Sulphate attack on concrete in the Oslo region. *Journal of the American Concrete Institution*, **56**, 257–264.

Murad, E. and Rojik, P. 2003. Iron-rich precipitates in a mine drainage environment: Influence of pH on mineralogy. *American Mineralogist*, **88**, 1915–1918, https://doi.org/10.2138/am-2003-11-1234

Murowchick, J.B. and Barnes, H.L. 1987. Effects of temperature and degree of supersaturation on pyrite morphology. *American Mineralogist*, **72**, 1241–1250.

Nixon, P.J. 1978. Floor heave in buildings due to use of pyritic shales as fill materials. *Chemistry and Industry*, **4**, 160–164.

Nordic Concrete Federation 2018. Impact of sulphide minerals (pyrrhotite) in concrete behaviour. *Workshop Proceedings No. 14*, 15–16 November 2018, Oslo, Norway, https://nordicconcrete.net/workshop-proceeding-no-14-impactof-sulphide-minerals-pyrrhotitein-concrete-aggregate-on-concretebehaviour/

Nordstrom, D.K. 1982. Aqueous pyrite oxidation and the consequent formation of secondary iron minerals. *Soil Society of America Special Publications*, **10**, 37–56.

Nordstrom, D.K. 2020. Geochemical modelling of iron and aluminium precipitation during mixing and neutralization of acid mine drainage. *Minerals*, **10**, https://doi.org/10.3390/min10060547

Penner, E. and Eden, W.J. 1972. *Expansion of Pyritic Shales, Canadian Building Digest*. Canadian Building Digest CBD-152. National Research Council Canada, Ottawa, Ontario, Canada.

Penner, E., Eden, W.J. and Gillott, J.E. 1973. Floor heave due to biochemical weathering of shale. *In*: *Proceedings of the 8th International Conference on Soil Mechanics and Foundation Engineering, Volume 2*. MIR Publication, Moscow, 151–158.

Pugh, C.E., Hossener, L.E. and Dixon, J.B. 1981. Pyrite and marcasite surface area as influenced by morphology and particle diameter. *Soil Science Society of America Journal*, **45**, 979–982, https://doi.org/10.2136/sssaj1981.03615995004500050033x

Pye, E.K. and Miller, J.A. 1990. Chemical and biochemical weathering of pyritic mudrocks in a shale embankment. *Quarterly Journal of Engineering Geology*, **23**, 365–381, https://doi.org/10.1144/GSL.QJEG.1990.023.04.11

Raiswell, R. and Berner, R.A. 1985. Pyrite formation in euxinic and semi-euxinic environments. *American Journal of Science*, **285**, 710–724, https://doi.org/10.2475/ajs.285.8.710

Reid, J.M. and Avery, K. 2009. *Measuring the Potential Reactivity of Sulfides in Aggregates*. TRL Limited Final Report for the Mineral Industry Sustainable Technology (MIST) Project MA/7/G/5/001.

Reid, J.M. and Cripps, J.C. 2019. Geochemical lessons from Carsington Dam failure of 1984 and reconstruction. *Quarterly Journal of Engineering Geology and Hydrogeology*, **52**, 414–424, https://doi.org/10.1144/qjegh2018-184

Reid, J.M., Czerewko, M.A. and Cripps, J.C. 2005. *Sulfate Specification for Structural Backfills*. TRL Report 447. Transport Research Laboratory (TRL), Crowthorne, UK.

Reid, J.M., Czerewko, M.A., Cripps, J.C. and Longworth, I. 2024. Sulfate and sulfides in soils and rocks. *In*: Brown, M., Burland, J., Chapman, T., Higgins, K., Skinner, H. and Toll, D. (eds) *ICE Manual of Geotechnical Engineering, Volume 1: Geotechnical Engineering Principles, Problematic Soils and Site Investigation*. 2nd edn. Emerald Publishing, Leeds, UK, 559–575.

Ricard, D.T. 2015. *Pyrite: A Natural History of Fool's Gold*. Oxford University Press, Oxford, UK.

Rimstidt, J.D. and Vaughan, D.J. 2003. Pyrite oxidation: a state-of-the-art assessment of the reaction mechanism. *Geochimica et Cosmochimica Acta*, **67**, 873–880, https://doi.org/10.1016/S0016-7037(02)01165-1

Ritsema, C.J. and Groenenberg, J.E. 1993. Pyrite oxidation, carbonate weathering, and gypsum formation in a drained potential acid sulphate soil. *Soil Science Society of America Journal*, **57**, 968–976, https://doi.org/10.2136/sssaj1993.03615995005700040015x

Sasaki, K., Tsunekawa, M., Ohtsuka, T. and Konno, H. 1998. The role of sulphur oxidizing bacteria *Thiobacillus thiooxidans* in pyrite weathering. *Colloids and Surfaces A: Physicochemical and Engineering Aspects*, **133**, 269–278, https://doi.org/10.1016/S0927-7757(97)00200-8

SBC 2007. *SBC 304: Structural – Concrete Structures*. National Committee for the Saudi Building Code (SBC), Riyadh.

Shayan, A. 1988. Deterioration of a concrete surface due to the oxidation of pyrite contained in pyritic aggregate. *Cement and Concrete Research*, **18**, 723–730, https://doi.org/10.1016/0008-8846(88)90095-6

Silverman, M.P. 1967. Mechanism of bacterial pyrite oxidation. *Journal of Bacteriology*, **94**, 1046–1051, https://doi.org/10.1128/jb.94.4.1046-1051.1967

Snedker, E.A. and Temporal, J. 1990. *M40 Motorway Banbury IV Contract – Lime Stabilisation*. Institution of Highways and Transportation, London.

Steward, H.E. and Cripps, J.C. 1983. Some engineering implications of chemical weathering of pyritic shale. *Quarterly Journal of Engineering Geology*, **16**, 281–289, https://doi.org/10.1144/GSL.QJEG.1983.16.04.05

St. John, T.W. 2020. Geotechnical characterization of sulfur species in UK Jurassic mudrocks. *Quarterly Journal of Engineering Geology and Hydrogeology*, **53**, 598–608, https://doi.org/10.1144/qjegh2019-148

Strogen, P. 2021. Lithological controls on the expansion of unbound aggregates in Dublin, eastern Ireland: two contrasting cases. *Quarterly Journal of Engineering Geology and Hydrogeology*, **55**, https://doi.org/10.1144/qjegh2021-054

Sutton, D., McCabe, B., O'Connell, A. and Cripps, J.C. 2013. A laboratory study of the expansion of an Irish pyritic mudstone/siltstone fill material. *Engineering Geology*, **152**, 194–201, https://doi.org/10.1016/j.enggeo.2012.10.011

Tacker, R.C. 2020. A review of 'pyrite disease' for paleontologists, with potential focused interventions. *Palaeontologia Electronica*, **23**, a45, https://doi.org/10.26879/1044

Taylor, R.K. 1988. Coal Measures mudrocks: composition, classification and weathering processes. *Quarterly Journal of Engineering Geology*, **21**, 85–99, https://doi.org/10.1144/GSL.QJEG.1988.021.01.06

Taylor, R.K. and Cripps, J.C. 1984. Mineralogical controls on volume change. *In*: Attewell, P.B. and Taylor, R.K. (eds) *Ground Movements and their Effects on Structures*. Surrey University Press, Guildford, UK, 268–302.

Temple, K.L. and Delchamps, E.W. 1953. Autotrophic bacteria and the formation of acid in bituminous coal mines. *Applied Microbiology*, **1**, 255–258, https://doi.org/10.1128/am.1.5.255-258.1953

Thomas, M.D.A., Kettle, R.J. and Morton, J.A. 1989. Expansion of cement-stabilized minestone due to the oxidation of pyrite. *Transportation Research Record*, **1219**, 113–120, https://onlinepubs.trb.org/Onlinepubs/trr/1989/1219/1219-011.pdf

Tuohy, B., Carroll, N. and Edger, M. 2012. *Report of the Pyrite Panel*. Department of the Environment, Community and Local Government, Government of Ireland, Dublin, https://www.pyriteboard.ie/Pyrite/media/Pyrite/Updated/Report-of-Pyrite-Panel-June-2012.pdf

Vaughan, D.J. (ed.) 2006. *Sulfide Mineralogy and Geochemistry*. Reviews in Mineralogy & Geochemistry, **61**. Mineralogical Society of America, Washington, DC, https://doi.org/10.1515/9781501509490

Vear, A. and Curtis, C. 1981. A quantitative evaluation of pyrite weathering. *Earth Surface Processes and Landforms*, **6**, 191–198, https://doi.org/10.1002/esp.3290060214

Watts, K. and Charles, A. 2015. *Building on Fill: Geotechnical Aspects*. 3rd edn. HIS BRE Press, Bracknell, UK.

Wiese, R.G., Jr, Powell, M.A. and Fyfe, W.S. 1987. Spontaneous formation of hydrated iron sulfates on laboratory samples of pyrite and marcasite bearing coals. *Chemical Geology*, **63**, 29–38, https://doi.org/10.1016/0009-2541(87)90071-4

Wilson, E.J. 1987. Pyritic shale heave in the Lower Lias at Barry, Glamorgan. *Quarterly Journal of Engineering Geology*, **20**, 251–253, https://doi.org/10.1144/GSL.QJEG.1987.020.03.07

Winter, M.G., Butler, A.M., Brady, K.C. and Stewart, W.A. 2002. Investigation of corroded stainless steel reinforcing elements in spent oil shale backfill. *Proceedings of the Institution of Civil Engineers – Geotechnical Engineering*, **155**, 35–46, https://doi.org/10.1680/geng.2002.155.1.35

Yamanaka, T., Miyasaka, H., Aso, I., Tanigawa, M. and Shoji, K. 2002. Involvement of sulfur- and iron-transforming bacteria in heaving of house foundations. *Geomicrobiology Journal*, **19**, 519–528, https://doi.org/10.1080/01490450290098487

A semi-quantitative geotechnical visualization tool for ground engineering projects

Thomas Berry[1], Evie Barratt[1]* and Chrysothemis Paraskevopoulou[2]

[1]Jacobs UK Ltd, 1 City Walk, Holbeck, Leeds LS11 9DX, UK
[2]School of Earth and Environment, University of Leeds, Woodhouse, Leeds LS2 9JT, UK
TB, 0000-0003-4540-9005; EB, 0009-0002-5750-5440; CP, 0000-0002-7063-5592
*Correspondence: evie.barratt@jacobs.com

Abstract: This paper presents a technique for the visualization and semi-quantification of various ground engineering criteria, in the form of a spider graph.

There are currently few techniques for effectively presenting multiple ground engineering and engineering complexity criteria in a single graphic. The technique described proposes a tool that can do so for technical and non-technical audiences alike. The input criteria can be adapted to suit any project in soil, rock or a combination of both.

The methodology and example presented is for a current Jacobs' rock engineering project. The geotechnical parameters used include unconfined compressive strength, Geological Strength Index, *in situ* stress and groundwater head. Engineering complexity is also considered and an additional criterion to allow for significant influences not accounted for in the other criteria. The input criteria must be set at the beginning of the project to ensure that comparisons can be made both between terrains (areas with similar ground engineering characteristics) and throughout the life cycle of a project.

The tool aims to communicate ground conditions and potential 'ground risk' with the usefulness of the tool being its simplicity and adaptability to meet the project needs at all stages of the project. A preliminary version of an 'app' is available to trial the tool.

As ground engineers, our primary task is to manage ground risk for our clients and our organizations. This paper presents a preliminary technique for the visualization and semi-quantification of ground conditions and engineering complexity, and provides a communication tool to help the working engineer or geoscientist to:

- organize and categorize the interpretation of data; and then
- share their findings with others.

The technique provides a structured methodology for current industry practices to semi-quantify and communicate ground conditions, uncertainty and potential risk to show relative differences in ground conditions between 'terrains' across and between projects and over time. The term 'terrain' is used throughout this paper and refers to areas with similar ground engineering characteristics.

The usefulness of the tool is its simplicity and adaptability to engineering projects in soil and/or rock, as the operator can set the most significant criteria for the project and apply the technique at all stages of the project. As such, results can be compared between terrains as the project matures and new information is available.

Geotechnical parameters are entered into the tool to provide a 'measure' of how favourable the ground conditions are in relation to proposed engineering works. In addition to the properties of the ground, the approach enables engineering judgement to be applied in terms of engineering complexity and a criterion to allow for any additional significant influence not accounted for in the other categories. Using six criteria gives an overview of the conditions and engineering in a specific terrain for comparison across the project. The approach has a method of weighting the actual criteria assessed and should be reflected to meet the requirements of the project.

The methodology and worked example, given later in this paper, are for a rock engineering project; however, the technique has been applied to projects wholly in soils, as well as mixed soil and rock conditions, by the adaption of the criteria to suit the specific needs of the project.

Background

Pictograms and schematic representations of ground engineering information have long been used to visualize and communicate engineering principles, characteristics and behaviour, to great effect. Selected examples for rock include the following:

- Hudson's (1989) rock mechanics–rock engineering interaction matrix;
- diagrams in Martin *et al.* (1999) showing tunnel instability and brittle failure as a function of rock mass rating and the ratio of the maximum far-field stress to the unconfined compressive strength modified from Hoek *et al.* (2000);

From: Griffiths, J. S., Giles, D. P., Davis, J. A. and Dunlop, J. J. (eds) 2025. *The Peter Fookes Engineering Geological Legacy in Geomodels, Geomaterials and Geomorphology*. Geological Society, London, Engineering Geology Special Publications, **30**, 179–189,
First published online 27 May 2025, https://doi.org/10.1144/egsp30-2024-2

- the simplified representation of the influence of scale on rock mass behaviour for designing underground excavations or rock slopes (Hoek 1983); and
- a table describing rock mass behaviour in underground excavations (Stille and Palmström 2008) (see Fig. 1).

Proposed semi-quantitative geotechnical visualization tool for ground engineering projects

The original intention was to design a tool that would allow comparison of ground conditions and engineering complexity across a large-scale excavation. The aim was to present the variations in ground conditions and complexity of the engineering in order to demonstrate the necessity for additional investigations or additional analytical consideration in the design process. However, practice has shown the technique is equally well suited to represent conditions for specific locations (i.e. at an individual footing for a particular structure) or for areas of similar ground ('terrains') over larger areas (e.g. sections of embankments, cuttings or tunnels).

The scale of the parameters used in the spider graphs are set at the beginning of the project so comparison between locations can be made and changes over time can be assessed as the project matures. The scale is set such that the less favourable conditions are on the outside of the spider graph web and the more favourable conditions are towards the centre. As such, the smaller the shaded area shown on the spider graph, the 'worse' the ground conditions and/or the more onerous the engineering (Fig. 2). As the ground model is developed and additional phases of ground investigations (GI) are incorporated, the area may increase or decrease over time depending on the outcome of the investigations.

Criteria

In general terms, there are three basic elements of the ground you would need to be able to design most foundation solutions. These parameters are the material properties, the mass properties and groundwater. In addition to the above, an appreciation of the *in situ* stress, the complexity and scale of the proposed engineering, and the influence or impact of the structure on the ground, and vice versa, are also required to gain a holistic understanding of the ground to define and solve engineering problems.

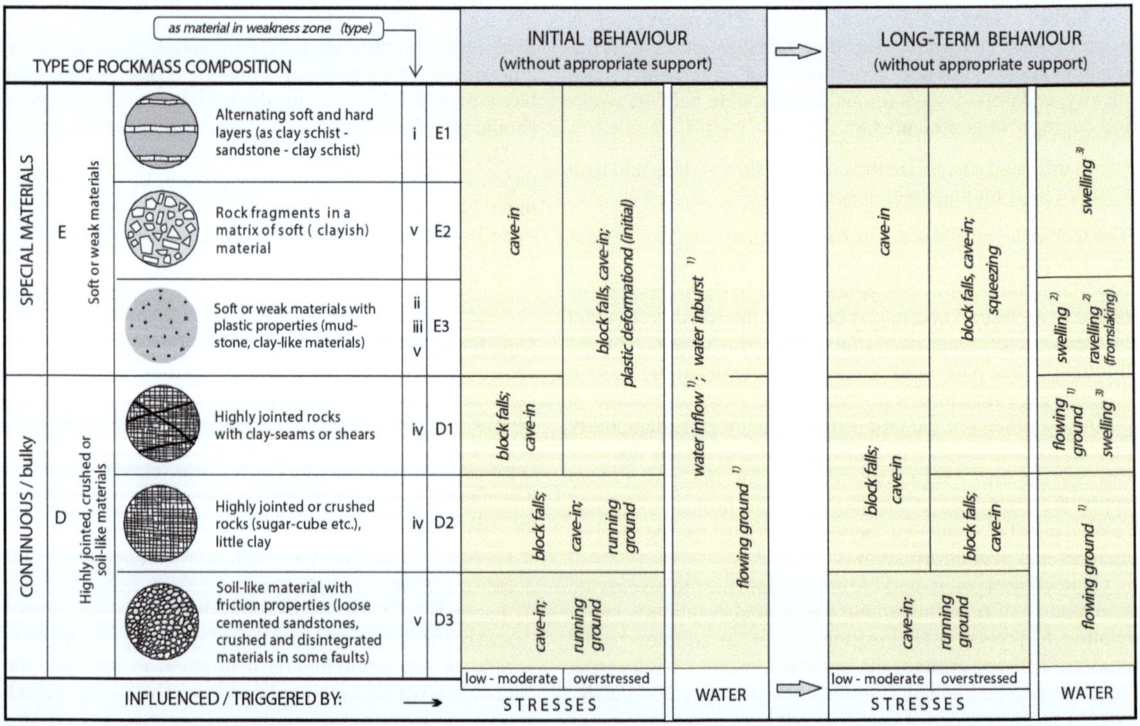

Fig. 1. Example of pictograms and schematic representations of ground conditions. [1]Will take place in porous materials and where there are channels (open joints); [2]requires materials with swelling minerals (smectite, anhydrite); [3]requires content of swelling clay in seams and clay zones. Source: modified from Stille and Palmström (2008).

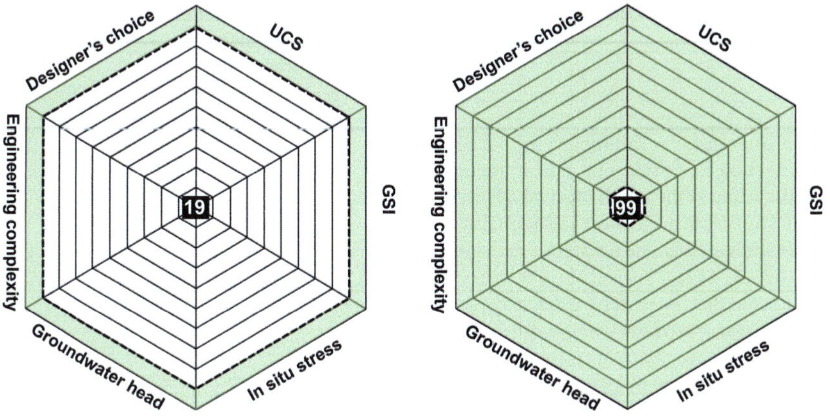

Fig. 2. Spider graphs showing examples of (left) poor and (right) good conditions. The green shaded area is referred to as the spider graph web.

In addition to understanding the fundamentals of the ground noted above, communication of technical information, associated geohazards and possible 'risks' in any discipline is important, and in rock engineering it is often crucial. The 'spider graphs' technique described below is our crude attempt to resolve the issue of representing semi-quantitative engineering and project information in a clear and concise manner to a technical and/or non-technical audience.

The spider graph (Fig. 3) is composed of a hexagon split into six triangles, each representing a ground parameter or element of the project. The subdivisions chosen for this example (for a rock engineering project) are as follows:

- the upper right sector represents the rock and is composed of the material properties, unconfined compressive strength (UCS) and Geological Strength Index (GSI);
- the lower central sector represents parameters that, broadly speaking, act on the rock, and is composed of *in situ* stress and groundwater head; and
- the upper left sector represents the project or engineering complexity and an additional criterion to allow for an additional significant influence not accounted for in the other criteria (i.e. designer's choice).

The six criteria selected for the spider diagrams are discussed in more detail below.

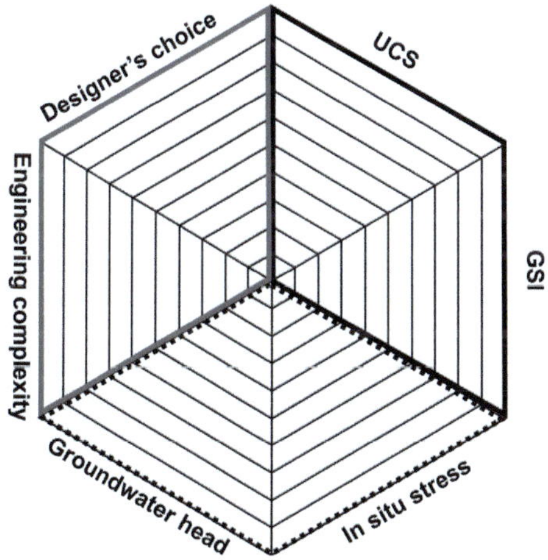

Fig. 3. Spider graph input parameter subdivisions. The black outline highlights the parameters that represent the rock material and mass properties; the black dotted outline highlights the parameters that act on the rock; and the grey outline highlights the parameters that represent the engineering project.

Unconfined compressive strength

The material properties of the rock are represented by the unconfined compressive strength (UCS) of the rock, measured in MPa. UCS was chosen as it is a well-understood parameter that is widely used and regularly tested for with standardized procedures, including the ISRM guidelines (Bieniawski and Bernede 1979; Fairhurst and Hudson 1999) or through well-understood correlations such as the point load test (PLT) (Bieniawski 1975) or Schmidt hammer rebound hardness test (Aydin 2015). In addition, UCS can be correlated with other material properties such stiffness (Goodman 1989).

Geological Strength Index

Discontinuities are represented by the unitless Geological Strength Index (GSI) (Hoek and Brown 1980; Hoek 1994; Marinos and Hoek 2000). GSI addresses 'two principal factors considered to have important influences on the mechanical properties of a rock mass, i.e. the structure (or blockiness) and the condition of the joints' (Hoek and Brown 2019, p. 449). GSI was chosen as it is a well -understood parameter that is widely used and regularly recorded in rock engineering projects.

In situ stress

The *in situ* stress can often have a significant influence on behaviour when engineering is carried out, and is included in our assessment as the value at the depth or level of interest measured in MPa. We suggest that the most influential stress is used and if this is not known, then the greater of the vertical or horizontal stresses (σ_v, σ_h or σ_H) is used. *In situ* stress is a well-understood parameter that is widely used and, if not tested for, can be estimated based on simple formulas or from a review of the available literature such as Baptie (2010). It should be noted that the authors considered replacing the *in situ* stress criterion with time-dependent behaviour. However, from an overall project perspective, *in situ* stress was considered more appropriate.

Groundwater head

Hydrogeology is a significant parameter in rock engineering projects and can be represented in a number of different ways such as average head, potential inflow and permeability but we suggest the highest groundwater head (in m) for the terrain or location should be used. The initial assumption is that the lower the groundwater head the better (we accept that this is not always the case, and the spider graph can be amended to suit this design case). Groundwater head was chosen as it is a well-understood parameter that is widely used and relatively easy to calculate from regularly measured groundwater readings.

Engineering complexity

Engineering complexity and scale can be quite subjective and as such we have deferred to the Eurocode 7: Geotechnical Design – Part 1: General Rules (BS EN 1997-1:2024: BSI 2024) definitions of Geotechnical Category. The initial assumption is that the simpler the engineering the more straightforward are the design considerations. The Eurocode 7: Geotechnical Design – Part 1: General Rules definitions of Geotechnical Category are summarized below:

(1) Geotechnical Category 1 should only include small and relatively simple structures with no excavation below the water table. Comparable local experience would lead the designer to understand the project to be sufficiently straightforward, thus requiring routine methods for foundation design and construction.

(2) Geotechnical Category 2 should include conventional types of structure and foundation with no exceptional risks, difficult ground conditions, high seismicity, unusual loading conditions or persistent ground movements. Examples of conventional structures complying with Geotechnical Category 2 include shallow and deep foundations, retaining or supporting structures, excavations, bridge piers and abutments, earthworks, ground anchors, and tunnels in hard, non-fractured rock and not subjected to special water tightness or other requirements.

(3) ‘Geotechnical Category 3 should include structures or parts of structures, which fall outside the limits of Geotechnical Categories 1 and 2’ (BSI 2024, p. 22), and includes:

- very large or unusual structures;
- structures involving abnormal risks, or unusual or exceptionally difficult ground or loading conditions;
- structures in highly seismic areas;
- structures in areas of probable site instability or persistent ground movements that require separate investigation or special measures

(BSI 2024, p. 22).

We suggest that engineers allocate Geotechnical Category 1 structures an engineering complexity value between 0 and 30, Geotechnical Category 2 structures a value between 30 and 60, and Geotechnical Category 3 structures a value between 60 and 100, thus affording the designer some autonomy to use their judgement to influence the significance of the complexity of the structure on the project. For example, it may be that all terrains within a project can be attributed to a Geotechnical Category 2; however, based on project-specific considerations, some terrains may have subtle differences in their complexity. Providing a range of values still allows for comparison between terrains. To support with the selection of parameters, discussion is provided in the methodology as to the sensitivity of each of the input values and the impact of these on the semi-quantitative output.

Designer’s choice

An undesignated criterion – the designer’s choice – is proposed to allow for consideration of, as part of the assessment, any additional significant influence not accounted for in the other criteria. These may include third-party restrictions (vibration limits, pumping restrictions, etc.), accessibility restrictions, the quality of GI information, the presence of residual geotechnical uncertainty, and terrain- or location-specific geohazards (mining risk, seismicity, etc.).

In addition, it may be used to represent other factors such as the aptitude or attitude of the client organization, the aggressivity of the programme and budget restrictions that will influence the engineer’s ability to fully characterize the ground for the proposed engineering. The assumption is that the more informed the client, the higher the budget, the longer the programme, and so on, the better.

It is important to note that designer’s choice criterion is not intended to facilitate ‘fixing’ of the outcomes to justify an opinion. This category is designed to allow for the visualization tool to be bespoke to the project and reflect the project’s key ground engineering considerations. A worked example is presented later in this paper to illustrate how this criterion can be adopted.

Methodology

Workflow

A suggested workflow for using the semi-quantitative geotechnical visualization tool is presented in Figure 4 and is described below:

(1) Define the problem: the first task is to define the project 'problem', ground conditions (including identifying the number of terrains) and the reliability of the project data available (the top row of Fig. 4).

(2) Set the criteria: the second task is to set the assessment criteria, ranges of values and weightings (the second row of Fig. 4).

(3) Carry out analysis: the third task is to create the spider graphs for each terrain within a project and analyse the results. There are two potential feedback loops to help to refine the technique during the initial use on the project.

Analytical procedure

There are several options for the way in which the criteria described above could be presented on the spider graph. For example, a simple linear scale could be used, and the shaded area of the spider graph used to visually represent and communicate the conditions. However, the methodology below proposes to employ a semi-quantitative option using a weighted inverse triangle technique to represent the site-specific data.

The area of each triangle representing a parameter in the spider graph is calculated from the outside in, resulting in the larger increases to the shaded triangle area (the spider graph web) occurring where the parameters are low or unfavourable. This approach has been adopted with the understanding that the impact of changing parameters on ground conditions is non-linear, with the greatest impact being where the ground conditions are poor. In other words, small increases or improvements to parameters representing 'poor' ground conditions have a more significant impact compared with small increases or improvements to parameters representing 'good' ground conditions. This logarithmic relationship is illustrated in Figure 5, with the equivalent height (h) of the spider graph web on the x-axis and the percentage of the total triangle area on the y-axis.

To generate a spider graph the following parameters need to be determined:

- Project-wide ranges for each input parameter: this information sets the scale of each of the six triangles of the spider graph and must be kept consistent throughout the project to allow comparisons to be made across different project stages and between different terrains within a project.
- Terrain-specific characteristic values for each input parameter: this value may be determined through statistical analysis; however, the parameter selection will be dependent on the data available and will be specific to the project. This value may change as the project progresses but will be plotted relative to the consistent project wide ranges, as discussed above.

Consideration should be given to each input parameter as to whether the minimum or maximum value is favourable or unfavourable. For example, a high GSI would be favourable for limiting the lining requirements for an underground excavation but would be unfavourable for the penetration of ground by a piling rig (assuming that it is difficult to achieve the design penetration).

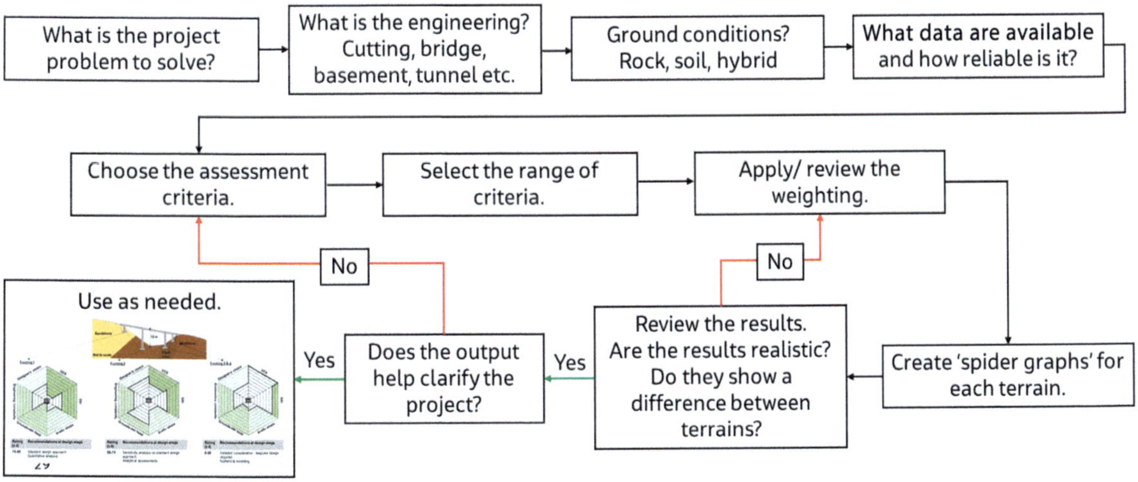

Fig. 4. Suggested workflow for using the semi-quantitative geotechnical visualization tool.

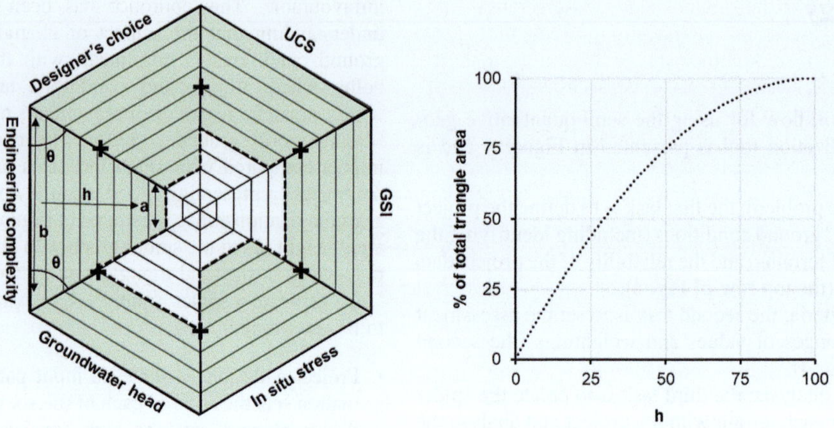

Fig. 5. Calculation of the semi-quantitative output based on the spider graph web.

A function is included in the generation of the spider graph to select whether the minimum value for each parameter is favourable, and the equivalent value result is then plotted to consistently show unfavourable conditions as a smaller shaded area on the web and favourable conditions shown as a larger area.

Determination of the input values for the criteria should be documented clearly to ensure that a consistent approach is adopted at each project stage.

Following the selection of the criteria, ranges and weightings, the area of the spider graph web is calculated as a proportion of the total spider graph area, as detailed below:

(1) Each of the six criteria that form the spider graph is represented by an equilateral triangle. The height of the triangle is nominally taken to be 100. The shaded area of the spider graph is based on an area of a trapezium, as detailed in equation (1) and Figure 5, where a is the roof of the trapezium, b is the base of the trapezium and h is the height of the trapezium:

$$\text{area of spider graph web per parameter} = \frac{a+b}{2}h. \tag{1}$$

(2) The height of the trapezium (h) is determined based on the project-specific ranges for each parameter and the terrain-specific characteristic value. The characteristic value is plotted as a proportion of the project-specific parameter ranges as detailed in equation (2), where TC is the terrain-specific characteristic value, PR is the project-specific parameter range and $\mathrm{PR_{min}}$ is the project-specific parameter range minimum:

$$h = \frac{\mathrm{TC} - \mathrm{PR_{min}}}{\mathrm{PR}} \times 100. \tag{2}$$

(3) The length of the roof of the trapezium (a), presented in Figure 5, is dependent on the height of the trapezium and the slope angle of the trapezium sides (see equation 3). In this case where six criteria are selected, the slope angle of the trapezium sides (θ) is taken to be 60° as the trapezium represents the shaded area within an equilateral triangle. θ will differ should there be more or less than six criteria to be commensurate with the total spider graph area (calculated in stage 5):

$$a = b - 2\frac{h}{\tan(\theta)}. \tag{3}$$

(4) The significance of each parameter may vary from project to project; therefore, an equal weighting across all parameters for the semi-quantitative output may not be appropriate. If a criterion is determined to be of higher significance to the project then a weighting of 1.25 might be applied, whereas if a criterion is determined to be of lower significance to the project a weighting of 0.75 may be more appropriate. The weighting adopted should be kept consistent across the project to allow for a consistent comparison across different locations or terrains. The total weightings of each parameter must equate to the total number of criteria displayed on the spider graph. The weighting is displayed visually on the spider graph, with a darker green shading illustrating where more significance has been placed on a parameter, and a lighter shading illustrating where less significance has been placed on a parameter.

For an example of how the weightings can be applied, consider a high strength rock environment where the variation of the UCS is between 155 and 175 MPa. The impact of variation in the parameter is small because of its narrow range and high strength. One could consider reducing the influence of this parameter by 25% (i.e. a weighting of 0.75).

Furthermore, if the groundwater pressure range were high due to seasonal fluctuations in precipitation and the operational requirements of the structure demanded a dry facility, one might consider increasing the weighting for groundwater pressure to 1.25. A worked example explaining the application of parameter weightings is provided later in this paper.

(5) The weighted shaded area of each parameter is totalled and a percentage of the total area of the spider graph is calculated. The spider graph semi-quantitative output is calculated as:

$$\% \text{ of total spider graph area} = \frac{\sum \text{weighted shaded area of each parameter}}{\text{total spider graph area}} \times 100. \quad (4)$$

Applications

The semi-quantitative geotechnical visualization tool can be used for soil, rock and hybrid ground conditions for most, if not all, engineering projects, at any stage of a project and throughout the life cycle of a project as a means of comparing the ground conditions, engineering complexity and designer's choice across geotechnical domains.

The numerical output can be used to communicate requirements for further geotechnical work, the potential extent and complexity analytical solutions required, the robustness of the engineering required, and the construction supervision requirements.

The figures can be used in reports, and added to plans, maps and cross-sections, as well as for proposed engineering design drawings.

The semi-quantitative geotechnical visualization tool can be used to develop ground models and help to optimize future ground engineering requirements for a project. Carter (1992) acknowledged that an appropriate database containing relevant information is key in the development of good predictive geological models and that geological risk can be minimized by increasing expenditure on site investigations, where it is prudent to do so. The development of a suitable predictive geological model helps to optimize the geotechnical design and is particularly important in view of claims regarding unforeseen ground conditions relating to the lack of, or quality of, geological data available at the time of tender (Carter 1992).

To illustrate this point, high-level recommendations based on the semi-quantitative output and the stage of the project are summarized in Figure 6. The rating intervals provided are non-linear, cognisant of the logarithmic relationship between the criteria and the semi-quantitative output. The recommendations provided can be adapted to suit project-specific requirements but provide an initial outline to communicate unfavourable ground conditions.

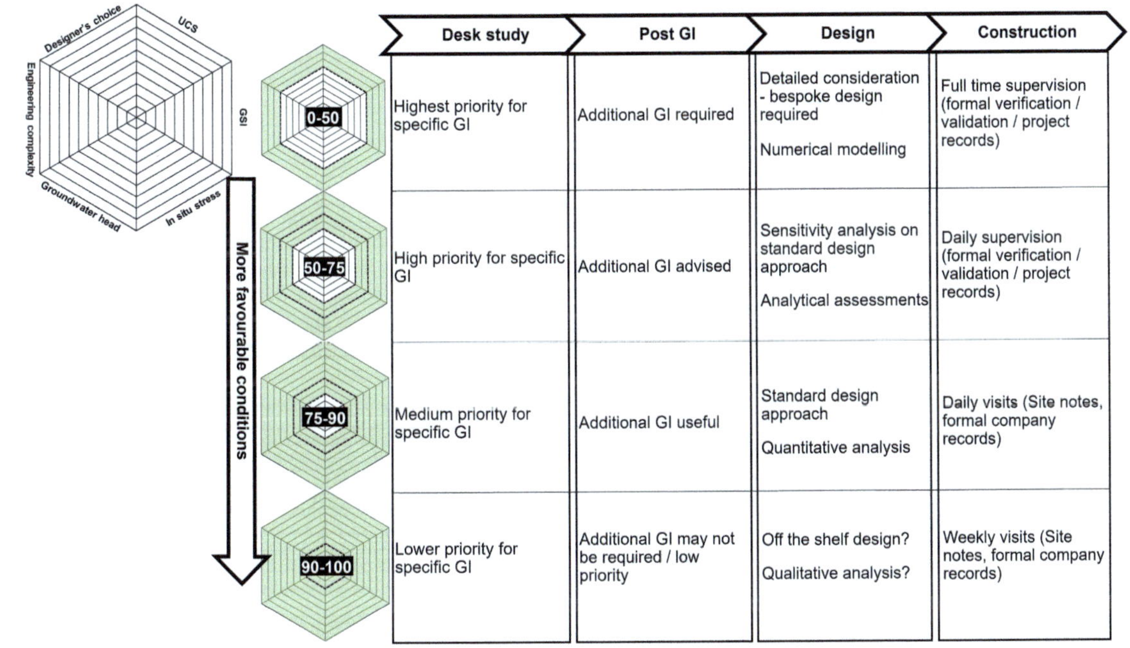

Fig. 6. Spider graph rating matrix.

Project case study

A worked example demonstrating the use of the spider graphs on a rock engineering project is presented below.

Project description

The project comprises the construction of a new bypass, the structure considered is an overbridge and the project is at the detailed design stage. Site details are anonymized.

The ground model for the project comprises a conformable sequence of Silurian-age sandstone and mudstones, which is presented in Figure 7. Based on the available information, including published information and project-specific GI data, it was concluded that the footings of the overbridge were to be founded on lithologies of different material and/or mass properties.

Based on the ground model and the engineering proposed, three terrains were identified:

- Terrain 1: Footing 1
  'Medium strong, orange brown sandstone, moderately, locally highly weathered. With very closely spaced randomly orientated planar, smooth, heavily iron-stained discontinuities.'
- Terrain 2: Footing 2
  'Medium strong, locally weak, laminated, dark grey mudstone, moderately, locally highly weathered. With extremely to very closely spaced randomly orientated planar, smooth, heavily iron-stained discontinuities.'
- Terrain 3: Footings 3 and 4
  'Very weak to extremely weak, indistinctly laminated, dark grey mudstone. With extremely to very closely spaced aligned to randomly orientated planar, smooth, polished, striated heavily iron-stained to gleyed discontinuities – fault zone.'

Spider graph development

Criteria, ranges and weighting

Criteria for each terrain are summarized in Table 1. The rock material and mass properties, the *in situ* stress, and

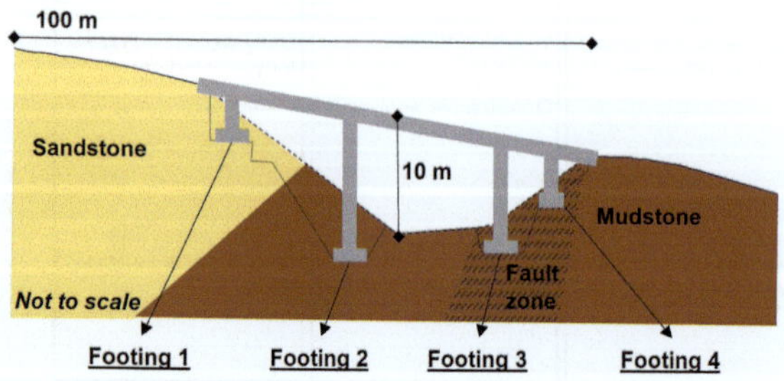

Fig. 7. Schematic cross-section illustrating the proposed overbridge and the ground conditions.

Table 1. Summary of terrain-specific values adopted for the project worked example

Terrain	Parameter	Units	Value
Footing 1	UCS	MPa	40
	GSI		50
	In situ stress	MPa	0.3
	Groundwater head	m	0.5
	Engineering complexity		50
	Designer's choice		50
Footing 2	UCS	MPa	25
	GSI		40
	In-situ stress	MPa	0.3
	Groundwater head	m	10.5
	Engineering complexity		50
	Designer's choice		30
Footing 3 and Footing 4	UCS	MPa	1
	GSI		20
	In situ stress	MPa	0.3
	Groundwater head	m	3
	Engineering complexity		50
	Designer's choice		50

Project-specific inputs are shown in the blue highlighted cells.

groundwater head were determined based on statistical analysis of available site-specific GI data.

For the purposes of this assessment, a conservative appraisal of the groundwater regime was adopted. Groundwater monitoring for the project recorded the maximum groundwater level across the site to be generally 1–2 m below the existing ground level within the sandstone and mudstone units. The groundwater head assumed for the project was based on a consistent depth below ground level in relation to the elevation of the footings. An engineering complexity of 50 was selected for each terrain which equates to Geotechnical Category 2 (BSI 2024).

The designer's choice criterion was used to describe the uncertainty associated with the proximity to a fault. Available GI data had not been able to confirm the extent of the fault zone and, therefore, the extent of the disturbed rock mass. The uncertainty in the ground conditions and the possible added complexity in the design are captured by the lower designer's choice value applied to the Footing 2 terrain due its potential proximity to the fault and the consequential uncertainty in the material strength and GSI.

Footings 3 and 4 were given a higher designer's choice rating compared with Footing 2 as the footings were understood to be within the fault zone and this is reflected in the reduced values of material strength and GSI (accepting that there was still some uncertainty in the magnitude of the impact of the faulting on the material strength and GSI). Footing 1 is furthest away from the fault but there is uncertainty associated with the presence or otherwise of accommodation faults impacting the footing.

Ranges for each of the parameters were set based on the available data for the project as summarized in Table 2. The project ranges adopted for each parameter were selected to

Table 2. General project input values for the project worked example

Parameter	Units	Minimum	Maximum	Favourable minimum value	Weighting
UCS	MPa	0.5	50	N	1.75
GSI		0	100	N	1.25
In situ stress	MPa	0	1	Y	0.50
Groundwater head	m	0	15	Y	1
Engineering complexity		0	100	Y	1
Designer's choice		0	100	N	0.5

Project-specific inputs are shown in the blue highlighted cells.

encompass the expected range of values across the project as this input information sets the scale for the spider graphs.

The weighting applied to each of the parameters was updated based on the ground model assumed and the engineering proposed. In this case, the significance and, therefore, the weighting applied to the *in situ* stress and the designer's choice were reduced as it was determined that these parameters would have less of an impact on the design of the over-bridge than the UCS and GSI, where the weighting was increased. Weightings for the groundwater and engineering complexity were set at unity.

Spider graph results

The spider graphs generated from the inputs summarized in Tables 1 and 2 are presented in Figure 8, with the recommendations based on the semi-quantitative output described in Figure 6.

Discussion

A simple tool to communicate ground conditions and engineering complexity, and thereby inference the overall possible geohazards and potential ground 'risk', has been proposed. The technique allows the use of the most significant ground parameters to be selected for the project (be it soil, rock or both) and weighted according to their significance to the project. It is recommended that the geotechnical criteria chosen are well understood and ubiquitously collected. Project complexity (indicative of ground–structure–

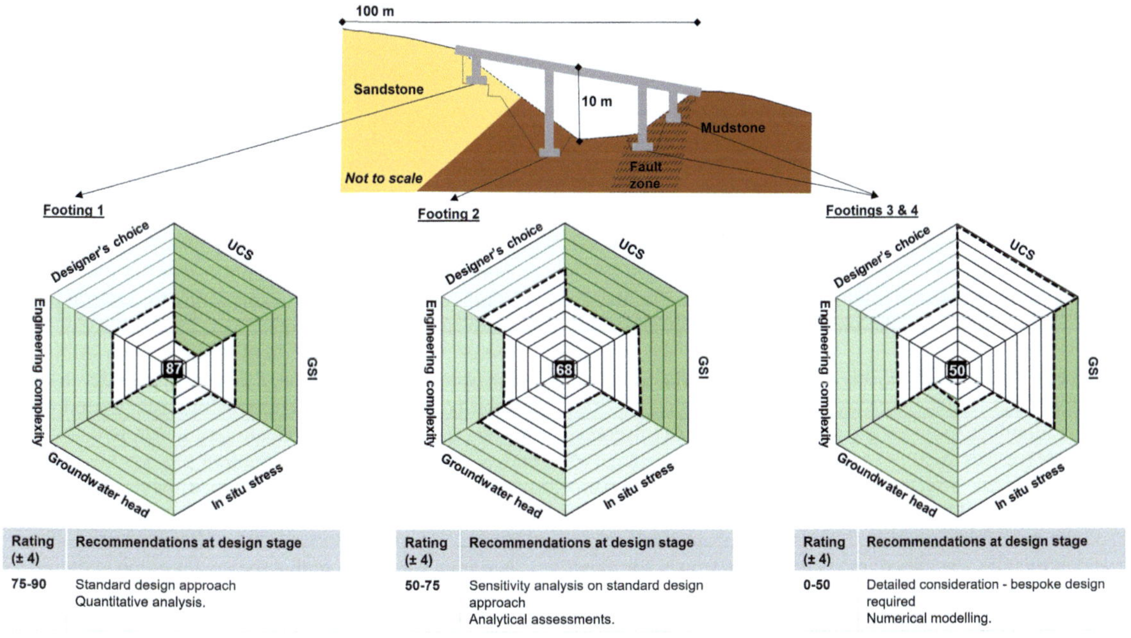

Fig. 8. Spider graph visual and semi-quantitative output for the example project.

ground interactions) and an additional criterion to allow for an additional significant influence not accounted for in the other criteria are also included.

The simplicity of the proposed technique is acknowledged; however, its simplicity is also what makes it effective. Ground engineers should adapt the proposed criteria to meet the needs of the project by, for example:

- the replacement of one or both of the 'project engineering' criteria with additional technical parameters such as seismicity, mining risk or ground behaviour; etc.; and/or
- the addition of one or more parameters to meet the specific requirements of the project (e.g. slope height and angle for rock cuttings).

Adapting the initial scale can influence the impact of the parameter by virtue of the area covered by the input. For example, if the anticipated range of the UCS is between 100 and 140 MPa and the parameter range is nominally set at 0 MPa (physically impossible) to 150 MPa, then the difference in UCS is less impactful than if the range was set at 100–150 MPa. The tool also provides opportunities to weight the influence of the parameters to reflect their significance to the project.

Initial applications for projects in soil/superficial deposits have used the following expanded criteria successfully with the specific parameters used based on the soil's engineering behaviour and whether the application is based on short- or long-term considerations:

- Material properties: the material properties of the soil represented by the undrained and drained shear strength, selected as appropriate for long- or short-term applications, granular or cohesive soils. A range of values for the project and terrain-specific characteristic values were selected based on project-specific geotechnical laboratory test data:
  - undrained strength: a strength parameter adopted when the rate of loading is faster than the dissipation of pore-water pressures (i.e. rapid loading of clay soils);
  - angle of shearing resistance: a shear strength parameter from the Mohr–Coulomb failure criterion, used to describe the friction shear resistance of soils together with the normal effective stress; and
  - cohesion: a shear strength parameter from the Mohr–Coulomb failure criterion, derived from interparticle bonds.
- Weathering: based on the grade of the classification system used.
- Discontinuities/fissures: the presence of fissures (discontinuities) reduces the soil strength of the material. A range of values for the project and terrain-specific characteristic values were selected based on engineering descriptions of the material encountered across the project following the principles described by Hoek and Brown for their GSI parameter (Hoek and Brown 1980; Hoek 1994).
- Hydrogeology: groundwater head was chosen as it is a well-understood parameter that is widely used and calculated from regularly measured groundwater readings.
- Engineering complexity: based on the Geotechnical Categories presented in the Eurocode 7: Geotechnical Design – Part 1: General Rules (BSI 2024).
- Designer's choice: an example of an additional consideration prudent to the soil engineering-based projects may include the influence of periglacial processes or climates.

Limitations

Although we believe that the semi-quantitative geotechnical visualization tool is useful, we acknowledge that it is not perfect and has several limitations including (but not limited to):

- An oversimplification of engineering problems that does not represent the full range or variability of the ground conditions in a terrain (or even in a smaller location), and the criteria do not fully describe the subtlety of the influence of the parameter on the engineering (or the coupled process or cumulative impact).
- The criteria, ranges and, specifically, weighting could be used for nefarious intent. The tool offers the engineer the ability to 'influence' the result to steer the answer towards an outcome that might suit them. This is not the intention or the purpose of these facilities.
- As far as possible, the same engineer should use the tool throughout the project life cycle to ensure that a consistent approach is adopted for the selection of criteria, ranges and weighting at the start of the project and for input values as the project progresses.
- The tool presented is intended to be used as a visualization tool only to aid communication to technical and non-technical audiences. The output of the tool is only as good as the input.

Adoption and future development

Our initial proposal for the tool has been outlined earlier in this paper, we will continue to use and refine it based on our own experience plus any feedback from industry. A preliminary version of an 'app' is available to trial the spider graph technique on your projects: https://jacobs.eu2.viktor.ai/public/spidergraph. The 'app' offers the functionality to select the number of criteria and define the criteria presented in the form of a spider graph.

Future works include adapting the criteria of the spider graph for use in establishing design classes for rock slope instability mitigation and support for underground openings in rock. Furthermore, applications for 'ground truthing' as part of the construction verification process are anticipated, and if ranges are applied it is hoped they could be used to help communicate assumptions of ground conditions in geotechnical baseline reports (GBRs) (Davis *et al.* 2023).

It is important to note that the spider graph technique does not aim to minimize the geological uncertainty but aims to communicate the designer's understanding of the ground conditions in different terrains and at different project stages and to communicate recommendations for the next steps.

Finally, it is intended to investigate how we can communicate the uncertainty in the geotechnical parameters used to generate the plot and, hence, the geological uncertainty overall. Early work suggests presenting minimum and maximum characteristic terrain-specific values on the spider graph, with the wider range presented representing parameters that have a greater uncertainty.

Conclusions

The semi-quantitative geotechnical visualization tool presented aims to communicate ground conditions and geohazards to a wider (non-geotechnical) audience to support the justification for further geotechnical work, such as GI, the nature of the analysis that may be suitable for the design and possible construction supervision requirements.

The usefulness of the tool is its simplicity and adaptability to meet the engineers and/or project needs. Experience has shown that the visualizations help to optimize future ground engineering input to projects, manage ground risks and maximize the value of our services to our clients. It is hoped that industry will find the proposed technique useful, adopt it, refine it and use it on projects for the benefit of their clients.

Acknowledgements The authors would like to thank the Jacobs Geodigital Team and Edward Dunn for support with the development of the 'app' for the semi-quantitative geotechnical visualization tool, and Peter Gilbert for his support and comments on the manuscript. We thank the reviewers for their constructive and helpful comments, which have improved the quality and clarity of our manuscript.

Competing interests The authors declare that they have no known competing financial interests or personal relationships that could have appeared to influence the work reported in this paper.

Author contributions **TB**: conceptualization (lead), formal analysis (lead), funding acquisition (lead), methodology (equal), writing – original draft (equal), writing – review & editing (supporting); **EB**: conceptualization (supporting), formal analysis (supporting), methodology (equal), writing – original draft (equal); **CP**: writing – review & editing (lead).

Funding The research was wholly funded by Jacobs UK Ltd (https://www.jacobs.com) and their support is gratefully acknowledged.

Data availability Data sharing is not applicable to this article as no datasets were generated during the current study.

References

Aydin, A. 2015. ISRM suggested method for determination of the Schmidt hammer rebound hardness: Revised version. *In*: Ulusay, R. (ed.) *The ISRM Suggested Methods for Rock Characterization, Testing and Monitoring: 2007–2014*. Springer International Publishing, Cham, Switzerland, 25–34, https://doi.org/10.1007/978-3-319-07713-0_2

Baptie, B. 2010. Seismogenesis and state of stress in the UK. *Tectonophysics*, **482**, 150–159, https://doi.org/10.1016/j.tecto.2009.10.006

Bieniawski, Z.T. 1975. The point-load test in geotechnical practice. *Engineering Geology*, **9**, 1–11, https://doi.org/10.1016/0013-7952(75)90024-1

Bieniawski, Z.T. and Bernede, M.J. 1979. Suggested methods for determining the uniaxial compressive strength and deformability of rock materials. *In*: Ulusay, R. and Hudson, J.A. (eds) *The Complete ISRM Suggested Methods for Rock Characterization, Testing and Monitoring: 1974–2006*. ISRM Turkish National Group and the ISRM, Ankara, 151–156.

BSI 2024. *BS EN 1997-1:2024 Eurocode 7 – Geotechnical Design – Part 1: General Rules*. CEN/British Standards Institution (BSI), London.

Carter, T.G. 1992. Prediction and uncertainties in geological engineering and rock mass characterization assessments. *In*: *Proceedings of the 4th Italian Rock Mechanics and Rock Mechanics Conference*, December, Torino, Italy. Italian Geotechnical Society (AGI), 1.1–1.22.

Davis, J., Essex, R., Farooq, I. and Drake, A. 2023. *Geotechnical Baseline Reports: A Guide to Good Practice*. CIRIA Report C807. Construction Industry Research and Information Association (CIRIA), London.

Fairhurst, C.E. and Hudson, J.A. 1999. Draft ISRM suggested methods for the complete stress-strain curve for intact uniaxial compression. *International Journal of Rock Mechanics and Mining Sciences*, **36**, 281–289.

Goodman, R.E. 1989. *Introduction to Rock Mechanics*. 2nd edn. John Wiley & Sons, New York.

Hoek, E. 1983. Strength of jointed rock masses. *Géotechnique*, **33**, 187–223, https://doi.org/10.1680/geot.1983.33.3.187

Hoek, E. 1994. Strength of rock and rock masses. *International Society for Rock Mechanics News Journal*, **2**, 4–16.

Hoek, E. and Brown, E.T. 1980. Empirical strength criterion for rock masses. *Journal of the Geotechnical Engineering Division*, **106**, 1013–1035, https://doi.org/10.1061/AJGEB6.0001029

Hoek, E. and Brown, E.T. 2019. The Hoek–Brown failure criterion and GSI – 2018 edition. *Journal of Rock Mechanics and Geotechnical Engineering*, **11**, 445–463, https://doi.org/10.1016/j.jrmge.2018.08.001

Hoek, E., Kaiser, P.K. and Bawden, W.F. 2000. *Support of Underground Excavations in Hard Rock*. CRC Press, Boca Raton, FL.

Hudson, J.A. 1989. *Rock Mechanics Principles in Engineering Practice*. Construction Industry Research and Information Association (CIRIA), London.

Marinos, P. and Hoek, E. 2000. GSI: A geologically friendly tool for rock mass strength estimation. Paper ISRM-IS-2000-035 presented at the *ISRM International Symposium*, November 19–24, 2000, Melbourne, Australia.

Martin, C.D., Kaiser, P.K. and Mccreath, D.R. 1999. Hoek–Brown parameters for predicting the depth of brittle failure around tunnels. *Canadian Geotechnical Journal*, **36**, 136–151, https://doi.org/10.1139/t98-072

Stille, H. and Palmström, A. 2008. Ground behaviour and rock mass composition in underground excavations. *Tunnelling and Underground Space Technology*, **23**, 46–64, https://doi.org/10.1016/j.tust.2006.11.005

Arabia Felix: Fookes in the desert

E. Mark Lee[1]*, Mike Sweeney[2‡] and Jim Clarke[3‡]

[1]Ebor Geoscience Ltd, 15 Whernside Avenue, York YO31 0QB, UK
[2]BP Upstream Technical Authority Geotechnics, Chertsey Road, Sunbury-on-Thames, Middlesex TW16 7LN, UK
[3]BP Geotechnics, Chertsey Road, Sunbury-on-Thames, Middlesex TW16 7LN, UK
EML, 0000-0002-5872-7654
‡Retired
*Correspondence: mark.lee@eborgeo.com

Abstract: Peter Fookes' background in chemistry, allied with his understanding of geo-materials and desert terrain, allowed him to make an important contribution in the development of hot salty desert concrete technology from the 1970s onwards. Through his consultancy work across the Middle East, Peter recognized that widespread premature concrete deterioration was found, at least in part, to be associated with the presence of chloride salts, either introduced by contaminated aggregates or mix water, or due to salt weathering. In 1992 Peter was made an Honorary Fellow of the Institute of Concrete Technology.

In the 1970s he was instrumental in setting up the Bahrain Surface Materials Resources Survey, opening the door for a generation of desert geomorphologists to contribute to the development boom that followed in the Middle East. On the award of an Honorary Fellowship of the Royal Geographical Society in 2003 he was described by Professor Sir Ron Cooke as 'a true father of engineering geomorphology'. In later years he helped the oil and gas industry address the geohazard challenges posed by routing major pipelines through deserts across the Sahara and in Arabia.

'Cracks in concrete are just cracks in concrete' the lead author (ML) thought naively when accompanying Peter Fookes to look at an abandoned oil refinery site in Algeria in 2003. It was a little paradise for a concrete specialist like Peter. By the end of the day, ML had learned to distinguish between cracks associated with expansive soils, sulfate attack, poor workmanship and thermal expansion. It came as a shock to be told that a 10 m length of concrete in hot desert environments can expand/contract by 5 mm when subjected to a rise or fall of 50°C. As an engineering geomorphologist, it gradually dawned on him that studying the performance of concrete in existing structures can offer important clues to the ground-related risks in an area. Peter had known this for decades, being a key part of the revolution in understanding concrete problems in the Middle East since the 1960s and 1970s.

Peter also encouraged and inspired many people, and was a great supporter of the use of geomorphology on desert projects. In 1974, he was responsible for establishing the Bahrain Surface Materials Resources Survey that introduced a generation of academic geomorphologists to commercial projects in the Middle East. On the award of Honorary Fellowship of the Royal Geographical Society in 2003, he was described by Professor Sir Ron Cooke as the 'father of engineering geomorphology'.

This paper takes selected events and projects from his more than 40 years of working in hot deserts and attempts to provide a structured narrative to what was an important element of his consultancy practice and research. Co-authors JC and MS worked in BP's Engineering Department, BP's technical support group for major projects and, being inside a client organization, were privileged to see the performance of major projects worldwide, often down years of service. The biggest learning of all was that the best specialists deliver the best outcomes. Project performances showed that Peter was the best of the best.

The selection of examples, of course, reflects our own biases towards terrain evaluation, the relationship between the salt cycle and concrete attack, and risk assessment. Others would undoubtedly have placed much more focus on his contribution to the understanding of aggregates and desert roads. We have also taken the liberty of extending the geographical coverage of the paper to include Iran and the Sahara (Libya and Algeria), and not just Arabia.

An early desert terrain model: Iran

Whilst at Imperial College, Peter worked on a number of engineering projects in Iran for Sir Alexander Gibb and Partners, Rendel, Palmer and Tritton (RPT), and Irendco (Tehran). His experience of the varied landscapes of SW

From: Griffiths, J. S., Giles, D. P., Davis, J. A. and Dunlop, J. J. (eds) 2025. *The Peter Fookes Engineering Geological Legacy in Geomodels, Geomaterials and Geomorphology*. Geological Society, London, Engineering Geology Special Publications, **30**, 191–207,
First published online 3 February 2025, https://doi.org/10.1144/egsp30-2023-33

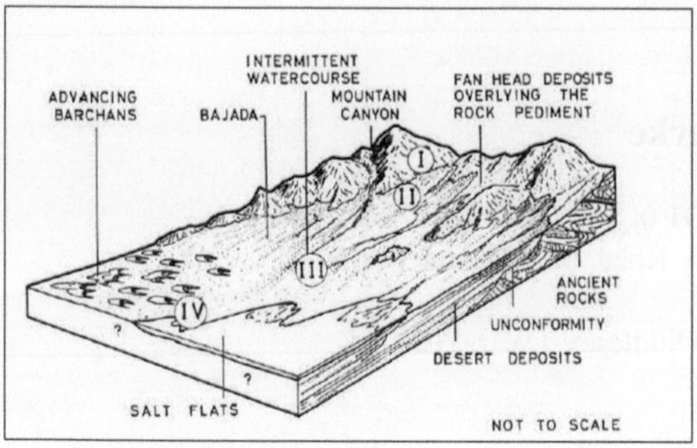

Fig. 1. Terrain model of mountain and plain landscapes of Iran. Source: from Fookes and Knill (1969).

central Iran allowed him to produce what could be described as the first engineering terrain model of a desert landscape to support road and rail planning (Fookes and Knill 1969). Four broad terrain zones were defined, each with a characteristic range of engineering soil and geotechnical conditions: Zone I – the mountain slopes; Zone II – the apron of alluvial fans (bajada); Zone III – the alluvial plains; and Zone IV – the base-level plains (Fig. 1; Table 1). The model is useful for predicting aggregate sources: rock quarries can be opened in Zone I, gravel borrow pits in Zone II and Zone III; and sand borrow in Zone III and Zone IV.

The Fookes and Knill model found widespread application beyond the area for which it was originally developed and was included in the classic text on desert geomorphology (Cooke *et al.* 1993) as a 'mountain-piedmont model'. Further adaptations made it far more generic and appropriate for a broad range of desert conditions whilst still retaining the basic four-fold landscape subdivision (e.g. Lee and Fookes 2005; Fookes *et al.* 2007).

Concrete problems in the Middle East

The Middle East construction boom in the 1960s and 1970s was accompanied by a surge in demand for concrete and aggregate (see Box 1). As there were no 'hot salty desert' standards, the construction industry, which was dominated by western design engineers and contractors, fell back on the codes and standards developed in temperate countries such as the UK, the USA and Canada. What followed was widespread, premature concrete deterioration, primarily due to rapid reinforcement corrosion.

Investigations of affected structures revealed that the concrete was often a high-quality mix, well compacted and cured, and free from alkali–aggregate reaction and sulfate attack. Many of these problems were found, at least in part, to be associated with the presence of chloride salts, either introduced by contaminated aggregates or mix water (internal), or due to salt weathering (external). At that time, chloride-related corrosion of the reinforcement (Box 2) was not generally recognized as a problem, with designers more focused on sulfate attack (Simm and Fookes 1989). The problems tended to be exacerbated by the rapid pace of construction and the use of unskilled labour.

For Peter, the concrete problems began to emerge at a time when he had just started his consultancy business. Investigating the problems provided some 40% of his workload at that time, travelling frequently and extensively across the region, and acting as a specialist consultant to companies such as Messrs Sandbergs, Rendel, Palmer and Tritton (RPT), and Halcrow. His background in chemistry, allied with his understanding of materials, allowed him to make an important contribution in the development of hot salty desert concrete technology. Peter and Laurence Collis (an Associate with Messrs Sandbergs) were the first to publicize the significance of chloride-related corrosion in the Middle East, in a series of articles in the concrete industry magazine *Concrete*, based on their regional experience (Fookes and Collis 1975*a*, *b*, 1976).

Table 1. Desert engineering zones in central Iran (from Fookes and Knill 1969)

Zone	I	II	III	IV
Principal engineering soil types		Rock fans	Silty, stony desert and sandy stony desert. Some evaporites	Sand dunes, loess and evaporites
Slope angle of desert surface (°)		2–12	0.5–2	0–0.5
Principal transporting agent of the environment		Gravity and as Zone III	Intermittent stream flow and sheet floods	Wind and evaporation
Geotechnical features		Good for foundation and fill	Generally very good foundation and fill material. Saline. May be pervious in foundations	Erratic behaviour to load bearing. Migrating dunes. Metastable loess. Saline. Absence of coarse material

Note that no descriptions were provided for Zone I.

Box 1. Concrete and aggregate: a quick introduction

Concrete is an 'artificial' breccia (angular aggregate) or conglomerate (rounded aggregate: e.g. Fookes 1980). It is a composite material comprising a cement matrix (the binding medium, a mix of cement and water: i.e. 'cement paste') within which are embedded fragments of relatively inert (i.e. 'sound') rock filler (the aggregate, generally natural sands and gravels or crushed rock graded to approximate a Fuller curve). Each particle of aggregate should be completely coated with cement paste, and the space (pores) between aggregate particles minimized and, ideally, completely filled with the paste.

In modern Portland cement concrete (the name was given in the mid-1800s because of the supposed resemblance of the resultant set concrete to the colour and appearance of Portland Stone) the relative proportions are typically 6–18% Portland cement, 14–22% water and 60–80% aggregate. The high percentage of aggregate to paste reflects the economics of concrete production; aggregate is cheaper than paste (i.e. the aggregate is often used for economic, rather than mechanical, reasons). However, cost is not the only reason for using aggregate: it confers considerable technical advantages on concrete, which has a higher volume stability and better durability than hydrated cement paste alone (Neville 2011), and helps to control the shrinkage cracking that occurs with hydrated cement alone.

The unconfined compressive strength of concrete typically used in building is in the range 20–50 MPa. However, the tensile strength is significantly lower (*c.* 10% of the compressive strength), 1–5 MPa, which is why structural concrete is usually reinforced by, for example, steel bars (rebars) or wires.

They recognized that the Middle East was the most severe environment for concrete anywhere in the world and more severe than typical hot-weather concrete as defined, for example, in American Concrete Institute Standard 305 (ACI Committee 305 1970). The severe environment reflected a combination of the climate and geology, notably contaminated aggregates and salt-rich ground conditions (see Fookes and Lee 2019).

Fookes and Collis (1976) were the first to make a general recommendation that strict limits on chloride levels in fresh concrete be set for the Middle East. They specified a total level of acid-soluble chlorides (as NaCl) of 0.5% by weight of cement (equivalent to 0.3% chloride ion by weight of cement), as well as limits on individual constituents of the concrete. The following years saw major changes to design practice; an amendment was issued to British Standard CP110 (Code of Practice for the Structural use of Concrete) in which maximum chloride levels in cement of 0.35% by weight were specified for ordinary Portland cement; and the American Concrete Institute issued a Guide to Durable Concrete in which a maximum chloride level of 0.1% by weight was specified (ACI Committee 201 1977). By the end of the 1970s most large contracts specified salt limits, although the improved practices had yet to filter down to smaller works (Fookes 1993).

A further series of articles in *Concrete* magazine with Ted Kay and Derek Pollock (both of Halcrow) described their project experience of the use of crack mapping to support the planning of remedial works (Pollock *et al.* 1981), rates of concrete deterioration (Fookes *et al.* 1981), and the use of laboratory testing and field trials to evaluate reinforcement corrosion (Kay *et al.* 1981). The authors had developed and used the techniques in assessing the condition of concrete structures across the Middle East. Concrete performance was measured in terms of deterioration rates using a five-fold classification based on field observations (Table 2) (Fookes *et al.* 1981). They found that on or near the coast structures would reach terminal condition (Class 4) within 12–15 years of construction.

Peter's work on concrete deterioration and aggregates was influential in the development of the Construction Industry Research and Information Association (CIRIA) Guide to concrete construction in the Gulf region (CIRIA 1984), which was important in the improvement in standards across the region (Fig. 2). Peter continued to be involved in the development of hot desert concrete technology in the following decades, notably on the Great Man-Made River Project in Libya (see later) and acted as an expert witness in construction disputes (e.g. Shaikh Isa Airbase in Bahrain). In 1992 Peter was made an Honorary Fellow of the Institute of Concrete Technology in recognition of his 'outstanding expertise involving a wide-ranging or substantial responsibility in fields relating to concrete technology'. He was also a contributing author to the updated CIRIA/Concrete Society guide to the construction of reinforced concrete in the Arabian Peninsula (Walker 2002).

Of particular significance to the construction industry was the recognition that the degree of salt contamination of aggregate sources and the salt attack hazard varied with the terrain setting (Fig. 3) (Fookes and Collis 1975*a*). The highest salt concentrations tend to be associated with areas of intense ground or surface water evaporation (i.e. playas, salt playas, sabkhas and salinas) (Fig. 3). These locations present significant salt attack and reinforcement corrosion hazards to concrete structures sited in these locations. Peter recognized that the height above the water-table surface that water can be drawn by capillary action (the 'capillary fringe') was of particular significance in defining the spatial extent of the hazard (these zones were further refined by Fookes and French 1977 in the context of desert road construction). He was also able to link the ground conditions expected in these different landforms to the engineering significance

Box 2. Reinforcement corrosion: a quick introduction

Reinforcement corrosion is the main cause of reinforced concrete damage in hot, salty desert environments. When reinforcing steel corrodes, the resulting rust (iron oxides and hydroxides) occupies a greater volume (up to 600%) than the original steel and creates tensile stresses in the concrete, which can eventually cause cracking, delamination and spalling. Corrosion of steel reinforcement will only occur in the presence of moisture, oxygen and non-alkaline or chloride-contaminated conditions. The highly alkaline environment of concrete (pH of 12–13) normally provides the embedded steel with corrosion protection ('passivation'). At a high pH, a thin 'passive' oxide layer forms on the steel and prevents metal ions from dissolving and reduces the corrosion rate to an insignificant level. However, this passive state can be lost by:

(1) Carbonation: this occurs when CO_2, which enters the concrete from the air, reacts with the hydroxides in the cement matrix to form carbonates:

$$Ca(OH)_2 + CO_2 \rightarrow CaCO_3 + H_2O.$$

This reaction reduces the pH of the concrete to as low as 8.5. This lower pH 'carbonation front' penetrates progressively further into the concrete, usually over a few years, through pores and along cracks, until it reaches the reinforcement and the passive layer is thus lost. If water (moisture) and oxygen are present, then general corrosion of the reinforcement can occur.

The rate of carbonation is, in rough terms, a function of the strength of the concrete. However, this is only because it reflects a number of properties of the concrete, including the porosity of the hardened concrete (reflecting the water/cement ratio and the aggregates used) and the relative humidity (moisture content) of the concrete. The highest rates of carbonation occur when the relative humidity is maintained between 50 and 75%. Below 25% relative humidity, the degree of carbonation that takes place is considered insignificant. Above 75% relative humidity, moisture in the pores restricts or slows CO_2 penetration (ACI Committee 201 1992).

(2) Exposure to chloride ions: chloride ions may be present in the original concrete mix (e.g. within the aggregates or water) or can penetrate the surface from the external environment. When the chloride ion concentration at the reinforcement reaches a 'threshold level' the passive layer is destroyed, initiating significant corrosion if water and oxygen are present. Rates of chloride ingress can be from a few months to a few years, as found in the Middle East. The presence of chlorides increases the risk of the corrosion of metals embedded in the concrete, especially to thin prestressing wires (Fookes 1995).

The 'threshold level' is controlled by the quality and carbonation state of the concrete, and the exposure conditions (i.e. the terrain setting). The most severe settings are moist with access to atmospheric oxygen (e.g. cracks in the concrete) and alternating wet/dry conditions, and rebar rusting may occur within a few weeks; if concrete is fully saturated, then oxygen is not available for the corrosion process. Very little happens when it is dry, the exception being where there are large daily swings of humidity (e.g. >15%) or fogs.

The carbonation of concrete lowers the amount of chloride ions needed to promote corrosion. For example, in new concrete with a strongly alkaline pH of 12–13, about 7000–8000 ppm of chlorides are required to start corrosion of the embedded steel. If, however, the pH is lowered to a range of 10–11, the chloride threshold for corrosion is significantly lower – at or below 100 ppm.

Table 2. Deterioration of reinforced concrete structures (after Fookes *et al.* 1981)

Class	Degree of deterioration
0	No defects
1	Rust staining and minor cracking
2	Onset of corrosion cracking
3	Moderately severe cracking with severe spalling
4	Severe cracking and spalling at a level indicating a terminal condition

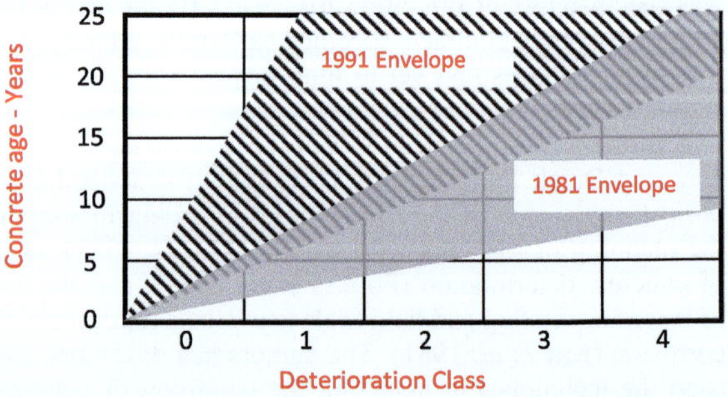

Fig. 2. Survey performance trend envelopes for reinforced concrete structures near the Gulf coast, 1981 and 1991. Source: adapted from Fookes (1993).

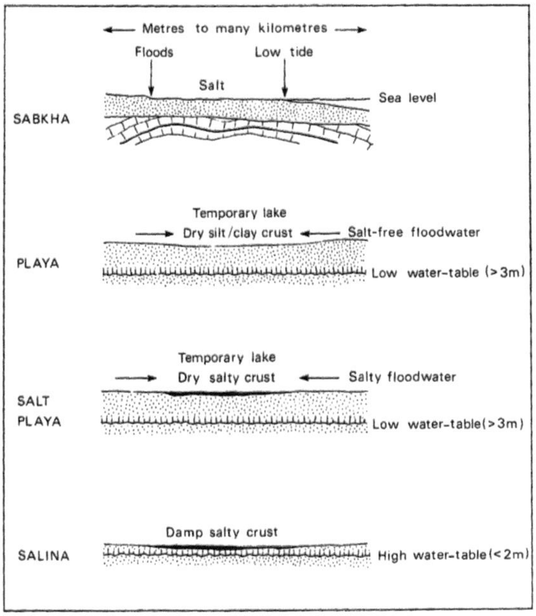

Fig. 3. Terrain settings associated with high salt concentrations. Source: from Fookes and Collis (1975*a*).

and construction techniques (Table 3); this can be viewed as a precursor to what would now be known as 'engineering terrain evaluation'. However, Peter only offered schematic ground models and basic cross-sections to delimit the approximate spatial distribution of these features. It would take the collaboration with up-and-coming young university-based geomorphological mappers – Denys Brunsden, Ron Cooke, John Doornkamp and David Jones – to develop what became 'engineering geomorphology' (a term first used by Dearman and Fookes 1974).

The Bahrain Surface Materials Resources Survey and the father of engineering geomorphology

In the early 1970s, Bahrain was undergoing a major expansion of infrastructure and population. The demand for aggregates was very high. However, quality aggregate resources were very limited and local producers struggled to meet the escalation in demand. The easiest source was the surface gravels on the back slopes of rocky escarpments and an even easier source were the Dilmun burial mounds, which are considered to be one of the most spectacular burial sites in the world. Unfortunately, these sources were contaminated with gypsum and attapulgite, and concrete problems soon became apparent. For example, the marble-covered concrete floors of the new parliament building began to swell and crack because of rusting of the reinforcing rods and expansion.

At that time, Peter was working as a consultant with Messers Sandberg, advising on the concrete problems in Bahrain. He saw that there was an urgent need to find 'clean', sound aggregates, especially sand and gravels. However, little geological information was available at that time. In 1974 Sandberg commissioned 1:20 000-scale aerial photographs of the island and prepared an interpretive geological and resource map (Sandberg 1974). Peter recognized that this work needed ground truthing by specialist geomorphologists with terrain mapping and desert experience. Already convinced by the value of geomorphological mapping, Peter saw that this would be an ideal task for the team of university-based geomorphologists that he had worked with in South Wales and Nepal (for details see Lee and Fookes 2015). In December 1974, John Doornkamp, Denys Brunsden, David Jones, Peter Bush and Graham Evans took leave from their universities and went to Bahrain on a preliminary geomorphological mapping investigation. They determined that there were a number of key problems: to map the sedimentary deposits, the solid geology, and the saline groundwater water and the gypsum in all the rocks and sediments.

Following this visit, Peter persuaded Sandberg, who negotiated with the Bahrain government, to commission a full survey of the geology, geomorphology and pedology (because agriculture was also affected by salts) to look for clean aggregate sources, especially sand and gravels. The brief given to the survey was to provide a rapid reconnaissance survey of:

(1) solid geology and structure;
(2) geomorphology and surface materials;
(3) agricultural soils;
(4) land capability for agriculture; and
(5) hazards to construction.

An 18-strong team of geologists, engineering geologists (a young John Charman), pedologists and geomorphologists (with Ron Cooke and Andrew Goudie providing desert expertise) went to Bahrain in April 1975. The team was co-led by Doornkamp, Brunsden and Jones, with Doornkamp responsible for the liaison with the Bahrain Government. Peter was not part of the survey team but provided a guiding voice throughout the project.

The focus of the survey was mapping the terrain and ground conditions:

(1) Three separate 1:10 000-scale map sheet series of the northern islands on: (i) geology, (ii) geomorphology and superficial materials; and (iii) soils for agriculture.
(2) Geomorphology and superficial materials of the Hawar Islands at a scale of 1:30 000.
(3) Synoptic 1:50 000 map sheets showing: (i) geology; (ii) geomorphology and superficial materials (except Hawar); (iii) drainage (except Hawar); (iv) soils (except Hawar); (v) land capability for agriculture (except Umm Na'san and Hawar); (vi) hazardous ground conditions for foundations (northern Bahrain only).

Table 3. An outline of some salty desert terrains and their engineering characteristics (from Fookes and Collis 1975*a*)

Name	Terrain	Groundwater table	Salts	Special engineering significance	Construction techniques
Sabkha	Coastal flat, inundated by seawater either tidally or during exceptional floods	Very near the surface	Thick surface salt deposits from evaporating sea brines. Salts usually include carbonates, sulfates and chlorides	Can be exceptionally aggressive to all types of foundations by salt weathering of stone and concrete and/or sulfate attack on cement-bound materials	Tank concrete foundations, use sulfate-resistant cement. Use inert aggregate. Capillary break layer or positive cut-off below sub-bases
Playa	Zone IV desert. Inland shallow centrally draining basin – of any size	Too deep for the capillary moisture zone to reach the ground surface but will be a temporary lake during floods	None if temporary lake is salt-free water	Non-special. Ground surface may be silt/clay or covered by windblown sands	Non-special
Salt playa	Ditto but often smaller than a playa	Ditto but lake of salty water	Surface salt deposits from evaporating temporary salty lake waters. Salts usually include carbonates, sulfates, chlorides and nitrates	Can be slightly to moderately aggressive to all types of foundations by salt weathering and sulfate attack. More severe near the water table	As sabkha
Salina	Ditto	Near surface; the capillary moisture zone from salty groundwater can reach the surface	Surface deposits from evaporating salty groundwater. Salts usually include carbonates, sulfates and chlorides	Can be slightly to exceptionally aggressive to all types of foundations by salt weathering and sulfate attack	As sabkha

The emphasis on mapping marks a significant advance on previous desert studies for engineering and planning purposes, which had relied on generic terrain models and cross-sections. The systematic mapping of geomorphology provided the spatial framework for understanding the nature and relative intensity of hazards, and the available of resources to be drawn from field evidence of landforms and near-surface deposits (e.g. Doornkamp *et al.* 1979).

The survey reported in 1976 in a six-volume manuscript report (Bahrain Materials Resources Survey 1976) and later consolidated into a single published volume: *Geology, Geomorphology and Pedology of Bahrain* (Doornkamp *et al.* 1980). The volume remains essential reading for anyone working on the geology and ground conditions of Bahrain. A summary of the survey and its findings can be found in Brunsden *et al.* (1979).

Bahrain was 'ground zero' for the development of engineering geomorphology in hot deserts, at least amongst UK practitioners. The civil engineering industry quickly became aware of the team of geomorphologists and how they could contribute to deciphering ground conditions both for engineering and planning purposes in the Middle East (e.g. Doornkamp *et al.* 1979). After Bahrain, they undertook aggregate resource surveys in Ras al Khaimah (Halcrow Middle East 1977*b*) and Dubai (Brunsden *et al.* 1976), aggressive ground hazard assessments at the site of Mina Jebel Ali airport, Dubai (Halcrow Middle East 1977*a*; Doornkamp *et al.* 1979) and for the Suez City development (Egypt Ministry of Housing and Reconstruction 1978), and wadi flood hazards in North Yemen (the Wadi Surdud Irrigation Scheme in 1979) and Saudi Arabia (the Wadi Dhamad Irrigation Scheme in 1982). Peter provided encouragement and support, often making potential clients aware of what the team could produce.

The rapid growth of engineering geomorphology in the desert led to two important developments. First, the establishment of a specialist consultancy firm (Geomorphological Services Ltd: GSL) by Brunsden, Cooke, Doornkamp and Jones.

John Doornkamp successfully combined his lecturing career at Nottingham University with the role of managing director of GSL. This company provided input into a number of major projects including the Channel Tunnel (Griffiths *et al.* 1995), applied Earth science mapping of the Torbay area (Doornkamp 1988), thc creation of a national landslide database for Great Britain (GSL 1986–87; Jones and Lee 1994), and the development of hazard maps and planning guidance for the Ventnor Undercliff (Lee and Moore 1991). Between 1979 and 1989 the company provided the initial career opportunity for a number of the next generation of engineering geomorphologists, including Jim Griffiths (Plymouth University, now retired), Roger Moore (Halcrow and now Jacobs), Peter Phipps (Mott McDonald) and Mark Lee.

The second development was the preparation of a report for the United Nations Resources Programme, 'An Assessment of Geomorphological Problems in Urban Areas of Drylands'. The report formed the basis of the book *Urban Geomorphology in Drylands* (Cooke *et al.* 1982). Unlike most desert geomorphology texts that tend to have a process domain focus (e.g. landforms shaped by wind, water or mass movement), this was engineering and planning problem focused, with chapters on identifying aggregate sources, aggressive ground hazards, flash flooding and sediment movement, and sand and dust hazards. The book also presents a variety of case studies based on the authors' field experiences. It was probably the first engineering geomorphology textbook and remains essential reading for professional geoscientists working in hot deserts. The book was dedicated to Peter Fookes.

In 2003 Peter was made an Honorary Fellow of the Royal Geographical Society (RGS). In making the presentation, Professor Sir Ron Cooke (at that time the RGS president) stated that the award was for leadership and development of engineering geomorphology:

> At academic meetings and in engineering consultancies he was a passionate advocate for the use of geomorphological maps of form, processes and history as an essential basis for engineering design. He has stayed loyal to this belief for forty years and will now insist on what he calls 'Total Geology' carried out by a Geo-Team as the foundation of a successful engineering project. Fookes should be regarded as a true Father of Engineering Geomorphology in Britain and a friend to Geography of very great value. There can be few people from outside the subject who have been more important or consistent advocates for geographers and for geomorphology
>
> (Cooke 2003).

The Great Man-Made River Project and corrosion risk

By the 1980s Peter had written a number of important papers that drew upon his desert experience, including road design (Fookes 1976), ground-related problems (e.g. Fookes 1978) and aggregates (Fookes and Higginbottom 1980). He was an internationally renowned expert and much sought after to advise on major projects, including settlement problems at Dubai Dry Dock caused by dissolution of gypsum beds in the soils (e.g. Fookes *et al.* 1985).

Perhaps the largest project he advised on at this time was the Great Man-Made River, which was described by Colonel Gaddafi as the 'eighth wonder of the world' (the cost has been of the order of $25 billion, funded by the Libyan government). Peter was a review consultant for Phase I and Phase II of the project, advising Price Brothers on both concrete durability and aggregate sourcing. The project involved the transportation of water from wells drilled into aquifers deep in the Sahara Desert to the coastal region of Libya (e.g. Fookes *et al.* 1991, 1993; Stoner and Fookes 1991). In Phase I, ground water was conveyed through more than 1900 km of prestressed concrete cylinder pipes (PCCP: *c.* 250 000 individual pipe sections; 2 Mm^3 of water/day) from Sarir and Tazerbo to a holding reservoir at Ajdabiya, from which the water was piped in two directions: west to Surt and north to Benghazi. Phase I became operational in 1991. Phase II, in western Libya, began supplying Tripoli with drinking water from the Murzuq area in 1996 (1275 km of pipeline). Phase III involved an expansion of the eastern network to wells further south around Kufra (700 km of pipeline), and the supply of water to Tobruk from wells in the Al-Jaghbub oasis (500 km of pipeline).

The pipes mainly had an inner diameter of 4.0 m, were 7.5 m long and weighed more than 70 t, and were laid in 7 m-deep trenches. The concrete pipe consisted of a 225 mm-thick concrete core within an embedded thin steel cylinder and externally wrapping prestressed wires. The prestressed wires were, in turn, covered by a 19 mm-thick sprayed-on layer of cement mortar. Tests conducted on the mortar suggested that it would take about 1 kyr for carbonation to penetrate to the prestressed wires (Stoner and Fookes 1991).

PCCP was selected because it had a good record for corrosion-free performance (Fookes *et al.* 1993). However, the threat from chloride-related corrosion (Box 2) and salt attack (Box 3) was recognized. In areas where the pipes were to be buried below the elevation of the maximum capillary rise (Table 4), they were coated with a spray-on coal tar epoxy paint; these protected pipes were known as 'black pipes', in contrast to the unprotected 'white pipes'; the recommended and final criteria chosen by the owners for corrosion protection are shown in Table 5. Phase I pipelines consisted of 60% (1176 km) uncoated pipes and 40% (750 km) black pipes. However, the majority of ground along the conveyance system was found to have high chloride concentrations and locally high sulfate concentrations, even when the water table and capillary rise were well below the pipe depth (Fookes *et al.* 1993).

The Phase I study saw Peter beginning to express his experience of chloride-related corrosion and salt attack in risk terms (i.e. a measure of the likelihood of an unacceptable outcome). A qualitative risk matrix approach was used that defined the risk associated with different hazards (physical

Box 3. Salt attack on concrete

Hot, salty drylands can experience some of the fastest rates of physical weathering in the world. This is because many salts undergo significant volume changes as a result of hydration during rainfall, fogs or other ground-wetting events and the subsequent dehydration by the high desert temperatures. Fogs and dews may contribute to salt weathering by adding significant quantities of moisture to the surface of materials. The salts crystallize within the pores and set up growth (tensile) stresses that often exceed the strength of concrete and most rocks (i.e. physical salt weathering). This process causes cracking and spalling of concrete and is repeated many times, and slowly the concrete surface is broken down to a salty powder. This is commonly called 'salt attack'.

Some salts can readily hydrate and dehydrate in response to the changes in temperature and humidity that can occur in hot drylands. For example, anhydrite ($CaSO_4$) hydrates to gypsum ($CaSO_4 \cdot 2H_2O$) with an associated volume increase of up to 63%, whereas gypsum dehydration results in a reversal back to anhydrite that leads to a 39% volume decrease (Zanbak and Arthur 1986). Pressures above 250 MPa have been measured during anhydrite hydration (Goudie 1998), which is significantly higher than the tensile strength of concrete (hydration pressures of up to 50 MPa have been measured for magnesium sulfate (Epsom salts) and sodium sulfate (Glauber's salt): Goudie 1998).

Sulfate attack is caused by chemical reactions between the Portland cement and sulfate ions present in solutions in contact with the concrete; sulfate ions can react with calcium hydroxide in the cement paste to form gypsum ($Ca\ SO_4 \cdot 2H_2O$) or with calcium aluminium hydrate to form ettringite ($((CaO)_3(Al_2O_3)(CaSO_4)_3 \cdot 32H_2O)$). Both can cause severe cracking and deterioration of the concrete, as the crystal growth exerts a pressure within the concrete (e.g. Whittaker and Black 2015). Yu *et al.* (2013) calculated a crystallization pressure of up to 21 MPa from a solution supersaturated with ettringite.

The factors controlling these sulfate reactions include the permeability of the concrete, the composition of the groundwater to which the concrete is exposed (see Fookes and Lee 2018) and the nature of the exposure (concrete exposed to cyclic wetting and drying will be more vulnerable than concrete continually immersed in sulfate-rich groundwater). The rate of sulfate attack increases with the sulfate content of the groundwater; BS EN 206-1: 2000 defined highly aggressive solutions as having a more than 3000 ppm concentration of water-soluble sulfates expressed as SO_3. The hygroscopicity (the capacity to react to the moisture content of the air by absorbing or releasing water vapour) of certain salts, such as sodium chloride (NaCl), to attract moisture into the pores within concrete can be an important factor in salt attack, especially in hyperarid environments (Goudie 1998).

salt attack, chemical salt attack and chloride-related corrosion) in a series of capillary rise/groundwater settings (Table 6). The consequence was not defined, although it was probably pipe failure. As the cumulative effects of hazards are progressive, potentially becoming more severe over time, three timescales were considered: short term (0–20 years), medium term (20–30 years) and long term (50 years). Four levels of risk were considered: no risk, low, medium and maximum.

In capillary rise/groundwater settings B–D the risk was mitigated by the use of coal tar epoxy paint ('black pipe'). However, in setting A no additional protection was provided.

Table 4. Height of capillary rise adapted for the corrosion condition classifications (from Fookes *et al.* 1993)

Material	Height of capillary rise (m)
Calcarenite	2.5
Sand and silty sand	4.5
Fine-grained soils	6.0
Fine-grained soils of high plasticity	10.0

Note that these include a significant factor of safety.

Table 5. Corrosion protection criteria (after Gibb 1986; quoted in Fookes *et al.* 1993)

	Recommended corrosion protection criteria	Final corrosion protection criteria
If groundwater from all sources* is ≤10 m from the surface, then any one of the following criteria will apply:		
Chloride concentration 2–10 m depth	≥500 ppm	≥5000 ppm
Sulfate concentration		≤3000 ppm
Soil resistivity; surface to 10 m	≥300 ohm cm	≥500 ohm cm
Soil sulfate concentration: Total SO_3		≥2%
Soil sulfate concentration: Soluble SO_3	≥500 ppm	≥5.6%
Soil pH	≤6.0	≤6.0

*The sources include wadi crossings, rainfall, leakage from adjacent conveyance or storage systems, spillage from structures in the conveyance system, changes in groundwater arising from irrigation systems, and soils with a natural moisture content independent of the capillary rise.

Table 6. Corrosion risk matrix (after Price Brothers 1985; quoted in Fookes *et al.* 1993)

Risk	Hazard scenario			
	A. Water table and capillary rise below the pipe	B. Capillary rise reaches the pipe	C. Water table and capillary rise reach the pipe	D. Pipe always below the water table
Physical salt attack	1–2	2–3	4–8	2–3
Chemical salt attack	1	1–2	3	3
Chloride-related corrosion	2–3	3–4	7–10	3

Risk classes:
1, No risk; 2, low risk in the long term; 3, low risk in the medium term; 4, low risk in the short term; 5, medium risk in the long term; 6, medium risk in the medium term; 7, medium risk in the short term; 8, maximum risk in the long term; 9, maximum risk in the medium term; 10, maximum risk in the short term.

Peter recognized that there would be a residual risk of pipe failure, especially from chloride-related corrosion (low risk in the medium and long term; the shaded cell in Table 6).

The significance of this residual risk became apparent within the first 10 years of operation of the Phase I system. Five catastrophic failures in 'white' pipes occurred between 1999 and 2001 (Kuwairi 2006; Elfergani *et al.* 2013). In August 1999, a pipe failed in the Sarir–Sirt line. One month later another pipe failed in the parallel Tazerbo–Benghazi line and, between September 1999 and April 2001, three more failures occurred. Investigations determined that the failures were caused by corrosion of the prestressed wires in the pipes due to attack by the chloride ions from the soil.

The pipe failures resulted in emergency unplanned shutdowns, which were not acceptable by the Great Man-Made River Authority (GMRA) as the project supplied all domestic water requirements for the entire population of the Libyan coastal strip. To minimize the risk the of occurrence of further rupture events and to maintain the uninterrupted flow of water, GMRA undertook a rehabilitation programme based on non-destructive inspections, aimed at identifying and treating all affected pipes. In total, 6960 pipes (around 3%) were found to be distressed in the 1655 km surveyed section of Phase I (Amaitik and Amaitik 2010). All of these distressed pipes were uncoated (white) pipes. After the system was back in operation, GMRA installed acoustic monitoring equipment to detect broken wires and track any deterioration on the remaining pipe sections (Lenghi *et al.* 2008). Only 'black' coated pipes were used on subsequent phases of the project.

Peter further developed a generic system for concrete damage risk using a risk scoring system (Fookes 1993). The approach attempted to define the relative influence of common factors on the performance of concrete in three settings: marine, inland and internal to a building or structure. Durability problems were divided into factors associated with (A) those stemming from construction materials (i.e. within the concrete); and (B) those stemming from external sources (i.e. provided by the environment but tempered by workmanship).

A risk score for each factor was calculated as the product of a hazard rating (1–10 scores; the severity of the factor in promoting or contributing to the deterioration of concrete) and a risk rating (1–5 scores the significance of the factor in promoting deterioration). The risk score ranges from 1 (low importance) to 50 (high importance). Table 7 presents the risk scores for reinforced concrete. Peter's risk model indicates that the highest risk scores were associated with contaminated aggregates and mix water, concrete exposed to saline water (seawater or groundwater), curing with salty water, and exposed poor-quality concrete.

In Salah Gas, Algeria and an engineering terrain evaluation

BP originally had its roots in the Middle East but by about 1980, in line with many of the industry majors, BP had handed production back to the host national oil companies. Their focus shifted to the Alaskan oilfields and the Forties Field in the North Sea. So, by the time the company started to plan opening up the large gas fields in central Algeria in the mid-1990s they were lacking in desert pipeline experience. They turned to Peter to advise them on the challenges they faced working in a very remote region.

During the front-end engineering design (FEED) for the In Salah Gas Project in Algeria, BP required a cost estimate within ±15% of the final cost, in order to support the decision to sanction the development with confidence. In order to achieve this level of precision, it was necessary to identify the potential terrain issues and take them into account during the route selection for the 500 km-long export pipeline (a 48 inch-diameter pipe buried in a 2.3 m-deep trench). In 1998 ML was asked to work with Peter (at the suggestion of Denys Brunsden) to develop an engineering terrain evaluation for the FEED contractor (Kellogg), having had previous desert mapping experience in Libya and Oman.

At the start of the FEED process, the pipeline engineers knew very little about the terrain conditions within the area, reflecting a combination of a lack of awareness of geological

Table 7. Deterioration of concrete risk scoring (after Fookes 1993; note the original table includes scores for mass concrete)

Factors	Scores: reinforced concrete								
	Marine			Inland			Interior		
	Hazard score	Risk rating	Risk score	Hazard score	Risk rating	Risk score	Hazard score	Risk rating	Risk score
(A) Corrosion of steel from salts (mainly chlorides) in or on:									
Fine aggregate	9	5	45	9	5	45	7	4	28
Coarse aggregate	2	2	4	2	2	4	2	2	4
Cement	1	1	1	1	1	1	1	1	1
Additives	2	2	4	2	2	4	2	2	4
Mix water	9	3	27	9	3	27	9	3	27
Contaminated reinforcement	6	3	18	6	3	18	6	3	18
(B) Deterioration of concrete, including that leading to reinforcement corrosion, from workmanship/environment									
Poorer quality concrete	7	5	35	7	4	28	6	3	18
Curing with salty water	8	3	24	8	3	24	6	3	18
Plastic shrinkage cracks	3	3	9	3	2	6	3	2	6
Plastic settlement cracks	5	2	10	5	2	10	5	2	10
Early thermal cracks	2	2	4	3	2	6	3	1	3
Large drying shrinkage cracks	4	3	12	5	4	20	4	1	4
Temperature movement cracks	3	3	9	4	3	12	3	1	3
Penetration of concrete by saline water (above sea level)	9	5	45	N/A			N/A		
Penetration of concrete by saline water (groundwater)	3	4	12	3	4	12	N/A		
Long-term carbonation	6	3	18	6	3	18	5	2	10
Physical salt weathering (ground level)	6	3	18	6	3	18	N/A		
Wetting/drying zone	5	4	20	5	4	20	3	2	6
Sulfate attack	4	2	8	4	1	4	N/A		
Alkali–aggregate reaction (AAR)	7	1	7	5	1	5	5	1	5

Risk score = Hazard score × Risk rating.
N/A, not applicable.

data sources and previous research (primarily published in French or German), and the military security problems in the country at that time. Indeed, the FEED team had been advised that in many areas the desert was pancake flat and devoid of evidence for trenching conditions. This proved not to be the case. The route (around 450–500 km long) passed through a landscape that includes, from the south, a sequence of NE–SW-trending plateaux and steep scarps (the Tademait Plateaux), a broad, low-lying depression covered, in part, by a sand sea (the Grand Erg Occidental) and, in the north, an almost flat plateau surface, bounded by steep scarp slopes (the Hassi R'Mel Plateau).

We identified a number of terrain issues that needed to be addressed, at a general level, in order to develop an overall cost estimate for pipeline construction. For once, the focus was not on concrete and aggregate sources, rather on geohazards and engineering cost drivers, including (for more details see Fookes *et al.* 2001*b*, 2005; Shilston *et al.* 2005):

- the potential impact of strong bedrock and duricrusts on trenching rates (e.g. Fookes *et al.* 2013);
- the feasibility of pipeline construction and operation through a major sand sea (the Grand Erg Occidental), which provided the shortest route for the export pipeline, taking account of the potential for loading or exposure of the pipe during periods of dune mobility; and
- the nature, extent and significance of other geohazards, notably wadi flooding/scour, land instability, subsidence and heave, and aggressive soils (e.g. Lee *et al.* 2013).

It was necessary to characterize the terrain ground conditions within a broad corridor containing a preferred pipeline route but, nevertheless, robust enough to cater for inevitable route changes, which did happen. The basis for the preliminary terrain evaluation was a combination of a desk study review and interpretation of a 1:100 000-scale composite (Satellite pour l'Observation de la Terre (SPOT) and Landsat™) satellite image of the route corridor (*c.* 100 km either side of the preferred route). This evaluation was carried out in the UK and was used to identify locations for a limited number of boreholes and trenches.

The results from the boreholes and trenches were then used, together with a limited terrain evaluation 'ground-truthing' programme in Algeria, to refine the terrain evaluation and provide typical engineering geological and geomorphological characteristics for each terrain unit. The ground truthing involved overflights in a Twin Otter (at *c.* 1000 ft) and a drive-over survey of accessible sections of the corridor. A helicopter-based reconnaissance of ground conditions was made within the sand sea to determine whether it was possible to define potential routes that maximize the use of interdune flats and slacks, with the crossing of active dunes limited to *c.* 10–30% of the selected alignment (the final route avoided crossing the sand sea).

The findings were presented as:

- 1:100 000-scale terrain unit maps of the route corridor, with detailed maps produced for sections of the sand sea (Fig. 4); and
- simple terrain models to demonstrate of the relationships between topography, terrain units and particular terrain characteristics: flood hazard, duricrusts, slope instability, and gypsum mobility and heave (Fig. 5).

The work combined both rapid engineering geomorphology mapping and a site-based engineering geological assessment. For example, the rate of trench excavation is a major factor in determining the overall construction costs. As a result, the project required a preliminary estimate of the expected trenching conditions along possible route alignments (i.e. the materials likely to be encountered within *c.* 3 m of the ground surface) and the potential rates of progress of different types of trenching equipment.

Peter's work with Geoff Pettifer (Pettifer and Fookes 1994) had demonstrated that bulk excavatability is controlled by the nature of the rock/duricrust (notably strength and fracture spacing) and the equipment used (size, type and method of working). We mapped the expected nature and extent of duricrusts, and made field assessments of the characteristic rock/duricrust strength (by hammer blows) and fracture spacing (by visual observation) for each terrain unit and subunit. This information was supplemented by point load test data collected during the borehole investigation. Progress rates were estimated for different types of trenching equipment:

(1) Conventional excavation methods (e.g. digging, rock breaking and blasting): the strength and fracture spacing information for each terrain unit was plotted on an excavatability chart (developed from Pettifer and Fookes 1994) to estimate the method of operation within each terrain unit.

(2) Trenching machines: a progress chart was developed using a combination of published data (Verhoef 1997; from the Vermeer T850 chain trencher) and data on the progress rates of a large rock trencher (Trencor 1860 HD trencher) working on a desert pipeline elsewhere in Algeria. For further details see Sweeney *et al.* (2005).

Over the next 15 years or so, Peter would work together with the authors on a series of desert pipeline projects in Egypt (across the Sinai), central Oman and in elsewhere in Algeria. He was also heavily involved in the terrain assessment of the BTC pipeline, which runs for almost 1800 km from the Caspian Sea through Azerbaijan and Georgia to the Turkish Mediterranean coast. As Peter grew older and less able to travel to remote regions he would act as a home-based advisor on projects in Tunisia and Libya (Phase III of the Great Man-Made River Project). Through this collaboration we would further refine what we came to call 'engineering terrain evaluation' (see Box 4).

It is clear that what's in the pipelines, that 'world of fossil fuels', will become a thing of the past but Peter's methods and his body of published case records will stand in future as a model of how to do all types of linear transportation routes. Much of his pipeline contribution was summarized in the ICE Conference on 'Terrain and Geohazard Challenges Facing Onshore Oil and Gas Pipelines', held in London in 2004 (Sweeney 2005).

As Mike Sweeney wrote in his 2017 Glossop Lecture:

> [T]he engineering terrain evaluation (ETE) process of dividing the landscape up into types of terrain with distinct engineering characteristics works well in pipelining.... From experience, pipeliners, and project and business managers, understand Ground Models presented in the ETE format of clear well-constructed block models, maps and sections (for examples, see also Fookes *et al.* 2015), and tied into engineering options and costs for each of the terrain types
>
> Sweeney (2017, p. 26).

Peter was also heavily involved in the Geological Society Working Party Report on hot deserts (Walker 2012).

Fookes in practice

Given the remoteness and challenging environment of most hot deserts, Peter's method of work relied heavily on observation and expert judgement. He had an uncanny ability to 'read the ground'. Judgement is the forming of an estimate or conclusion from the information available at that time, in what could be described as the intelligent use of experience. Peter used inductive reasoning that goes from the specific to the general by means of associative inference or analogy and lessons learnt from past precedent, case histories and experience. The method found its formal expression in the concept of 'total geology models' (Fookes 1997; Fookes *et al.* 2001*a*).

A favourite approach would be to record terrain characteristics using a 1–5 scale that he developed early in a field visit, with class 5 representing the most onerous conditions in each case (for examples see Tables 2 and 8). The objective was for each classification to be engineer friendly, with the minimum of geo-jargon, to be made by an experienced eye, and needing no reliance on sampling and testing.

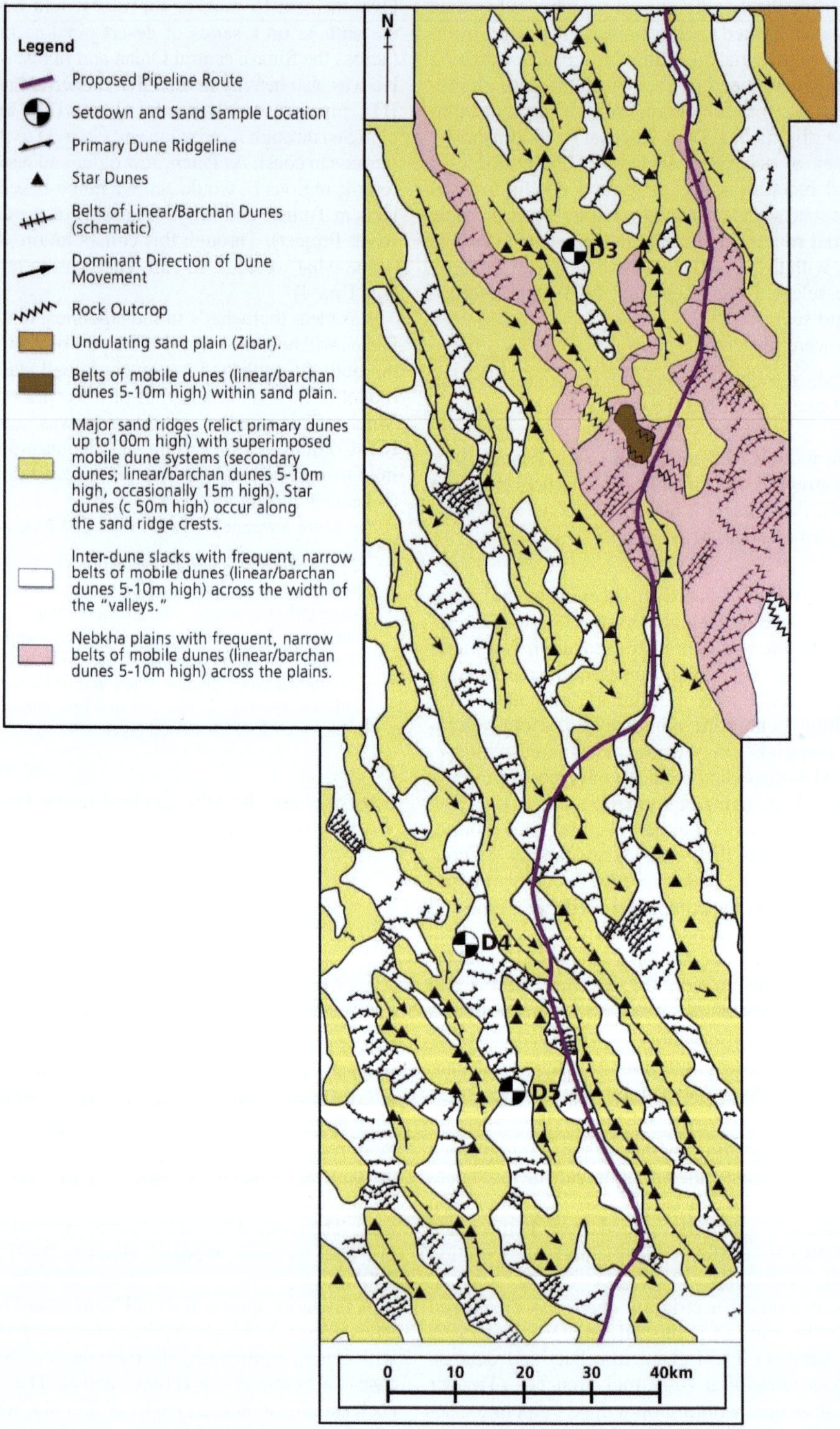

Fig. 4. Sample section of the detailed terrain map of the Grand Erg Occidental, Algeria. Source: from Lee (1998).

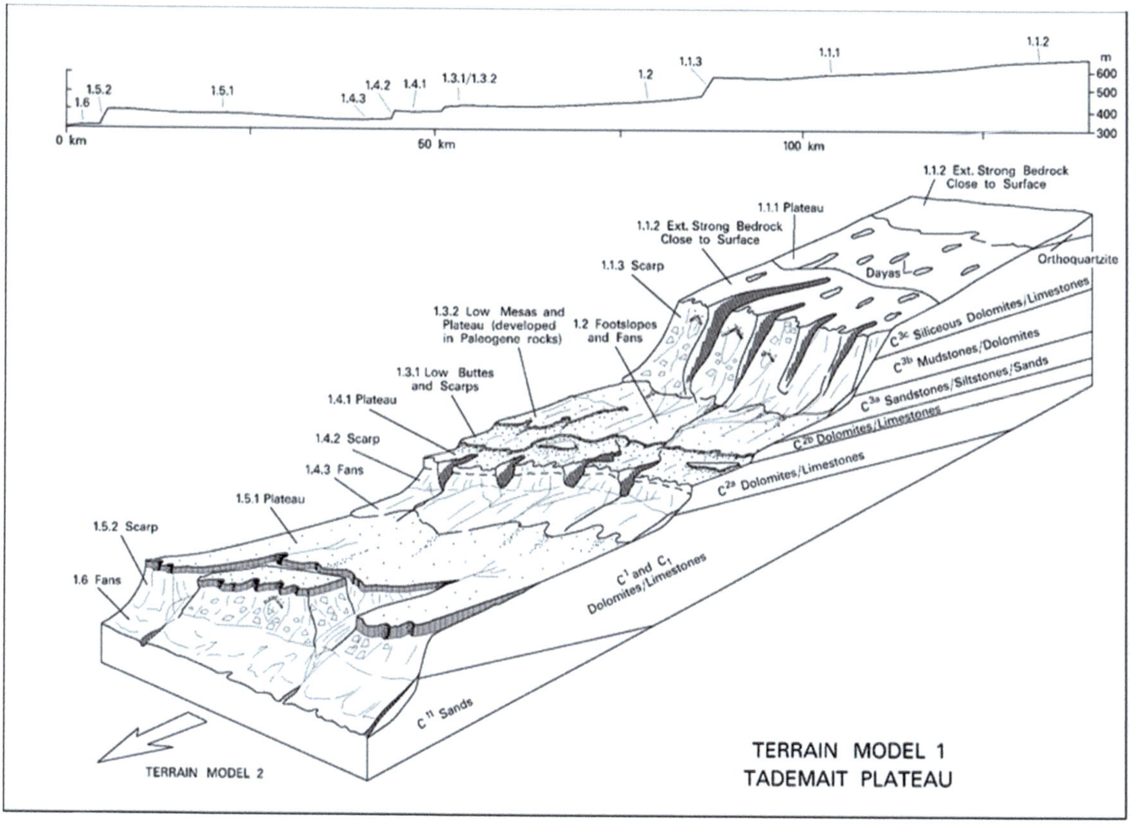

Fig. 5. Engineering terrain model: Tademait Plateau, Algeria. Source: from Lee (1998).

Peter had a talent for articulating the accuracy of the judgements made by the experienced eye, in a way that project engineers understood. In practice, these judgements proved to be of huge value. For example, at an early stage of the In Salah project, the judgements were always an order of magnitude better than anything the project had, and often stood as

Box 4. A brief history of terrain evaluation

Terrain evaluation has its origins in the need to organize and communicate specific Earth science information or intelligence in a way that is of direct relevance to the end user (e.g. civil and military engineers, land-use planners, agriculturalists, and foresters). An interesting summary of the development of the approach can be found in Mitchell (1973). Perhaps the most well-known examples of the use of terrain evaluation have been:

- the Commonwealth Scientific and Industrial Research Organization (CSIRO) applications in Australia between the 1940s and 1960s, to support agricultural development (the Land Research and Regional Survey division) and engineering (the Division of Soil Mechanics); and
- the Military Engineering Experimental Establishment (MEXE) and Soil Science Laboratory, Oxford University system for military terrain evaluation, developed in the 1960s. The military need for terrain information includes both strategic (e.g. the gross distribution of terrains and transport corridors) and tactical (e.g. visibility, cross-country mobility, reaction of the terrain to deformation, and the availability of water and construction resources).

It has tended to be associated with rapid assessments of remote, inaccessible regions.

Terrain evaluation methods have been used by the pipeline industry in Alaska and other Arctic regions, as described in Kreig and Reger (1976) and Kreig and Metz (1989).

Table 8. Typical Fookes 1–5 class system for recording terrain characteristics (adapted from Fookes *et al.* 2005)

	Fluvial systems	Flooding	Gypsum heave	Duricrust
Class 1	Sheet flow and infiltration	None	None seen or expected	None seen or expected
Class 2	As for Class 1 plus fluvial channels <5 m wide, 1 m deep	Minor, not significant	No or very little instability seen but in certain circumstances could be expected	Crust (i.e. BS 5930 (BSI 1999) up to moderately weathered: e.g. gypcrete)
Class 3	As for Class 2 but channels <20 m wide, 3 m deep	Likely but minor realignment training works required	Heave seen, with vertical movement up to 0.5 m	Floating crete (i.e. BS 5930 above moderately weathered: e.g. calcrete, siliceous calcrete)
Class 4	As for Class 3 but channels <100 m wide, 5 m deep	Very likely, training works/realignment required	Heave common, with vertical movement up to 2 m	Cap rock (i.e. surface-enriched bedrock)
Class 5	As for Class 4 but channels >100 m wide, 5 m deep	Very likely, best avoided	As for Class 4, but movement >2 m	Crete and cap rock

the key reference throughout. Feedback was immediate, along the lines 'we've never had anything like this before' and 'we didn't know you could do this'.

Many of the papers in this volume emphasize Peter's contribution to the development of the concept of ground models. However, Peter's ground models were not drafted for engineering geologists, as there were often none in the management teams associated with the projects on which he worked. They were developed to be understandable by engineers and project managers. Ground model outputs had to fit the objectives of the particular project management stage, for example, in terms of route and cost feasibility. That was a learning exercise for all. Peter and the team would present the ground models and an appreciation of the engineering implications. It worked, everyone understood. That gift of communicating to a non-specialist audience is rare.

His natural authority, humour and clarity were great assets in selling the messages. Whenever Peter and the team were due to deliver a trip debrief, the word would get out and the conference rooms at the field base camp, or back in London, would fill to overflowing.

In the pipeline industry, colleagues referred to Peter as 'the Professor', and he liked that. But in one sense it wasn't fair. He didn't emerge from his study with the next great thought. He didn't do academic abstraction or research; it didn't usually end up back in a laboratory. His laboratory was the project, his work was driven by the project needs and his contribution was giving them the tools to solve their problems. Above all, working in the desert with Peter was always huge fun. He had a mischievous sense of humour that will be much missed.

Arabia Felix

Peter loved being in the desert and was an avid reader of books about desert exploration, especially the exploits of Wilfred Thesiger in the Empty Quarter. However, it was Bertram Thomas (wazir, first minister, to the Sultan of Oman) who had been the first European to cross the Empty Quarter in 1931–32, travelling from Salalah to Dohar, accompanied by tribal guides who steered him from well to well. His account of the journey was published in 1932, *Arabia Felix: Across the Empty Quarter of Arabia* (Thomas 1932). Arabia Felix was one of three regions into which the Romans divided the Arabian peninsula: Arabia Felix in the east, Arabia Deserta (deserted Arabia) and Arabia Petraea (the Roman Province of Petraea) in the west.

Arabia Felix can be translated as 'fecund, fertile' and 'happy, fortunate, blessed'. Thomas' use of the term is in marked contrast to the feelings of one of the earliest travellers across western Arabia, Charles Doughty, who wrote in 1888 'I passed one good day in Arabia, and all the rest were evil'. This was not Peter's experience and he certainly felt fortunate to have worked for more than 40 years in a region that challenged his curiosity and inspired some of his most important work. Fertile Arabia indeed.

Acknowledgements The author would like to thank the reviewers for their helpful comments.

Competing interests The authors declare that they have no known competing financial interests or personal relationships that could have appeared to influence the work reported in this paper.

Author contributions **EML**: writing – original draft (lead); **MS**: writing – review & editing (supporting); **JC**: writing – review & editing (supporting).

Funding This research received no specific grant from any funding agency in the public, commercial, or not-for-profit sectors.

Data availability Data sharing is not applicable to this article as no datasets were generated or analysed during the current study.

References

ACI Committee 201 1977. Guide to Durable Concrete. *Journal of the American Concrete Institute*, **74**, 573–609.

ACI Committee 201 1992. *201.2R-92: Guide to Durable Concrete*. American Concrete Institute (ACI), Farmington Hills, MI.

ACI Committee 305 1970. *ACI 305-72: Recommended Practice for Hot Weather Concreting*. American Concrete Institute (ACI), Farmington Hills, MI.

Alexander Gibb and Partners 1986. *Reports 3, 5 and 6 on Pipeline Corrosion*. Unpublished reports. Alexander Gibb and Partners, Reading, UK.

Amaitik, N.M. and Amaitik, S.M. 2010. Prediction of pipe failures in water mains using artificial neural network models. *In*: *Proceedings of the 11th International Arab Conference of Information Technology (2010 ACIT)*, University of Garyounis, Benghazi, 1–8.

British Standards Institution (BSI - BS) 1972 CP 110-1. *Code of Practice for The structure use of concrete Part 1. Design, materials and workmanship*. British Standards Institution, London.

British Standards Institution (BSI) 1999. *BS 5930:1999 Code of practice for site investigations*. British Standards Institution, London.

British Standards Institution (BSI) 2000. *BS EN 206-1:2000. Concrete. Specification, performance, production and conformity*. British Standards Institution, London.

Brunsden, D., Doornkamp, J.C. and Jones, D.K.C. 1976. *Report on a Geomorphological Interpretation of the available 1:15 000 Monochromatic Aerial Photography, in State of Dubai Aggregate Survey*. Wimpey Laboratories Ltd.

Brunsden, D., Doornkamp, J.C. and Jones, D.K.C. 1979. The Bahrain Surface Materials Resources Survey and its application to regional planning. *The Geographical Journal*, **145**, 1–35, https://doi.org/10.2307/633073

CIRIA 1984. *The CIRIA Guide to Concrete Construction in the Gulf Region*. Report SP 31. Construction Industry Research and Information Association (CIRIA), London.

Cooke, R.U. 2003. Annual General Meeting 2 June 2003. *The Geographical Journal*, **169**, 283–290, https://doi.org/10.1111/1475-4959.00090

Cooke, R.U., Brunsden, D., Doornkamp, J.C. and Jones, D.K.C. 1982. *Urban Geomorphology in Drylands*. Oxford University Press, Oxford, UK.

Cooke, R.U., Warren, A. and Goudie, A.S. 1993. *Desert Geomorphology*. University College London Press.

Dearman, W.R. and Fookes, P.G. 1974. Engineering geological mapping for civil engineering practice in the United Kingdom. *Quarterly Journal of Engineering Geology*, **7**, 223–256, https://doi.org/10.1144/GSL.QJEG.1974.007.03.01

Doornkamp, J.C. (ed.) 1988. *Applied Earth Science Mapping for Planning and Development. Torbay*. M1 Press, Nottingham, UK.

Doornkamp, J.C., Brunsden, D., Jones, D.K.C., Cooke, R.U. and Bush, P.R. 1979. Rapid geomorphological assessments for engineering. *Quarterly Journal of Engineering Geology*, **12**, 189–204, https://doi.org/10.1144/GSL.QJEG.1979.012.03.06

Doornkamp, J.C., Brunsden, D., Jones, D.K.C. and Cooke, R.U. 1980. *Geology, Geomorphology and Pedology of Bahrain*. GeoBooks, Norwich, UK.

Egypt Ministry of Housing and Reconstruction 1978. *Suez Master Plan*. Report prepared by Sir William Halcrow and Partners, etc., with contributions from D. Brunsden, R. Cooke, J.C. Doornkamp, D.K.C. Jones and P.R. Bush.

Elfergani, H.A., Pullin, R. and Holford, K.M. 2013. Damage assessment of corrosion in prestressed concrete by acoustic emission. *Construction and Building Materials*, **40**, 925–933, https://doi.org/10.1016/j.conbuildmat.2012.11.071

Fookes, P.G. 1976. Road geotechnics in hot deserts. *Highway Engineer*, **23**, 11–29.

Fookes, P.G. 1978. Middle East – inherent ground problems. *Quarterly Journal of Engineering Geology*, **11**, 33–49, https://doi.org/10.1144/GSL.QJEG.1978.011.01.04

Fookes, P.G. 1980. An introduction to the influence of natural aggregates on the performance and durability of concrete. *Quarterly Journal of Engineering Geology*, **13**, 207–229, https://doi.org/10.1144/GSL.QJEG.1980.013.04.02

Fookes, P.G. 1993. Concrete in the Middle East – past, present and future: a brief review. *Concrete*, **27**, 14–20.

Fookes, P.G. 1995. Aggregates: a review of prediction and performance. *In*: Glanville, J. and Neville, A. (eds) *Prediction of Concrete Durability. Proceedings of STATS 21st Anniversary Conference*. E & FN Spon, London, 91–115.

Fookes, P.G. 1997. Geology for engineers: the geological model, prediction and performance. The First Glossop Lecture. *Quarterly Journal of Engineering Geology*, **30**, 290–424, https://doi.org/10.1144/GSL.QJEG.1997.030.P4.02

Fookes, P.G. and Collis, L. 1975*a*. Problems in the Middle East. *Concrete*, **9**, 12–17.

Fookes, P.G. and Collis, L. 1975*b*. Aggregates and the Middle East. *Concrete*, **9**, 14–19.

Fookes, P.G. and Collis, L. 1976. Cracking and the Middle East. *Concrete*, **10**, 14–19.

Fookes, P.G. and French, W.J. 1977. Soluble salt damage to surfaced roads in the Middle East. *Journal of the Institution of Highway Engineers*, **24**, 10–20.

Fookes, P.G. and Higginbottom, I.E. 1980. Some problems of construction aggregates in desert areas, with particular reference to the Arabian peninsula 1: Occurrence and special characteristics. *Proceedings of the Institution of Civil Engineers, Part 1*, **68**, 39–67, https://doi.org/10.1680/iicep.1980.2504

Fookes, P.G. and Knill, J.L. 1969. The application of engineering geology to the regional development of northern and central Iran. *Engineering Geology*, **3**, 81–120, https://doi.org/10.1016/0013-7952(69)90001-5

Fookes, P.G. and Lee, E.M. 2018. The engineering geology of playas, salt playas and salinas. *Quarterly Journal of Engineering Geology and Hydrogeology*, **51**, 287–298, https://doi.org/10.1144/qjegh2017-084

Fookes, P.G. and Lee, E.M. 2019. The engineering geology of concrete in hot drylands. *Quarterly Journal of Engineering Geology and Hydrogeology*, **52**, 257–279, https://doi.org/10.1144/qjegh2018-185

Fookes, P.G., Pollock, D.J. and Kay, E.A. 1981. Middle East concrete (2) – Rates of deterioration. *Concrete*, **15**, 12–19.

Fookes, P.G., French, W.J. and Rice, S.M.M. 1985. The influence of ground and groundwater geochemistry on construction in the Middle East. *Quarterly Journal of Engineering Geology*, **18**, 101–128, https://doi.org/10.1144/GSL.QJEG.1985.018.02.01

Fookes, P.G., Stoner, J.R. and Mackintosh, J.H. 1991. The Libyan Great Man-Made River Project Phase I. Part 2: concrete technology of prestressed concrete pipe manufacture. *Proceedings of the*

Institution of Civil Engineers, Part 1, **90**, 881–919, https://doi.org/10.1680/iicep.1991.15387
Fookes, P.G., Stoner, J.R. and Mackintosh, J.H. 1993. Great Man-Made River Project, Libya, Phase I: a case study on the influence of climate and geology on concrete technology. *Quarterly Journal of Engineering Geology*, **26**, 25–60, https://doi.org/10.1144/GSL.QJEG.1993.026.01.04
Fookes, P.G., Baynes, F.J. and Hutchinson, J.N. 2001*a*. Total geological history: a model approach to understanding site conditions (Technical Note). *Ground Engineering*, **34**, 42–47.
Fookes, P.G., Lee, E.M. and Sweeney, M. 2001*b*. Pipeline route selection and ground characterization, Algeria. *Geological Society, London, Engineering Geology Special Publications*, **18**, 115–121, https://doi.org/10.1144/GSL.ENG.2001.018.01.17
Fookes, P.G., Lee, E.M. and Sweeney, M. 2005. In Salah Gas Project, Algeria – part 1: terrain evaluation for desert pipeline routing. *In*: Sweeney, M. (ed.) *Terrain and Geohazard Challenges Facing Onshore Oil and Gas Pipelines*. Thomas Telford, London, 144–161.
Fookes, P.G., Lee, E.M. and Griffiths, J.S. 2007. *Engineering Geomorphology: Theory and Practice*. Whittles Publishing, Dunbeath, UK.
Fookes, P.G., Hart, A.B. and Lee, E.M. 2013. Some near-surface desert features of significance in engineering geology evaluations. *Quarterly Journal of Engineering Geology and Hydrogeology*, **46**, 259–266, https://doi.org/10.1144/qjegh2012-083
Fookes, P.G., Pettifer, G. and Waltham, A. 2015. *Geomodels in Engineering Geology: An Introduction*. Whittles Publishing, Dunbeath, UK.
Goudie, A.S. 1998. The salt weathering hazard in deserts. *In*: Kalvoda, J. and Rosenfeld, C.L. (eds) *Geomorphological Hazards in High Mountain Areas*. Kluwer Academic, Dordrecht, The Netherlands, 107–120.
Griffiths, J., Brunsden, D., Lee, E.M. and Jones, D.K.C. 1995. Geomorphological investigations for the Channel Tunnel terminal and portal. *Geographical Journal*, **161**, 275–284, https://doi.org/10.2307/3059832
GSL 1986–87. *Review of Research into Landsliding in Great Britain*. Report to the Department of the Environment. Geomorphological Services Limited (GSL), London.
Halcrow Middle East 1977*a*. *Report on the Geomorphological Investigations in Dubai, April 1977*. Report prepared by J.C. Doornkamp, D. Brunsden, D.K.C. Jones, P. Bush, M.J. Gibbons and R.U. Cooke.
Halcrow Middle East 1977*b*. *Khor Khwair Reconnaissance Geomorphological Investigation*. Report prepared by D.K.C. Jones, D. Brunsden, J.C. Doornkamp, P.R. Bush and M.J. Gibbons.
Jones, D.K.C. and Lee, E.M. 1994. *Landsliding in Great Britain*. HMSO, London.
Kay, E.A., Fookes, P.G. and Pollock, D.J. 1981. Middle East concrete (3) – Deterioration related to chloride ingress. *Concrete*, **15**, 22–28.
Kreig, R.A. and Metz, M.C. 1989. Recent advances in route selection and cost estimating methodology for pipelines in Arctic and Subarctic regions. Presented at the *International Symposium on Geocryological Studies in Arctic Regions*, 1–10 August 1989, Yamburg, USSR.
Kreig, R.A. and Reger, R.D. 1976. Preconstruction terrain evaluation for the Trans-Alaska Pipeline Project. *In*: Coates, D.R. (ed.) *Geomorphology and Engineering*. Routledge, London, 55–76.
Kuwairi, A. 2006. Water mining: the Great Man-made River, Libya. *Proceedings of the Institution of Civil Engineers – Civil Engineering*, **159**, 39–43, https://doi.org/10.1680/cien.2006.159.5.39
Lee, E.M. 1998. *In Salah Gas: Terrain Evaluation Report*. Unpublished report to BP.
Lee, E.M. and Fookes, P.G. 2005. Hot Drylands. *In*: Fookes, P.G., Lee, E.M. and Milligan, G. (eds) *Engineering Geomorphology*. Whittles Publishing, Dunbeath, UK, 419–453.
Lee, E.M. and Fookes, P.G. 2015. A note on the origins of engineering geomorphology in the UK. *Quarterly Journal of Engineering Geology and Hydrogeology*, **48**, 147–156, https://doi.org/10.1144/qjegh2014-048
Lee, E.M. and Moore, R. 1991. *Coastal Landslip Potential Assessment: Isle of Wight Undercliff, Ventnor*. Department of the Environment, London.
Lee, E.M., Fookes, P.G. and Hart, A.B. 2013. Observations on the impact of climate change on landform behaviour and geohazards in the Algerian Sahara. *Quarterly Journal of Engineering Geology and Hydrogeology*, **46**, 107–116, https://doi.org/10.1144/qjegh2012-058
Lenghi, A., Amaitik, N. and Wrigglesworth, M. 2008. Expansion of existing monitoring system on Great Man-Made River Project using acoustic fibre optic technology. *Water Practice and Technology*, **3**, https://doi.org/10.2166/wpt.2008.072
Mitchell, C.W. 1973. *Terrain Evaluation – The World's Landscapes*. Longman, London.
Neville, A. 2011. *Properties of Concrete*. 5th edn. Pearson Education, Harlow, UK.
Pettifer, G.S. and Fookes, P.G. 1994. A revision of the graphical method for assessing the excavatability of rock. *Quarterly Journal of Engineering Geology*, **27**, 145–164, https://doi.org/10.1144/GSL.QJEGH.1994.027.P2.05
Pollock, D.J., Kay, E.A. and Fookes, P.G. 1981. Crack mapping for investigation of Middle East concrete. *Concrete*, **15**, 12–18.
Price Brothers 1985. *Preliminary Corrosion Report for the Great Man-Made River Project*. Unpublished report. Price Brothers, Dayton, OH.
Sandberg, M. 1974. *A Photogeological Interpretation of Bahrain, Adjacent Islands Near-shore Shallows for Materials Resources and Engineering Purposes*. Unpublished report for the Government of Bahrain.
Shilston, D.T., Lee, E.M., Fookes, P.G. and Standish, I. 2005. In Salah Gas, Algeria – part 2: terrain evaluation for detailed site design. *In*: Sweeney, M. (ed.) *Terrain and Geohazard Challenges Facing Onshore Oil and Gas Pipelines*. Thomas Telford, London, 162–174.
Simm, J.D. and Fookes, P.G. 1989. Improving reinforced concrete durability in the Middle East during the period 1960–1985: an analytical review. *Proceedings of the Institution of Civil Engineers, Part 1*, **86**, 333–358, https://doi.org/10.1680/iicep.1989.1629
Stoner, J.R. and Fookes, P.G. 1991. The Libyan Great Man-Made River Project Phase I. Part 1: manufacture of prestressed concrete cylinder pipe. *Proceedings of the Institution of Civil Engineers, Part 1*, **90**, 853–879, https://doi.org/10.1680/iicep.1991.15386
Sweeney, M. (ed.) 2005. *Terrain and Geohazard Challenges Facing Onshore Oil and Gas Pipelines*. Thomas Telford, London.
Sweeney, M. 2017. Terrain and geohazard challenges for remote region onshore. pipelines: risk management, Geoteams and Ground Models. *Quarterly Journal of Engineering Geology and Hydrogeology*, **50**, 13–52, https://doi.org/10.1144/qjegh2016-074

Sweeney, M., Pettifer, G.S., Shilston, D.T., Bel-Ford, P. and Stockbridge, M. 2005. In Salah Gas, Algeria – part 3: prediction and performance of large chain trenchers on a desert pipeline project. *In*: Sweeney, M. (ed.) *Terrain and Geohazard Challenges Facing Onshore Oil and Gas Pipelines.* Thomas Telford, London, 529–548.

Thomas, B. 1932. *Arabia Felix: Across the Empty Quarter of Arabia.* Scribners, New York.

Verhoef, P.N.W. 1997. *Wear of Rock Cutting Tools: Implications for the Site Investigation of Rock Dredging Projects.* Balkema, Rotterdam, The Netherlands.

Walker, M. (ed.) 2002. *Guide to the Construction of Reinforced Concrete in the Arabian Peninsula.* CIRIA Publication, C577/ Concrete Society Special Publications, **CS136**. Construction Research and Information Association, London/Concrete Society, Crowthorne, UK.

Walker, M.J. (ed.) 2012. *Hot Deserts: Engineering Geology and Geomorphology: Geological Society Working Party Report.* Geological Society, London, Engineering Geology Special Publications, **25**, https://doi.org/10.1144/EGSP25.00

Whittaker, M. and Black, L. 2015. Current knowledge of external sulfate attack. *Advances in Cement Research*, **27**, 532–545, https://doi.org/10.1680/adcr.14.00089

Yu, C., Sun, W. and Scrivener, K. 2013. Mechanism of expansion of mortars exposed in sodium sulphate solutions. *Cement and Concrete Research*, **43**, 105–111, https://doi.org/10.1016/j.cemconres.2012.10.001

Zanbak, C. and Arthur, R.C. 1986. Geochemical and engineering aspects of anhydrite/gypsum phase transitions. *Environmental & Engineering Geoscience*, **XXIII**, 419–433, https://doi.org/10.2113/gseegeosci.xxiii.4.419

Rock coast geomodel: process–response dynamics, resilience and vulnerability

Cherith A. Moses

Geography and Geology, Edge Hill University, Ormskirk, Lancashire L39 4QP, UK
0000-0001-7222-9486
Correspondence: mosesc@edgehill.ac.uk

Abstract: Geology and geomorphology are linked starkly in coastal environments, exemplifying the importance of Fookes' geomodel approach to understanding site dynamics. Cliffs, and the shore platforms that front them, are subject to both marine and subaerial processes. Throughout his work, Fookes emphasizes the role of climate as a weathering agent, influencing material properties and requiring consideration in engineering projects. Comparatively, more is known about the operation of rock weathering processes on shore platforms than on cliffs. Yet, with the changing climate, it is becoming apparent that winter salt and frost weathering, and summer salt and wetting–drying weathering, contribute to more frequent occurrences of coastal rockfalls. Increasing storm activity and rising sea-levels are enhancing marine and subaerial processes. Rock coasts are amongst the most rapidly eroding coastlines in Europe. New applications of technologies, including seismometers, LiDAR and InSAR, provide a fuller understanding of rock coast process–response geomodel behaviour and the resilience of rock coasts on engineering timescales. Communicating scientific understanding of rock coast dynamics to policy-makers, planners and the public remains a challenge. Assessments of the hazard, risk, resilience and vulnerability of rock coasts associated with climate change provide useful communication tools. Databases such as the British Geological Survey GeoCoast geohazard data product and EMODnet Geology (European Marine Observation and Data Network) coastal behaviour data products integrate and visualize available data and information to communicate situational awareness to a broad audience.

Much of Peter Fookes' work is of direct relevance to rock coasts in relation, for example, to geomaterial properties (Fookes 1991), rock slope stability (Fookes and Sweeney 1976), the durability of rock and both reinforced and unreinforced concrete materials for breakwaters and coastal protection works (Fookes and Poole 1981). In his Glossop Lecture he notes that characterizing near-surface hazard events and short-term occurrences, including coastal change, is an important part of the work of engineering geology in helping to reduce engineering risk (Fookes 1997). Throughout his work, the importance of geomorphology in engineering geology is emphasized together with the value of the geomodel for integrating and communicating geological and geomorphological understanding for engineering projects (Fookes and Vaughan 1986; Fookes 1997; Griffiths 2014; Hearn 2021).

Engineering Geomorphology – Theory and Practice (Fookes *et al.* 2007), Part 4, consisting of nine chapters, is dedicated to coasts, including the hazards associated with cliff retreat on rock coasts and their management. Coasts feature again throughout *Geomodels in Engineering Geology – an Introduction* (Fookes *et al.* 2015). A key value of the geomodel is that it facilitates the study of process–response systems (Fig. 1). Using this approach, geomorphological and geological data, relevant to understanding ground conditions, are categorized and may be quantified to model climate change impacts on the coastal system or to establish design life for coastal defence measures (Griffiths and Stokes 2008). The geomodel approach has been applied to identify the Coastal Behaviour Systems and sediment cells that underpin the development and review of UK Shoreline Management Plans (Hosking and McInnes 2002; McInnes *et al.* 2011). It is also applied to risk appraisal and analysis (McInnes and Moore 2014) and cliff instability and management (Sellwood *et al.* 2000; Lee and Clark 2002; Moore *et al.* 2002, 2010, 2016; Halcrow Group Ltd 2009; Moore and McInnes 2011; Moore and Davis 2015; Saroglou and Alexandrou 2016). Further applications of the geomodel approach are to be found in guidance for both decision-makers and practitioners (e.g. Moore and McInnes 2011, 2021; McInnes and Moore 2014; Blanco *et al.* 2019; Environment Agency 2020).

This paper focuses on rock coasts, where the geomodel approach has been successfully applied to understand ground behaviour to inform the design of engineering solutions addressing coastal management needs, e.g. at Clayton Bay and Filey Bay, Yorkshire, NE England (Moore *et al.* 2002; Halcrow Group Ltd 2009) and Lyme Regis, Dorset, SW England (Sellwood *et al.* 2000; Moore *et al.* 2016). The importance of climate, and climate change, to engineering projects by influencing the morphology and magnitude-frequency of weathering and erosion processes may be captured in the geomodel (Fookes 1997). To inform development and application of rock coast geomodels (Fig. 1) at

From: Griffiths, J. S., Giles, D. P., Davis, J. A. and Dunlop, J. J. (eds) 2025. *The Peter Fookes Engineering Geological Legacy in Geomodels, Geomaterials and Geomorphology*. Geological Society, London, Engineering Geology Special Publications, **30**, 209–235,
First published online 26 March 2025, https://doi.org/10.1144/egsp30-2023-32

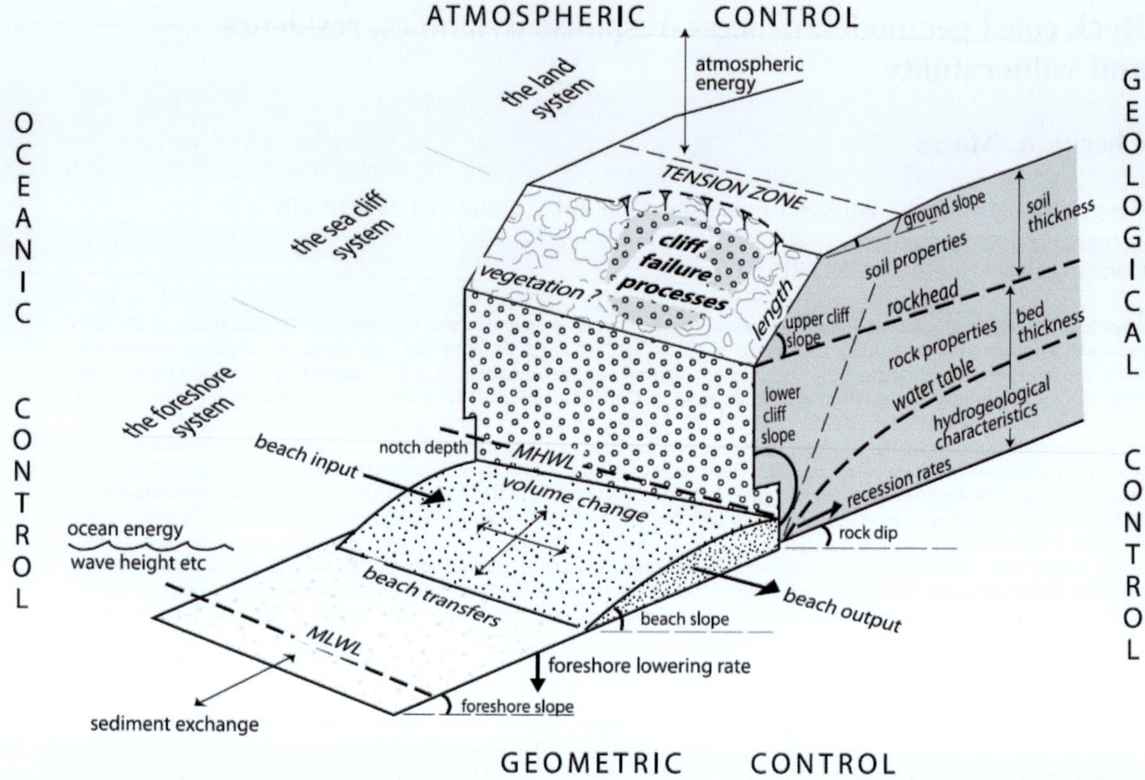

Fig. 1. Key components of the rock coast geomodel. The conceptual model shows interlinked systems of foreshore (shore platform); cliff and land; energy inputs from oceanographic (marine) and atmosphere (climate); geology and geomorphology of the cliff; geomorphological processes. Additional factors such as vegetation or human interference, e.g. presence of sea walls, can be built into the model. Source: modified by Griffiths and Stokes (2008) from Lee and Brunsden (2001) and Brunsden (2002).

specific sites and, more broadly, for decision-making at the coast (Blanco *et al.* 2019; Moore and McInnes 2021), this paper reviews recent developments in understanding the response of key components of the rock coast geomodel to climate change and sea-level rise.

Rock coasts are cliffed and composed of consolidated material irrespective of its hardness (Sunamura 1992). They are typically classified as either ‘soft’ or ‘hard’ depending upon the materials in which they are developed. Soft rock coasts range from those developed in weak surficial deposits through to bedded and jointed weak rock (Lee 2002; Lee and Clark 2002). Hard rock coasts are considered to range from those developed in weak, sedimentary rocks through to resistant, igneous rocks (Trenhaile 1987; Sunamura 1992). Rock coasts make up more than half the world’s coastline, with an earlier, widely used, estimate of 80% (Emery and Kuhn 1982) refined to 52% using remote sensing methods (Young and Carilli 2019). Approximately half the coastline of Europe is rock coast (Prémaillon *et al.* 2018; Regard *et al.* 2022), of which 80% is hard rock, 13% soft rock and 7% intermediate (calculated from the EMODnet coastal typology data product; Regard *et al.* 2022). In the UK, 60% of the coastline is rock coast (18% soft, 42% hard rock) and in Ireland 57% (1% soft, 56% hard rock; Salman *et al.* 2004). Cliff retreat rates are generally highly variable: hard, 0.029 ± 0.029 m a^{-1}; intermediate, 0.10 ± 0.04 m a^{-1}; soft, 0.23 ± 0.13 m a^{-1} (median and variance between 17% and 83% quantiles; Prémaillon *et al.* 2018; Regard *et al.* 2022). These values, calculated from a global database, mask the stochastic nature of coastal cliff retreat and site-specific variations. For example, on the soft rock, London Clay coastline, Isle of Sheppey, Kent, UK, an historical average of 1.02 ± 0.08 m a^{-1} has been measured, with a maximum rate of 1.93 ± 0.08 m a^{-1} at Warden Point (Nicholls *et al.* 2000). On the Cretaceous Chalk coasts of NW Europe, historical cliff retreat rates average 0.49 ± 0.38 m a^{-1} with site-specific rates, measured over short periods of a few years, of >1.2 m a^{-1} (Moses and Robinson 2011).

Rock coasts are dynamic on engineering timescales of concern to engineering geologists, their behaviour driven by the interaction of lithology with marine and subaerial processes. Improving understanding of the response of rock coasts to climate change is of importance for climate-resilient project design. With UK predictions of warmer, wetter winters and hotter, drier summers with increased intensity of heavy summer rainfall events and increased rates of sea-level rise (Kendon *et al.* 2022), the need to understand the dynamics of the rock coast geomodel is ever more pressing. This is particularly the case for predictive modelling of cliff behaviour in future climate scenarios, the topic of Fookes' (1997) Glossop Lecture. This paper, therefore, uses the geomodel approach to review current understanding of the influence of climate, via marine and subaerial processes, on the process–response dynamics of rock coasts. A range of rock types is included, but the focus is largely on the Cretaceous Chalk coasts of SE England and northern France and the London Clay coast of SE England, which are two of the most intensively studied globally (Fig. 2). Following a brief introduction to the key characteristics of rock coasts, emphasis is placed on more recent new applications of a variety of technologies to improve understanding of rock coast dynamics. Although still some way from full understanding and quantification of all parameters of the rock coast geomodel to allow future changes to be modelled in a deterministic way, as noted by Griffiths and Stokes (2008), this paper highlights the significant research progress being made. It also addresses the increasing emphasis placed on understanding coastal risk, resilience and vulnerability for decision-making (Blanco *et al.* 2019; Moore and McInnes 2021) by highlighting some of the ways these are represented to draw attention to rock coasts.

1. South Tyneside, NE England, Magnesian Limestone
2. Yorkshire, NE England, Jurassic sandstone and shale, glacial till
3. Norfolk, E England, glacial till
4. Suffolk, E England, composite Pleistocene, Pliocene, Paleogene and Cretaceous deposits
5. Isle of Sheppey, Kent, SE England, London Clay
6. Kent and East Sussex, SE England, Cretaceous Chalk
7. Normandy, N France, Cretaceous Chalk
8. Glamorgan, S Wales, Carboniferous limestone
9. North Devon, SW England, Carboniferous sandstone
10. Burren, Co. Clare, W Ireland, Carboniferous limestone

Fig. 2. Location and geology of study sites, UK, Ireland and northern France.

This paper considers only the components of rock coast geomodels that are present on the shoreline, namely the cliff and foreshore (Fig. 1). It is important to note that the geomodel should also consider regional geological and geomorphological history as well as current ground conditions (Fookes 1997). Rock coast geomodels, therefore, need to be considered in the context of broader inland, nearshore and offshore ground conditions. This is particularly the case in engineering projects in the energy industry, including the siting of offshore energy developments and infrastructure required to bring energy onshore (Guinan *et al.* 2020; Nyberg *et al.* 2021; Yeomans *et al.* 2023). Advances continue to be made in mapping the geology of seabed environments to inform geomodels for ground engineering from offshore to onshore (Mellett *et al.* 2015; Moses and Vallius 2021; Asch *et al.* 2022).

Rock coast geomodel characteristics

The two key morphological components of the rock coast geomodel are cliffs and the foreshore, or shore platform (Fig. 1). Two types of rock coast are recognized: (1) plunging cliffs, with vertical or subvertical cliffs that descend far below sea-level; and (2) cliffs with shore platforms where vertical or subvertical cliffs are fronted by gently sloping or horizontal to subhorizontal platforms. Their geomorphology is described by Trenhaile (1987) and Sunamura (1992) and by Bird (2016), who also addresses their management. Two types of shore platform are differentiated based on morphology: Type A dips gently seawards with no discernible break of slope; and Type B is horizontal or subhorizontal, terminating at the seaward end with a marked scarp (Sunamura 1992, 2015). It is often difficult to differentiate the two types of shore platform in the field because the edge of the platform is seldom exposed at low tide, making it difficult to measure its full horizontal extent, or width, which is an important parameter in the long-term modelling of cliff retreat (Kennedy 2015). Plunging cliffs are generally considered stable, with no appreciable retreat (Trenhaile 1987; Sunamura 1992, 2015). Research interest has, therefore, focused on rock coasts with cliffs fronted by platforms that are formed as the cliffs retreat. In comparison with soft coastlines, such as beaches, dunes and salt marshes which have uniform substrates and morphologies dominated by a few processes, rock coasts present more complex environments. They may be inherited from former sea-levels and climates and have

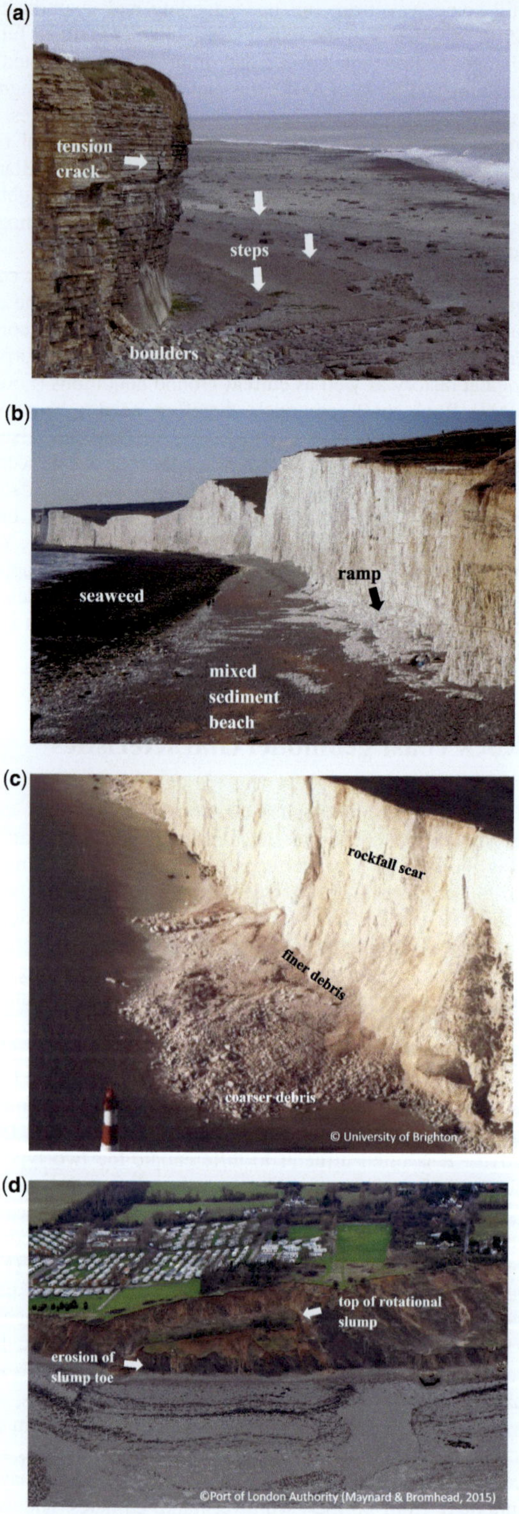

Fig. 3. Cliff and shore platform characteristics. (**a**) Limestone cliff and platform, Nash Point, South Wales. Note the boulder accumulation at the back of the platform, the steps the distribution of boulders across the platform and the tension crack on the upper cliff. (**b**) Chalk cliff and platform, Birling Gap, East Sussex. Note the exposed ramp at the cliff-platform junction, the varying depth of beach sediment and the extent of seaweed cover. (**c**) Beachy Head rockfall, East Sussex. Note the scar on the chalk cliff and the larger boulders on the outer, seaward, edge of the fall debris with finer material adjacent to the cliff. (**d**) Cliff and platform, Warden Point, Isle of Sheppey, Kent. Note the rotational slump with the top of the slump rotated backwards; erosion at the toe of the slump has created a lower, secondary, cliff. Source: (c) © University of Brighton; (d) Maynard and Bromhead (2015), © Port of London Authority.

complex geological settings with varying degrees of resistance to a range of processes including weathering, marine, biological and mass movements (Stephenson and Naylor 2010). In the UK, because of their relatively rapid cliff retreat rates, there has been a particular interest in the erosion and management of soft rock coasts, principally London Clay and glacial till (Lee 2002; Lee and Clark 2002; Brew 2004, 2007), but also Cretaceous Chalk, which may be included in both categories of 'soft' and 'hard' rock coast (Moses and Robinson 2011). Many UK coastal cliffs have national and international conservation designations because of their geological, geomorphological and biological significance (May and Hansom 2003) and these also require consideration in engineering projects.

Detailed reviews of cliff and shore platform development in Great Britain are provided in Kennedy *et al.* (2014), *Rock Coast Geomorphology: a Global Synthesis*, which is the most comprehensive review of rock coast geomorphology to date (cliffs, Lim 2014; shore platforms, Moses 2014). Typically, cliffs developed in well-jointed rocks are vertical to subvertical, with rockfall debris accumulated periodically on the platform below the cliff (Fig. 3a–c). Cohesive clay cliffs generally exhibit a crenulated and irregular cliff line and cliff face owing to rotational slumps being a dominant mode of retreat. In both cases, fall and slump debris protect the toe of the cliff from marine attack for a period until the debris is removed by wave processes (Fig. 3d). At the junction of the cliff and platform there is often either a ramp or a step (Fig. 3a, b). Ramps are associated with protection of the junction from erosion, either by rockfall debris or by deep beach sediment (Trenhaile 1987; Kirkby 2023), while steps are considered to form where there is a harder, more resistant, bed present at the foot of the cliff (Trenhaile 1987). The rate of removal of rockfall debris is related to the exposure of the shoreline to wave action and also the size of debris. On well-jointed lithologies, larger rockfall debris travels furthest from the cliff and may form a protective apron (Fig. 3c) with the finer debris being removed first, as recorded on the chalk coasts of SE England and northern France (Hénaff *et al.* 2006). The size of larger clasts that remain intact during the fall is related to the pattern of discontinuities – bedding, jointing and fractures – in the cliff (Mortimore *et al.* 2004*a*; Lim 2014).

Discontinuity patterns also exert a control on foreshore development (Moses 2014). Across shore platforms individual beds may be stripped back by wave impact removing joint- or fracture-bounded boulders (Figs 3a, 4a & b). These boulders usually reside on the middle to lower section of the platform, while boulders derived from rockfall debris tend to remain on the upper platform close to the toe of the cliff. Movement of boulders across the platform may cause impact abrasion (Fig. 4c). Where boulders or cobbles remain static they may protect the platform surface from abrasion (Fig. 5a). Downwearing of shore platforms occurs both at a micro-scale (millimetres to centimetres) on the exposed rock surface and at a meso-scale (tens of centimetres to metres) when individual steps are eroded. Step removal is stochastic and may be related to the occurrence of resistant layers (Fig. 5b). At low tide, shore platforms are exposed to a range of subaerial weathering processes including solution, wetting–drying, salt weathering and frost weathering. When the platform is inundated as the tide rises, then marine abrasion processes also operate. Where platforms are backed by hard engineering structures there may be enhanced downwearing owing to wave reflection, and this may be compounded by the use of heavy plant in construction (Fig. 5c, d). In addition to removing rock when moving, heavy machinery may compress and weaken the surface so that it is more susceptible to marine erosion. The surface depressions created then act as a trap for beach sediment, exacerbating erosion beyond the life of the construction project as seen on the chalk shore platforms of East Sussex (Dornbusch *et al.* 2007). Organisms that inhabit shore platforms may also cause erosion, e.g. by abrasion on and into the rock surface. Seaweed cover is thought to mainly provide protection from marine abrasion and mitigate the impact of subaerial weathering. The presence of beach material precludes biological colonization of the platform surface and there is often an abraded area close to the toe of the beach where sediment regularly moves back and forth (Fig. 5e). Generally, subaerial weathering processes are considered to be most effective on the middle to upper parts of the platform that are exposed for the longest time during the tidal cycle. The middle and lower parts of the platform may be dissected by shore-normal runnels (Fig. 5f, g). Relatively little is known about their initiation, formation and contribution to platform development, but they are thought to be linked to abrasion by sediment movement in the incoming and outgoing tide (Moses 2014).

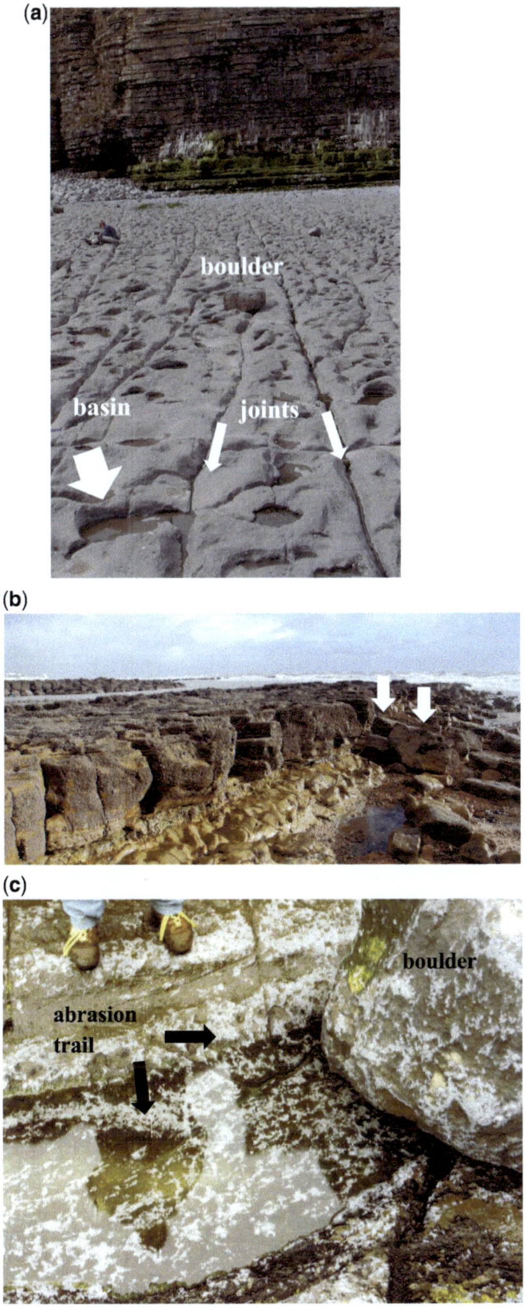

Fig. 4. Meso- and micro-morphologies typical of platforms formed in well jointed and bedded sedimentary rocks. (**a**) Discrete solution basins on the mid-upper platform, limestone, South Wales. Note the solutionally widened joint sets. (**b**) Step backwearing by the removal of joint-bounded blocks, sandstone, Bexhill, East Sussex. (**c**) Boulders moved by storm waves, limestone, Burren, Western Ireland. The light-coloured areas are where the rock surface has been fractured off by abrasion.

Rock coast geomodel dynamics in response to climate change

Cliff retreat rates: decadal

The most common approach to assess changes in cliff retreat rate and behaviour in response to climate change is to measure cliff line movement, from first edition Ordnance Survey

(a)
(b)
(c)
(d)
(e)
(f)
(g)

maps, published in the 1800s, to the present day, using a combination of data sources including maps, aerial photographs, LiDAR surveys and often also direct measurements from field surveys. The most intensively studied lithologies are those that erode most rapidly and thus are potentially the most sensitive to variations in climate and relative sea-level rise (RSLR). Studies considered here include Cretaceous Chalk, glacial till, London Clay and composite materials (Table 1).

The results are variable with some studies finding an increase in cliff retreat rates in recent decades, from approximately 1950 in most cases, whereas others indicate a decrease in recent cliff retreat rates when compared with longer-term historical rates, usually measured over approximately 130–150 years. This is the case even within the same lithology. For example, studies on the Cretaceous Chalk of SE England show a recent decrease in cliff retreat rates while in contrast, studies conducted in northern France show a recent increase. These are considered in detail in Moses and Robinson (2011) and, in summary, linking measured rates to factors known to influence cliff retreat (geological structure, rainfall, the occurrence of frost, wave climate, beach width and volumes and platform height and width) has proved difficult, mainly owing to the very different scales at which the data are recorded. These and other studies (Table 1) identify a range of, now well-established, factors that influence cliff retreat rates at specific sites and across a range of soft rock cliffs (reviewed by Sunamura 2015). Increasing shore platform width may reduce wave attack at the toe of the cliff, although measuring shore platform width from maps and aerial photos is problematic (Dornbusch *et al.* 2008*c*). Other factors include changes in beach volume and beach level related to variations in longshore drift owing to the presence of coastal defences, groynes and harbour breakwaters and occurrence of mass movement material with long residence times. For example, large-scale rockfalls on Cretaceous Chalk cliffs may have a residence time of 10–40 years (Mortimore *et al.* 2004*a*) and the erosion cycle for large-scale rotational slumps on London Clay coasts may be 30–40 years up to 100 years where exposure to wave energy is low (Hutchinson 1973; Dixon and Bromhead 2002). Bray and Hooke (1997) suggest that cliff height, through its control on the volume of debris deposited, influences the cycle of erosion from the initial mass movement to removal of debris from the toe of the slope: low, simple cliffs of <20 m height, cut in soft glacial sediment, 5–10 years; medium cliffs of *c.* 20–40 m height and composed of consolidated clay, or soft sedimentary rocks, 30–40 years; and for high cliffs of >40 m and up to as much as 200 m height in harder sedimentary rocks, 100–150 years. Removal of coastal defences may trigger a period of rapid cliff retreat, as has been the case at Happisburgh, Norfolk, UK, where a 1 km stretch of cliff retreated by *c.* 140 m over a 20 year period following the removal of sea defences (Payo *et al.* 2018).

Despite these complications, increased rainfall, water levels and wave energy associated with storm activity and variations in tidal cycles are key influences on cliff retreat rates on soft rock cliffs (Brooks *et al.* 2012; Pye and Blott 2015; Burningham and French 2017). Yet it remains difficult to establish the role of RSLR, particularly the rate of RSLR, in driving coastal cliff retreat. To address this, there is increasing interest in investigating centennial and millennial timescale cliff retreat using a combination of cosmogenic isotope dating and numerical modelling (see 'Cliff retreat rates: Holocene'). At annual to seasonal timescales, new applications of a range of technologies are yielding insights into cliff and shore platform response to climate-related driving factors (see 'Cliff retreat: annual and seasonal' section).

Cliff retreat rates: Holocene

Over approximately the last 10 years, the dating of accumulated cosmogenic isotopes, such as ^{10}Be, combined with numerical modelling, has emerged as a tool to understand cliff and shore platform erosion over millennia (Table 2; Fig. 6). The aim, providing context for historical and current rates of cliff erosion, is to facilitate calibration of long-term dynamic models of cliff and shore platform erosion. This is considered central to the development of predictive models of cliff retreat that account for climate change (Hurst *et al.* 2016, 2017; Trenhaile 2018; Shadrick *et al.* 2021, 2022, 2023*a*). The value of modelling millennia-scale, Holocene, cliff retreat rates is posited to be that such long-term retreat rates (1) incorporate long return period mass wasting events characteristic of coastal cliff retreat, (2) are unaffected by human intervention on the coast and (3) include prehistoric rates of RSLR that are comparable with projected future rates (Hurst *et al.* 2016, 2017; Shadrick *et al.* 2021, 2022, 2023*a*). Shadrick *et al.* (2022) consider this to represent an advancement over the well-established and widely used Soft-Cliff and Platform Erosion (SCAPE) model (Walkden and Hall 2005), which is reliant on historical cliff retreat rates and designed primarily for use on soft rock cliffs. More recent developments to SCAPE, however, incorporate rock hardness, enabling its application to hard rock coasts (e.g. Walkden *et al.* 2020; SCAPE+, National Network of

Fig. 5. Visual evidence of platform erosion. (**a**) Large cobble protecting the platform surface from sand abrasion, glacial till, Holderness, Yorkshire, NE England. (**b**) Flint band protecting the chalk shore platform, Seven Sisters, East Sussex, England. (**c**) Erosion of the platform beneath a concrete groyne, chalk, Saltdean, East Sussex. Arrows indicate the original platform surface. (**d**) enhanced abrasion in front of seawall, chalk shore platform, Saltdean, East Sussex, exacerbated by using heavy plant in construction as described by Dornbusch *et al.* (2007). (**e**) Abrasion zone adjacent to the beach toe where sediment frequently moves across the platform surface. (**f**) Runnels on the middle and lower platform, chalk, East Sussex. The darker colour on the lower platform is seaweed cover. (**g**) Runnels on the lower platform, London Clay, Isle of Sheppey, Kent, SE England.

Table 1. Cliff retreat and sea-level rise: decadal-scale recent and historic cliff retreat rates

Recent rates relative to historic rates	Lithology	Location	Key drivers	Data sources	Total years	Authors and notes
Decreasing	Cretaceous Chalk	East Sussex	Reduction in beach width Increase in shore platform width	First edition map; 1973 aerial photos; 2005 LiDAR survey; 2006 beach survey	130	Dornbusch *et al.* (2008*b*) 1873–1925/1928: 0.37 m a^{-1} 1925/1928–73: 0.35 m a^{-1} 1973–2001: 0.27 m a^{-1}
Increasing	Cretaceous Chalk	Northern France	Sediment starvation: large rockfalls with long residence time + harbour breakwaters and groynes	Historical maps; aerial photos	300	Prêcheur (1960) 300 years: 0.08–0.8 m a^{-1} Hénaff *et al.* (2002) 1824–1986: 0.05–0.07 m a^{-1} 1947–95: 0.14 m a^{-1} Costa *et al.* (2003, 2006) 1966–95: 0.21 m a^{-1} Costa *et al.* (2004) 1966–95 0.26 m a^{-1}
Increasing	Glacial till	Yorkshire	Sea defences Sea-level rise Beach levels 18.6 year nodal tidal cycle	Historical maps; direct measurements; biannual beach surveys	160 50	Pye and Blott (2015) Epoch 1: 1852–1908 Epoch 2: 1908–51 Epoch 3: 1951–89 Epoch 4: 1989–2013 Retreat rate in Epoch 4 > other Epochs on undefended sections Lee (2011) Early 1950s: 1.2 m a^{-1} 2000: 1.5 m a^{-1}
Decreasing	London Clay	Kent	Historic human interference (beach levels) Occurrence of large rotational slumps with long residence time	Historical maps; field survey; differential global positioning systems survey	131	Nicholls *et al.* (2000) 1867–97 : 3.29 ± 0.33 m a^{-1} 1897–1998: 1.67 ± 0.1 m a^{-1} Hutchinson (1965, 1968), Bromhead (1979), Dixon and Bromhead (2002) Aerial photos and field survey
Increasing	Composite (Pliocene + early–mid Pleistocene marine deposits overlying Paleogene and Cretaceous basement)	Suffolk	Higher rainfall and surges associated with positive phases of the NAO Complex interaction of • met-ocean forcing, • inherited geological and geomorphological controls, • Evolving anthropogenic intervention	Historical maps; aerial photos; Environment Agency annual beach survey; differential global positioning systems survey; geomorphological mapping	127	Brooks *et al.* (2012) 1883–2010: 3.49 ± 0.4 m a^{-1} 1993–2010: 4.66 ± 0.55 m a^{-1} Burningham and French (2017) Cluster-based segmentation of shoreline, graphically indicating shoreline change, 1881–2015; 1990–2015.

Table 2. Cliff retreat and sea-level rise: millennial scale Holocene and historical cliff retreat rates

Cliff retreat rate ($cm\ a^{-1}$)		Lithology	Location	Modelling approach	Authors
Holocene	Historic (measurement period, years)				
2–6	22–32 (150)	Cretaceous Chalk	East Sussex, SE England	Steady-state geometric model	Hurst *et al.* (2016)
5–9	10–18 (29)	Cretaceous Chalk	Normandy, NW France	Steady-state geometric model	Duguet *et al.* (2021)
11–13	15 (45)	Cretaceous Chalk	Normandy, NW France	Steady-state geometric model	Regard *et al.* (2012)
6–18 (2000 years BP) 15–110 (7000 years BP)	22–40 (130)	Cretaceous Chalk	East Sussex and Kent, SE England	Morphodynamic process-based model	Shadrick *et al.* (2023*a*)
4.5 ± 0.63	2.7 ± 2.9 (7)	Jurassic shale and sandstone	Yorkshire, NE England	Steady-state geometric model	Swirad *et al.* (2020)
7.5–17.5 (2400 years BP) 25–100 (7000 years BP)	2–25 (130)	Carboniferous and Jurassic sandstone	Devon, SW England; Yorkshire, NE England	Morphodynamic process-based model	Shadrick *et al.* (2021, 2022)

Regional Coastal Monitoring Programme of England 2023). The use of cosmogenic isotopes is considered to have transformed the ability to investigate coastal cliff retreat over centennial to millennial timescales (Hurst *et al.* 2016; Trenhaile 2018; Shadrick *et al.* 2022), potentially informing predictions of future rock coast geomodel behaviour.

Cosmogenic isotopes are produced by the interaction of cosmic rays with the rock surface. Their concentration indicates the period of exposure of the surface, providing a geochronometer to quantify how long the surface has been exposed or to interpret its rate of weathering and/erosion (Fig. 6). A range of isotopes may be used, dependent on rock type, and ^{10}Be, used in rock coast studies published to date, requires the presence of quartz (Moses *et al.* 2014). On chalk shore platforms it is therefore the indigenous siliceous flint that is sampled rather than the chalk. At the simplest level, as the cliff retreats and the foreshore is exposed, concentrations of ^{10}Be increase across the shore platform with the most recently exposed, upper platform, accumulating lower concentrations compared with the longer-exposed middle and lower platform. This simple distribution, which underpins the studies of Holocene cliff retreat, is complicated by a range of additional factors. For example, seawater shields the rock surface and ^{10}Be concentrations decrease with water depth. Local water levels, influenced by tides and RSLR may, therefore, reduce the concentration on the lower platform. In addition, downwearing of the shore platform removes the richest ^{10}Be from the surface, exposing lower concentrations. The result is a 'humped' distribution of ^{10}Be across the platform that is dependent on both the rate of cliff retreat and shore platform downwearing (Fig. 6). Higher rates of cliff retreat result in lower ^{10}Be concentrations. Increasing tidal ranges may cause the ^{10}Be peak to migrate seaward; rising sea-level may result in higher ^{10}Be content because of a suggested associated reduction in platform downwearing. More detailed explanations of the principles and the methodological approach are provided by Hurst *et al.* (2016, 2017).

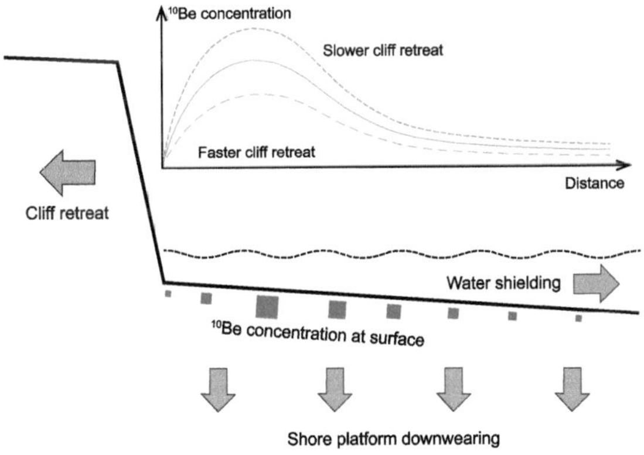

Fig. 6. Schematic representation of expected 'humped' relationship between distance from the cliff and ^{10}Be concentration, assuming equilibrium retreat of the cliff–platform system. Source: after Hurst *et al.* (2016).

Cosmogenic isotope geochronology, combined with numerical modelling, has been applied on sandstone and chalk rock coasts to ascertain Holocene rates of cliff retreat (Table 2). In studies to date, two different modelling approaches are used: steady-state geometric or morphodynamic process-based modelling. For the Cretaceous Chalk cliffs of SE England and NW France, studies using a steady-state geometric modelling approach indicate either comparable (Regard *et al.* 2012) or increasing cliff retreat rates from the Holocene to modern, historical period (Hurst *et al.* 2016; Duguet *et al.* 2021). Use of a morphodynamic process-based modelling approach indicates declining cliff retreat rates from the Holocene to the modern, historical period (Shadrick *et al.* 2023*a*). For sandstone cliffs both modelling approaches yield a decline in cliff retreat rates from Holocene to modern (Swirad *et al.* 2020; Shadrick *et al.* 2021, 2022). This is considered to be due to lower rates of platform downwearing on sandstone when compared with chalk (Shadrick *et al.* 2023*a*). Studies of Holocene cliff retreat rates using cosmogenic isotopes remain small in number and methodological approaches are in ongoing development. The dynamic process-based modelling approach allows for transient cliff retreat rather than Holocene-averaged cliff retreat rates (Shadrick *et al.* 2023*a*; Table 2). There are also variations across the studies in the modelling assumptions, in both approaches, that may influence the results. Parameters such as cliff morphology, platform morphology, width and gradient, downwearing rate and presence/absence of beach or mass movement sediment may be represented in a variety of ways. For example, platform morphology may be fixed or dynamic through model simulation time (Shadrick *et al.* 2023*a*); after initial exposure the platform may be assumed never to have been reburied or shielded by sediment cover (e.g. Swirad *et al.* 2020); either geology (e.g. Swirad *et al.* 2020) or physical processes (e.g. Shadrick *et al.* 2022) may be factored as the dominant control. Hurst *et al.* (2017, p. 82) investigate a range of factors that may influence the modelled results, concluding that 'cliff retreat rates derived from cosmogenic ^{10}Be might be considered as maximum estimates'. Nonetheless, this approach has been used to forecast cliff retreat rates for the next century suggesting an increase of up to one order of magnitude by 2100 on historically stable coastlines (Shadrick *et al.* 2022). While there is discussion on the validity of such predictions, they highlight some of the key process–response issues that require further research and refinement to improve predictive capability for future rock coast responses to climate change and SLR (Dickson *et al.* 2023; Shadrick *et al.* 2023*b*; Matsumoto *et al.* 2024).

All of these cosmogenic isotope studies are predicated on the assumption that active cliff retreat has occurred throughout the Holocene, an assumption that is open to challenge, at least for the chalk and sandstone cliffs of East Sussex, SE England (Dornbusch 2022). Combining evidence, including historical cliff retreat rates calculated over 400 years, the presence of mid-Holocene terrestrial habitats fronting contemporary cliffs and a geomorphological landscape-type interpretation, Dornbusch (2022) suggests that cliff erosion can only have started in the last few centuries. The main line of evidence is that remnants of Eemian (last interglacial 130–115 ka BP) cliffs are still present in places. In addition, it is suggested that, despite sea-level rise, there has been a persistent decrease in erosion rates to less than half the maximum rate observed in the late nineteenth to early twentieth century along this coastline. While millennial-scale studies of cliff retreat based on cosmogenic isotopes may offer potential insights into cliff response to RSLR, the wide range of assumptions of the various modelling approaches and the assertion that 'we cannot use historical rates to assess the risk associated with rock coasts' (Shadrick *et al.* 2022, p. 7) require further examination (Dickson *et al.* 2023; Shadrick *et al.* 2023*b*; Acharya-Chowdhury *et al.* 2024; Matsumoto *et al.* 2024). A better understanding of (1) shore platform downwearing and equilibrium state, (2) lithological controls on rock coast evolution, (3) linkages between shore platform downwearing and cliff retreat and (4) inheritance is required.

Cliff retreat: annual and seasonal

Studies on the chalk coast SE England and NW France show that cliff retreat rates are higher in winter, associated with a higher frequency of rockfalls linked to subaerial rather than marine processes: higher number of rain days and frost days, heavier winter rainfall and sequential wet years that elevate groundwater levels. The longer the period of antecedent rainfall, the deeper and larger the slide. It has also been noted that rockfalls may occur during the summer when dry periods are interrupted by rainfall (Mortimore and Duperret 2004; Duperret *et al.* 2005; Moses and Robinson 2011; Lim 2014). On cohesive clay coasts, higher cliff retreat rates are associated with higher rainfall levels and wave energies during the winter period (Hutchinson 1973; Dixon and Bromhead 2002; Brooks *et al.* 2012; Pye and Blott 2015; Burningham and French 2017). Annual and seasonal variations in cliff retreat rates are strongly influenced by the moderating effects of rockfall debris, episodically protecting the cliff toe from wave attack, and fluctuating beach levels with higher summer beach levels protecting the cliff toe, as is well demonstrated on the Holderness tills (Pringle 1985). Cliff retreat rates on decadal timescales are well documented, although this has been predominantly in terms of cliff top retreat (Moses and Robinson 2011; see 'Cliff retreat rates: decadal' section). More recently annual and seasonal cliff erosion dynamics have received increasing attention associated, in part, with noticeable changes in seasonal weather. Recent summer heat waves in the UK have sparked concern for a potential increase in the frequency of summer rockfalls associated with higher rock temperatures (Ground Engineering 2022*a*, *b*). There is a particular research interest, therefore, in gaining a more detailed understanding of the influence of lithological, marine and subaerial factors on cliff retreat rates at annual and seasonal timescales.

Studies have focused on collecting higher temporal and spatial resolution data and new applications of technologies and of mathematical analytical techniques. For example, using a team of volunteers to collect a 7-year database of rockfall location and volume over 37.5 km of chalk cliffs in northern France, Letortu *et al.* (2014*a*, *b*) investigate the influence of triggering factors on location and volume of fall material. High rainfall was found to trigger massive rock falls, with material lost from across the whole cliff face, of 200–1400 m^3 and up to more than 10 000 m^3. Marine processes were found to trigger rockfalls predominantly located only at the cliff foot and of smaller volumes <200 m^3. Similar findings are reported for interbedded claystone–sandstone cliffs in California (Young *et al.* 2021). Winter frost action triggered small debris falls, also <200 m^3, on either the cliff top or cliff foot rather than from across the whole cliff face. Deciphering the mechanisms associated with these triggering factors, however, remains difficult in the absence of high resolution spatial and temporal data on lithological, meteorological and marine parameters (Letortu *et al.* 2015*a*). A further study at the same site, using repeat terrestrial laser scanning (TLS) surveys, every 4–5 months over a 3-year period, indicated that the key precursor for larger rockfalls is the formation of a notch at the base of the cliff (Letortu *et al.* 2015*b*). While it has frequently been argued that notch development at the toe of the cliff is largely responsible for initiating rockfalls on jointed rock coasts (Hutchinson 1971; Trenhaile and Bryne 1986; Sunamura 1992), evidence for this on the chalk coasts of southern England and northern France has been sparse (Moses and Robinson 2011). It is more common for the base of chalk cliffs to project seawards as a ledge and some slight notching has been identified above the height of wave attack where it is related to subaerial weathering of weaker beds (Costa *et al.* 2004; Hénaff *et al.* 2006; Lageat *et al.* 2006). Based on field evidence, notching has been considered to be of secondary importance for chalk cliff retreat where rock structure is held as the dominant control on the type and volume of rockfall with different types of failure attributed to different chalk units (Duperret *et al.* 2004; Mortimore *et al.* 2004*a*). Models, however, suggest that as the depth of the notch increases the removal of support at the toe of the slope results in the development of a tension crack within the cliff (Hutchinson 1971; Hoek and Bray 1981; Styles *et al.* 2011). Hoek and Bray (1981) also model the critical depth that the notch must reach to trigger failure. However, where cliff toe notches develop, they are usually localized and any relationship between their depth and the size of the resulting rockfall has not been investigated in the field. Few studies quantify the rate of erosion at the toe of the slope, although Ellis (1986) measured a rate of 7.2 mm a^{-1} at Friar's Bay, between Brighton and Newhaven in East Sussex, SE England, using the micro erosion meter (MEM). A comparable figure, of up to 3.5 mm over a 6-month period, has been recorded at an adjacent site using unmanned aerial vehicle (UAV) digital photogrammetry (Gilham *et al.* 2019), with authors noting that erosion at the toe of the slope is localized and highly variable. Mortimore *et al.* (2004*a*) have noted that notches are most commonly associated with harder chalks as they may be short-lived in softer rocks that collapse readily (Moses and Robinson 2011; Trenhaile 2015).

Other studies suggest that coastal cliffs behave in a similar manner to non-marine cliffs, as self-organizing systems where precursory rockfalls are important for the development of large scale rockfalls. For example, on composite cliffs of Jurassic limestones, shales and mudstones, capped by a sequence of Cretaceous sandstones and Quaternary glacial tills, in North Yorkshire, on the basis of 32 monthly TLS scans, Rosser *et al.* (2007) found no evidence of basal notching, limited seasonal control, limited control of wave action at the cliff toe and limited environmental controls. The key finding, that large rockfalls were preceded by a series of smaller debris and rock falls, suggested that the frequency of the largest events could be estimated by analysis of the smaller events. The high spatial and temporal resolution of TLS survey data can facilitate such analyses. On a complex cliff consisting of three types of limestone at Marsden Bay, NE England, Westoby *et al.* (2020) have investigated this further. Cliff fall data, collected by monthly TLS survey over a 24-month period for a 0.6 km stretch of coast, were analysed to examine magnitude-frequency and power law scaling used to predict rockfalls. Over the survey period, precursory rockfalls accounted for approximately 80% of the total eroded volume but less than 50% of the erosion scars on the cliff face; erosion rates were higher in winter, associated with short-term increases in rainfall and wave activity from winter storms; varying sensitivity of the different limestones to environmental forcing was dependent on jointing patterns. The authors suggest that it may be possible, using power law scaling and high-resolution datasets, for coastal managers to predict rockfall occurrence from marine forecasts or real-time offshore wave buoy data and wave transformation modelling. Gilham *et al.* (2019) have experimented with this, using power law scaling to predict rockfall activity in relation to significant wave height (H_s). Finding statistically significant correlations between H_s and power law scaling coefficients, they applied Monte Carlo simulations to predict erosion scenarios based on H_s probabilities and sea-level forecasts derived from the UKCP09 medium emission climate model. They use these to estimate a *c.* 11% chance that the A259 road, located 42.1 m behind a cliff, would be breached by coastal erosion by 2090. In this case the cliff erosion dataset was collected by UAV as it is cheaper and quicker than TLS. Even more rapid and cost-effective high-resolution data collection is now possible using LiDAR-equipped smartphones (Luetzenburg 2022; Tavani *et al.* 2022), which are being tested for rapid assessment of cliff behaviour (Bessin *et al.* 2023).

To further investigate the role of marine processes in cliff retreat, a growing number of studies deploy seismometers, often combined with TLS surveys, to detect cliff micro-displacements in response to wave impact (Table 3). Usually, seismometers are emplaced at or close to the cliff top, which means that ground movements are detected only as waves impact on the toe of the cliff and not as they travel across the platform where they are out of the instrument detection

Table 3. Cliff response to marine processes: insights from seismograph studies

Key findings	Site location and geology	Monitoring techniques	Monitoring period	Authors
– Vertical cliff-top ground motions increase with increasing Hs and tidal elevation – Peak seismic activity: high tide; wind direction aligned with max fetch – Secondary peak: when waves reach ramp	Staithes, north Yorkshire, UK interbedded mudstones, shale, siltstone, ironstone, and sandstone, are capped with glacial till	4.5 Hz integrated seismic system triaxial geophone installed at top of cliff; LiDAR (airborne, at each low tide) + TLS (monthly) surveys	Eight months (September–April, year not specified)	Lim *et al.* (2011)
– Episodic displacements associated with extreme wave conditions cause failure (time lag) – Displacements under normal conditions unlikely to weaken rock mass		Güralp 6TD broadband seismometer installed at top of cliff; weather data (rainfall, atmospheric pressure, wind direction and wind velocity), tidal data (predicted tidal height and observed tidal residual), oceanographic data (significant wave height and peak wave period).	2 × 16 day periods, summer (July 2009) and winter (November–December 2009)	Brain *et al.* (2014)
– Under storm conditions ground motions become more uniform along the coast (where >2 m water depth) – Ground motions higher on headlands than bays – Platform width, elevation and topography dominant control on delivery of wave energy to cliff		7 × Güralp 6TD broadband seismometer installed at *c.* 100 m intervals, 10 m inland from cliff edge; weather and oceanographic data as above	Eight months (December 2013 to July 2014)	Jones *et al.* (2018)
– Type of wave impact more important than Hs: energy of waves breaking directly on cliff face > breaking/unbroken	Del Mar, California, USA Sandy claystone, sandstone, overlain by Pleistocene terrace deposits	Nanometrics Compact Trillium broadband seismometer, installed at top of cliff; high-resolution video to record impact time of each wave	Forty-five minutes (March 2016)	Thompson *et al.* (2019, 2022)
	Onaero Bay, Taranaki, New Zealand Sandstone, mudstone overlain by Quaternary deposits	As above + Geospace Technologies HS-1-LT geophone, water pressure transducer installed at cliff toe, offshore wave data, tidal level data	Thirty days (December 2017 to January 2018)	
– Failure events occurred most frequently overnight and during a wetter winter – Water content most important: (1) subsurface flow towards the cliff; (2) rain directly onto the cliff; (3) condensation of water vapour onto the cliffs at night. – Marine, wind and frost processes negligible	Jasmund peninsula, Rügen, Germany Weakly cemented Maastrichtian chalk	Nanometrics Trillium Compact 120 s seismometers and PE6/B 4.5 Hz geophones, installed at four cliff-top sites along 7 km as well as 10 UAV surveys, weather, sea-level and groundwater data. 81 failure events recorded	Twenty-five months (March 2017 to April 2019)	Dietze *et al.* (2020)

– Thermal amplitude greatest at cliff top (rockfalls caused by frost weathering most likely to develop preferentially at top of cliff) – Cliff top ground motion: displacement up to 50 μm (in line with previous studies but considered high for the relatively calm wave conditions and owing to elasticity of the chalk – Significant cliff face displacements usually during spring tides (time lag >1 day from highest tide to maximum displacement) – Cliff top displacement strongly linked to Hs – Most favourable conditions for displacement = medium H_s, low wind speed, high water level, high tide coefficient, low rainfall, medium temperature (recorded extension displacement on existing fractures) – Least favourable conditions for displacement = high H_s, high wind speed, high water level, medium tide coefficient, high rainfall, high temperature (recorded contraction displacement on existing fractures) contrary to what is expected	St Marguerete-sur-Mer, Normandy Upper Cretaceous Chalk, partly overlain with 10–15 m thick clay and sand Paleogene deposits	44 sensors of nine different types installed to measure: • cliff mechanical response (seismometers, geophones, extensometers, fibre optic); • marine forcing (pressure sensors, video camera); • subaerial forcing (weather station, temperature sensors, fibre optic devices, piezometer). TLS topographic surveys (3 monthly); 1 × Ground penetrating radar and 1 × electrical resistivity tomography survey to assess chalk homogeneity and internal structure and properties)	Thirteen months (November 2018 to January 2020)	Letortu *et al.* (2022)

H_s, significant wave height; TLS, terrestrial laser scanning.

range. As waves move across the platform their energy is dissipated, as indicated by pressure sensor studies, with platform elevation and width being dominant controls on energy reaching the cliff toe (Ogawa *et al.* 2016). Tidal range also influences wave energy dissipation and in macro-tidal environments waves of <2 m may travel unbroken across the platform at high tide (Stephenson *et al.* 2018). Although seismic studies indicate that vertical cliff-top ground motions increase with increasing H_s and tidal elevation, this is mediated by cliff–platform junction elevation and beach gradient. For example, during winter storms, Earlie *et al.* (2015, 2018) found that where H_s <2 m, elevation of the cliff–platform junction is an important control on energy delivered to the cliff but where H_s >2 m, the beach gradient exerts more influence on energy delivery. Ground motions become more uniform along the coast at water depths of >2 m, as may be associated with storms, at the foot of the cliff. Additionally, seismic studies indicate that the type of wave impact is more important than H_s with waves that break directly onto the cliff face delivering more energy than breaking or unbroken waves (Table 3).

Seismometers have also been deployed in a slightly different way to identify the timing of cliff falls and elucidate their trigger processes on the Maastrichtian chalk cliffs, Jasmund peninsula, Rügen, Germany (Dietze *et al.* 2020; Table 3). This weakly cemented chalk has an annual cliff retreat rate of 25 cm a^{-1}, comparable with the Cretaceous Chalk cliffs of SE England and northern France. Of 81 failure events recorded, 65 occurred during a wet winter, 11 during a dry winter, and across the entire study period 50 occurred overnight between 8 p.m. and 8 a.m. Dietze *et al.* (2020) suggest that overnight accumulation of dew increases the cliff water content, ultimately driving cliff retreat at this site. Other studies on the Cretaceous Chalk cliffs also identify the water content of the cliffs as an important trigger for rockfalls, although usually associated with higher winter rainfall (Duperret *et al.* 2004; Moses and Robinson 2011) rather than overnight fluctuations as identified by Dietze *et al.* (2020). Also, although the number of frost days has been linked to a greater occurrence of winter rockfalls (e.g. Hutchinson 1971; Prémaillon *et al.* 2018), there are no direct field measurements on cliffs. Frost shattering on shore platforms has been measured, however, indicating that a minimum of 10 frost cycles below seawater freezing temperature (−2.5°C) are required and for durations of 2–3.5 h in order for rock breakdown to occur (Dewez *et al.* 2015). Field observations suggest that, following a night of freezing, as temperatures rise in the morning a rain of debris is released from cliff faces (Mortimore *et al.* 2004*a*), indicating that frost action is superficial.

Despite the constant exposure of coastal cliffs to wave wash, splash and spray, salt weathering, which – like freeze–thaw weathering – is a superficial weathering process, has received little attention. More recent research, however, indicates that seawater may weaken chalk cliffs at depth (Lawrence *et al.* 2013, 2018). Salt weathering is usually associated with granular disintegration and surface spalling and flaking to create small-scale surficial features such as alveoli, although larger-scale tafone may also develop (Pye and Mottershead 1995). To be effective it requires a salt supply and environmental conditions that cause evaporation or facilitate temperature fluctuations, both of which are required for salt crystals to develop and grow. Rock breakdown occurs owing to salt crystal growth, differential thermal expansion and hydration. Porous rocks are particularly susceptible to salt weathering (Robinson and Moses 2011). Duperret *et al.* (2005) investigated the occurrence of salt weathering on the chalk cliffs of northern France, suggesting that it is mainly surficial because the thermal conditions within the chalk do not allow the salts to crystallize out of solution. However, solution of carbonate rocks may occur in seawater and chemical solution at depth in chalk, brine-saturated, oil reservoirs has been found to result in a reduction in rock strength (Heggheim *et al.* 2005). Seawater may further contribute to the weakening of chalk when magnesium (Mg^{2+}) and sulfate (SO_4^{2-}) from seawater concentrate in the rock. Sulfate may act as a catalyst for the substitution of calcium (Ca^{2+}) from the chalk by Mg^{2+}. When this dolomitization process occurs at grain boundaries, the different size of the Mg^{2+} compared with Ca^{2+} may cause stress and, consequently, a decrease in the mechanical strength of chalk (Austad *et al.* 2008; Nermoen *et al.* 2015; Minde and Hiorth 2019).

This ‘salt water weakening’ has been investigated on the Cretaceous Chalk cliffs of East Sussex, SE England, where saturation by meteoric waters is already known to contribute to cliff retreat through a reduction in rock strength and thus also the angle of friction in cliffs (Mortimore *et al.* 2004*b*). Lawrence *et al.* (2013) tested samples collected both from the cliff face, exposed to seawater saturation, and also from inland quarry sites unaffected by salt from seawater. On the coast, samples were collected from two different zones, (1) in direct contact with the sea at high tide and (2) not in direct contact with the sea but in a zone affected by waves, splash and spray. In addition, samples collected from the inland quarry were subjected to salt wetting–drying experimental cycles to assess the impact on rock strength. Triaxial strength testing indicated that samples that had experienced salt-water saturation were up to 55% weaker than those that had not. The authors then investigated how far into the cliff face this salt-water weakening effect extends (Lawrence *et al.* 2018). Three *c.* 9 m long cores were extracted horizontally into the cliff at two contrasting sites, both in Newhaven Chalk: an engineered cliff protected by a sea wall and a natural cliff where the lower section is in contact with the sea at high tide. Cores were drilled using air flushing to ensure that the pore water content and chemistry were not affected. Laboratory tests assessed the impact of salt water on chalk strength and extent to which any weakening effects penetrated the cliff. Chloride and sulfate concentrations provided a proxy for the presence of seawater. The sea wall site had a seawater penetration depth of only 0.6 m, whereas the natural cliff site showed seawater present up to and beyond 6 m depth into the cliff. Cores from both sites showed a decrease in rock strength at the cliff face, increasing with depth into the cliff and with the decline being greater at the natural cliff site. The authors note that there is not a straight correlation between chalk weakening and salt content and that there are other processes that also weaken chalk close to the cliff face, including heterogeneity within the chalk, e.g. in relation to sedimentation, bioturbation, diagenesis and notch development and oversteepening of the cliff face. Nonetheless, the results demonstrate that chalk is weaker at the cliff face and for several metres into the cliff and the effect that this has on cliff stability merits consideration. That salt can penetrate to depths of more than 7 m into the cliff face has not previously been demonstrated or considered. Assessment of the mechanisms requires further investigation to identify whether any reductions in rock strength are greater in the presence of seawater compared with meteoric water.

Most recently, efforts are being made to shed further light on high-resolution spatial and temporal responses of cliffs to lithological, marine and subaerial parameters leading to cliff fatigue and cliff failure. Letortu *et al.* (2022) report results from the first 4 months of a 13-month continuous, multiparameter monitoring programme on Cretaceous Chalk cliffs in northern France (Table 3). The study site, St Marguerete-sur-Mer, Normandy, has previously been monitored and there is general consensus that the main drivers of erosion are: lithological control (Costa *et al.* 2004); intense rainfall events (Duperret *et al.* 2004; Lageat *et al.* 2006; Letortu *et al.* 2015*b*); and marine factors (Letortu *et al.* 2015*a*, *b*, 2019). Letortu *et al.* (2022) report some unexpected findings which analyses of the full 13-month dataset, yet to be published, may help to elucidate.

Foreshore: shore platform downwearing

Shore platforms may play a key role in influencing temporal variations in cliff retreat rates. For example, on cohesive clay shorelines in both coastal and lakeshore settings, it is generally agreed the primary control on the long-term rate of cliff retreat is the rate of lowering of the shore platform, and beach if present, fronting the cliff toe and the water level at the cliff toe (Moses and Robinson 2021). Understanding shore platform downwearing is important, therefore, for a number of reasons: firstly, shore platforms are dynamic, and the level of the shore platform can determine the height of the platform–cliff junction, the level and behaviour of the beach and thus the extent to which wave action and associated spray impact on and up the cliff face (e.g. Brew 2004, 2007); secondly, removal of sediment by platform downwearing not only leaves the cliff face more exposed but also removes weight from the toe of the slope and may therefore contribute to mass movement and retreat of the cliff (e.g. Hutchinson 1973; Dixon and Bromhead 2002); thirdly, weathering and erosion of the areas around engineering structures such as groynes and sea walls can reduce their engineering lifespan (e.g. Dornbusch *et al.* 2007; Dornbusch 2010); fourthly, shore platforms provide sediment to the coastal sediment budget (e.g. Dornbusch *et al.* 2006; Payo *et al.* 2018; Moses and Robinson 2021).

There is an ongoing 'wave v. weathering' debate, concerning the relative roles of marine and subaerial processes as the dominant processes in their formation (e.g. Stephenson and Kirk 1998; Trenhaile and Kanyaya 2007; Kennedy *et al.* 2011; Dickson *et al.* 2013; Matsumoto *et al.* 2016, 2018; Trenhaile 2019, 2020). Present-day shore platforms may be inherited features that predate the relatively stable sea-levels of the late Holocene over the last 2000 years (Regard *et al.* 2012; Thébaudeau *et al.* 2013; Moses 2014). Because of the links between shore platforms and cliff retreat, understanding the key drivers for shore platform development is important in informing models of cliff retreat and their response to future climate change. Platform width is influenced by erosion rates of the cliffs at the inner edge relative to the outer, seaward edge of the platform. If the cliff retreat rate is greater than that of the outer platform edge, then the platform width may increase until waves no longer reach the cliff. This is known as static equilibrium, whereby the platform width remains constant but downwearing continues and so the platform slope declines. If the cliff retreat rate equals the platform edge retreat, then the platform width and gradient remain constant over time, known as dynamic equilibrium. Platform elevation, width, gradient and level of the cliff–platform junction are influenced by rock hardness and tidal range, and so can vary locally and regionally. Rock structure, via discontinuities such as joints, fractures and bedding planes, influences platform morphology and the production of boulders and coarse debris. Boulders may be present on shore platforms and are derived either from cliff falls or platform erosion or a combination of the two (Moses 2014).

A dominant area of research has involved measurement of shore platform downwearing rates, incorporating both subaerial weathering and marine erosion. These studies, reviewed in detail in Moses and Robinson (2011) and Moses (2014), provide insights into the role of lithology, annual and seasonal climatic variations and wave energy. The most intensively studied shore platforms in the British Isles are on Blue Lias limestone in South Wales and Cretaceous Chalk in SE England, with a limited number of studies on other rock types. Platform downwearing rates vary according to lithology, with the most rapid downwearing on softer cohesive clays and the lowest rates on harder limestones: glacial till > London Clay > chalk > schist > limestone > basalt and granite (Table 4). It is also now well established that there are variations in platform downwearing rates across platforms from the cliff toe to the platform seaward edge: upper > middle > lower (Moses 2014). Higher rates of downwearing on the upper platform are associated with abrasion caused by beach material being drawn across the platform by wave action (Fig. 5e). Where the beach depth is greater than the layer mobilized by wave action (active layer), then it protects the platform from abrasion and this can often be seen as a ramp at the cliff–toe junction when it is revealed following storm activity. The upper platform is exposed to subaerial processes for the longest time period during the tidal cycle and so this means that subaerial weathering processes such as wetting–drying, salt weathering and frost weathering

Table 4. Summary of shore platform downwearing rates for the British Isles (based on data from Moses and Robinson 2011, 2021; Moses 2014)

Lithology	Rate (mm a^{-1})	Measurement period (years)	Measurement method
Glacial till	42	1	TEB
London Clay	18	10	TEB
Chalk	4	2–28	MEM, laser scanner, photogrammetry
Schist	1	1	MEM
Limestone	0.05	1	MEM, laser scanner
Basalt & granite	<0.05	NA[a]	MEM

TEB, Traversing erosion beam; MEM, micro erosion meter.
[a]Basalt and granite have proved too resistant to record downwearing using MEM.

can operate effectively (Moses and Robinson 2011, 2021; Moses 2014). Similarly, these processes can operate effectively on the mid-platform while the lower platform has less opportunity to fully dry out between tides and is often covered with a thick layer of seaweed that may protect the platform surface from weathering and erosion (Coombes *et al.* 2013; Gowell *et al.* 2015). The presence of seaweeds on hard engineering structures may also protect them from weathering and erosion (Baxter *et al.* 2022).

The extent to which the presence of organisms may erode or protect shore platforms is debated. Bioerosion is associated with, for example, limpets and sea urchins, which abrade the rock surface, and bivalve molluscs that bore directly into the rock to depths of 10–20 cm (Trudgill 1987; Trudgill *et al.* 1987; Andrews and Williams 2000; Brew 2004, 2007; Naylor *et al.* 2012). Bivalve molluscs, known as piddocks, such as *Pholas dactylus* and *Hiatella arctica*, occur most frequently on the lower shore where densities of more than 700 individuals per square metre to depths of 15 cm have been recorded on cohesive clay platforms (Brew 2004, 2007). On the middle to upper shore platform, other organisms, such as Mud Shrimp and limpets, may contribute to downwearing. For example, on the cohesive clay shore platform at Warden Point, Kent, SE England, high densities of Mud Shrimp, which were recorded up to depths of 1 cm on the upper platform, are considered to have contributed to higher rates of platform downwearing. Their role, however, was not so visible on the glacial till platform, at Easington, Yorkshire, NE England, where the upper part of the platform is frequently covered by sand bars, preventing colonization (Brew 2004). The middle part of the platform, on chalk for example, often has a high density of limpets where their grazing and abrasion of the rock surface may contribute as much as 35% of platform downwearing (Andrews and Williams 2000). Although, as is noted above, the presence of a dense cover of seaweed may protect the lower platform from weathering and erosion, this may not be the case where the coverage is of lower density. Low-density seaweed cover may contribute to platform downwearing via a whiplash effect whereby fronds sweep over and abrade the platform surface as they are agitated by wave action (Trudgill 1985, 1988). The relative roles of organisms in protecting or eroding shore platforms and their contribution to seasonal, annual and longer-term downwearing remain an area of discussion. Seasonal variations in shore platform downwearing rates are generally higher in the winter than in the summer, usually attributed to frost weathering and wave-driven abrasion rather than any biological processes (Moses and Robinson 2011; Moses 2014).

Elevation of the shore platform, and the height of sediments accumulated on it, relative to mean sea-level (msl) is an important control on cliff toe erosion. If the elevation is above msl then waves will not directly impact the toe of the cliff, even at high tide, although wave splash and spray may do so. If platform elevation is at or below msl then waves may, depending on the depth of beach material, reach the toe of the cliff directly, causing erosion and saturating the toe area (e.g. Hutchinson 1973). Even if waves are prevented from reaching the toe of the cliff by the presence of beach material, saturation may still occur owing to fluctuations in the water level within the beach (Duperret *et al.* 2005). Variations in platform elevation and the level of the cliff–platform junction have been related to geology, tidal range, exposure to wave action and, locally, protection by deep beach or rockfall sediment (Moses 2014).

Until recently, understanding of shore platform dynamics has been informed largely through direct measurement of downwearing rates combined with field survey. Shore platform downwearing rates have typically been measured directly, at discrete points on the platform, usually of no more than 30–50 cm^2 area using an MEM, laser scanner or traversing erosion beam. Aerial coverage is achieved by arranging the sites in a transect across the platform and duplicating measurement sites. Usually, however, the total measurement area remains relatively small because of time constraints for recording measurements within the tidal range. Details of the measurement methods and a discussion of their relative merits is provided in Moses *et al.* (2014). Although they have the advantage of providing directly measured rates, these are usually measured over short timescales of 1–3 years. Also, these instruments measure surface downwearing at a micro-scale of millimetres and so cannot record meso-scale downwearing caused by the removal of individual beds. This has previously been addressed using photogrammetry combined with differential global positioning systems survey. For example, Dornbusch and Robinson (2011) measured step backwearing by block removal over a 28-year period on chalk shore platforms in Sussex, SE England. When averaged across the platform, the downwearing rate of 0.6–5 mm a^{-1} is comparable with rates measured directly using the MEM (Moses 2014). Although quantifying shore platform downearing rates from aerial photographs using photogrammetry facilitates coverage over larger areas and longer, decadal, timescales, it can be difficult to identify suitable fixed points for analysis meaning that the accuracy decreases across the platform from upper (more accurate) to lower (less accurate; Dornbusch *et al.* 2008*a*). Dornbusch and Robinson (2011) address this discrepancy by also including detailed field survey data in the calculation.

The accuracy of measuring longer-term shore platform downwearing over decades has been improved by combining historical aerial photographs with contemporary UAV photomosaics, analysed using structure-from-motion (SfM) photogrammetry. This method enables downwearing rates to be calculated on decadal timescales and across hundreds of metres of platform. For example, Buchanan *et al.* (2020) cover an area of platform 500 m alongshore and 200 m across shore, measuring downwearing over a 78-year period on Blue Lias limestone, Glamorgan, Wales. In addition to calculating an overall average platform downwearing rate, this method also facilitates the calculation of annual and seasonal changes and identifies areas of intense erosion associated with the backwearing of individual beds. The rates, 1.25–5.3 mm a^{-1}, are a similar order of magnitude but higher than those measured using only the MEM or laser scanner at the same site (0.042 ± 0.005–0.196 mm a^{-1}; Swantesson *et al.* 2006) and on the Carboniferous limestone platforms of nearby Gower (0.033 ± 0.009 mm a^{-1}; Shakesby and Walsh 1986). This is most likely because meso- and macro-scale erosion is included in the calculation, which is also averaged over a much longer time period and area of platform when compared with direct measurement methods. Structure-from-motion photogrammetry has also been used at a more local scale to measure the amount of erosion caused by boulder movement impact across platforms during individual storm events (Cullen and Bourke 2018). Swirad *et al.* (2019) use SfM to investigate shore platform downwearing on Jurassic shale and sandstone platform in north Yorkshire, NE England. Measurements were recorded over 15 sites, each 0.5 × 0.5 m, across the platform measured 13 times over a 1-year period from April 2016 to April 2017. The mean erosion rate of 0.528 mm a^{-1} is lower than the average of 3.21 ± 4.76 mm a^{-1} measured by Robinson (1977*a*, *b*) at the same site using the MEM but shows the same cross-platform variations being higher closer to the cliff and also where the duration of tidal inundation is highest. In contrast, on mudstone and limestone platforms of the Kaikoura Peninsula, New Zealand, downwearing rates calculated using SfM are much higher (10.299 mm a^{-1}) than those measured at the same site using the MEM (2.244 mm a^{-1}; Omidiji *et al.* 2023). The authors further refine the SfM method to discern micro- and macro-scale down downwearing rates, both of which are higher than MEM-measured rates. These higher values are attributed to the inclusion of intense granular disintegration, flaking, micro- and polygonal cracking and biological activities which are better captured using SfM than MEM.

Even more extensive areas of platform can be measured from satellite imagery. Mider *et al.* (2020) have measured platform downwearing along a 30 km stretch of coast, for a 3.5-year period, on the Cretaceous Chalk of East Sussex, SE England. They utilize persistent scatterer interferometry to assess synthetic aperture radar (SAR) data acquired by the Sentinel-1 SAR satellite. Using radar images has the advantage, over aerial photography, that radar waves can penetrate most clouds. Interferometric synthetic aperture radar (InSAR) techniques can provide reliable ground surface measurements over relatively long time periods with millimetre precision. Although it has been used by geologists to measure ground movement since the 1990s, this is its first application to the measurement of shore platform downwearing. Interestingly on this softer lithology, the published average of 0.36 mm a^{-1}, across the study area (Mider *et al.* 2020) is approximately an order of magnitude lower than the average from studies at the same field sites where it was measured directly using the MEM (1.5, Foote *et al.* 2006, to 3.5 mm a^{-1}, Ellis 1986). This may be due to the relatively coarse spatial resolution of SAR data which, at 30 m, can capture seawall, beach and platform together and so does not always exclusively measure platform lowering. It may also partly be a function of the methodology. Satellite altimetry provides net ground surface movement over time and so must also consider net vertical ground motion (VGM) when quantifying surface erosion. East Sussex VGM is 1.363 mm a^{-1} (Yuan *et al.* 2022). The corrected figure of 1.7 mm a^{-1} is comparable with MEM-measured rates for the same sites. While it merits further investigation, this new application of InSAR has enabled a comparison of downwearing rates on platforms in front of natural, unprotected, cliffs (2.5 mm a^{-1}) with those protected by hard engineering structures (1.7 mm a^{-1}), and also comparison of downwearing rates on individual units of chalk with different strength properties (weaker, Seaford Chalk, 2.02 mm a^{-1}; stronger, Lewes Nodular Chalk, 1.98 mm a^{-1}; all figures from Mider *et al.* 2020, corrected for VGM).

The new application of InSAR to the study of shore platform erosion dynamics may address a particular knowledge gap, namely information on seasonal–annual timeframes over hundreds of metres to kilometre scale, which is considered potentially critical in understanding the future evolution of rock coasts and their response to climate change (Kennedy *et al.* 2017). Also critical to this is the vexatious issue of understanding how shore platform width and gradient change over time (Stephenson 2001). This is important for modelling longer term cliff retreat rates, to inform modelling approaches dependent on whether platform width increases or remains constant as cliffs retreat, resulting in static and dynamic equilibrium respectively (Trenhaile 2020). A variety of methods have been used to assess changes in platform width over time, including tidal elevation, the distance between mean high and mean low water lines (for example, derived from historical maps, LiDAR profile and differential global positioning systems survey, Dornbusch *et al.* 2008*b*), morphology, sedimentology, biology and wave processes (Kennedy 2015). More accurate delimitation of platform width can be facilitated with increasing availability and the use of seamless terrestrial and marine topographic, LiDAR and bathymetric, datasets in which mapping is not limited by physical access (Kennedy 2015).

Rock coast geomodel resilience, risk and vulnerability

Understanding coastal risk, resilience and vulnerability informs decisions taken by planners and policymakers (Blanco *et al.* 2019; Moore and McInnes 2021). From a management perspective, in comparison with unconsolidated sediment beaches that may recover following an erosion event (e.g. Brooks *et al.* 2017), rock coasts cannot be restored once they are eroded (Sunamura 1992; Blanco *et al.* 2019). Unlike salt marshes, that can respond dynamically to sea-level rise by adjusting their elevation by sedimentation and biogenic growth, rock coasts can only respond by erosion and the redistribution of sediment along- and offshore (Naylor *et al.* 2010). Erosion management efforts on rock coasts therefore focus on mitigating the risk of erosion, for example by maintaining beach volumes (Moses and Williams 2008) or by installing sea defences, reprofiling, draining and slope stabilization measures (Palmer 2002; Windle *et al.* 2015; Moore *et al.* 2016). The resilience of the rock coast geomodel depends on its ability to withstand major disruption, as defined by Aven (2011), caused for example by high rainfall or high wave energy associated with winter storms, and to recover within an acceptable time. Softer rock coasts, with higher cliff retreat rates, may therefore be considered to have lower resilience. Although harder rock coasts are usually considered to be resilient, they have more limited capacity for a dynamic response to climate change pressures (Naylor *et al.* 2010).

Where rock coasts are densely populated, as is the case in many parts of the UK coast, particularly in SE England, understanding their resilience to climate change is important for effective coastal management. In some cases, this has been approached by assessing the future risk of cliff collapse using probabilistic approaches. Such studies are usually site specific and require there to be a good understanding of the scale or magnitude of the cliff erosion hazard (e.g. Mortimore *et al.* 2004*a*), a database of historical frequency statistics (e.g. Lee 2011, 2016) or high resolution cliff erosion datasets (e.g. Rosser *et al.* 2007; Gilham *et al.* 2019; Westoby *et al.* 2020), and where quantitative data are not available, an expert judgement of the likelihood of occurrence (Fookes *et al.* 2007). In addition, such assessments are concerned chiefly with lithological and process controls without considering social and economic parameters. To address this, there is a growing interest in the use of coastal vulnerability indices that incorporate physical, social and economic parameters and can be mapped at local, regional and national scales using GIS. The earliest coastal vulnerability index, which was developed primarily for use in the US, used only physical parameters, including geology, geomorphology, elevation, shoreline change rate, wave and tide regime and sea-level rise (Gornitz and Kanciruk 1989; Gornitz 1990, 1991) before social and economic factors were incorporated (Gornitz *et al.* 1991, 1994; Gornitz and White 1992). Although Pethick and Crooks (2000) use an entirely geomorphology-based approach in their consideration of UK coastal vulnerability, others include geology, geomorphology and the degree to which the coast is exposed to wave action. For example, regional coastal slope, geomorphology, mean tidal range, relative sea-level rise, wave height and shoreline change, aspect and cliff type are used to calculate and map coastal vulnerability for parts of the coastline of Ireland in Counties Wicklow and Dublin (Caloca-Casado 2018; GSI 2023). For the coast of mainland Great Britain, the British Geological Survey (BGS) has produced a database of a range of geological and geomorphological properties that influence coastal vulnerability, including geology, geomorphology, behaviour and coastal context, e.g. shape, profile, height. Analysis of key variables and combinations of data available within the BGS GeoCoast product include lithology, cliff strength, backshore buffer and geomorphology (BGS GeoCoast 2023). For the coast of Northern Ireland, coastal vulnerability has been assessed and mapped by combining data on coastal characteristics, coastal forcing and the principal socio-economic elements for which coastal erosion poses a threat (McLaughlin *et al.* 2002; McLaughlin and Cooper 2010; Table 5). The measurement and mapping of coastal vulnerability provides important information for policymakers of the underlying causes and potential for ameliorating vulnerability (Pethick and Crooks 2000), assisting national level assessments on the basis of which resource allocation is often prioritized. The further application of the index at regional and local scales may facilitate the identification of setback lines or hazard zones to control planning, to identify infrastructure at risk and inform development or management plans (McLaughlin and Cooper 2010). At a European level, EMODnet Geology data products visualize coastal migration data from both field and satellite data and coastal type, including lithology and geomorphology. These products can be used to identify areas of coastal retreat associated with eroding rock coasts, alongside other coastal types. A new product integrates data on coastal resilience and vulnerability from scientific papers and published/unpublished government reports providing the first pan-European assessment of coastal vulnerability (EMODnet Geology 2023; Moses *et al.* 2025). All data products are available on the EMODnet Map Viewer in the catalogue for EMODnet Geology, Coastal Behaviour (https://emodnet.ec.europa.eu/geoviewer/).

Table 5. Coastal characteristics, coastal forcing and socio-economic parameters commonly used in the assessment of coastal vulnerability

Subindex	Variable	1	2	3	4	5
Coastal characteristics	Shoreline type	High cliff >40 m	Medium cliff; 20–<40 m	Low cliff; 10–<20 m	Shingle ridge or bar	Shingle beach or dune
	Rivers	Absent				present
	Solid geology	Plutonic Volcanic (lava) High- to medium-grade metamorphics	Low-grade metamorphics; sandstone and conglomerate well cemented	Most sedimentary rocks	Coarse and/or poorly sorted unconsolidated sediments	Fine unconsolidated sediment, volcanic ash
	Drift geology	Bedrock Urban	Till or boulder clay		Raised beach deposits	Alluvium Blown sand Peat Glacial sands and gravels Glacial outwash sands Recent marine
	Elevation (m)	>30	20–<30	10–<20	5–<10	<5
	Orientation	Not relevant		Easterly		Northerly
	Inland buffer (metres from MHW)	500–1000 m inland				0–<500 m inland
Coastal forcing	Significant wave height (m)	0–<0.74 N 0–<0.24 E	0.74–<1.49 N 0.24–<0.48 E	1.49–<2.23 N 0.48–<0.72 E	2.23–<2.98 N 0.72–<0.96 E	>2.98 N >0.96 E
	Tidal range (m)	>5	3.5–<5	2–<3.5	1–<2	<1
	Difference in modal and storm waves (m)	<0.10 N <0.10 S	0.10–<1.70 N 0.10–<0.25 S	1.70–<3.30 N 0.25–<0.40 S	3.30–<4.90 N 0.40–<0.55 S	>4.9 N >0.55 S
	Frequency of onshore storms (%)	0–<2.8	2.8–<5.6	5.6–<8.4	8.4–<11.2	>11.2
Socioeconomic	Settlement	No settlement	Village	Small town	Large town	City
	Cultural heritage	Absent				Present
	Roads	Absent		A class		Motorway Dual carriageway
	Railways	Absent				present
	Land use	Water bodies Marsh/bog and moor Sparsely vegetated areas Bare rocks	Natural grasslands Coastal areas	Forest	Agriculture	Urban and industrial Infrastructure
	Conservation designation	Absent		International		National

A range of variables are ranked, and each subindex is given equal weighting in the calculation of a coastal vulnerability index. The parameters used for each subindex may vary according to the scale of application of the coastal vulnerability index – local, regional and national (McLaughlin *et al.* 2002; McLaughlin and Cooper 2010). MHW, mean high water.

Rock coast geomodel and climate change

The purpose of this chapter has been to review, from a predominantly UK and northern European perspective, current understanding of rock coast erosion dynamics and response to climate change as viewed through the Fookes geomodel. Recent research developments informing understanding of process–response and impacts of climate change are highlighted. The geomodel is readily applied to categorize and quantify ground conditions in relation to the key components, cliffs and foreshore, providing the foundation of site investigation (Fookes 1997; Fookes *et al.* 2000, 2001, 2007, 2015). With improving data availability, it may also be extended to nearshore and offshore environments for offshore and coastal infrastructure projects (see the Introduction). In addition to lithology the geomodel integrates geomorphology, facilitating description and analyses of dynamic process–response mechanisms and rates in a site investigation (Sellwood *et al.* 2000; Brunsden 2002; Griffiths and Stokes 2008; Griffiths 2014; Moore *et al.* 2016; Hearn 2021). It provides a valuable means of communicating that information and its relevance to engineering priorities, identified as a priority for geoengineering and environmental management (Griffiths and Culshaw 2004; Griffiths and Lee 2022). The use of the geomodel also forms the basis for quantifying risk, resilience and vulnerability and establishing the degree of uncertainty that exists in the interpretation of ground conditions (Griffiths and Stokes 2008). The value of the geomodel is dependent on the level of detail that is built into it in relation to the different parameters associated with geological, geometrical, atmospheric and oceanic controls (Fig. 1; Griffiths and Stokes 2008). This chapter highlights some of recent research advances that not only enhance the level of detail that can be built into geomodels but also raise further questions and identify gaps in our understanding of the rock coast geomodel and its response to climate change.

Predictions of the response of rock coasts to climate change and RSLR are informed by understanding of present and past responses to marine and subaerial controls. Although the climate response is complicated by a range of factors, it is recognized that increased rainfall, water levels and wave energy associated with storms drive higher rates of cliff retreat dependent on the frequency of storm occurrence (sections 'Cliff retreat rates: decadal' and 'Cliff retreat: annual and seasonal'). Although sea-level rise, particularly increasing rates of RSLR, is generally expected to increase rates of cliff retreat (section 'Cliff retreat rates: Holocene'), it is also recognized that this may not be the case and that the future behaviour of cliffs may differ from their historical patterns of change (Lee 2002; Lee and Clark 2002; Shadrick *et al.* 2022). It is possible that rapid RSLR level rise may reduce rates of cliff retreat – deeper water levels and less subaerial exposure of foreshores may reduce shore platform downwearing rates through reduced abrasion and weathering, and if waves remain unbroken then the delivery of wave energy to the cliff toe will be reduced (see sections 'Cliff retreat: annual and seasonal' and 'Foreshore: shore platform downwearing'). Higher sea-levels will, though, assist in raising groundwater levels within the cliff mass and increase the vertical extent of wetting of the cliff face, both of which reduce rock mass strength, rendering the cliff more susceptible to mass movements (see section 'Cliff retreat: annual and seasonal'). Numerical, geometric and process-based models that are used to simulate rock coast behaviour under future climate scenarios are underpinned by the understanding that cliffs and shore platforms are linked subsystems (see section 'Cliff retreat rates: Holocene'). Although understanding of the modes, mechanisms and rates of cliff retreat and shore platform downwearing is continually improving, establishing a clear quantitative link between the two remains elusive: firstly, models often focus on retreat at the toe of the cliff, assuming that the cliff top retreat rate is a reflection of this; secondly, studies of historical cliff retreat focus on the cliff top rather than the toe and there are very few measurements of cliff toe retreat, so the relationship between the two remains to be tested; thirdly cliff retreat and shore platform downwearing are measured on different temporal and spatial scales, making it difficult to relate the two; fourthly, measurements of shore platform downwearing remain incomplete, relying predominantly on short-term, micro-scale measurements. The use of InSAR, as discussed in the section 'Foreshore: shore platform downwearing', presents an important opportunity to address this knowledge gap. It has been used to date to separately measure shore platform downwearing (Mider *et al.* 2020) and coastal landslide ground movement (O'Connor *et al.* 2021). Combining both cliffs and shore platforms over tens of kilometres and several decades to quantify concurrent rates of retreat and downwearing respectively would provide new insights into quantitative links between the two to inform simulation model development and validation. Meanwhile the increasing volume of high-resolution data from the application of a range of techniques, including S-f-M photogrammetry, LiDAR, UAV and seismic sensors, alongside InSAR and other more traditional data collection methods, continues to improve the level of detail that can be captured in the geomodel. Combined with robust analytical methods and the expert judgement espoused by Fookes and others, rock coast geomodel ground conditions, risk, resilience and vulnerability and potential future responses climate change are increasingly better captured to inform engineering projects, policy and planning.

Acknowledgements The author would like to thank Colin Serridge for insightful comments on an earlier draft and Arran Haigh for cartography. Jim Griffiths and Andrew Hart are thanked for their very helpful reviews.

Competing interests The authors declare that they have no known competing financial interests or personal relationships that could have appeared to influence the work reported in this paper.

Author contributions **CM**: conceptualization (lead), writing – original draft (lead), writing – review & editing (lead).

Funding This research received no specific grant from any funding agency in the public, commercial, or not-for-profit sectors.

Data availability All data generated or analysed during this study are included in this published article (and, if present, its supplementary information files).

References

Acharya-Chowdhury, L., Dickson, M., Norton, K., Rowland, J., Hall, B. and Stephenson, W. 2024. Reconciling short-and long-term measurements of coastal cliff erosion rates. *Engineering Geology*, **341**, 107703, https://doi.org/10.1016/j.enggeo.2024.107703

Andrews, C. and Williams, R.B.G. 2000. Limpet erosion of chalk shore platforms in Southeast England. *Earth Surface Processes and Landforms*, **25**, 1371–1381, https://doi.org/10.1002/1096-9837(200011)25:12<1371::AID-ESP144>3.0.CO;2-#

Asch, K., Kitazato, H. and Vallius, H. (eds) 2022. From continental shelf to slope: mapping the oceanic realm. *Geological Society, London, Special Publications*, **505**, https://doi.org/10.1144/SP505-2022-96

Austad, T., Strand, S., Madland, M.V., Puntervold, T. and Korsnes, R.I. 2008. Seawater in chalk: an EOR and compaction fluid. *SPE Reservoir Evaluation and Engineering*, **11**, 648–654, https://doi.org/10.2118/118431-pa

Aven, T. 2011. On some recent definitions and analysis frameworks for risk, vulnerability, and resilience. *Risk Analysis: An International Journal*, **31**, 515–522, https://doi.org/10.1111/j.1539-6924.2010.01528.x

Baxter, T., Coombes, M. and Viles, H. 2022. No evidence that seaweed cover enhances the deterioration of natural cement-based mortar in intertidal environments. *Earth Surface Processes and Landforms*, **7**, 3453–3464, https://doi.org/10.1002/esp.5467

Bessin, Z., Jaud, M., Letortu, P., Vassilakis, E., Evelpidou, N., Costa, S. and Delacourt, C. 2023. Smartphone structure-from-motion photogrammetry from a boat for coastal cliff face monitoring compared with Pléiades tri-stereoscopic imagery and unmanned aerial system imagery. *Remote Sensing*, **15**, 3824, https://doi.org/10.3390/rs15153824

BGS GeoCoast 2023. British Geological Survey GeoCoast data product. https://www.bgs.ac.uk/geology-projects/hazard-and-resilience-modelling/coastal-vulnerability-index-cvi/

Bird, E. 2016. *Coastal Cliffs: Morphology and Management.* Springer, https://doi.org/10.1007/978-3-319-29084-3

Blanco, B.L.D.S.R., Ballard, B.W., Simm, J., Brampton, A., Sutherland, J., Gouldby, B. and Rossington, K. 2019. *Coastal morphological modelling for decision-makers.* Report, **SC090036/R**. Environment Agency, Bristol, https://www.gov.uk/flood-and-coastal-erosion-risk-management-research-reports/coastal-morphological-modelling-for-decision-makers

Brain, M.J., Rosser, N.J., Norman, E.C. and Petley, D.N. 2014. Are microseismic ground displacements a significant geomorphic agent? *Geomorphology*, **207**, 161–173, https://doi.org/10.1016/j.geomorph.2013.11.002

Bray, M.J. and Hooke, J.M. 1997. Prediction of soft-cliff retreat with accelerating sea-level rise. *Journal of Coastal Research*, **13**, 453–467.

Brew, D.S. 2004. *Understanding and predicting beach morphological change processes associated with the erosion of cohesive foreshores: scoping report.* DEFRA Flood Management Division Technical Report, **FD1915/TR**.

Brew, D.S. 2007. *Understanding and predicting beach morphological change processes associated with the erosion of cohesive foreshores.* DEFRA Flood Management Division Technical Report, **FD1926/TR**.

Bromhead, E.N. 1979. Factors affecting the transition between the various types of mass movement in coastal cliffs consisting largely of overconsolidated clay, with special reference to Southern England. *Quarterly Journal of Engineering Geology*, **12**, 291–300, https://doi.org/10.1144/gsl.qjeg.1979.012.04.06

Brooks, S.M., Spencer, T. and Boreham, S. 2012. Deriving mechanisms and thresholds for cliff retreat in soft-rock cliffs under changing climates: rapidly retreating cliffs of the Suffolk coast, UK. *Geomorphology*, **153**, 48–60, https://doi.org/10.1016/j.geomorph.2012.02.007

Brooks, S.M., Spencer, T. and Christie, E.K. 2017. Storm impacts and shoreline recovery: mechanisms and controls in the southern North Sea. *Geomorphology*, **283**, 48–60, https://doi.org/10.1016/j.geomorph.2017.01.007

Brunsden, D. 2002. Geomorphological roulette for engineers and planners: some insights into an old game. *Quarterly Journal of Engineering Geology and Hydrogeology*, **35**, 101–142, https://doi.org/10.1144/1470-92362001-40

Buchanan, D.H., Naylor, L.A., Hurst, M.D. and Stephenson, W.J. 2020. Erosion of rocky shore platforms by block detachment from layered stratigraphy. *Earth Surface Processes and Landforms*, **45**, 1028–1037, https://doi.org/10.1002/esp.4797

Burningham, H. and French, J. 2017. Understanding coastal change using shoreline trend analysis supported by cluster-based segmentation. *Geomorphology*, **282**, 131–149, https://doi.org/10.1016/j.geomorph.2016.12.029

Caloca-Casado, S. 2018. *Coastal vulnerability assessment of Co. Dublin and Co. Wicklow to impacts of sea-level rise.* Doctoral dissertation, National University of Ireland, Maynooth.

Coombes, M.A., Naylor, L.A., Viles, H.A. and Thompson, R.C. 2013. Bioprotection and disturbance: seaweed, microclimatic stability and conditions for mechanical weathering in the intertidal zone. *Geomorphology*, **202**, 4–14, https://doi.org/10.1016/j.geomorph.2012.09.014

Costa, S., Lageat, Y., Henaff, A., Delahaye, D. and Plessis, E. 2003. Origine de la variabilité spatiale du recul des falaises crayeuses du nord-ouest du Bassin de Paris. L'exemple du littoral haut-normand (France). *Hommes et Terres du Nord Numero Spécial*, **1**, 22–31, https://doi.org/10.3406/htn.2003.2832

Costa, S., Delahaye, D., Frieré-Diaz di Nocera, L., Davidson, R. and Plessis, E. 2004. Quantification of the Normandy and Picardy chalk cliff retreat by photogrammetric analysis. *Geological Society, London, Special Publications*, **20**, 139–148, https://doi.org/10.1144/gsl.eng.2004.020.01.11

Costa, S., Hénaff, A. and Lageat, Y. 2006. The gravel beaches of north-west France and their contribution to the dynamic of the coastal cliff-shore platform system. *Zeitschrift für Geomorphologie*, **144**, 199–214.

Cullen, N.D. and Bourke, M.C. 2018. Clast abrasion of a rock shore platform on the Atlantic coast of Ireland. *Earth Surface Processes and Landforms*, **43**, 2627–2641, https://doi.org/10.1002/esp.4421

Dewez, T.J.B., Regard, V., Duperret, A. and Lasseur, E. 2015. Shore platform lowering due to frost shattering during the 2009 winter at Mesnil Val, English channel coast, NW France. *Earth Surface Processes and Landforms*, **40**, 1688–1700, https://doi.org/10.1002/esp.3760

Dickson, M.E., Ogawa, H., Kench, P.S. and Hutchinson, A. 2013. Sea-cliff retreat and shore platform widening: steady-state equilibrium? *Earth Surface Processes and Landforms*, **38**, 1046–1048, https://doi.org/10.1002/esp.3422

Dickson, M.E., Matsumoto, H., Stephenson, W.J., Swirad, Z.M., Thompson, C.F. and Young, A.P. 2023. Sea-level rise may not uniformly accelerate cliff erosion rates. *Nature Communications*, **14**, 8485, https://doi.org/10.1038/s41467-022-34386-3

Dietze, M., Cook, K.L., Illien, L., Rach, O., Puffpaff, S., Stodian, I. and Hovius, N. 2020. Impact of nested moisture cycles on coastal chalk cliff failure revealed by multiseasonal seismic and topographic surveys. *Journal of Geophysical Research: Earth Surface*, **125**, e2019JF005487, https://doi.org/10.1029/2019jf005487

Dixon, N. and Bromhead, E.N. 2002. Landsliding in London Clay coastal cliffs. *Quarterly Journal of Engineering Geology and Hydrogeology*, **35**, 327–343, https://doi.org/10.1144/1470-9236/2000-53

Dornbusch, U. 2010. The role of water and beach levels in seawall abrasion in the macro-tidal high energy environment of Southeast England. *In*: *Coasts, Marine Structures and Breakwaters: Adapting to Change: Proceedings of the 9th International Conference Organised by the Institution of Civil Engineers and held in Edinburgh on 16 to 18 September 2009*. Thomas Telford, 708–719, https://doi.org/10.1680/cmsb.41301.0062

Dornbusch, U. 2022. Holocene and historic rates of rock coast erosion – a discussion focussed on Southeast England. *Marine Geology*, **449**, 106817, https://doi.org/10.1016/j.margeo.2022.106817

Dornbusch, U. and Robinson, D.A. 2011. Block removal and step backwearing as erosion processes on rock shore platforms: a preliminary case study of the chalk shore platforms of south-east England. *Earth Surface Processes and Landforms*, **36**, 661–671, https://doi.org/10.1002/esp.2086

Dornbusch, U., Robinson, D.A., Moses, C.A. and Williams, R.B.G. 2006. Chalk coast erosion and its contribution to the shingle budget in East Sussex. *Zeitschrift für Geomorphologie*, **144**, 215–230.

Dornbusch, U., Robinson, D.A., Williams, R.B. and Moses, C.A., 2007. Chalk shore platform erosion in the vicinity of sea defence structures and the impact of construction methods. *Coastal Engineering*, **54**, 801–810, https://doi.org/10.1016/j.coastaleng.2007.05.012

Dornbusch, U., Moses, C., Robinson, D.A. and Williams, R. 2008*a*. Soft copy photogrammetry to measure shore platform erosion on decadal timescales. *Journal of Coastal Conservation*, **11**, 193–200, https://doi.org/10.1007/s11852-008-0025-8

Dornbusch, U., Robinson, D.A., Moses, C.A. and Williams, R.B. 2008*b*. Temporal and spatial variations of chalk cliff retreat in East Sussex, 1873 to 2001. *Marine Geology*, **249**, 271–282, https://doi.org/10.1016/j.margeo.2007.12.005

Dornbusch, U., Williams, R.B., Moses, C.A. and Robinson, D.A. 2008*c*. Foreshore narrowing along the coast of southeast England, UK – a reevaluation. *Journal of Coastal Research*, **24**, 14–24, https://doi.org/10.2112/06-0751.1

Duguet, T., Duperret, A., Costa, S., Regard, V. and Maillet, G. 2021. Coastal chalk cliff retreat rates during the Holocene, inferred from submarine platform morphology and cosmogenic exposure along the Normandy coast (NW France). *Marine Geology*, **433**, 106405, https://doi.org/10.1016/j.margeo.2020.106405

Duperret, A., Genter, A., Martinez, A. and Mortimore, R.N. 2004. Coastal chalk cliff instability in NW France: role of lithology, fracture pattern and rainfall. *Geological Society, London, Special Publications*, **20**, 33–55, https://doi.org/10.1144/gsl.eng.2004.020.01.03

Duperret, A., Taibi, S., Mortimore, R.N. and Daigneault, M. 2005. Effect of groundwater and sea weathering cycles on the strength of chalk rock from unstable coastal cliffs of NW France. *Engineering Geology*, **78**, 321–343, https://doi.org/10.1016/j.enggeo.2005.01.004

Earlie, C.S., Young, A.P., Masselink, G. and Russell, P.E. 2015. Coastal cliff ground motions and response to extreme storm waves. *Geophysical Research Letters*, **42**, 847–854, https://doi.org/10.1002/2014gl06254

Earlie, C., Masselink, G. and Russell, P. 2018. The role of beach morphology on coastal cliff erosion under extreme waves. *Earth Surface Processes and Landforms*, **43**, 1213–1228, https://doi.org/10.1002/esp.4308

Ellis, N. 1986. *Morphology, process and rates of denudation on the chalk shore platforms of East Sussex*. PhD thesis, Brighton Polytechnic (now University of Brighton).

Emery, K.O. and Kuhn, G.G. 1982. Sea cliffs: their processes, profiles, and classification. *Geological Society of America Bulletin*, **93**, 644–654, https://doi.org/10.1130/0016-7606(1982)93<644:sctppa>2.0.co;2

EMODnet Geology 2023. European Marine Observation Data Network, Geology Portal, https://emodnet.ec.europa.eu/en/geology

Environment Agency 2020. Assessment of coastal erosion and landsliding for the funding of coastal risk management projects, https://www.gov.uk/government/publications/assess-coastal-erosion-and-landsliding-for-fcerm-funding

Fookes, P.G. 1991. Geomaterials. *Quarterly Journal of Engineering Geology and Hydrogeology*, **24**, 3–15, https://doi.org/10.1144/gsl.qjeg.1991.024.01.02

Fookes, P.G. 1997. Geology for engineers: the geological model, prediction and performance. *Quarterly Journal of Engineering Geology and Hydrogeology*, **30**, 293–424, https://doi.org/10.1144/gsl.qjeg.1997.030.p4.02

Fookes, P.G. and Poole, A.B. 1981. Some preliminary considerations on the selection and durability of rock and concrete materials for breakwaters and coastal protection works. *Quarterly Journal of Engineering Geology and Hydrogeology*, **14**, 97–128, https://doi.org/10.1144/gsl.qjeg.1981.014.02.03

Fookes, P.G. and Sweeney, M. 1976. Stabilization and control of local rock falls and degrading rock slopes. *Quarterly Journal of Engineering Geology and Hydrogeology*, **9**, 37–55, https://doi.org/10.1144/gsl.qjeg.1976.009.01.03

Fookes, P.G. and Vaughan, P.R. 1986. *A Handbook of Engineering Geomorphology*. Chapman and Hall.

Fookes, P.G., Baynes, F. and Hutchinson, J.N. 2000. Total geological history: a model approach to the anticipation, observation, and understanding of site conditions. *In*: *Proceedings of GeoEng2000, an International Conference on Geotechnical & Geological Engineering*, 19–24 November, Melbourne. Technomic, Warsaw, 370–460.

Fookes, P.G., Baynes, F. and Hutchinson, J.N. 2001. Total geological history: a model approach to the anticipation, observation, and understanding of site conditions. *Ground Engineering*, **34**, 42–47.

Fookes, P.G., Lee, E.M. and Griffiths, J.S. 2007. *Engineering Geomorphology: Theory and Practice*. Whittles.

Fookes, P., Pettifer, G. and Waltham, T. 2015. *Geomodels in Engineering Geology-An Introduction*. Whittles.

Foote, Y., Plessis, E., Robinson, D.A., Hénaff, A. and Costa, S. 2006. Rates and patterns of downwearing of chalk shore platforms of the Channel: comparisons between France and England. *Zeitschrift für Geomorphologie*, **144**, 93–115.

Gilham, J., Barlow, J. and Moore, R. 2019. Detection and analysis of mass wasting events in chalk sea cliffs using UAV photogrammetry. *Engineering Geology*, **250**, 101–112, https://doi.org/10.1016/j.enggeo.2019.01.013

Gornitz, V. 1990. Vulnerability of the East Coast, U.S.A. to future sea level rise. *Journal of Coastal Research*, **9**, 201–237.

Gornitz, V. 1991. Global coastal hazards from future sea level rise. *Palaeogeography, Palaeoclimatology, Palaeoecology*, **89**, 379–398, https://doi.org/10.1016/0031-0182(91)90173-o

Gornitz, V. and Kanciruk, P. 1989. *Assessment of Global Coastal Hazards from Sea Level Rise (No. CONF-8907104-1)*. Oak Ridge National Laboratory, Oak Ridge, TN.

Gornitz, V. and White, T.W. 1992. *A Coastal Hazards Database for the U.S. West Coast: ORNL/CDIAC-81, NDP-043C*. Oak Ridge National Laboratory, Oak Ridge, TN.

Gornitz, V., White, T.W. and Cushman, R.M. 1991. *Vulnerability of the US to Future Sea Level Rise (No. CONF-910780-1)*. Oak Ridge National Laboratory, Oak Ridge, TN.

Gornitz, V.M., Daniels, R.C., White, T.W. and Birdwell, K.R. 1994. The development of a coastal risk assessment database: vulnerability to sea-level rise in the US Southeast. *Journal of Coastal Research*, **12**, 327–338.

Gowell, M.R., Coombes, M.A. and Viles, H.A. 2015. Rock-protecting seaweed? Experimental evidence of bioprotection in the intertidal zone. *Earth Surface Processes and Landforms*, **40**, 1364–1370, https://doi.org/10.1002/esp.3736

Griffiths, J.S. 2014. Feet on the ground: engineering geology past, present and future. *Quarterly Journal of Engineering Geology and Hydrogeology*, **47**, 116–143, https://doi.org/10.1144/qjegh2013-087

Griffiths, J.S. and Culshaw, M.G. 2004. Seeking the research frontiers for UK engineering geology. *Quarterly Journal of Engineering Geology and Hydrogeology*, **37**, 317–325, https://doi.org/10.1144/1470-9236/04-056

Griffiths, J.S. and Lee, E.M. 2022. Geomorphology in environmental management 1965–2000. *Geological Society, London, Memoirs*, **58**, 135–149, https://doi.org/10.1144/m58-2020-22

Griffiths, J.S. and Stokes, M. 2008. Engineering geomorphological input to ground models: an approach based on Earth systems. *Quarterly Journal of Engineering Geology and Hydrogeology*, **41**, 73–91, https://doi.org/10.1144/1470-9236/07-010

Ground Engineering 2022*a*. Heatwave could increase rockfall risk, BGS warns. https://www.geplus.co.uk/news/heat wave-could-increase-rockfall-risk-bgs-warns-18-07-2022/

Ground Engineering 2022*b*. Devon cliff collapses for second time during heatwave. https://www.geplus.co.uk/news/devon-cliff-collapses-for-second-time-during-heatwave-11-08-2022/

GSI 2023. Geological Survey of Ireland, Coastal Vulnerability Index (CVI) mapping to evaluate impacts of sea-level rise, https://www.gsi.ie/en-ie/programmes-and-projects/marine-and-coastal-unit/projects/Pages/Coastal-Vulnerability-Index.aspx

Guinan, J., McKeon, C., O'Keeffe, E., Monteys, X., Sacchetti, F., Coughlan, M. and Nic Aonghusa, C. 2020. INFOMAR data supports offshore energy development and marine spatial planning in the Irish offshore via the EMODnet Geology portal. *Quarterly Journal of Engineering Geology and Hydrogeology*, **54**, qjegh2020-033, https://doi.org/10.1144/qjegh2020-03

Halcrow Group Limited 2009. Cayton Bay Cliff Stability Assessment Ground Investigation and Appraisal of Engineering Stabilisation Options, https://www.northeastcoastalobservatory.org.uk/data/smp_&_strategy_reports/

Hearn, G.J. 2021. Getting the basic ground model right in engineering practice. *Quarterly Journal of Engineering Geology and Hydrogeology*, **54**, qjegh2020-055, https://doi.org/10.1144/qjegh2020-055

Heggheim, T., Madland, M.V., Risnes, R. and Austad, T. 2005. A chemical induced enhanced weakening of chalk by seawater. *Journal of Petroleum Science and Engineering*, **46**, 171–184, https://doi.org/10.1016/j.petrol.2004.12.001

Hénaff, A., Lageat, Y., Costa, S. and Plessis, E. 2002. Le recul des falaises crayeuses du Pays de Caux: determination des processus d'érosion et quantification des rythmes d'évolution. *Géomorphologie*, **2**, 107–118, https://doi.org/10.3406/morfo.2002.1132

Hénaff, A., Lageat, Y. and Costa, S. 2006. Geomorphology and shaping processes of chalk shore platforms of the Channel coasts. *Zeitschrift für Geomorphologie*, **144**, 61–91.

Hoek, E. and Bray, J.D. 1981. *Rock Slope Engineering*. CRC Press, https://doi.org/10.1201/9781482267099

Hosking, A. and McInnes, R. 2002. Preparing for the impacts of climate change on the central south coast of England: a framework for future risk management. *Journal of Coastal Research*, **36**, 381–389, https://doi.org/10.2112/1551-5036-36.sp1.381

Hurst, M.D., Rood, D.H., Ellis, M.A., Anderson, R.S. and Dornbusch, U. 2016. Recent acceleration in coastal cliff retreat rates on the south coast of Great Britain. *Proceedings of the National Academy of Sciences*, **113**, 13336–13341, https://doi.org/10.1073/pnas.1613044113

Hurst, M.D., Rood, D.H. and Ellis, M.A. 2017. Controls on the distribution of cosmogenic ^{10}Be across shore platforms. *Earth Surface Dynamics*, **5**, 67–84, https://doi.org/10.5194/esurf-5-67-2017

Hutchinson, J.N. 1965. *Survey of the Coastal Landslides of Kent*. Building Research Station, Note 1965 EN35/65.

Hutchinson, J.N. 1968. Field meeting on the coastal landslides of Kent. *Proceedings of the Geologists' Association*, **79**, 227–237, https://doi.org/10.1016/s0016-7878(68)80039-2

Hutchinson, J.N. 1971. Field and laboratory studies of a fall in Upper Chalk cliffs at Joss Bay, Isle of Thanet. *Proceedings of the Roscoe Memorial Symposium*, **1971**, 692–706.

Hutchinson, J.N. 1973. The response of London Clay cliffs to differing rates of toe erosion. *Geologia Applicata Idrogeologia*, **8**, 221–239.

Jones, E.V., Rosser, N.J. and Brain, M.J. 2018. Alongshore variability in wave energy transfer to coastal cliffs. *Geomorphology*, **322**, 1–14, https://doi.org/10.1016/j.geomorph.2018.08.019

Kendon, M., McCarthy, M., Jevrejeva, S., Matthews, A., Sparks, T., Garforth, J. and Kennedy, J. 2022. State of the UK Climate 2021. *International Journal of Climatology*, **42**, 1–80, https://doi.org/10.1002/joc.7787

Kennedy, D.M. 2015. Where is the seaward edge? A review and definition of shore platform morphology. *Earth-Science Reviews*, **147**, 99–108, https://doi.org/10.1016/j.earscirev.2015.05.007

Kennedy, D.M., Paulik, R. and Dickson, M.E. 2011. Subaerial weathering v. wave processes in shore platform development: reappraising the Old Hat Island evidence. *Earth Surface Processes and Landforms*, **36**, 686–694, https://doi.org/10.1002/esp.2092

Kennedy, D.M., Stephenson, W.J. and Naylor, L.A. (eds) 2014. *Rock Coast Geomorphology: a Global Synthesis*. Geological Society, London, Memoirs, **40**, https://doi.org/10.1144/M40

Kennedy, D.M., Coombes, M.A. and Mottershead, D.N. 2017. The temporal and spatial scales of rocky coast geomorphology: a commentary. *Earth Surface Processes and Landforms*, **42**, 1597–1600, https://doi.org/10.1002/esp.4150

Kirkby, M.J. 2023. Cliffs and ramparts: persistent steep slopes in the landscape. *Geomorphology*, **463**, 108776, https://doi.org/10.1016/j.geomorph.2023.108776

Lageat, Y., Hénaff, A. and Costa, S. 2006. The retreat of the chalk cliffs of the Pays de Caux (France): erosion processes and patterns. *Zeitschrift für Geomorphologie*, **144**, 183–197.

Lawrence, J.A., Mortimore, R.N., Stone, K.J. and Busby, J.P. 2013. Sea saltwater weakening of chalk and the impact on cliff instability. *Geomorphology*, **191**, 14–22, https://doi.org/10.1016/j.geomorph.2013.02.022

Lawrence, J.A., Spence, R., Mortimore, R.N., Eade, M. and Bottrell, S.H. 2018. Coastal cliff rock mass weakening of Chalk and the impact of salt water. *Proceedings of the Institution of Civil Engineers-Geotechnical Engineering*, **171**, 545–555, https://doi.org/10.1680/jgeen.18.00040

Lee, E.M. 2002. *Soft Cliffs: Prediction of Recession Rates and Erosion Control Techniques*. R&D Project FD2403/1302. DEFRA/Environment Agency, London.

Lee, E.M. 2011. Reflections on the decadal-scale response of coastal cliffs to sea-level rise. *Quarterly Journal of Engineering Geology and Hydrogeology*, **44**, 481–489, https://doi.org/10.1144/1470-9236/10-063

Lee, E.M. 2016. Landslide risk assessment: the challenge of communicating uncertainty to decision-makers. *Quarterly Journal of Engineering Geology and Hydrogeology*, **49**, 21–35, https://doi.org/10.1144/qjegh2015-066

Lee, E.M. and Brunsden, D. 2001. Sediment budget analysis for coastal management, west Dorset. *Geological Society, London, Special Publications*, **18**, 181–187, https://doi.org/10.1144/gsl.eng.2001.018.01.25.

Lee, E.M. and Clark, A.R. 2002. *Investigation and Management of Soft Rock Cliffs*. Thomas Telford.

Letortu, P., Costa, S., Bensaid, A., Cador, J.-M. and Quénol, H. 2014*a*. Vitesses et modalités de recul des falaises crayeuses de Haute-Normandie (France): méthodologie et variabilité du recul. *Géomorphologie: Relief, Processus, Environnement*, **20**, 133–144, https://doi.org/10.4000/geomorphologie.10588

Letortu, P., Costa, S. and Bonnet, E. 2014*b*. Spatial analysis of coastal chalk cliff falls in Upper Normandy (France) from Veules-les-Roses to Le Treport (2002–2009). *Revue Internationale de Géomatique*, **24**, 335–354, https://doi.org/10.3166/rig.24.335-354

Letortu, P., Costa, S., Cador, J.-M., Coinaud, C. and Cantat, O. 2015*a*. Statistical and empirical analyses of the triggers of coastal chalk cliff failure. *Earth Surface Processes and Landforms*, **40**, 1371–1386, https://doi.org/10.1002/esp.3741

Letortu, P., Costa, S. *et al*. 2015*b*. Retreat rates, modalities and agents responsible for erosion along the coastal chalk cliffs of Upper Normandy: the contribution of terrestrial laser scanning. *Geomorphology*, **245**, 3–14, https://doi.org/10.1016/j.geomorph.2015.05.007

Letortu, P., Costa, S., Maquaire, O. and Davidson, R. 2019. Marine and subaerial controls of coastal chalk cliff erosion in Normandy (France) based on a 7-year laser scanner monitoring. *Geomorphology*, **335**, 76–91, https://doi.org/10.1016/j.geomorph.2019.03.005

Letortu, P., Le Dantec, N. *et al*. 2022. Experimental field study on the fatigue and failure mechanisms of coastal chalk cliffs: implementation of a multi-parameter monitoring system (Sainte-Marguerite-sur-Mer, France). *Geomorphology*, **408**, 108211, https://doi.org/10.1016/j.geomorph.2022.108211

Lim, M. 2014. The rock coast of the British Isles: cliffs. *Geological Society, London, Memoirs*, **40**, 19–38, https://doi.org/10.1144/m40.3

Lim, M., Rosser, N.J., Petley, D.N. and Keen, M. 2011. Quantifying the controls and influence of tide and wave impacts on coastal rock cliff erosion. *Journal of Coastal Research*, **27**, 46–56, https://doi.org/10.2112/jcoastres-d-09-00061.1

Luetzenburg, G. 2022. Investigating coastal change with smartphone LiDAR. *Nature Reviews Earth & Environment*, **3**, 104–104, https://doi.org/10.1038/s43017-022-00265-0

Matsumoto, H., Dickson, M.E. and Kench, P.S. 2016. An exploratory numerical model of rocky shore profile evolution. *Geomorphology*, **268**, 98–109, https://doi.org/10.1016/j.geomorph.2016.05.017

Matsumoto, H., Dickson, M.E. and Kench, P.S. 2018. Modelling the relative dominance of wave erosion and weathering processes in shore platform development in micro- to mega-tidal settings. *Earth Surface Processes and Landforms*, **43**, 2642–2653, https://doi.org/10.1002/esp.4422

Matsumoto, H., Dickson, M.E., Stephenson, W.J., Thompson, C.F. and Young, A.P. 2024. Modelling future cliff-front waves during sea level rise and implications for coastal cliff retreat rates. *Scientific Reports*, **14**, 7810, https://doi.org/10.1038/s41598-024-57923-0

May, V.J. and Hansom, J.D. 2003. *Coastal Geomorphology of Great Britain*. Geological Conservation Review Series No. 28. JNCC, Peterborough.

Maynard, D. and Bromhead, E.N. 2015. Landslide activity in London Clay at Warden Point, Isle of Sheppey, SE England. *Quarterly Journal of Engineering Geology and Hydrogeology*, **48**, 159–162, https://doi.org/10.1144/qjegh2015-048

McInnes, R.G. and Moore, R. 2014. *Living with Ground Instability and Landslides: an International Good Practice Guide*. CH2M HILL.

McInnes, R.G., Cope, S.N., Moore, R., Bradbury, A.P. and Millerchip, C.J. 2011. Adapting to Coastal Change along England's Southern Shorelines (ACCESS). Report for SCOPAC, https://southerncoastalgroup-scopac.org.uk/wp-content/uploads/2018/11/ACCESS_web_copy.pdf

Mclaughlin, S. and Cooper, J.A.G. 2010. A multi-scale coastal vulnerability index: a tool for coastal managers? *Environmental Hazards*, **9**, 233–248, https://doi.org/10.3763/ehaz.2010.0052

McLaughlin, S., McKenna, J. and Cooper, J.A.G. 2002. Socio-economic data in coastal vulnerability indices: constraints and opportunities. *Journal of Coastal Research*, **36**, 487–497, https://doi.org/10.2112/1551-5036-36.sp1.487

Mellett, C., Long, D., Carter, G., Chiverell, R. and Van Landeghem, K. 2015. *Geology of the seabed and shallow subsurface: the Irish Sea*. British Geological Survey Commissioned Report, **CR/15/057**.

Mider, G., Lawrence, J., Mason, P. and Ghail, R. 2020. Monitoring littoral platform downwearing using differential SAR interferometry. *Remote Sensing*, **12**, 3243, https://doi.org/10.3390/rs12193243

Minde, M.W. and Hiorth, A. 2019. Compaction and fluid–rock interaction in chalk insight from modelling and data at pore-, core-, and field-scale. *Geosciences*, **10**, 6, https://doi.org/10.3390/geosciences10010006

Moore, R. and Davis, G. 2015. Cliff instability and erosion management in England and Wales. *Journal of Coastal Conservation*, **19**, 771–784, https://doi.org/10.1007/s11852-014-0359-3

Moore, R. and McInnes, R.G. 2011. *Cliff instability and coastal erosion management in Great Britain: Good practice guide*. Halcrow Group Ltd, https://www.researchgate.net/publication/323150427_Cliff_Instability_and_Erosion_Management_in_Great_Britain_A_Good_Practice_Guide

Moore, R. and McInnes, R. 2021. *Coastal erosion and climate change: guidance for policymakers, planners and stakeholders.* University of Sussex. Report, https://hdl.handle.net/10779/uos.23491373.v1

Moore, R., Barter, P., McConnell, K., Riby, J. and Matthews, C. 2002. Cliff behaviour assessment and management strategy, Filey Bay, Yorkshire, UK. *In*: McInnes, R.G. and Jakeways, J. (eds) *Instability: Planning and Management.* Thomas Telford, 641–648.

Moore, R., Rogers, J., Woodget, A. and Baptiste, A. 2010. Climate change impact on cliff instability and erosion. *Flood and Coastal Risk Management, FCRM>10, Defra and Environment Agency Annual Conference*, Telford, UK, paper 079.

Moore, R., Davis, G., Stannard, M. and Browning, N. 2016. Stabilising Lyme Regis – a strategic approach. *Proceedings of the Institution of Civil Engineers – Civil Engineering*, **170**, 63–70, https://doi.org/10.1680/jcien.16.00008

Mortimore, R.N. and Duperret, A. (eds) 2004, Coastal chalk cliff instability. *Geological Society, London, Special Publications*, **20**, https://doi.org/10.1144/gsl.eng.2004.020.01.01

Mortimore, R.N., Lawrence, J., Pope, D., Duperret, A. and Genter, A. 2004*a*. Coastal cliff geohazards in weak rock: the UK Chalk cliffs of Sussex. *Geological Society, London, Special Publications*, **20**, 3–31, https://doi.org/10.1144/gsl.eng.2004.020.01.02

Mortimore, R.N., Stone, K.J., Lawrence, J. and Duperret, A. 2004*b*. Chalk physical properties and cliff instability. *Geological Society, London, Special Publications*, **20**, 75–88, https://doi.org/10.1144/gsl.eng.2004.020.01.05

Moses, C.A. 2014. The rock coast of the British Isles: shore platforms. *Geological Society, London, Memoirs*, **40**, 39–56, https://doi.org/10.1144/m40.4

Moses, C. and Robinson, D. 2011. Chalk coast dynamics: implications for understanding rock coast evolution. *Earth-Science Reviews*, **109**, 63–73, https://doi.org/10.1016/j.earscirev.2011.08.003

Moses, C. and Robinson, D.A. 2021. Intertidal cohesive foreshores: erosion rates and processes illustrated by a shore platform at Warden Point, Kent, UK. *Marine Geology*, **442**, 106658, https://doi.org/10.1016/j.margeo.2021.106658

Moses, C.A. and Vallius, H. 2021. Mapping the geology and topography of the European seas (European Marine Observation and Data Network, EMODnet). *Quarterly Journal of Engineering Geology and Hydrogeology*, **54**, qjegh2020-131, https://doi.org/10.1144/qjegh2020-131

Moses, C.A. and Williams, R.B. 2008. Artificial beach recharge: the South East England experience. *Zeitschrift für Geomorphologie, Supplementary Issues*, **52**, 107–124, https://doi.org/10.1127/0372-8854/2008/0052s3-0107

Moses, C., Robinson, D. and Barlow, J. 2014. Methods for measuring rock surface weathering and erosion: a critical review. *Earth-Science Reviews*, **135**, 141–161, https://doi.org/10.1016/j.earscirev.2014.04.006

Moses, C., Chopra, T. *et al.* 2025. EMODnet Geology: pan-European assessment of coastal resilience and vulnerability. *Quarterly Journal of Engineering Geology and Hydrogeology*, **58**, qjegh2024-119, https://doi.org/10.1144/qjegh2024-199

National Network of Regional Coastal Monitoring Programme of England 2023. SCAPE+, https://coastalmonitoring.org/ccoresources/iCOASST/SCAPE/ (accessed 9 June 2023).

Naylor, L.A., Stephenson, W.J. and Trenhaile, A.S. 2010. Rock coast geomorphology: recent advances and future research directions. *Geomorphology*, **114**, 3–11, https://doi.org/10.1016/j.geomorph.2009.02.004

Naylor, L.A., Coombes, M.A. and Viles, H.A. 2012. Reconceptualising the role of organisms in the erosion of rock coasts: a new model. *Geomorphology*, **157**, 17–30, https://doi.org/10.1016/j.geomorph.2011.07.015

Nermoen, A., Korsnes, R.I., Hiorth, A. and Madland, M.V. 2015. Porosity and permeability development in compacting chalks during flooding of nonequilibrium brines: insights from long-term experiment. *Journal of Geophysical Research: Solid Earth*, **120**, 2935–2960, https://doi.org/10.1002/2014jb011631

Nicholls, R.J., Dredge, A. and Wilson, T. 2000. Shoreline change and fine-grained sediment input: Isle of Sheppey Coast, Thames Estuary, UK. *Geological Society, London, Special Publications*, **175**, 305–315, https://doi.org/10.1144/gsl.sp.2000.175.01.22

Nyberg, J., Zillén-Snowball, L. and Strömstedt, E. 2021. Spatial characterization of seabed environmental conditions and geotechnical properties for the development of marine renewable energy in Sweden. *Quarterly Journal of Engineering Geology and Hydrogeology*, **55**, qjegh2021-091, https://doi.org/10.1144/qjegh2021-09

O'Connor, W., Mider, G., Lawrence, J.A., Agar, S., Mason, P.J., Ghail, R. and Scoular, J. 2021. An investigation into ground movement on the Ventnor landslide complex, UK using persistent scatterer interferometry. *Remote Sensing*, **13**, 3711, https://doi.org/10.3390/rs13183711

Ogawa, H., Dickson, M.E. and Kench, P.S. 2016. Generalised observations of wave characteristics on near-horizontal shore platforms: synthesis of six case studies from the North Island, New Zealand. *New Zealand Geographer*, **72**, 107–121, https://doi.org/10.1111/nzg.12121

Omidiji, J., Stephenson, W. and Norton, K. 2023. Cross-scale erosion on shore platforms using the micro-erosion meter and structure-from-motion (SfM) photogrammetry. *Geomorphology*, **108736**, https://doi.org/10.1016/j.geomorph.2023.108736

Palmer, M.J. 2002. Coastal cliff retreat and instability in a weak rock, Fairlight Cove, East Sussex, UK. *In*: McInnes, R.G. and Jakeways, J. (eds) *Instability: Planning and Management.* Thomas Telford, 535–542.

Payo, A., Walkden, M. *et al.* 2018. A quantitative assessment of the annual contribution of platform downwearing to beach sediment budget: Happisburgh, England, UK. *Journal of Marine Science and Engineering*, **6**, 113, https://doi.org/10.3390/jmse6040113

Pethick, J.S. and Crooks, S. 2000. Development of a coastal vulnerability index: a geomorphological perspective. *Environmental Conservation*, **27**, 359–367, https://doi.org/10.1017/s0376892900000412

Prêcheur, C. 1960. *Le littoral de la Manche, de Sainte-Adresse à Aulte. Étude morphologique.* Norois.

Prémaillon, M., Regard, V., Dewez, T.J. and Auda, Y. 2018. GlobR2C2 (Global Recession Rates of Coastal Cliffs): a global relational database to investigate coastal rocky cliff erosion rate variations. *Earth Surface Dynamics*, **6**, 651–668, https://doi.org/10.5194/esurf-6-651-2018

Pringle, A.W. 1985. Holderness coast erosion and the significance of ords. *Earth Surface Processes and Landforms*, **10**, 107–124, https://doi.org/10.1002/esp.3290100204

Pye, K. and Blott, S.J. 2015. Spatial and temporal variations in soft-cliff erosion along the Holderness coast, East Riding of Yorkshire, UK. *Journal of Coastal Conservation*, **19**, 785–808, https://doi.org/10.1007/s11852-015-0378-8

Pye, K. and Mottershead, D.N. 1995. Honeycomb weathering of Carboniferous sandstone in a sea wall at Weston-super-Mare, UK.

Quarterly Journal of Engineering Geology, **28**, 333–347, https://doi.org/10.1144/gsl.qjegh.1995.028.p4.03

Regard, V., Dewez, T. *et al.* 2012. Late Holocene seacliff retreat recorded by ^{10}Be profiles across a coastal platform: theory and example from the English Channel. *Quaternary Geochronology*, **11**, 87–97, https://doi.org/10.1016/j.quageo.2012.02.027

Regard, V., Prémaillon, M. *et al.* 2022. Rock coast erosion: an overlooked source of sediments to the ocean. Europe as an example. *Earth and Planetary Science Letters*, **579**, 117356, https://doi.org/10.1016/j.epsl.2021.117356

Robinson, D.A. and Moses, C.A. 2011. *Rock Surface and Weathering: Process and Form. The SAGE Handbook of Geomorphology.* SAGE, London, 91–309, https://doi.org/10.4135/9781446201053.n17

Robinson, L.A. 1977*a*. Marine erosive processes at the cliff foot. *Marine Geology*, **23**, 257–271, https://doi.org/10.1016/0025-3227(77)90022-6

Robinson, L.A. 1977*b*. Erosive processes on the shore platform of northeast Yorkshire, England. *Marine Geology*, **23**, 339–361, https://doi.org/10.1016/0025-3227(77)90038-x

Rosser, N., Lim, M., Petley, D., Dunning, S. and Allison, R. 2007. Patterns of precursory rockfall prior to slope failure. *Journal of Geophysical Research: Earth Surface*, **112**, F4, https://doi.org/10.1029/2006jf000642

Salman, A., Lombardo, S. and Doody, P. 2004. *Living with Coastal Erosion in Europe: Sediment and Space for Sustainability*. Eurosion Project Reports, http://www.eurosion.org/reports-online/reports.html

Saroglou, C. and Alexandrou, D. 2016. Coastal cliff instability, a case study from the UK. *Bulletin of the Geological Society of Greece*, **50**, 949–958, https://doi.org/10.12681/bgsg.11799

Sellwood, M., Davis, G., Brunsden, D. and Moore, R. 2000. Ground models for the coastal landslides at Lyme Regis, Dorset, UK. *In*: Bromhead, E., Dixon, N. and Ibsen, M.L. (eds) *Landslides in Research, Theory and Practice*. Thomas Telford, London, **3**, 1361–1366, https://doi.org/10.1680/lirtapv1.34617

Shadrick, J.R., Hurst, M.D., Piggott, M.D., Hebditch, B.G., Seal, A.J., Wilcken, K.M. and Rood, D.H. 2021. Multi-objective optimisation of a rock coast evolution model with cosmogenic ^{10}Be analysis for the quantification of long-term cliff retreat rates. *Earth Surface Dynamics*, **9**, 1505–1529, https://doi.org/10.5194/esurf-9-1505-2021

Shadrick, J.R., Rood, D.H., Hurst, M.D., Piggott, M.D., Hebditch, B.G., Seal, A.J. and Wilcken, K.M. 2022. Sea-level rise will likely accelerate rock coast cliff retreat rates. *Nature Communications*, **13**, 7005, https://doi.org/10.1038/s41467-022-34386-3

Shadrick, J.R., Rood, D.H., Hurst, M.D., Piggott, M.D., Wilcken, K.M. and Seal, A.J. 2023*a*. Constraints on long-term cliff retreat and intertidal weathering at weak rock coasts using cosmogenic ^{10}Be, nearshore topography and numerical modelling. *Earth Surface Dynamics*, **11**, 429–450, https://doi.org/10.5194/esurf-11-429-2023

Shadrick, J.R., Rood, D.H. and Hurst, M.D. 2023*b*. Reply to: sea-level rise may not uniformly accelerate cliff erosion rates. *Nature Communications*, **14**, 8486, https://doi.org/10.1038/s41467-023-44149-3

Shakesby, R.A. and Walsh, R.P.D. 1986. Micro-erosion meter measurements of erosion of limestone, Oxwich Point, Gower: some technical considerations and preliminary results. *Cambria*, **13**, 213–234.

Stephenson, W. 2001. Shore platform width – a fundamental problem. *Zeitschrift für Geomorphologie*, **45**, 511–527, https://doi.org/10.1127/zfg/45/2001/511

Stephenson, W.J. and Kirk, R.M. 1998. Rates and patterns of erosion on inter-tidal shore platforms, Kaikoura Peninsula, South Island, New Zealand. *Earth Surface Processes and Landforms*, **23**, 1071–1085, https://doi.org/10.1002/(sici)1096-9837(199812)23:12<1071::aid-esp922>3.0.co;2-u

Stephenson, W.J. and Naylor, L.A. 2010. Rock coast geomorphology. *Geomorphology*, **114**, 1–100, https://doi.org/10.1016/j.geomorph.2009.02.013

Stephenson, W.J., Naylor, L.A., Smith, H., Chen, B. and Brayne, R.P. 2018. Wave transformation across a macrotidal shore platform under low to moderate energy conditions. *Earth Surface Processes and Landforms*, **43**, 298–311, https://doi.org/10.1002/esp.4245

Styles, T.D., Coggan, J.S. and Pine, R.J. 2011. Back analysis of the Joss Bay Chalk Cliff Failure using numerical modelling. *Engineering Geology*, **120**, 81–90, https://doi.org/10.1016/j.enggeo.2011.04.004

Sunamura, T. 1992. *Geomorphology of Rocky Coasts.* Wiley.

Sunamura, T. 2015. Rocky coast processes: with special reference to the recession of soft rock cliffs. *Proceedings of the Japan Academy, Series B*, **91**, 481–500, https://doi.org/10.2183/pjab.91.481

Swantesson, J.O.H., Moses, C.A., Berg, G.E. and Magnus Jansson, K. 2006. Methods for measuring shore platform micro erosion: a comparison of the micro-erosion meter and laser scanner. *Zeitschrift für Geomorphologie*, **144**, 1–17.

Swirad, Z.M., Rosser, N.J. and Brain, M.J. 2019. Identifying mechanisms of shore platform erosion using structure-from-motion (SfM) photogrammetry. *Earth Surface Processes and Landforms*, **44**, 1542–1558, https://doi.org/10.1002/esp.4591

Swirad, Z.M., Rosser, N.J., Brain, M.J., Rood, D.H., Hurst, M.D., Wilcken, K.M. and Barlow, J. 2020. Cosmogenic exposure dating reveals limited long-term variability in erosion of a rocky coastline. *Nature Communications*, **11**, 3804, https://doi.org/10.1038/s41467-020-17611-9

Tavani, S., Billi, A. *et al.* 2022. Smartphone assisted fieldwork: towards the digital transition of geoscience fieldwork using LiDAR-equipped iPhones. *Earth-Science Reviews*, **227**, 103969, https://doi.org/10.1016/j.earscirev.2022.103969

Thébaudeau, B., Trenhaile, A.S. and Edwards, R.J. 2013. Modelling the development of rocky shoreline profiles along the northern coast of Ireland. *Geomorphology*, **203**, 66–78, https://doi.org/10.1016/j.geomorph.2013.03.027

Thompson, C.F., Young, A.P. and Dickson, M.E. 2019. Wave impacts on coastal cliffs: do bigger waves drive greater ground motion? *Earth Surface Processes and Landforms*, **44**, 2849–2860, https://doi.org/10.1002/esp.4712

Thompson, C.F., Dickson, M.E. and Young, A.P. 2022. Seismic signatures of individual wave impacts on a coastal cliff. *Earth Surface Processes and Landforms*, **47**, 2833–2845, https://doi.org/10.1002/esp.5426

Trenhaile, A.S. 1987. *The Geomorphology of Rock Coasts*. Clarendon Press, Oxford.

Trenhaile, A.S. 2015. Coastal notches: their morphology, formation, and function. *Earth-Science Reviews*, **150**, 285–304, https://doi.org/10.1016/j.earscirev.2015.08.003

Trenhaile, A.S. 2018. Shore platform erosion and evolution: implications for cosmogenic nuclide analysis. *Marine Geology*, **403**, 80–92, https://doi.org/10.1016/j.margeo.2018.05.005

Trenhaile, A.S. 2019. Hard-rock coastal modelling: past practice and future prospects in a changing world. *Journal of Marine Science and Engineering*, **7**, 34, https://doi.org/10.3390/jmse7020034

Trenhaile, A. 2020. Modelling the development of dynamic equilibrium on shore platforms. *Marine Geology*, **427**, 106227, https://doi.org/10.1016/j.margeo.2020.106227

Trenhaile, A.S. and Bryne, M.L. 1986. A theoretical investigation of the Holocene development of rock coasts, with particular reference to shore platforms. *Geografiska Annaler: Series A, Physical Geography*, **68**, 1–14, https://doi.org/10.1080/04353676.1986.11880154

Trenhaile, A.S. and Kanyaya, J.I. 2007. The role of wave erosion on sloping and horizontal shore platforms in macro and mesotidal environments. *Journal of Coastal Research*, **23**, 298–309, https://doi.org/10.2112/04-0282.1

Trudgill, S. 1985. *Limestone Geomorphology*. Longman, London.

Trudgill, S.T. 1987. Bioerosion of intertidal limestone, Co. Clare, Eire – 3: zonation, process and form. *Marine Geology*, **74**, 111–121, https://doi.org/10.1016/0025-3227(87)90009-0

Trudgill, S. 1988. Integrated geomorphological and ecological studies on rocky shores in southern Britain. *Field Studies*, **7**, 239–279.

Trudgill, S.T., Smart, P.L., Freiderich, H. and Crabtree, R.W. 1987. Bioerosion of intertidal limestone, Co. Clare, Eire – 1: *paracentrotus lividus*. *Marine Geology*, **74**, 85–98, https://doi.org/10.1016/0025-3227(87)90007-7

Walkden, M.J.A. and Hall, J.W. 2005. A predictive mesoscale model of the erosion and profile development of soft rock shores. *Coastal Engineering*, **52**, 535–563, https://doi.org/10.1016/j.coastaleng.2005.02.005

Walkden, M., Johnson, A. and Longfield, S. 2020. Erosion due to sea level rise: a Glamorgan case study. *In*: Hardiman, N. and the Institute of Civil Engineers (eds) *Coastal Management 2019: Joining Forces to Shape our Future Coasts*. ICE Publishing, 565–574, https://doi.org/10.1680/cm.65147.565

Westoby, M., Lim, M., Hogg, M., Dunlop, L., Pound, M., Strzelecki, M. and Woodward, J. 2020. Decoding complex erosion responses for the mitigation of coastal rockfall hazards using repeat terrestrial lidar. *Remote Sensing*, **12**, 2620, https://doi.org/10.3390/rs12162620

Windle, J.C., Skipper, J., Beetham, L., Winter, M.G., Smith, D.M., Eldred, P.J.L. and Toll, D.G. 2015. Fairlight Cove landslip remediation: a case history of ongoing resident management. *In*: Winter, M.G., Smith, D.M., Eldred, P.J.L. and Toll, D.G. (eds) *Geotechnical Engineering for Infrastructure and Development: XVI European Conference on Soil Mechanics and Geotechnical Engineering*. ICE Publishing, 1927–1932.

Yeomans, C.M., Claridge, H., Hudson, A.J., Shail, R.K., Willems, C., Eyre, M. and Harker, C. 2023. A single multi-scale and multi-sourced semi-automated lineament detection technique for detailed structural mapping with applications to geothermal energy exploration. *Quarterly Journal of Engineering Geology and Hydrogeology*, **56**, qjegh2022-051, https://doi.org/10.1144/qjegh2022-051

Young, A.P. and Carilli, J.E. 2019. Global distribution of coastal cliffs. *Earth Surface Processes and Landforms*, **44**, 1309–1316, https://doi.org/10.1002/esp.4574

Young, A.P., Guza, R.T., Matsumoto, H., Merrifield, M.A., O'Reilly, W.C. and Swirad, Z.M. 2021. Three years of weekly observations of coastal cliff erosion by waves and rainfall. *Geomorphology*, **375**, 107545, https://doi.org/10.1016/j.geomorph.2020.107545

Yuan, R., Kennedy, D.M., Stephenson, W.J. and Finlayson, B.L. 2022. The precision and accuracy of measuring micro-scale erosion on shore platforms. *Marine Geology*, **443**, 06691, https://doi.org/10.1016/j.margeo.2021.106691

LiDAR for rapid geomorphological assessment in relict periglacial landsystems

David Peter Giles

Card Geotechnics Ltd (CGL), 4 Godalming Business Centre, Woolsack Way, Godalming, Surrey GU7 1XW, UK
0000-0001-8016-0538
Correspondence: drdpgiles@gmail.com

Abstract: The construction of the A21 Sevenoaks bypass in Kent is a seminal case study in engineering geology. In 1965 during the earthworks for the road, major slope failures occurred presenting a significant setback to the programme. No slope stability issues were foreseen before the groundworks commenced. Post-failure forensic investigations revealed multiple shear surfaces within lobe-like landforms that dissected the original route alignment. At that time, these landforms were not recognized, and their problematic nature not fully understood. Similarly, the Maidstone district has long been described as an area of relict periglacial landforms, specifically cambering and associated gulls, and valley bulge. This chapter presents key case studies, which detail the nature of the relict periglacial landsystem that gave rise to the encountered geohazards and demonstrates the advances in ground model assessment offered by airborne LiDAR (Light Detection and Ranging; laser scanning) datasets for engineering geomorphology. High-resolution digital elevation models derived from the LiDAR imagery (that became available for the first time in 2019) can be utilized to interpret and visualize the on-ground geomorphology. The available LiDAR imagery is described, and examples of their use presented from these classic case study sites.

Professor Peter Fookes had recognized at an early stage in his career that understanding aspects of the relict periglacial landsystem in the UK (Higginbottom and Fookes 1970) was critical in the development of engineering geology ground models (Fookes 1991) and to the understanding of ground behaviour and the associated risk of the legacy geological hazards to construction projects (for example Fookes 1997; Fookes *et al.* 2000, 2007). Fookes also recognized the importance of applied geological mapping and aerial imagery to provide a synoptic view of the terrain under engineering investigation (for example Dearman and Fookes 1974; Brunsden *et al.* 1975; Fookes *et al.* 1991). This chapter sets out to develop this '*Fookesian*' theme with examples of digital imagery that can enhance the ground model development process. It should be noted that since Peter Fookes' early papers there have been significant advances in the understanding of relict periglacial geomorphology, significantly supported by field-based study of contemporary periglacial landscapes. Remote sensing techniques have also developed to aid in the identification of these landforms, complementing the traditional methods of geomorphological mapping and aerial photographic interpretation that still remain a critical tool in landscape evaluation.

Kent is an area where intense periglacial conditions during the Quaternary have left a legacy of relict landforms that are problematic in engineering geology. This chapter presents a digital methodology for the rapid and initial appraisal of these geohazard landforms that could be encountered in this relict periglacial landsystem. Two case studies are presented to illustrate the use of LiDAR (Light Detection and Ranging) for this rapid terrain evaluation.

The A21 Sevenoaks bypass in Kent, UK is a classic case study in the development of engineering geology and, indeed, geotechnics in the UK. This site was one of the first investigated instances of a slope failure where (then) unrecognized former periglacial conditions had affected the overall stability of the slope. Traditionally the landforms and deposits at the site have been ascribed as products of solifluction with both lobate landforms and sheets being present. The term solifluction is used to include both frost creep and gelifluction (Murton and Ballantyne 2017). Both processes are forms of mass movement or wasting downslope in periglacial conditions. Frost creep describes downslope soil movement due to repeated cycles of volumetric expansion due to ice-lens growth during freezing and contraction (settling) during thaw of seasonally frozen ground. Gelifluction is the slow gravity-induced shear deformation of soil associated with generation of high porewater pressures during thaw of soil containing excess ice. Gelifluction involves winter freezing of water within a soil in the 'active' upper soil layer that subsequently melts in the warmer summer conditions. Where the melting water cannot drain away, the soil becomes saturated and the increased pore pressures and disturbed soil structure result in the weakening of soils and slow flow downslope.

Examples of similar contemporary processes and landforms in current permafrost environments of northern

From: Griffiths, J. S., Giles, D. P., Davis, J. A. and Dunlop, J. J. (eds) 2025. *The Peter Fookes Engineering Geological Legacy in Geomodels, Geomaterials and Geomorphology*. Geological Society, London, Engineering Geology Special Publications, **30**, 237–265,
First published online 21 March 2025, https://doi.org/10.1144/egsp30-2024-46

Canada, where a variety of slope failures triggered by permafrost thaw could be observed (Lewkowicz 1990; Harris and Lewkowicz 1993; Lewkowicz and Harris 2005), suggest that active layer failures are deemed to be more appropriate interpretations of the landforms and periglacial legacy observed at Sevenoaks. This chapter sets out to demonstrate how digital data presented by airborne LiDAR imaging can be used to further delineate, visualize and interpret these landforms in this relict periglacial landsystem.

The second case study considers an area of relict periglacial cambering with associated gull structures and how LiDAR imagery has been used to visualize and delineate these periglacial geohazards.

The sites described in these case studies are located on the footslope of the Lower Cretaceous Hythe Formation scarp (Fig. 1).

Laser scanning in site investigation and geomorphological assessment

The advent of terrestrial and airborne digital laser scanning has provided growing opportunities for engineering geology. Specifically, the advent and availability of LiDAR datasets, at engineering project ground resolutions of 1–2 m, has seen a significant growth in geohazard and geomorphological applications for identifying and delineating problematic ground conditions on and around a development site (e.g. Moore *et al.* 2022). Digital models of the ground surface for use in geological applications were initially generated by photogrammetric techniques (e.g. Moore 1974; Chandler and Moore 1989; Chandler 2001; Walstra *et al.* 2007). Developments in the use of ground-based laser-scanning systems for the study of terrain offered new tools for the rapid acquisition of digital elevation data for geomorphological analysis and for the derivation of 'bare earth' digital terrain models (DTM) that removed the vegetation cover in the images

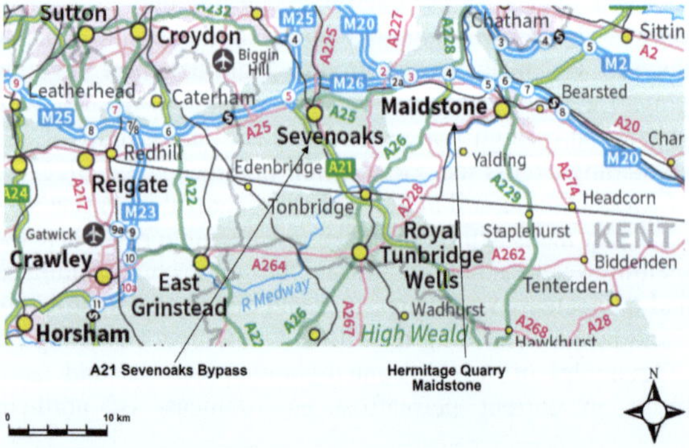

Fig. 1. Site location map. Source: reproduced from the Ordnance Survey 1:25 000 map with permission of the Controller of Her Majesty's Stationery Office, Crown Copyright, Licence No. 100012585.

(Whitworth *et al.* 2005, 2006*a*, *b*; Buckley *et al.* 2008). Terrestrial laser scanning has been used to study both coastal and inland landslides (e.g. Rowlands *et al.* 2003; Rosser *et al.* 2005; Jones 2006; Dunning *et al.* 2010; Ellis *et al.* 2011; Jaboyedoff *et al.* 2012) and rock slope stability (e.g. Kemeny and Donovan 2005; Mikoš *et al.* 2005; Chen *et al.* 2020), while other studies have taken advantage of the survey repeatability for change detection and temporal studies of landslide movement (Hsiao *et al.* 2004). Airborne-mounted systems that offered more synoptic views of the ground surface have also seen considerable expansion with the advent of wider data availability and software processing and the creation of elevation models. In 2010 the Geological Society produced a Special Publication that considered the use of such elevation models for geoscience (Fleming *et al.* 2010*a*, *b*). This volume dealt with more regional studies than engineering site-specific applications (e.g. Hall *et al.* 2010; Hobbs *et al.* 2010; McKenzie *et al.* 2010). The ready supply of high-quality elevation data provided by the UK Environment Agency and the access to open-source software such as QGIS (Baghdadi *et al.* 2018; https://qgis.org) for the data processing and presentation has presented an invaluable resource for rapid engineering site assessment and geohazard assessment.

Airborne laser scanning and LiDAR

LiDAR mapping is a method of generating precise and directly georeferenced spatial information about the shape and surface characteristics of the Earth (NOAA 2012). It is a system like radar that uses laser light instead of radio waves to generate a detailed elevation model of the Earth's surface. These data, in the form of many thousands of x, y, z coordinates (tuplets), can be collected from a laser rangefinder mounted on an aircraft, helicopter or satellite platform (Giglierano 2010). The system uses an active sensor, with a day/night and cloud-penetrating capability, to generate pulses of light and record their multiple surface reflections. These laser pulses, intense polarized and focused beams of light, are processed to measure the time taken between the emission of the beam and the reflection back by a ground surface object. Multiple reflections are recorded for the first return to the last or *bare-earth* return (Fig. 2a), with each pulse return representing increasing penetration beneath the surface vegetation canopy (Fig. 2b). GPS systems are used to record the position of the pulse emitting platform to calculate the absolute location of the sensor above the ground surface. By measuring the travel time difference between the laser pulse being emitted and returned, the accurate height of the ground surface can be calculated (NOAA 2012). These data are recorded as a dense x, y, z point cloud, measured from the typically 150 000 pulses emitted per second from the laser (150 kHz). This densely spaced network of highly accurate georeferenced elevation points can then be processed, via interpolation algorithms, to generate a digital

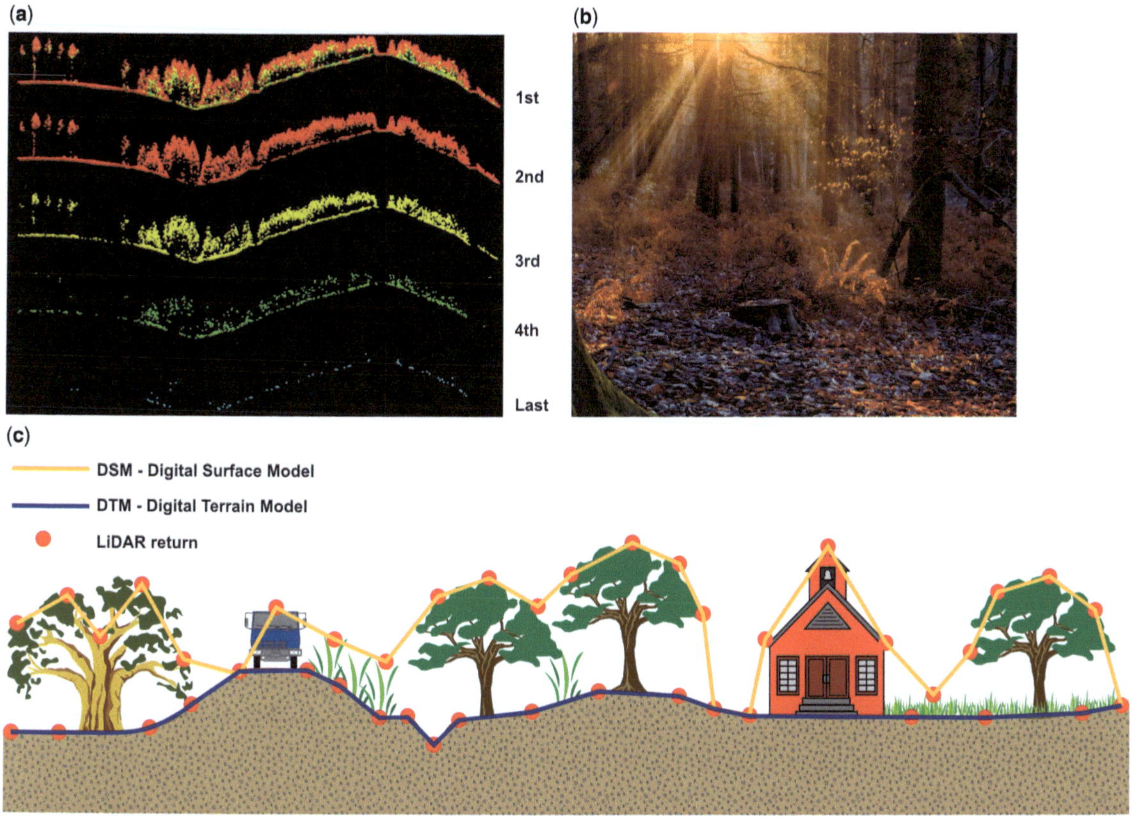

Fig. 2. (**a**) LiDAR return signals. (**b**) Illustration of LiDAR vegetation canopy penetration. (**c**) Derivation of digital surface and digital terrain models. Source: (b) photo courtesy of Elizabeth Tarren.

surface model (DSM) and DTM (Fig. 2c) of the Earth's surface (Privett 2019). The DSM represents data from the early signal returns (Fig. 3a) and the DTM from the last or bare-earth return (Fig. 3b). LiDAR systems normally operate in the near-infrared (NIR) wavelengths of the electromagnetic spectrum (905 µm). Some sensors have a water-penetrating capability in calm, non-turbid, conditions operating in a green wavelength. Intensity data are also recorded representing the strength (intensity) of the returned signal. Wehr and Lohr (1999) provide a more detailed account of the technical detail of the light detection and ranging processes. Table 1 outlines the terminology associated with airborne LiDAR data.

In the UK, the principal source of publicly available LiDAR data is provided by the Environment Agency Geomatics Survey Open Data programme with data provided from the Defra data download platform (Defra n.d.). Data have been regularly captured since 1998 and are available at ground resolutions from 2 m to 0.25 m pixel size, although this is an ongoing programme with the entirety of the UK yet to be fully captured. This ground resolution is significant as the data become focused on the requirements of site-specific project scales rather than more regional broad-scale assessments. The vertical accuracy has a Root Mean Square Error (RMSE) of ± 0.15 m (the error difference between the ground truth survey and the measured LiDAR height data). Several factors can influence the acquired LiDAR accuracy (Aerial Precision 2023).

(1) *Scan density and resolution*: the accuracy of LiDAR mapping is heavily dependent on the density of laser points captured per unit area. Higher scan density makes more accurate representations of complex surfaces, as fine details are better captured. The resolution of the LiDAR system also plays a crucial role in accurately representing more minor features.
(2) *Calibration and system parameters*: proper calibration of LiDAR systems is essential for accurate measurements. System parameters such as laser pulse frequency, beam divergence and scanner angular accuracy must be precisely configured to ensure accurate point cloud generation.
(3) *Terrain and surface characteristics*: the nature of the terrain and surfaces being scanned can influence accuracy. LiDAR works exceptionally well on surfaces with high reflectivity and minimal occlusion. Complex environments, like dense vegetation or reflective materials,

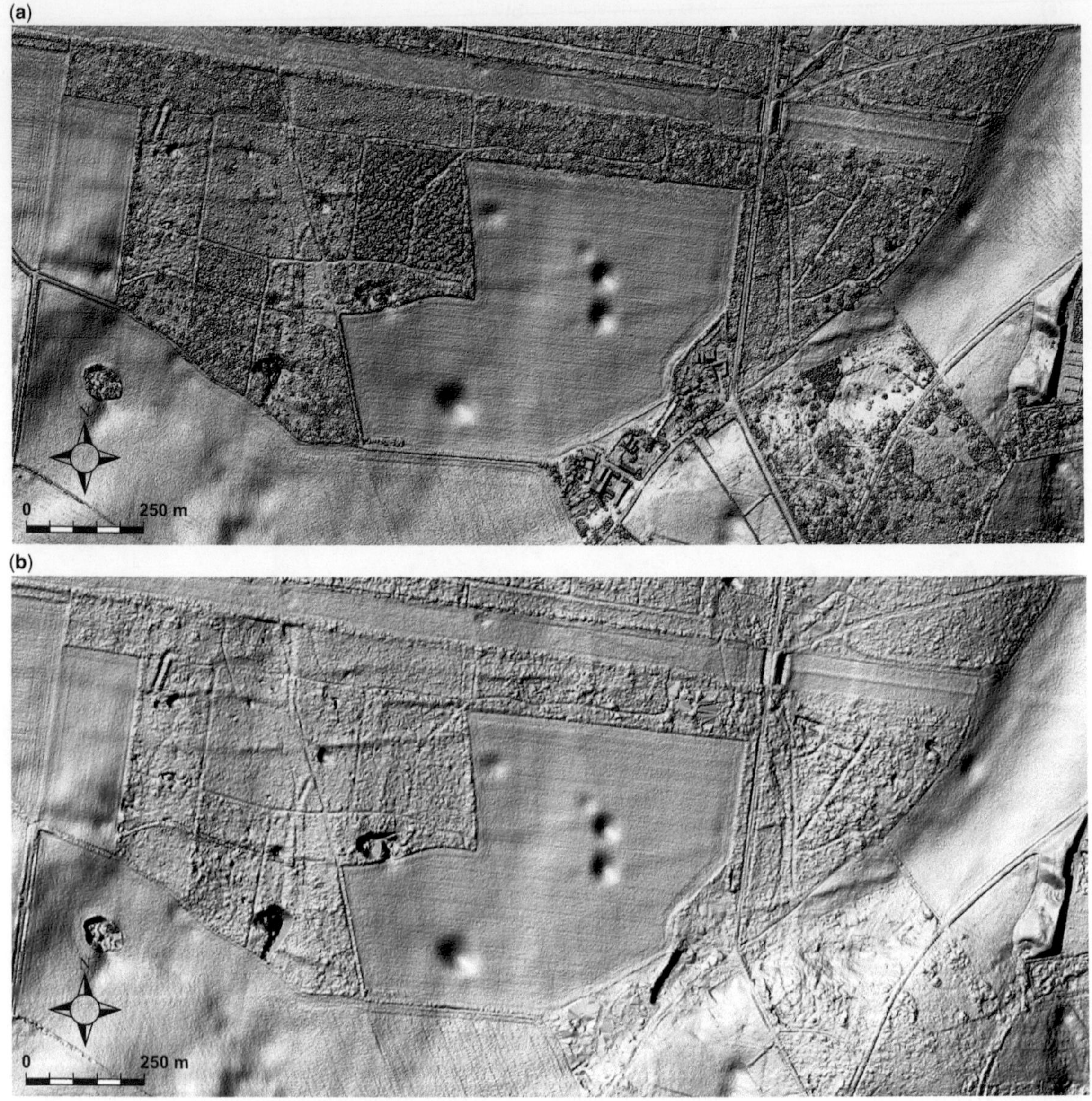

Fig. 3. (**a**) Digital surface model (DSM) with vegetation canopy. (**b**) Same scene with a digital terrain model (DTM) with vegetation canopy penetration. Example from Holme Farm, Rowlands Castle, Hampshire showing small chalk pits and chalk sinkholes.

can pose challenges and affect the accuracy of the collected data.

(4) *Georeferencing and GPS accuracy*: integrating LiDAR data with accurate GPS coordinates and georeferencing is crucial. Minor errors in GPS measurements can lead to misalignment between LiDAR point clouds and the real-world coordinates, affecting overall accuracy.

(5) *Atmospheric conditions*: atmospheric factors such as humidity, rain and fog can scatter and absorb laser pulses, leading to inaccurate distance measurements. Advanced algorithms can help compensate for these effects, but optimal conditions produce more accurate results.

One of the major advantages from LiDAR imaging for ground engineering and engineering geomorphological assessment is the ability of the system to penetrate the vegetation cover and canopy (Fig. 4).

Table 1. Terminology associated with airborne LiDAR data

Term	Definition
LAS	Laser file format. File format for the interchange of three-dimensional point cloud data between data users. Although developed primarily for the exchange of LiDAR point cloud data, this format supports the exchange of any three-dimensional x, y, z tuplet. LAS is a binary file format that maintains information specific to the LiDAR nature of the data while not being overly complex.
RMSE	Root Mean Square Error. Measure of the accuracy of the acquired data. Like the measure of standard deviation if there is no bias in the data.
FVA	Fundamental vertical accuracy. Measure of the accuracy of the data in open areas at a high level of confidence (95%).
SVA	Supplemental vertical accuracy. Like the FVA (95% accuracy), but it is a measure of the individual ground cover types, of which the bare earth is one. The value is typically calculated at the 95th percentile where errors do not follow a normal distribution, where 95% of elevation values have elevation errors equal to or less than the 95th percentile. For example, if 20 points are collected, the worst point (5%) can be removed and the 19th point provided as the SVA value. The SVA value is reported separately for each land cover type tested and, thus, provides more in-depth information than the consolidated vertical accuracy.
CVA	Consolidated vertical accuracy. Measure of the accuracy at the 95% level. It is a measure of the entire dataset; all the points collected in the various ground types are used together.
Classification	Data that have been processed to define the type of object that the pulses have reflected off. Can be as simple as unclassified (i.e. object not defined) to buildings and high vegetation. The most common is to classify the dataset for points that are considered 'bare earth' and those that are not (unclassified).
Return Number (First/Last Returns)	Many LiDAR systems can capture the first, second, third and, ultimately, the 'last' return from a single laser pulse. The return number can be used to help determine what the reflected pulse is from (e.g. ground, tree, understorey).
Point Spacing	How close the laser points are to each other, analogous to the pixel size of an aerial image. Also called 'posting density' or 'nominal point spacing.' The point spacing determines the resolution of derived gridded products.
Pulse Rate	Number of discrete laser 'shots' per second that the LiDAR instrument is firing. Data are captured at approximately 50 000 to 150 000 pulses per second.
Intensity Data	When laser return is recorded, the strength of the return is also recorded. The values represent how well the object reflected the wavelength of light used by the laser system (e.g. 1064 nm for most commercial topography sensors). These data resemble a black and white photo but cannot be interpreted in the same manner.
RTK GPS (Real Time Kinematic GPS)	Satellite navigation that uses the carrier phase (a waveform) that transmits (carries) the Global Positioning System (GPS) signal instead of the GPS signal itself. The actual GPS signal has a frequency of about 1 MHz, whereas the carrier wave has a frequency of 1500 MHz, so a difference in signal arrival time is more precise. The carrier phase is more difficult to use (i.e. the equipment is more costly); however, once it has been resolved, it produces a more accurate position in relation to the higher frequency.
DEM	Surface created from elevation point data to represent the topography. Often a DEM is more easily used in a geographical information system (GIS) or computer-aided design (CAD) application than the raw point data it is constructed from. The techniques used to create DEMs range from simple (e.g. nearest neighbour) to complex (e.g. kriging) gridding routines and can create slightly different surface types.
DSM	Digital surface model. Represents the mean surface level elevations of the reflective surfaces of trees, buildings, and other features elevated above the 'bare earth'.
DTM	Digital terrain model. Represents the mean surface elevations of the reflective surfaces of the 'bare earth'.

Source: adapted from (NOAA 2012).

Using these elevation models, surface visualizations can be generated, extenuating breaks of slope and elevation changes, allowing features such as scarps, back-tilted slope units, lobate accumulation zones and irregular topography to be enhanced to facilitate delineation and classification of landslides. A combination of shaded relief models and the application of slope, curvature and topographic roughness filters provides an optimum method for landslide mapping and other geohazard delineation (Fig. 5).

Fig. 4. Standard aerial photograph from the Weald of Kent with DTM overlay demonstrating the power of the LiDAR to penetrate the vegetation canopy and reveal the underlying topographic and geomorphological detail (LiDAR tile 10 km^2).

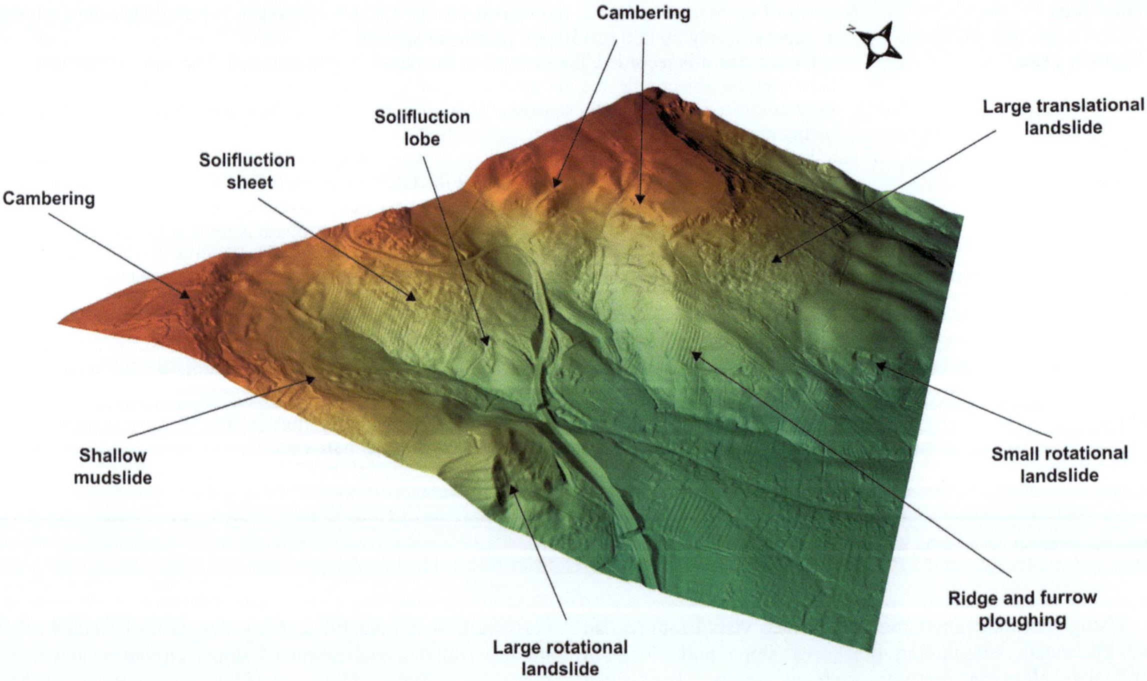

Fig. 5. Shaded relief models provide an optimum method for landslide mapping and other geohazard delineation. Example shown from a periglaciated Jurassic terrain in the Cotswolds with rotational and translational landslides, cambering and solifluction related landforms.

Circular depressions can be identified and once all manner of anthropogenic origins have been discounted using desk-study and local knowledge, potential dissolution features within soluble rocks can be identified for further on-ground investigation. Other hazards such as subsidence, evidence of former mineral extraction and quarrying and former water courses can also be identified. On the HS2 rail project LiDAR was successfully utilized to automate fault identification along the HS2 Chilterns tunnel (Sercombe *et al.* 2023).

Geological setting of case studies

The Lower Greensand escarpment, part of the eroded Wealden Anticline, comprises the Lower Cretaceous Hythe Formation overlying clays and mudstones of the Atherfield Clay and Weald Clay formations (Fig. 6). The Hythe Formation is typically a strong sandy limestone (locally termed *Ragstone*), interbedded with weaker glauconitic sandy mudstone (locally termed *Hassock*). The relatively lithologically competent Hythe Formation caps the underlying weaker mudstones and clays of the Atherfield Clay and Weald Clay formations, a classic geological setting for landsliding and slope instability (Radley 2006; Radley and Allen 2012*a*, *b*).

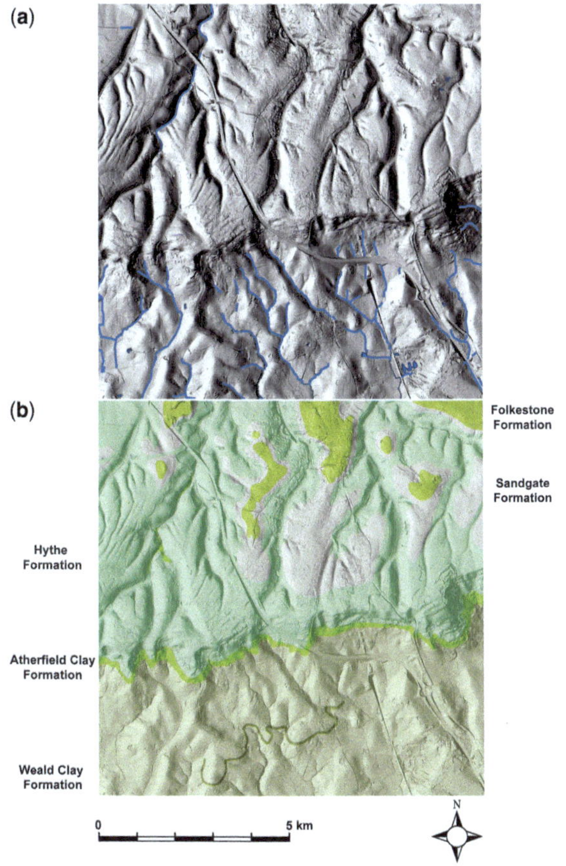

Fig. 6. (**a**) Regional hillshaded 1 m digital terrain model displaying active and dry river valley network (Ordnance Survey mapped river system overlaid in blue). (**b**) Regional 1 m digital terrain model with bedrock geology overlaid.

Geomorphological setting: the relict periglacial landsystem

Kent lies beyond the maximum (Anglian) limit of glaciation in England, which is mapped on the northern fringe of London (Lukas *et al.* 2017). The terrain in Kent, however, has experienced repeated episodes of periglaciation during the Quaternary (Murton and Giles 2016), some associated with the development of permafrost. Periglacial environments are characterized by cold climate nonglacial conditions and ground freezing.

The Kent case study sites presented are in a region that experienced these recurrent periglacial conditions throughout the Quaternary (Fig. 7). Pleistocene periglacial processes operating on frost-susceptible bedrock and soils in this area have had a major impact on the landscape geomorphology as well as the engineering geological properties of the *in situ* rocks and soils (Murton and Giles 2016). These processes include the growth of segregated ice promoting heave and fracture of the substrate, melt of excess ice with the resultant thaw consolidation of the ground generating excess porewater pressures and a reduction in material strength and downslope movement of rock or soil because of creep of ice-rich material and/or sliding or flow during seasonal or ultimate thaw. The area along the Hythe Formation escarpment was affected by large-scale rotational and translational landsliding, with many rotated, translated and back-tilted blocks in evidence. Within this landsystem other periglacial landforms and soils are also present, including relict solifluction-related landforms and active layer detachment slide remnants, described in detail below. Periglacial soils present include granular head deposits (angular to subangular clasts set in a silty or sandy matrix) and clay-rich head deposits (reworked, and partially remoulded) clayey debris. These more cohesive deposits may incorporate lithorelicts of the original clay fabric, fragments of more resistant rocks from upslope and sometimes non-local clasts derived from adjacent sediments. The accumulation of granular head deposits has generally been attributed to solifluction, although locally debris flows and active-layer detachment slides may have contributed debris, especially on steeper slopes (Harris 2013). Significantly for the engineering geology ground model, relict shear surfaces can be found within or beneath clay-rich head deposits. Figure 8 illustrates the general geomorphological setting of the area with the Wealden Clay Formation shallow slope underneath the steeper Hythe Formation escarpment.

Kent is an area where these former periglacial (permafrost) conditions have also left a landform legacy in the form of cambered strata with rotated blocks and gulls (Murton and

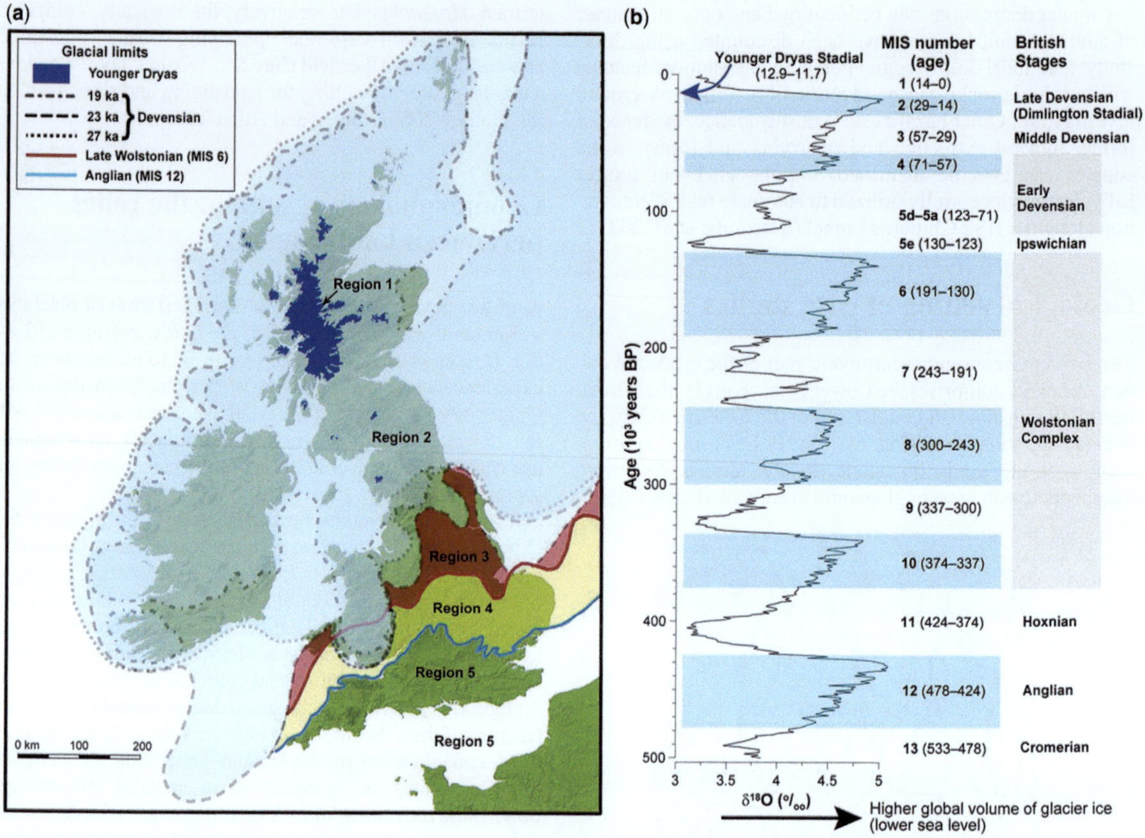

Fig. 7. Periglacial regions of the UK and Ireland. Region 1 (blue) was largely covered by glacier ice during the Younger Dryas Stadial, and periglacial activity within this zone has been confined to high ground during the Holocene. Region 2 (green) was covered by the last (Late Devensian) British–Irish Ice Sheet and experienced periglacial conditions during ice-sheet recession in the Late Devensian and during the Younger Dryas. Region 3 (brown and red) was subject to periglacial conditions during Wolstonian deglaciation and throughout the Devensian. Region 4 (yellow and olive green) experienced periglaciation during Anglian deglaciation and during the Wolstonian and Devensian. Region 5 (dark green) experienced recurrent periglacial conditions throughout the Quaternary. Source: Murton and Ballantyne (2017).

Fig. 8. Weald Clay Formation slope below Hythe Formation escarpment, Hubbard's Hill, Sevenoaks.

Giles 2016). Cambered strata display evidence of large-scale flexing and stretching of competent caprocks over the upper parts of valley-side slopes. The strata tend to show a local dip (camber) towards the valley floor, in contrast to the regional dip, and their angle of dip commonly exceeds the regional dip (Giles *et al.* 2017; Giles and Murton 2023). The caprock comprises a relatively brittle limestone, sandstone or ironstone that overlies an incompetent mudrock, usually clay (Fig. 9).

When evaluating the engineering geological periglacial legacy, two key elements of the periglacial–permafrost environment must be appreciated. The ground profile active layer represents the upper layer of ground subject to annual freezing and thawing in areas underlain by permafrost. Beneath the active layer is an ice-rich layer, commonly between 0.5 and 2 m thick, that is transitional between the seasonally thawing active layer and underlying deeper and more thermally stable permafrost (Murton and Ballantyne 2017).

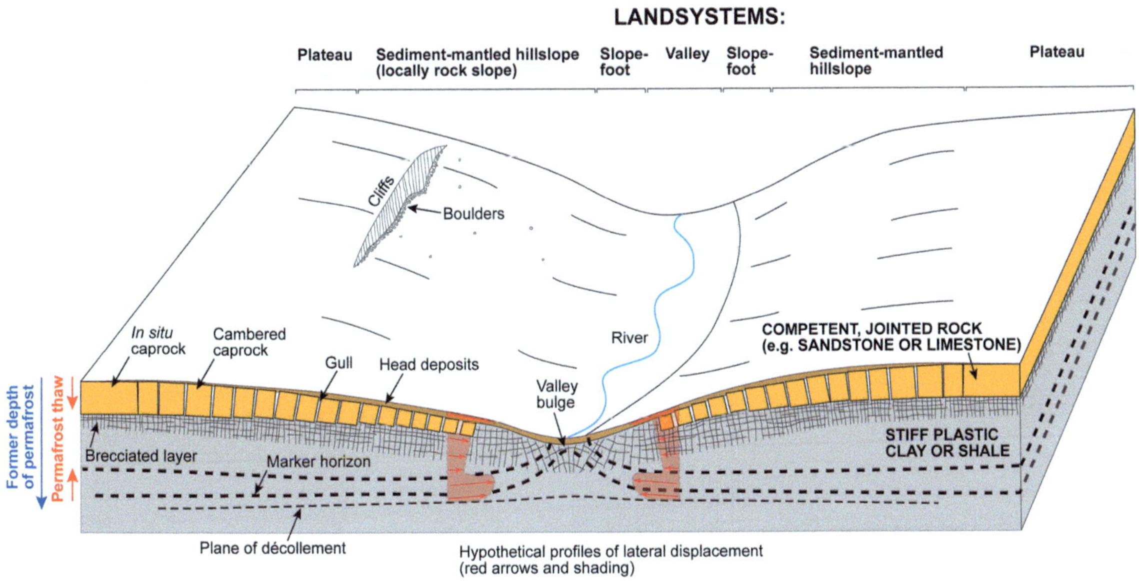

Fig. 9. Ground model of caprock plateau–mudstone valley landsystem, showing cambered caprock and gulls beneath valley sides and valley bulge structure in clay beneath the valley centre (Murton and Ballantyne 2017).

In terms of a landsystem framework the site of the Sevenoaks bypass and Maidstone can be considered as being a **Slope-foot landsystem** (Murton and Ballantyne 2017). Slope-foot landsystems abut the foot of rock slopes (in this case the escarpment of the stronger Hythe Formation) or grade upslope into sediment-mantled hillslopes. They functioned as depositional zones in the periglacial debris system, with sediment deposited by colluvial processes, including solifluction, and landslides including debris flow and overland flow (slopewash). Slope-foot sediments prograded onto adjacent lowlands, burying or interdigitating with sediments such as river gravels in valleys. Depositional landforms characteristic of slope-foot landsystems include individual lobes of sediment, isolated or coalescent alluvial fans and sediment aprons, and extensive sheets or remnant patches of granular and clayey head deposits (Murton and Ballantyne 2017). Sediment lobes may be present which can contain clayey head deposits up to 4 m thick, interrupted and significantly underlain by shear planes, and be overlain by silty clay slopewash deposits. Deposition of the head is attributed to periglacial mudflows or mudslides associated with thaw consolidation during summer thaw of an ice-rich layer at the base of the active layer or top of permafrost, akin to active-layer detachment failures observed in modern-day Arctic regions (Harris 1998) (Fig. 10).

Other significant relict periglacial geohazards

Relict periglacial shears have long been observed on the shallow hillslopes from deposits developed in southern England with documented examples within the Weald, Atherfield, Wadhurst and Gault formation clays (e.g. Weeks 1969; Pugh *et al.* 1991; Bromhead *et al.* 2000; Davies *et al.* 2003; Ellis *et al.* 2011; George *et al.* 2019). Relict periglacial shears are shear surfaces within cohesive bedrock or clay-rich head deposits that are no longer active under present-day environmental conditions, but which may be reactivated. Reactivation may result from excavation or loading of a hillslope, or from altering the drainage conditions such that porewater pressure within the soils increases.

Relict shears were observed and categorized during the ground investigation for the M25 motorway near Denham, Buckinghamshire, where several sheared and unsheared discontinuities were described within the London Clay Formation and Lambeth Group clays (Spink 1991). These discontinuities took a variety of forms, including low-angle near-surface shears (with associated random accommodation shears), which overlay and truncated high-angle shears that were thought to have formed through the collapse on thawing of the top of the permafrost layer (Fig. 11). Deeper low-angle shears were also encountered: one continuous subhorizontal and planar; and the other discontinuous, random and undulating. These have been attributed to shearing at the base of a permafrost layer at a thawing front. Also present were unsheared discontinuities considered to be thermal contraction cracks. Spink (1991) proposed a classification scheme recognizing seven different types of discontinuity seen in the Eocene clays (Table 2). Movement along any of these shear planes has the potential to be reactivated, so their anticipation and early identification on clay-dominated sites influenced by former periglacial conditions through careful choice of borehole technique, trial pitting and core logging is highly advisable.

(a) (b) (c) (d)

Fig. 10. Contemporary examples of (**a**) slickensides below a modern-day active-layer detachment slide; (**b**) modern retrogressive flow lobes; (**c**) retrogressive flowslides of various ages, All Banks Island, Canada; (**d**) detachment slides and mudflows, Longyeardalen, Svalbard. Source: (a)–(c) Giles *et al.* (2017), photo credit D.J.A. Evans ; (d) Hanne H. Christiansen, The University Centre in Svalbard (UNIS).

Shallow periglacial shears indicate shallow translational landsliding, like active-layer detachment slides in Arctic regions (Lewkowicz 1990; Hutchinson 1991; Harris and Lewkowicz 1993; Lewkowicz and Harris 2005; Harris 2013). Sliding on low-gradient slopes is attributed to build-up of high porewater pressures during thaw consolidation of ice-rich soil in the active layer. Pore pressures greater than hydrostatic pressures can reduce the shear strength of sediment on slopes as gentle as 4° to the Factor of Safety threshold for failure. Deeper relict shears have been attributed to landslides seated at the base of upward-thawing permafrost ('underthaw'), again with porewater pressures considerably more than hydrostatic pressures needed to induce failure (Spink 1991). Overall, relict periglacial shears are a major periglacial geohazard in the UK (Skempton *et al.* 1991; Culshaw *et al.* 2017).

Case study: the A21 Sevenoaks bypass, Kent

During the construction of the Sevenoaks bypass, which began in 1965, major slope failures occurred during the earthworks programme that led to a new understanding of the

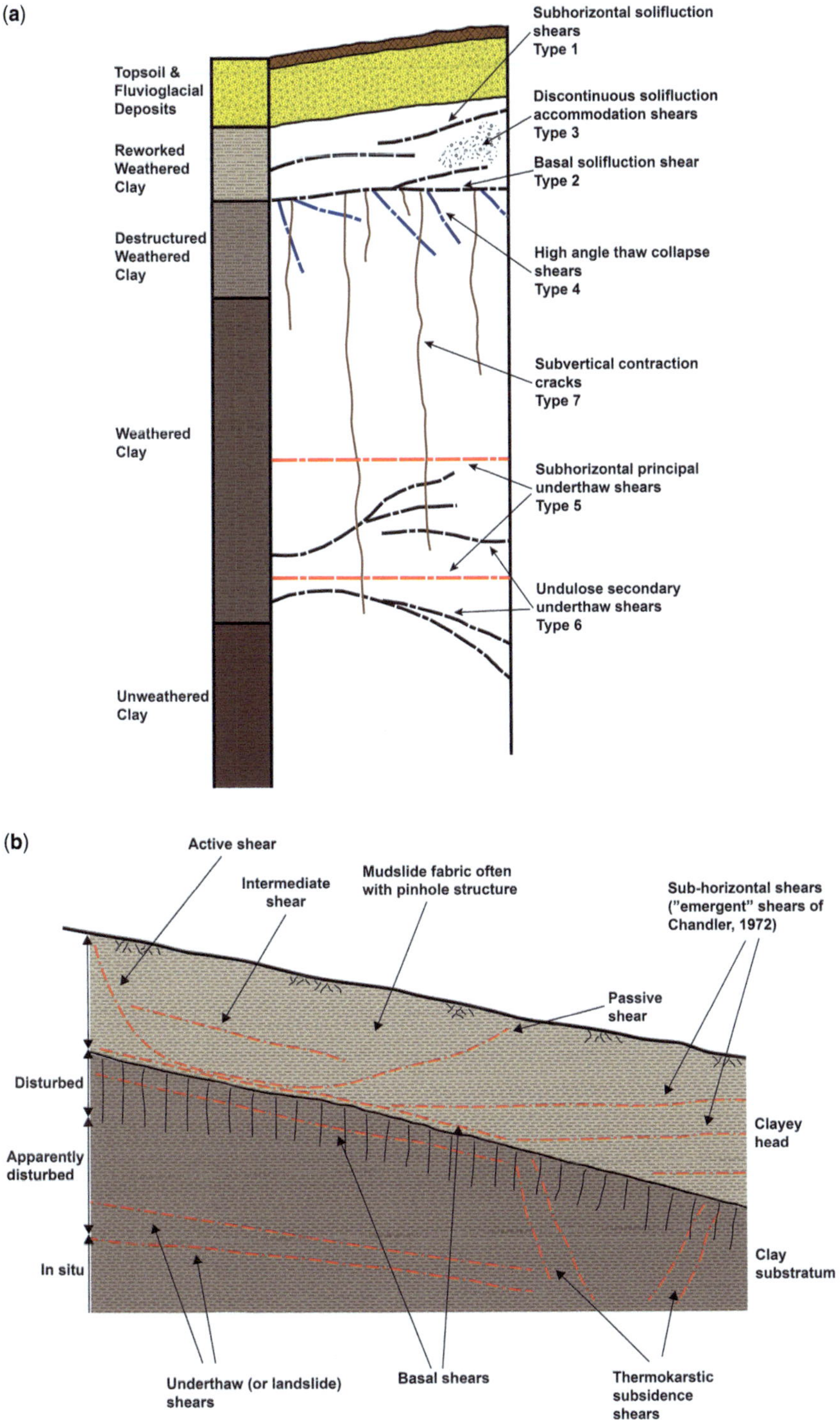

Fig. 11. Schematic profile through weathered Eocene clay illustrating periglacial discontinuity forms. Source: Griffiths and Giles (2017), modified from Spink (1991) and Hutchinson (1992).

Table 2. Different types of discontinuity seen in Eocene clays (Spink 1991)

Discontinuity category	Discontinuity style
Type 1	Subhorizontal solifluction shears
Type 2	Basal solifluction shears
Type 3	Discontinuous solifluction accommodation shears
Type 4	High-angle thaw collapse shears
Type 5	Subhorizontal principal underthaw shears
Type 6	Undulose secondary underthaw shears
Type 7	Subvertical contraction cracks

behaviour and geotechnical properties of clay-dominated slopes. The project is considered a seminal case study in engineering geology (Weeks 1969, 1970; Skempton and Weeks 1976; Martin *et al.* 2017). The failures occurred in the non-engineered ground during earthworks construction near the base of the south-facing scarp slope of the Lower Greensand Ridge that lies just to the south of the North Downs in Kent (Fig. 12), within the zone of the slope-foot as demonstrated in the conceptual model shown in Figure 9.

The nature of the landforms, their initial origin and associated geohazard were not part of the engineering knowledge base at the time of the road construction. Subsequent interpretations ascribed the landforms as relict solifluction lobes where flow could be expected as their mode of emplacement, yet clear evidence of shearing was also present within them. With current-day thinking, and the development of a site conceptual ground model in this relict periglacial terrain, we would now anticipate areas of both relict solifluction (flow) and active layer detachment slides (shear). These two key processes of **solifluction** and **active-layer detachment** are critical in the development and understanding of ground models in relict periglacial landsystems.

Solifluction sheets and aprons

Solifluction landforms are expanses of mobile or formerly mobile sediments on gentle to moderate slopes that have moved downslope by solifluction, and which often terminate

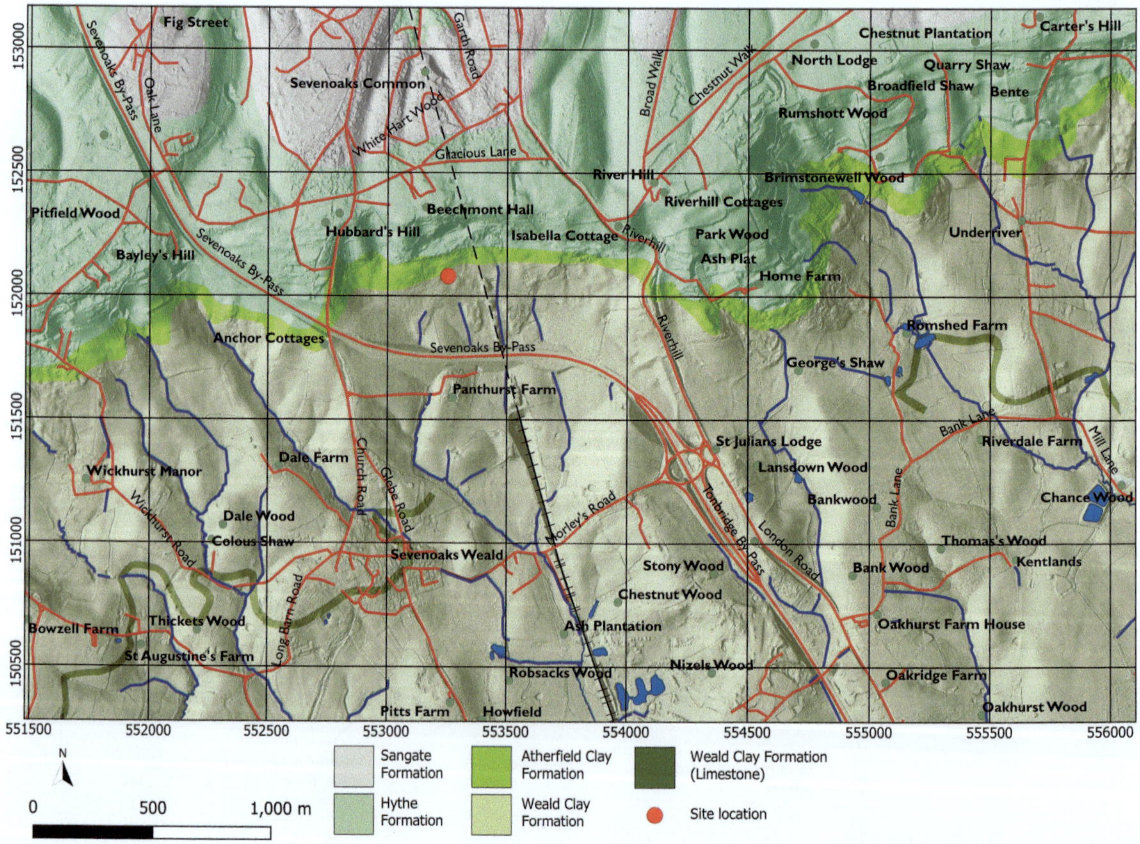

Fig. 12. Geological setting with bedrock stratigraphy overlaid on 1 m DTM.

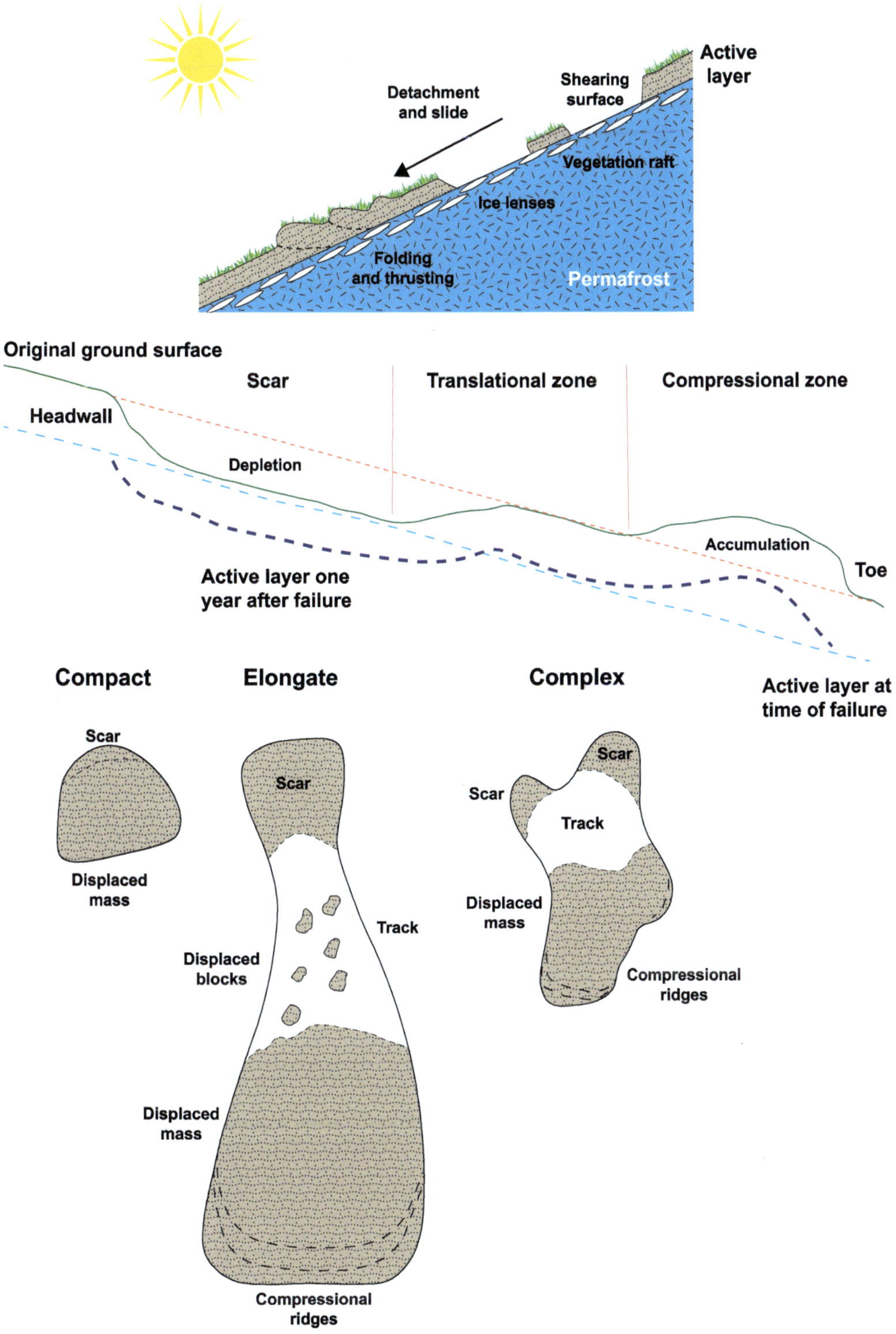

Fig. 13. Active layer detachment slide failure mechanisms and slide morphology. Source: adapted from Favero (2009).

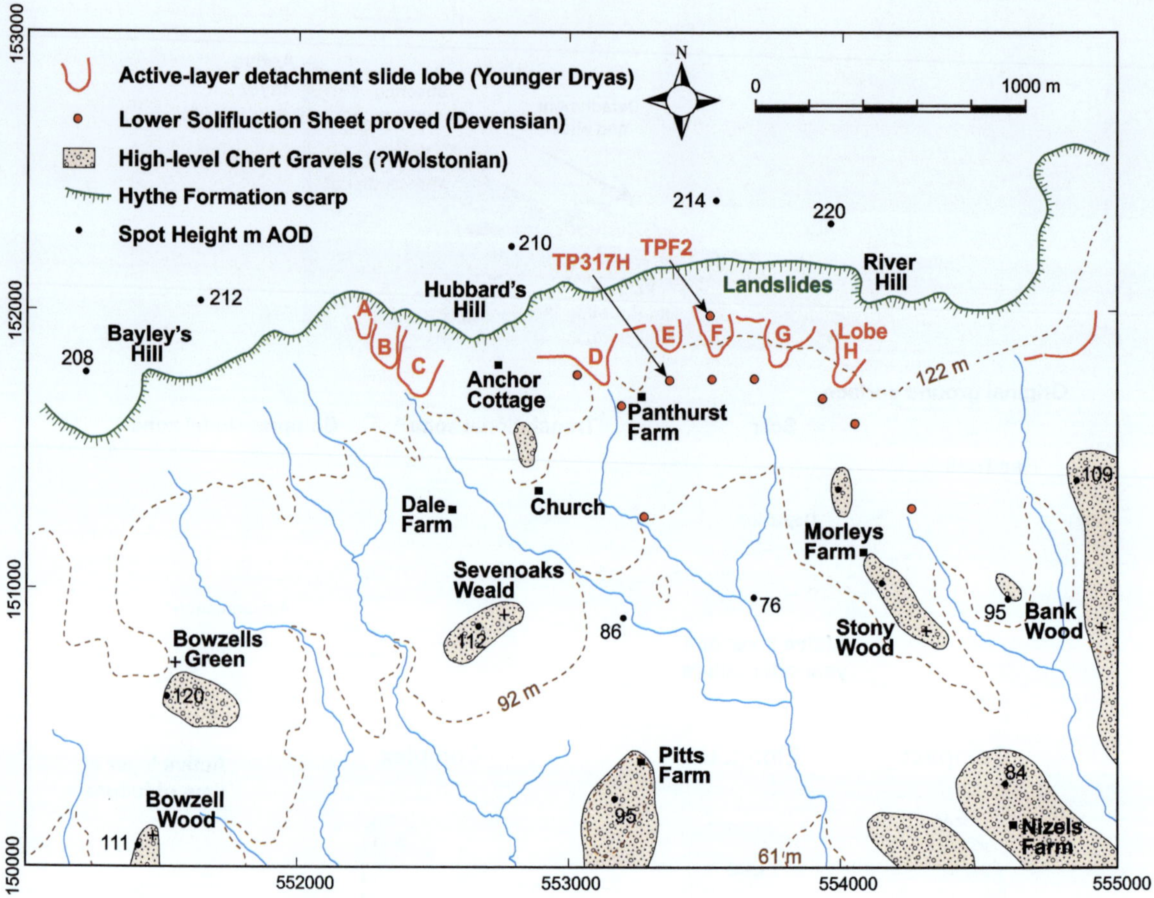

Fig. 14. Identified periglacial landforms and deposits near Hubbard's Hill, Sevenoaks at the site of the original A21 bypass scheme showing locations of post-slope failure trial pit ground investigation. Source: Murton and Giles (2016).

downslope at a steep or vertical face. Solifluction sheets generally have little or no surface expression. Morphologically, they are smooth convexo-concave slopes that can extend from several hundred metres to 3–4 km downslope. Low-strength horizons can be expected in clay-rich deposits. They are the predominant form of periglacial mass movement in active periglacial environments, and solifluction deposits and landforms are widespread and common in relict form. They result from the slow downslope movement of soil due to recurrent freezing and thawing of the ground and are emplaced on slopes with inclinations as low as 1–2° due to excessive porewater pressures generated in thaw (Giles *et al.* 2017; Ballantyne 2018).

Solifluction lobes

Solifluction lobes have a clear morphological expression on the ground and are defined by a steep vertical face, commonly developed at the downslope margin of a sheet that can rise to 1.5 m high. Lobes can often override solifluction sheets. Low-strength horizons can be expected in clay-rich deposits. Again, they result from the slow downslope movement of soil due to recurrent freezing and thawing of the ground (Giles *et al.* 2017; Ballantyne 2018).

Solifluction benches and terraces

Solifluction benches and terraces have a gently to moderately sloping land surface with a steeply sloping vertical face or riser up to several metres high at the downslope margin. The riser commonly runs parallel to the valley axis, delineating a terrace- or bench-like landform underlain by head deposits. Again low-strength horizons can be expected in clay-rich deposits. Solifluction benches and terraces tend to develop in valley-floor locations, where thick accumulations of granular or clay-rich head deposits have been incised by later fluvial activity to produce a terrace-like form (Giles *et al.* 2017; Ballantyne 2018).

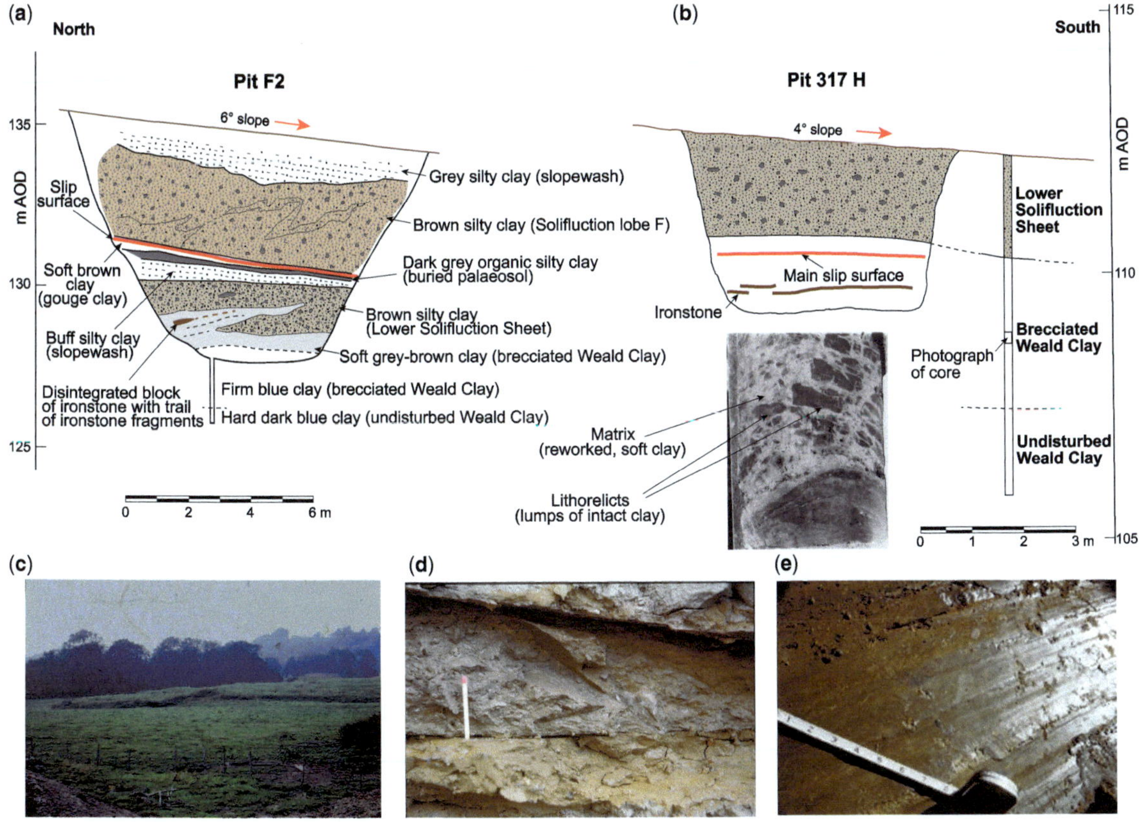

Fig. 15. (**a**) Vertical cross-section through trial pit F2 excavated in lobe F (Fig. 14), showing two clay-rich head deposits separated by a prominent slip surface. (**b**) Vertical cross-section through Trial Pit 317 H (Fig. 14) and adjacent borehole, 100 m south of solifluction lobe E with Brecciated Weald Clay, 3.5 m depth in borehole 317H. Here angular to subangular lithorelicts of clay are set within a matrix of soft, soliflucted, reworked clay. (**c**) Photograph of Lobe F, looking NW (dated 1967). A lower solifluction sheet extends beyond the edge of the lobe in the foreground. (**d**) Multiple shear planes visible in Trial Pit F2, match for scale. (**e**) Clearly polished and striated shear surface from Trial Pit F2. Source: Murton and Giles (2016), modified from Skempton and Weeks (1976); original photographs from Weeks (1970).

Active layer slides

Active layer slides are translational (surface-parallel) landslides that when initially developed usually involved the full depth of the permafrost active layer (Fig. 13). Relict forms are identifiable as a shallow translational planar slide. These slope failures are initially triggered by rapid thaw of ice at the base of the active layer or in the transition layer at top of permafrost. Two forms can be recognized in the field, active-layer detachment slides, where blocks of intact soil have moved downslope over permafrost on a slope-parallel shear plane or thin layer of liquefied sediment, and active-layer flowslides or skinflows where the sliding mass loses coherence and continues to move downslope as a viscous flow (Ballantyne 2018). Many active-layer failures are seen to exhibit hybrid characteristics. The nature of movement depends on moisture content and soil composition. Cohesionless sands or silty soils with low liquid limits will tend to form flowslides, whereas clay-rich soils with relatively high liquid limits will tend to move as translational slides (Giles *et al.* 2017; Ballantyne 2018). These translational slides maintain their coherence above a basal shear zone, especially where previous two-sided freezing of the active layer has resulted in drying and overconsolidation of the central part of the active layer (Lewkowicz and Harris 2005).

Figure 10 shows contemporary examples of modern day active-layer detachment slides, mudflows and retrogressive flow lobes with slickensides indicating the downslope shearing.

Ground engineering consequences

The legacy of the periglacial environment can be considered in terms of the geotechnical and geomorphological

Fig. 16. (**a**) Slope failure in Lobe G (Fig. 14) occurring during earthworks for the original bypass alignment. (**b**) Earthworks in area of Lobe G. (**c**) Head of embankment failure Lobe B. Note the downslope distortion of the installed drainage pipe. (**d**) Toe of embankment failure, Lobe B. Source: original photographs from Weeks (1970).

Fig. 17. Face of Trial Pit F2, excavated in Lobe F (Fig. 14), with Professors Skempton and Morgenstern inspecting the exposure. Source: original photographs from Weeks (1970).

modification of the ground conditions in those areas affected. The past permafrost conditions in the Sevenoaks area are very likely to have developed at depths typically investigated by engineering geologists and geotechnical engineers during ground investigations and at levels where they can affect the engineering works. Essential to the understanding of the engineering geology of periglaciated regions is the concept of frost susceptibility (Murton and Giles 2016; Murton and Ballantyne 2017). Frost-susceptible ground represents soil, rock, or the weathered profile in which segregated ice will form (causing frost heave) (van Everdingen 1998). Frost-susceptible ground eventually becomes ice-rich and the ground becomes increasingly vulnerable to thaw-weakening effects when its ice content converts back to free water. Under these conditions disturbance and/or redeposition of affected sediments can occur when permafrost degrades with seasonal summer thaw or wider climatic change. The legacy of these conditions leads to a variety of geohazards as relict problems within the ground profile, all with significant ground engineering implications. Berry *et al.* (2020)

provide a more detailed narrative on the ground engineering consequences that could be anticipated in such terrains.

Figure 14 maps out these significant periglacial landforms that were encountered during the Sevenoaks bypass construction. Figures 15–18 detail the highly problematic ground conditions that were present in the ground profile from this periglacial legacy with a variety of remoulded soils, numerous shears and active failure planes.

Fig. 18. Lobe E (Fig. 14), taken in spring 2002, showing signs of active slope movement with tension cracking, compressional ridges, hummocky ground and compression zones all in evidence, indicating the still metastable nature of these relict active-layer detachment slides. The winter of 2002 was particularly wet.

Geomorphological visualization using LiDAR

The availability of LiDAR data coverage for the UK has greatly improved in recent years and, in 2019, the area around the A21 Sevenoaks bypass became available for the first time at 1 m resolution, enabling site visualization and interpretation to be undertaken in some detail. LiDAR imagery allows for a first look at a landscape before any actual field-based walkover survey or geomorphological mapping is undertaken. It is important to note that the digital imagery analysis is not a replacement for these more traditional site assessment techniques, but it does provide targets for further ground truthing and reinforcement of image interpretations. Mapping out of identified features, breaks of slope and changes in surface texture can all support the overall ground model development. The 1 m composite DTM was used to generate both hillshade and topographic visualizations of this classic engineering geology site. Figure 19 shows the nature of the terrain around the area of the original alignment of the bypass works.

(a)

(b)

Fig. 19. (**a**) 1 m DTM hillshade image (×5 vertical exaggeration, sun illumination angle 270°) of the now-constructed A21 Sevenoaks bypass and Lower Greensand escarpment. (**b**) 3D visualization highlighting the escarpment with major rotational and translational landslides with the relict active-layer detachment slides below (escarpment width in lower image *c.* 500 m).

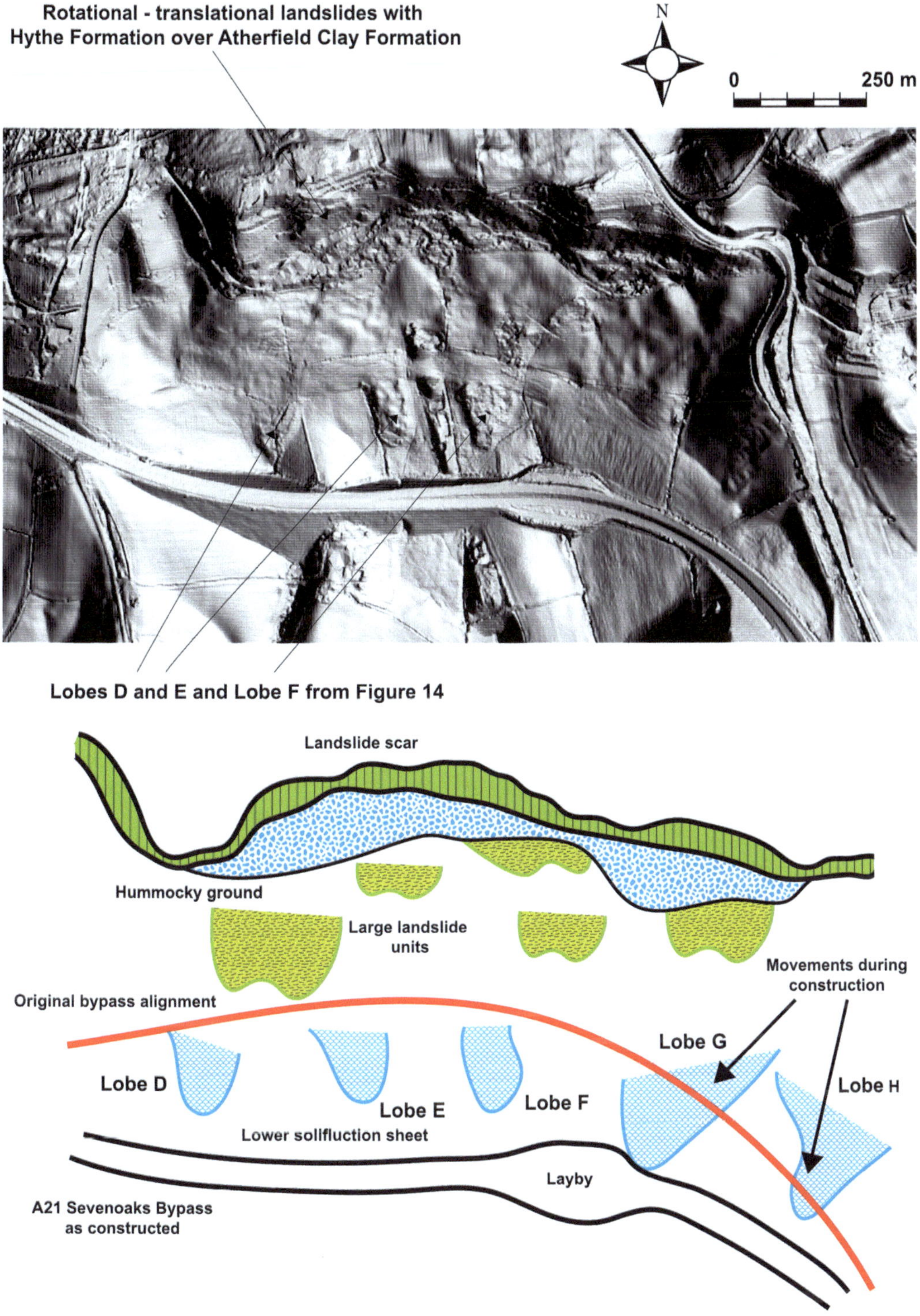

Fig. 20. LiDAR DTM scene of Lobes E and F (Fig. 14) which are clearly identifiable from the imagery. Diagrammatic interpretation and original road alignment below.

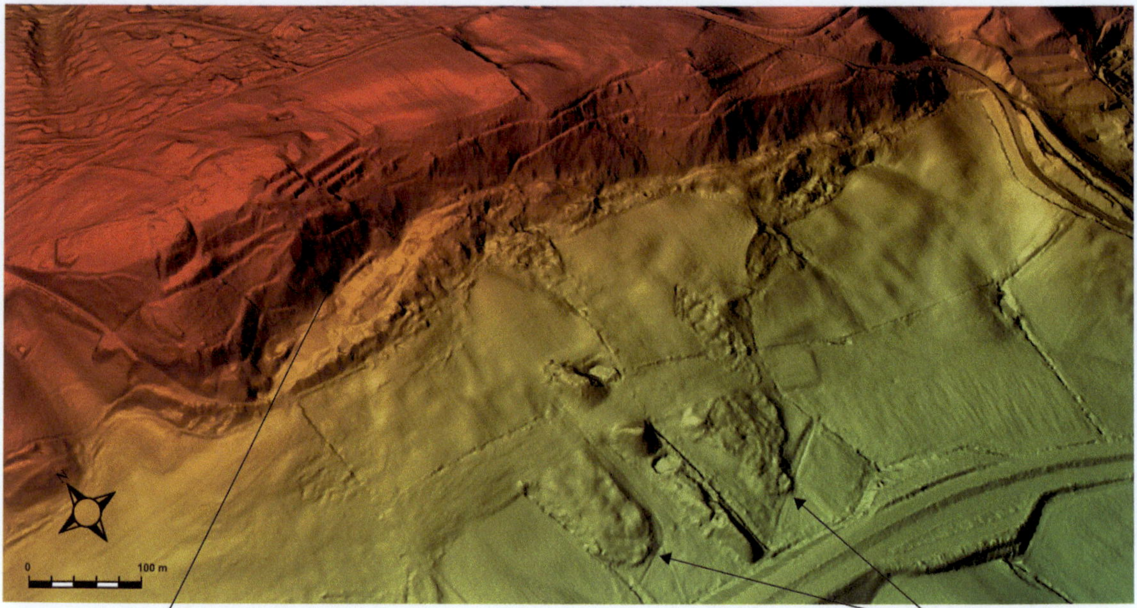

Fig. 21. 3D perspective model of Lobes E and F (Fig. 14).

Hillshade and topographic models

The 1 m LiDAR composite grid can be enhanced via a hillshade algorithm to provide a much clearer and higher impact image of the site for geomorphological interpretation (Figs 19 & 20). This can be further processed with a three-dimensional topographic visualization where the vertical height can be exaggerated to enhance the surface geomorphology (Fig. 21). Geological overlays can be added to explore the geomorphological–geological relationships. Figure 22 clearly picks out the morphology of the lobes fully utilizing the vegetation penetration capability of the LiDAR and DTM image. The wider topographic view delineates the significant rotational and translational landsliding along the Hythe Formation escarpment along with the periglacially derived landforms below. Texture analysis of smooth and rough surfaces facilitates the interpretation of the displaced masses.

Fig. 22. Lobe F (Fig. 14), taken in 2018, still showing signs of metastability with suspended (has moved within the last 12 months, but is not active at present) slope movement. Lobe height *c.* 1.5 m.

Fig. 23. Cambered strata with intermittent infilled gulls in the Hythe Formation at Hermitage Quarry, near Maidstone, Kent. Cliff height *c.* 15 m. Source: Giles *et al.* (2017).

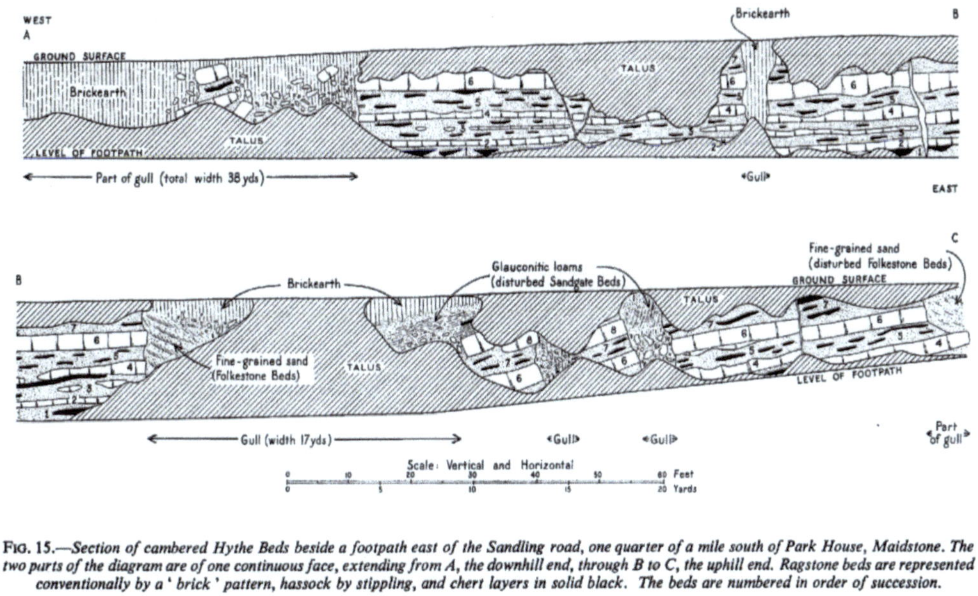

Fig. 24. Ground complexity on cambered strata of the Hythe Formation in the Maidstone area, Kent. Source: Worssam (1963).

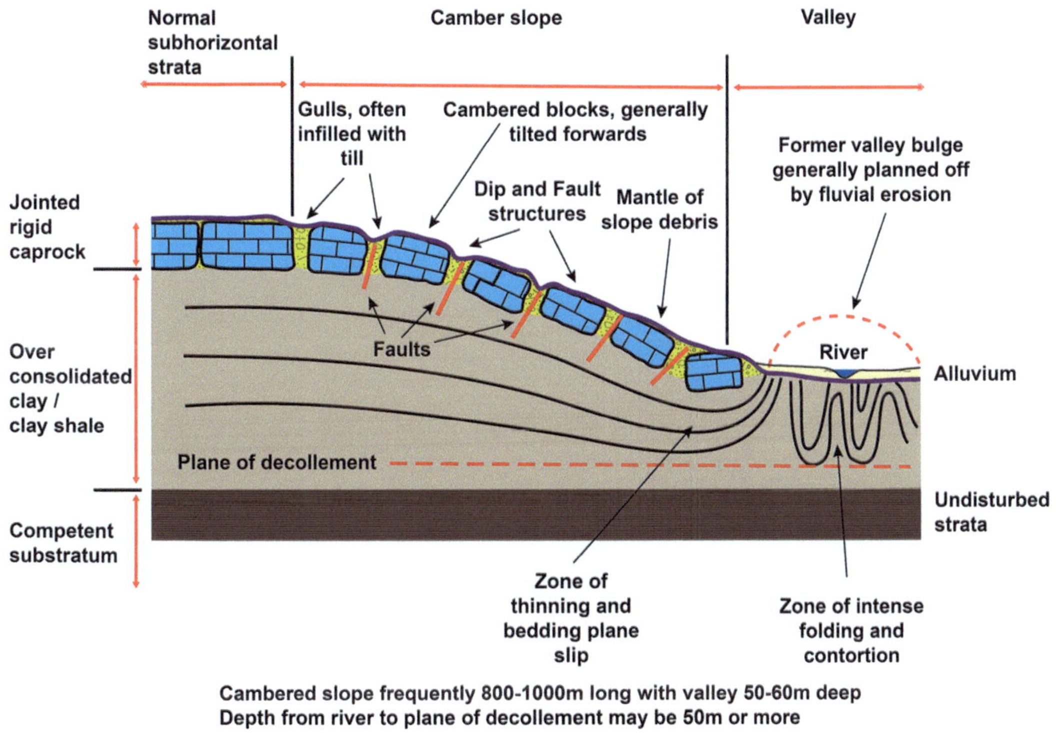

Fig. 25. Schematic section showing periglacial cambering, gulls and valley bulging. Source: Griffiths and Giles (2017).

Fig. 26. Gulls filled with silt and sand in the Hythe Formation at Hermitage Quarry, Maidstone, Kent (southern wall of quarry workings). Cliff height *c.* 15 m.

Case study: Maidstone, Kent

Relict evidence of cambering and associated gulls has been reported in the Maidstone area in Hermitage Quarry (Murton and Giles 2016; Giles *et al.* 2017) and at nearby Allington (Bell *et al.* 2003; Whiteman 2024) (Fig. 23). Wider ground complexity on cambered strata has also been recorded and described for many years (Worssam 1963) (Fig. 24).

Larger macrostructures can be identified in periglacially affected bedrock where valleys have been incised through a relatively competent caprock such as limestone or sandstone into underlying and incompetent stiff clay or mudrock. These structures include cambered strata, gulls, dip-and-fault structures, superficial valley disturbances and valley bulges (Fig. 25) (Giles and Murton 2023). They have a range of morphologies, some of which daylight at the ground surface and are clearly mappable, and others which are concealed at ground level and therefore show little sign of immediate surface expression.

The cambering process involves mass movement of bedrock, whereby the weight of the caprock drives the underlying clay towards the valley axis. The movement involves flexing, stretching, slumping downward and outward, and subsequent gradual breakup of the caprock into separate blocks. Individual blocks also tend to tilt towards the valley axis, with the result that they dip more steeply than the general slope of the valley-side camber, sometimes producing a

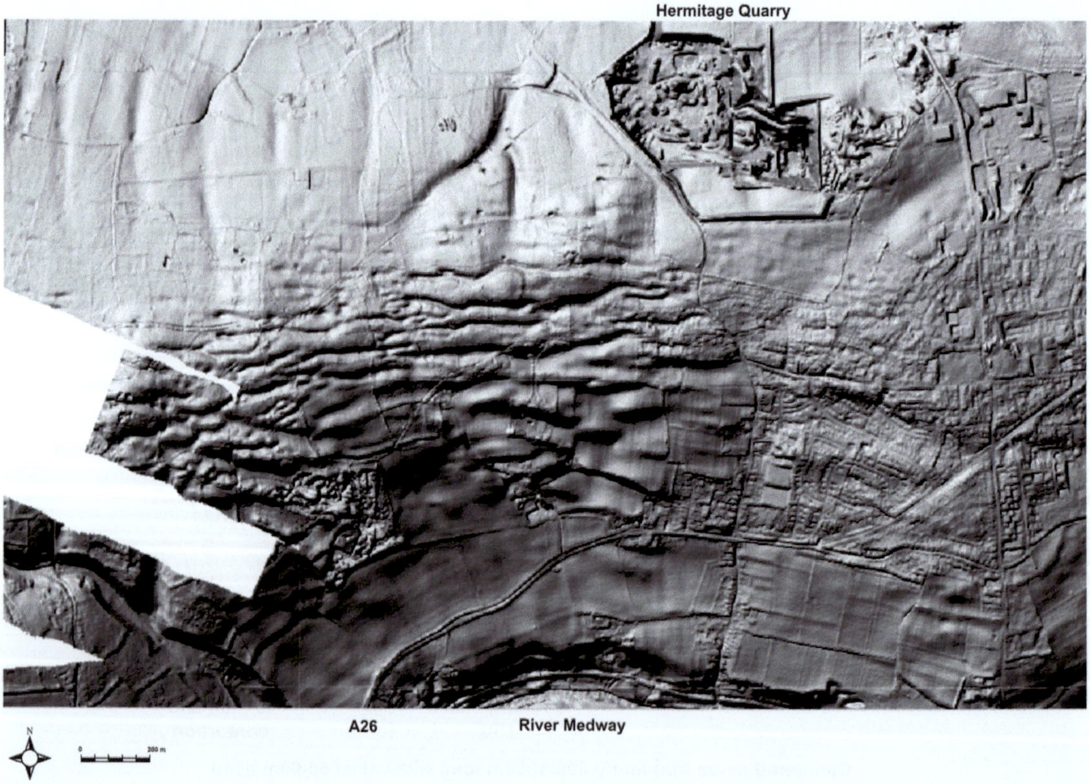

Fig. 27. Environment Agency 50 cm digital terrain model (DTM) LiDAR tile of Maidstone area displaying significant cambered landforms providing a clear visualization of the expected ground morphology in such a terrain.

distinctive dip-and-fault structure (Fig. 24). The relative importance of lateral movement v. subsidence varies.

Widened vertical joints termed gulls commonly separate cambered strata into blocks. Gulls vary in width from centimetres to a few metres, and in height from commonly about 1 m to at least 15 m. They are often aligned more or less parallel to valley-side slopes, but some gulls can run directly into the hillside.

Gulls are widespread in the Hythe Formation and have been noted as far as 550 m back from the Hythe escarpment. Some gulls can be large in size, up to 45 m wide has been recorded, and up to 500 m long (Edmonds 2018). Many remain open and hidden underground, whereas others are infilled partially or completely with clastic sediment derived from the sides and/or above, as observed near Maidstone as reported by Worssam *et al.* (1981) and Whiteman (2024). The gull-fill occurs in widened fissures (gulls) in the Hythe Formation (Fig. 26). The fractures or fissures in the cambered beds usually trend parallel to the elevation contour.

Worssam *et al.* (1981) originally suggested that gulls might have formed along fissures as a result of dissolution of the Hythe Formation, water being concentrated at a few widely spaced points.

Large segregations of ground ice may have accumulated in the gulls during the periglacial conditions that existed in late Devensian (i.e. Late Pleistocene) times although the role of ground ice in gull formation is uncertain. No modern analogues have been reported from modern permafrost regions, and knowledge of ground ice in bedrock is limited. It is possible that ground ice accumulated within gulls and contributed to their widening and/or infilling, but details are speculative. Modern ice wedges tend to form a polygonal pattern, which is inconsistent with the parallel, linear pattern of ridges and depression shown in Figures 27, 28 and 29. A possible analogy may be with dilation-crack ice, where dilation caused by e.g. pingo uplift stretches the frozen overburden, and meltwater later fills the cracks. At Hermitage Quarry, however, tension was not generated vertically by uplift but instead largely horizontally by cambering. It is possible, therefore, that meltwater from snow, possibly supplemented by hoar frost, filled or partly filled any dilation cracks (i.e. gulls) as they opened. Melt of the dilation-crack ice would have allowed infilling with material beside and/or above the ice (J. B. Murton, pers. comm.).

Some of the gulls, as revealed in quarries, may extend through the entire depth of the Hythe Formation to the

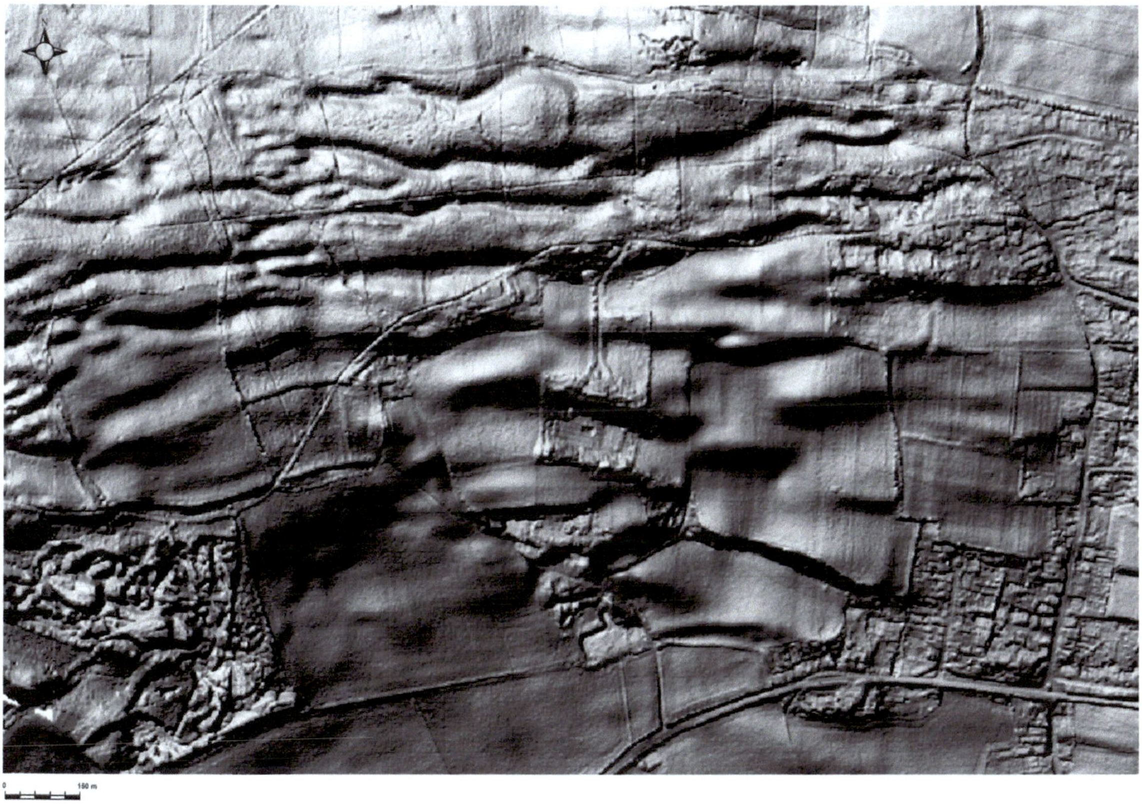

Fig. 28. Zoom detail of digital terrain model (DTM) LiDAR tile of Maidstone area displaying significant cambered landforms.

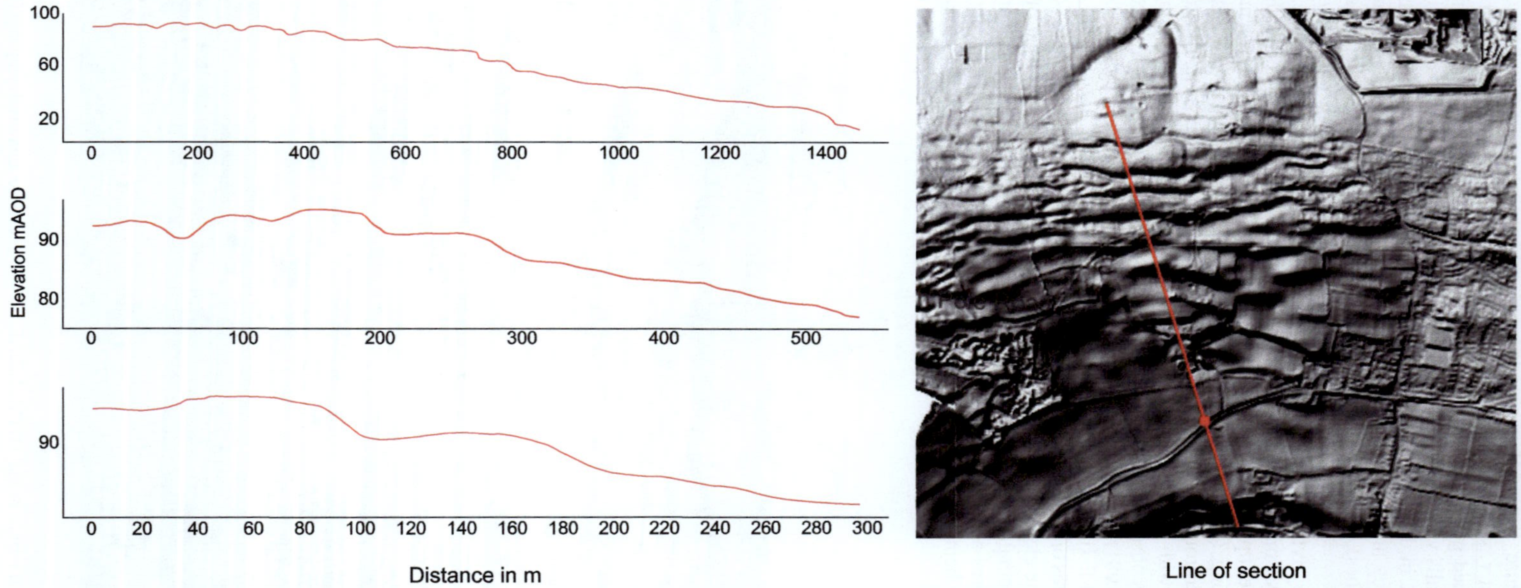

Fig. 29. Cross-section profiles showing amplitude and scale of cambering observed on LiDAR imagery (distances shown in metres).

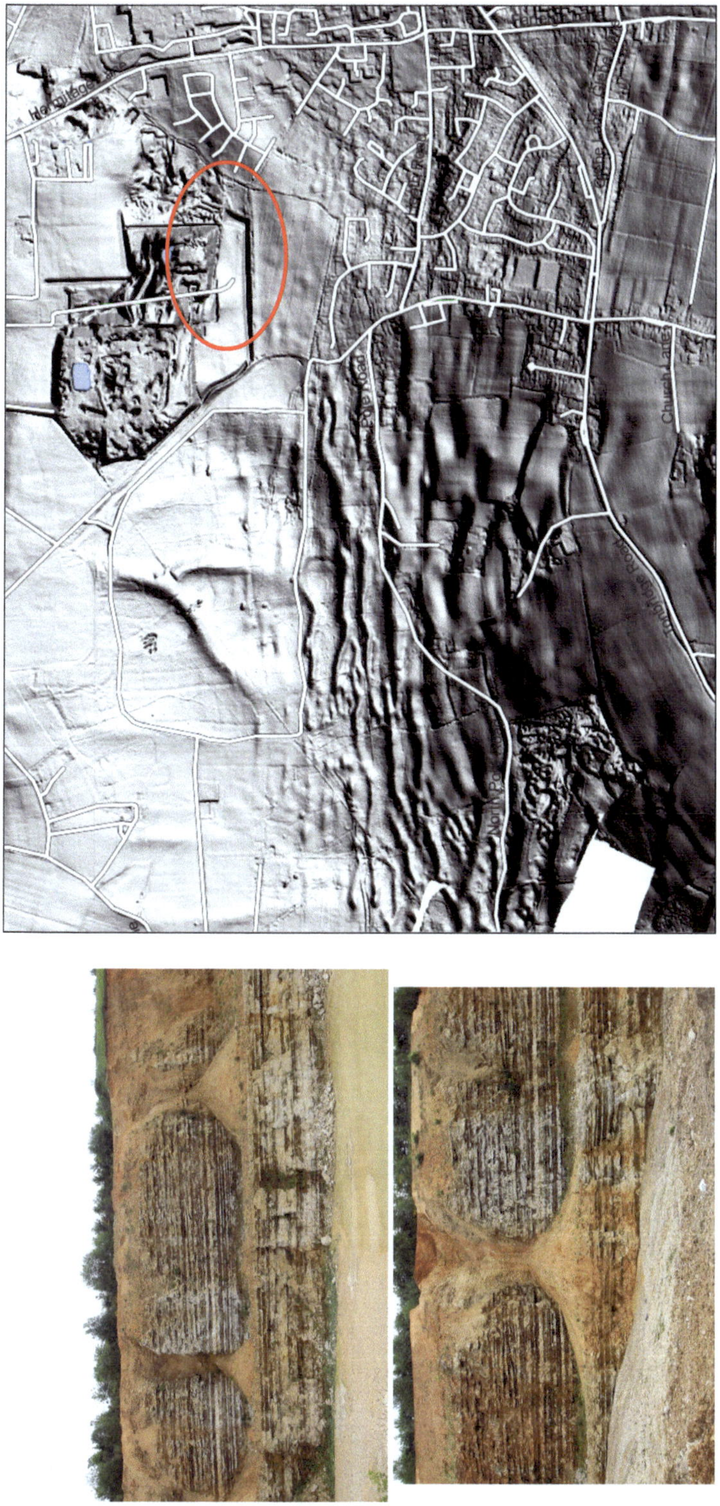

Fig. 30. Illustration of landforms as observed in Hermitage Quarry with respect to LiDAR digital terrain model (DTM) image.

underlying Atherfield Clay Formation and some that contain windblown sand or silt fill have been observed up to 45 m wide at the surface and half a kilometre in length. As the ground ice melted with the disappearance of periglacial conditions at the end of the Devensian, the fill deposits slowly subsided into the gulls (Edmonds 2018).

Ground engineering consequences

The gulls present a variety of ground engineering-related hazards including potential ground movement upon loading, shear surfaces, voids, low shear strength infill, anisotropic permeability, anisotropic stiffness and high-permeability zones. The gulls have been filled with a variety of loose materials. Those on the higher ground contain clays and sand from the Sandgate or Folkestone formations, or chert fragments from the Hythe Formation, while those on valley slopes generally contain river gravel or windblown sand or silt (*brickearth*) type material. The latter occurs as a thin layer of yellowish-brown loam or silt some 0.5–1.0 m in thickness above the Hythe Beds.

Geomorphological visualization using LiDAR

There is significant evidence from the LiDAR DTM imagery of this cambered terrain that can be viewed to the south of Hermitage Quarry located directly to the NW of the site (Fig. 27).

The image clearly illustrates the scale and distribution of these landforms and the problematic ground conditions that they present (Fig. 28). These features would present considerable challenges for on-ground mapping with the LiDAR providing that greater synoptic view. Profile cross-sections (Fig. 29) show the amplitude and scale of these cambered blocks, although offering no insight as to the gull hazard present below ground, which can only be predicted based on a suitable conceptual model (Murton and Ballantyne 2017).

Figure 30 provides a visual contrast as to the size and nature of the cambers and gulls in cross-section, and as to the distance that these features are from the valley floor, suggesting that even the LiDAR image cannot fully delineate the total extent of these relict periglacial geohazards.

Conclusions

The site at Sevenoaks in Kent is a classic location in the development of engineering geology, similarly with the site described at Maidstone. These sites present a variety of landforms and deposits as a legacy of the former periglacial environment in the region during the Quaternary. The availability of high-resolution LiDAR imagery allows high-quality DTMs to be used with hillshaded and topographic visualizations to greatly enhance the mapping and understanding of these complex sites in terms of their geomorphology and landforms. Digital LiDAR imagery presents the engineering geologist with a high-quality dataset at project-scale ground resolutions that can be digitally processed and enhanced to provide an important tool in ground assessment and geohazard identification and delineation. The use of LiDAR for large-scale geomorphological mapping, ground model development and engineering geological assessment provides a fitting and logical extension to the developmental work of Fookes in these areas.

Acknowledgements As part of the Engineering Group of the Geological Society Working Party on the engineering geology and geomorphology of glaciated and periglaciated terrains (Griffiths and Martin 2017), a field visit to the Sevenoaks and Hermitage Quarry sites (Murton and Giles 2016) was attended by both engineering geologists and Quaternary geoscientists. A significant aspect of this meeting was a discussion of the mechanics operating in the Quaternary that left the associated ground legacy seen at Sevenoaks and Maidstone. In this highly problematic landsystem with its associated geological and geotechnical hazards the engineering geologists were presented with a reinterpretation of the (engineering geological) orthodox views of soliflucted-related landforms. The mantra that the best geologist is the one who has seen the most rocks was perfectly illustrated by the Quaternary geoscientists who had field experience in current permafrost environments of northern Canada where a variety of slope failures triggered by permafrost thaw could be observed. Their input to this model development and the engineering geological understanding is greatly acknowledged. Specific thanks go to Professors Julian Murton and David Evans who helped with this new appreciation of the Sevenoaks site. Thanks also to the two reviewers of this chapter who provided excellent feedback and helpful comments.

Competing interests The authors declare that they have no known competing financial interests or personal relationships that could have appeared to influence the work reported in this paper.

Author contributions **DPG**: writing – original draft (lead).

Funding This research received no specific grant from any funding agency in the public, commercial, or not-for-profit sectors.

Data availability The datasets generated during and/or analysed during the current study are available in the Defra repository, https://environment.data.gov.uk/survey.

References

Aerial Precision 2023. *Unveiling the Precision of LiDAR Mapping: How Accurate is it Really?*, https://www.aerial-precision.com/

blog/exploring-the-precision-of-lidar-mapping-unveiling-its-accuracy

Baghdadi, N., Mallet, C. and Zribi, M. 2018. *QGIS and Generic Tools*. John Wiley & Sons, https://doi.org/10.1002/9781119457091

Ballantyne, C.K. 2018. *Periglacial Geomorphology*. Wiley Blackwell.

Bell, F.G., Culshaw, M.G. and Northmore, K.J. 2003. The metastability of some gull-fill materials from Allington, Kent, UK. *Quarterly Journal of Engineering Geology and Hydrogeology*, **36**, 217–229, https://doi.org/10.1144/1470-9236/02-005

Berry, T.W., Fish, P.R., Price, S.J. and Hadlow, N.W. 2020. Periglacial geohazards in the UK. *Geological Society, London, Engineering Geology Special Publications*, **29**, 259–289, https://doi.org/10.1144/EGSP29.10

Bromhead, E.N., Huggins, M. and Ibsen, M.L. 2000. Shallow landslides in Wadhurst clay at Robertsbridge, Sussex, UK. *In*: Bromhead, E., Dixon, N. and Ibsen, M.L. (eds) *Landslides in Research, Theory and Practice: Proceedings of the 8th International Symposium on Landslides*. Thomas Telford, Cardiff, 1–183, https://www.icevirtuallibrary.com/doi/abs/10.1680/lirtapv1.34617.0031

Brunsden, D., Doornkamp, J.C., Fookes, P.G., Jones, D.K.C. and Kelly, J.M.H. 1975. Large scale geomorphological mapping and highway engineering design. *Quarterly Journal of Engineering Geology and Hydrogeology*, **8**, 227–253, https://doi.org/10.1144/GSL.QJEG.1975.008.04.01

Buckley, S.J., Howell, J.A., Enge, H.D. and Kurz, T.H. 2008. Terrestrial laser scanning in geology: data acquisition, processing, and accuracy considerations. *Journal of the Geological Society, London*, **165**, 625–638, https://doi.org/10.1144/0016-76492007-100

Chandler, J.H. 2001. Terrain measurement using automated digital photogrammetry. *Geological Society, London, Engineering Geology Special Publications*, **18**, 13–18, https://doi.org/10.1144/gsl.eng.2001.018.01.02

Chandler, J.H. and Moore, R. 1989. Analytical photogrammetry: a method for monitoring slope instability. *Quarterly Journal of Engineering Geology and Hydrogeology*, **22**, 97–110, https://doi.org/10.1144/GSL.QJEG.1989.022.02.02

Chen, N., Cai, X., Li, S., Zhang, X. and Jiang, Q. 2020. Automatic extraction of rock mass discontinuity based on 3D laser scanning. *Quarterly Journal of Engineering Geology and Hydrogeology*, **54**, https://doi.org/10.1144/qjegh2020-054

Culshaw, M.G., Entwisle, D.C., Giles, D.P., Berry, T., Collings, A., Banks, V.J. and Donnelly, L.J. 2017. Material properties and geohazards. *Geological Society, London, Engineering Geology Special Publications*, **28**, 599–740, https://doi.org/10.1144/EGSP28.6

Davies, J.P., Loveridge, F.A., Perry, J., Patterson, D. and Carder, D. 2003. Stabilisation of a landslide on the M25 highway London's main artery. *Proceedings of 12th Pan-American Conference on Soil Mechanics and Geotechnical Engineering*, 22–25 June 2003, Boston, USA.

Dearman, W.R. and Fookes, P.G. 1974. Engineering geological mapping for civil engineering practice in the United Kingdom. *Quarterly Journal of Engineering Geology and Hydrogeology*, **7**, 223–256, https://doi.org/10.1144/GSL.QJEG.1974.007.03.01

Defra n.d. Defra survey data download. Retrieved from https://environment.data.gov.uk/DefraDataDownload/?Mode=survey

Dunning, S.A., Rosser, N.J. and Massey, C.I. 2010. The integration of terrestrial laser scanning and numerical modelling in landslide investigations. *Quarterly Journal of Engineering Geology and Hydrogeology*, **43**, 233–247, https://doi.org/10.1144/1470-9236/08-069

Edmonds, C. 2018. Five decades of settlement and subsidence. *Quarterly Journal of Engineering Geology and Hydrogeology*, **51**, 403–416, https://doi.org/10.1144/qjegh2018-089

Ellis, L.A., Harrison, E. and Bowden, A.J. 2011. Landslides on Gault: geomorphological identification and qualitative risk assessment. *Quarterly Journal of Engineering Geology and Hydrogeology*, **44**, 35–48, https://doi.org/10.1144/1470-9236/08-096

Favero, P. 2009. *Active Layer Detachment Morphology, Sedimentology, and Mechanisms, Fosheim Peninsula, Ellesmere Island*. Unpublished PhD thesis, University of Ottawa, Canada, https://doi.org/10.20381/ruor-12426

Fleming, C., Giles, J.R.A. and Marsh, S.H. (eds) 2010*a*. *Elevation Models for Geoscience*. Geological Society, London, Special Publications, **345**, https://doi.org/10.1144/SP345.0

Fleming, C., Marsh, S.H. and Giles, J.R.A. 2010*b*. Introducing elevation models for geoscience. *Geological Society, London, Special Publications*, **345**, 1–4, https://doi.org/10.1144/SP345.1

Fookes, P.G. 1991. Quaternary engineering geology. *Geological Society, London, Engineering Geology Special Publications*, **7**, 73–98, https://doi.org/10.1144/GSL.ENG.1991.007.01.04

Fookes, P.G. 1997. Geology for engineers: the geological model, prediction and performance. *Quarterly Journal of Engineering Geology and Hydrogeology*, **30**, 293–424, https://doi.org/10.1144/GSL.QJEG.1997.030.P4.02

Fookes, P.G., Dale, S.G. and Land, J.M. 1991. Some observations on a comparative aerial photography interpretation of a landslipped area. *Quarterly Journal of Engineering Geology and Hydrogeology*, **24**, 249–265, https://doi.org/10.1144/GSL.QJEG.1991.024.03.01

Fookes, P.G., Baynes, F.J. and Hutchinson, J.N. 2000. Total geological history: a model approach to the anticipation, observation and understanding of site conditions. *ISRM International Symposium*, Melbourne, Australia, November, Paper ISRM-IS-2000-010.

Fookes, P.G., Lee, E.M. and Griffiths, J.S. 2007. *Engineering Geomorphology: Theory and Practice*. Whittles

George, O., Grainger, R. and Conie, P. 2019. The influence of periglacial action on road construction: a case study from the A21 Tonbridge to Pembury dualling. *Quarterly Journal of Engineering Geology and Hydrogeology*, **52**, 320–325, https://doi.org/10.1144/qjegh2018-018

Giglierano, J.D. 2010. LiDAR basics for natural resource mapping applications. *Geological Society, London, Special Publications*, **345**, 103–115, https://doi.org/10.1144/SP345.11

Giles, D.P. and Murton, J.B. 2023. Bedrock disturbances. *In*: *Reference Module in Earth Systems and Environmental Sciences*. Elsevier, https://doi.org/10.1016/B978-0-323-99931-1.00086-6

Giles, D.P., Griffiths, J.S., Evans, D.J.A. and Murton, J.B. 2017. Geomorphological framework: glacial and periglacial sediments, structures and landforms. *Geological Society, London, Engineering Geology Special Publications*, **28**, 59–368, https://doi.org/10.1144/EGSP28.3

Griffiths, J.S. and Giles, D.P. 2017. Conclusions and illustrative case studies. *Geological Society, London, Engineering Geology Special Publications*, **28**, 891–936, https://doi.org/10.1144/EGSP28.9

Griffiths, J.S. and Martin, C.J. (eds) 2017. *Engineering Geology and Geomorphology of Glaciated and Periglaciated Terrains – Engineering Group Working Party Report*. Geological Society, London, Engineering Geology Special Publications, **28**, https://doi.org/10.1144/EGSP28-000

Hall, M., Cooper, A.H., Ford, J., Price, S. and Burke, H. 2010. The use of NEXTMap Britain for geological surveying in the Vale of York. *Geological Society, London, Special Publications*, **345**, 55–66, https://doi.org/10.1144/SP345.7

Harris, C. 1998. Periglacial slope deposits, Sevenoaks, Kent. *In*: Murton, J.B., Whiteman, C.A., Bates, M.R., Bridgland, D.R., Long, A.J., Roberts, M.B. and Waller, M.P. (eds) *The Quaternary of Kent and Sussex: Field Guide*. Quaternary Research Association, London, 7–8.

Harris, C. 2013. Slope deposits and forms. *In*: Elias, S.A. and Mock, C.J. (eds) *Encyclopedia of Quaternary Science*, **3**, 2nd edn. Elsevier, Amsterdam, 481–489, https://doi.org/10.1016/B978-0-444-53643-3.00101-1

Harris, C. and Lewkowicz, A.G. 1993. Form and internal structure of active-layer detachment slides, Fosheim Peninsula, Ellesmere Island, Northwest Territories, Canada. *Canadian Journal of Earth Sciences*, **30**, 1708–1714, https://doi.org/10.1139/e93-149

Higginbottom, I.E. and Fookes, P.G. 1970. Engineering aspects of periglacial features in Britain. *Quarterly Journal of Engineering Geology and Hydrogeology*, **3**, 85–117, https://doi.org/10.1144/GSL.QJEG.1970.003.02.02

Hobbs, P.R.N., Gibson, A., Jones, L., Pennington, C., Jenkins, G., Pearson, S. and Freeborough, K. 2010. Monitoring coastal change using terrestrial LiDAR. *Geological Society, London, Special Publications*, **345**, 117–127, https://doi.org/10.1144/SP345.12

Hsiao, K.H., Liu, J.K., Yu, M.F. and Tseng, Y.H. 2004. Change detection of landslide terrains using ground-based LiDAR data. *XXth ISPRS Congress, Istanbul, Turkey, Commission VII, WG*, **7**, 5, https://cartesia.org/geodoc/isprs2004/comm7/papers/122.pdf

Hutchinson, J.N. 1991. Periglacial and slope processes. *Geological Society, London, Engineering Geology Special Publications*, **7**, 283–331, https://doi.org/10.1144/gsl.eng.1991.007.01.27

Hutchinson, J.N. 1992. Engineering in relict periglacial and extraglacial areas in Britain. *In*: Gray, J.M. (ed.) *Applications of Quaternary Research, Quaternary Proceedings*, **2**. Quaternary Research Association, Cambridge, 49–65.

Jaboyedoff, M., Oppikofer, T., Abellán, A., Derron, M.H., Loye, A., Metzger, R. and Pedrazzini, A. 2012. Use of LIDAR in landslide investigations: a review. *Natural Hazards*, **61**, 5–28, https://doi.org/10.1007/s11069-010-9634-2

Jones, L.D. 2006. Monitoring landslides in hazardous terrain using terrestrial LiDAR: an example from Montserrat. *Quarterly Journal of Engineering Geology and Hydrogeology*, **39**, 371–373, https://doi.org/10.1144/1470-9236/06-009

Kemeny, J. and Donovan, J. 2005. Technical Note: rock mass characterisation using LiDAR and automated point cloud processing. *Ground Engineering*, November, 26–29, https://cdn.ca.emap.com/wp-content/uploads/sites/13/2005/11/2005-11_Pages_26-29.pdf

Lewkowicz, A.G. 1990. Morphology, frequency and magnitude of active-layer detachment slides, Fosheim Peninsula, Ellesmere Island, NWT. *Permafrost-Canada, Proceedings of the Fifth Canadian Permafrost Conference*, Université Laval, Nordicana, **54**, 111–118.

Lewkowicz, A.G. and Harris, C. 2005. Morphology and geotechnique of active layer detachment failures in discontinuous and continuous permafrost, northern Canada. *Geomorphology*, **69**, 275–297, https://doi.org/10.1016/j.geomorph.2005.01.011

Lukas, S., Preusser, F., Evans, D.J., Boston, C.M. and Lovell, H. 2017. The Quaternary. *Geological Society, London, Engineering Geology Special Publications*, **28**, 3–58, https://doi.org/10.1144/EGSP28.2

Martin, C.J., Morley, A.L. and Griffiths, J.S. 2017. Introduction to engineering geology and geomorphology of glaciated and periglaciated terrains. *Geological Society, London, Engineering Geology Special Publications*, **28**, 1–30, https://doi.org/10.1144/EGSP28.1

McKenzie, A.A., Rutter, H.K. and Hulbert, A.G. 2010. The use of elevation models to predict areas at risk of groundwater flooding. *Geological Society, London, Special Publications*, **345**, 75–79, https://doi.org/10.1144/SP345.9

Mikoš, M., Vidmar, A. and Brilly, M. 2005. Using a laser measurement system for monitoring morphological changes on the Strug rock fall, Slovenia. *Natural Hazards and Earth System Sciences*, **5**, 143–153, https://doi.org/10.5194/nhess-5-143-2005

Moore, J.F.A. 1974. Mapping major joints in the Lower Oxford Clay using terrestrial photogrammetry. *Quarterly Journal of Engineering Geology and Hydrogeology*, **7**, 57–67, https://doi.org/10.1144/GSL.QJEG.1974.007.01.04

Moore, R., Fish, P., Trinder, S., Czarnomski, C., Dabson, O. and Fitzgerald, R. 2022. Engineering geomorphology of HS2: management of geohazards. *Quarterly Journal of Engineering Geology and Hydrogeology*, **55**, qjegh2021-122, https://doi.org/10.1144/qjegh2021-122

Murton, J.B. and Ballantyne, C.K. 2017. Periglacial and permafrost ground models for Great Britain. *Geological Society, London, Engineering Geology Special Publications*, **28**, 501–597, https://doi.org/10.1144/EGSP28.5

Murton, J.B. and Giles, D.P. 2016. *The Quaternary Periglaciation of Kent. Field Guide*. Quaternary Research Association, London.

National Oceanic and Atmospheric Administration (NOAA) Coastal Services Center 2012. *Lidar 101: An Introduction to Lidar Technology, Data, and Applications*. NOAA Coastal Services Center, (November), 76, https://coast.noaa.gov/data/digitalcoast/pdf/lidar-101.pdf

Privett, K.D. 2019. The lines of evidence approach to challenges faced in engineering geological practice. *Quarterly Journal of Engineering Geology and Hydrogeology*, **52**, 141–172, https://doi.org/10.1144/qjegh2018-131

Pugh, R.S., Weeks, A.G. and Hutchinson, D.E. 1991. Landslip and remedial works in Wadhurst Clay. *Slope Stability Engineering Developments and Applications: Proceedings of the International Conference on Slope Stability, Isle of Wight*. Institution of Civil Engineers, Thomas Telford, London, 377–382, https://doi.org/10.1680/ssedaa.16606.0058

Radley, J.D. 2006. A Wealden guide I: the Weald Sub-basin. *Geology Today*, **22**, 109–118, https://doi.org/10.1111/j.1365-2451.2006.00563.x

Radley, J.D. and Allen, P. 2012*a*. The non-marine Lower Cretaceous Wealden strata of southern England. *Proceedings of the Geologists' Association*, **123**, 235–244, https://doi.org/10.1016/j.pgeola.2012.01.001

Radley, J.D. and Allen, P. 2012*b*. The Wealden (non-marine Lower Cretaceous) of the Weald sub-basin, southern England. *Proceedings of the Geologists' Association*, **123**, 245–318, https://doi.org/10.1016/j.pgeola.2012.01.003

Rosser, N.J., Petley, D.N., Lim, M., Dunning, S.A. and Allison, R.J. 2005. Terrestrial laser scanning for monitoring the process of hard rock coastal cliff erosion. *Quarterly Journal of Engineering Geology and Hydrogeology*, **38**, 363–375, https://doi.org/10.1144/1470-9236/05-008

Rowlands, K.A., Jones, L.D. and Whitworth, M. 2003. Landslide laser scanning: a new look at an old problem. *Quarterly Journal of Engineering Geology and Hydrogeology*, **36**, 155–157, https://doi.org/10.1144/1470-9236/2003-08

Sercombe, G., Paraskevopoulou, C., Bedi, A. and Vazaios, I. 2023. Automating fault identification along the HS2 Chilterns tunnel alignment from aerial LiDAR scanning. *In*: *Expanding Underground-Knowledge and Passion to Make a Positive Impact on the World*. CRC Press, 2519–2527. https://www.taylorfrancis.com/chapters/oa-edit/10.1201/9781003348030-303/automating-fault-identification-along-hs2-chilterns-tunnel-alignment-aerial-lidar-scanning-sercombe-paraskevopoulou-bedi-vazaios

Skempton, A.W. and Weeks, A.G. 1976. The Quaternary history of the Lower Greensand escarpment and Weald Clay vale near Sevenoaks, Kent. *Philosophical Transactions of the Royal Society of London*, **A283**, 493–526. https://www.jstor.org/stable/74699

Skempton, A.W., Norbury, D., Petley, D.J. and Spink, T.W. 1991. Solifluction shears at Carsington, Derbyshire. *Geological Society, London, Engineering Geology Special Publications*, **7**, 381–387, https://doi.org/10.1144/GSL.ENG.1991.007.01.34

Spink, T.W. 1991. Periglacial discontinuities in Eocene Clay near Denham, Buckinghamshire. *Geological Society, Engineering Geology Special Publications*, **7**, 389–396, https://doi.org/10.1144/GSL.ENG.1991.007.01.35

Van Everdingen, R. (ed.) 1998. *Multi-Language Glossary of Permafrost and Related Ground-ice Terms*. National Snow and Ice Data center / World Data Center for Glaciology, Boulder, Colorado, https://globalcryospherewatch.org/reference/glossary_docs/Glossary_of_Permafrost_and_Ground-Ice_IPA_2005.pdf [revised May 2005].

Walstra, J., Chandler, J.H., Dixon, N. and Dijkstra, T.A. 2007. Aerial photography and digital photogrammetry for landslide monitoring. *Geological Society, London, Special Publications*, **283**, 53–63, https://doi.org/10.1144/SP283.5

Weeks, A.G. 1969. The stability of natural slopes in south-east England as affected by periglacial activity. *Quarterly Journal of Engineering Geology*, **2**, 49–61, https://doi.org/10.1144/GSL.QJEG.1969.002.01.04

Weeks, A.G. 1970. *Stability of the Lower Greensand Escarpment in Kent*. PhD thesis, University of Surrey, https://epubs.surrey.ac.uk/id/eprint/848552

Wehr, A. and Lohr, U. 1999. Airborne laser scanning – an introduction and overview. *ISPRS Journal of Photogrammetry and Remote Sensing*, **54**, 68–82, https://doi.org/10.1016/S0924-2716(99)00011-8

Whiteman, C.A. 2024. Periglacial landforms and landscape development in southern England. *Proceedings of the Geologists' Association*, https://doi.org/10.1016/j.pgeola.2024.04.006

Whitworth, M.C.Z., Giles, D.P. and Murphy, W. 2005. Airborne remote sensing for landslide hazard assessment: a case study on the Jurassic escarpment slopes of Worcestershire, UK. *Quarterly Journal of Engineering Geology and Hydrogeology*, **38**, 285–300, https://doi.org/10.1144/1470-9236/04-057

Whitworth, M.C.Z., Giles, D.P., Anderson, I. and Clewett, M. 2006*a*. Terrestrial laser scanning for applied geoscience and landslide studies in the urban environment. *Engineering Geology for Tomorrow's Cities. The Tenth IAEG Congress*, 6–10 September 2006, Nottingham, United Kingdom, Paper No. 252, https://media.geolsoc.org.uk/iaeg2006/

Whitworth, M.C.Z., Giles, D.P. and Anderson, I. 2006*b*. Landslide imaging techniques for urban geoscience reconnaissance. *Engineering Geology for Tomorrow's Cities. The Tenth IAEG Congress*, 6–10 September 2006, Nottingham, United Kingdom, Paper No. 245, https://media.geolsoc.org.uk/iaeg2006/

Worssam, B.C. 1963. *Geology of the country around Maidstone* (explanation of one-inch geological sheet 288, new series).

Worssam, B.C., Neale, J. and Fleney, J. 1981. Pleistocene deposits and superficial structures, Allington Quarry, Maidstone, Kent. *In*: *The Quaternary in Britain*. Pergamon Press, Oxford, 221–229.

Endpiece: reflections on Peter Fookes' legacy

David T. Shilston[1], E. Mark Lee[2] and James S. Griffiths[3]*

[1]AtkinsRéalis, Woodcote Grove, Ashley Road, Epsom KT18 5BW, UK
[2]Ebor Geoscience Ltd, 15 Whernside Avenue, York YO31 0QB, UK
[3]School of Geography, Earth and Environmental Sciences, University of Plymouth, Drake Circus, Plymouth PL4 8AA, UK

EML, 0000-0002-5872-7654; JSG, 0000-0002-6431-3992
*Correspondence: jim.griffiths@plymouth.ac.uk

Abstract: This volume is about the work of Professor Peter G. Fookes and the legacy he has left to the present and future generations of engineering geologists. In this endpiece the nature of this legacy is reflected through the experiences of scientists and engineers who worked with him, his emphasis on case studies in his published work, and how Peter's ground model concept espoused in his 1997 Glossop Lecture can be used to accommodate landscape changes and changes to ground conditions that may occur during the lifetime of a project.

> Legacy: the long-lasting impact of particular events, actions, etc., that took place in the past, or of a person's life (*Oxford English Dictionary* definition).

Professor Peter George Fookes' professional life and work is described in this volume by John Charman, with the other papers describing some examples of how his work has formed the foundation for many developments in engineering geology. However, what do we, as three authors who each knew Peter for significant parts of our professional careers, regard as his legacy to engineering geology and how might his legacy influence the future of the discipline? These thoughts are explored in this final chapter under three distinct themes: Peter's role in furthering the profession through his links with, and support for, fellow professionals, notably those in their early careers and including those in allied disciplines; the emphasis in his publications on case studies; and, looking ahead, how Peter's ground model concept has the potential to be adapted to situations where landscape change (and changes in ground conditions) can be significant over the lifetime of a project.

Furthering the profession

In a series of papers presented to the Association of Engineering Geologists in the USA, Tepel (2002*a*, *b*, 2003, 2004*a*, *b*, *c*, *d*) put forward a vision for the future of engineering geology. A key element of the vision was a concept he described as 'stewardship'. Stewardship requires leadership and commitment by individuals, companies and professional bodies. Tepel (2004*c*, p. 16) stated:

> Engineering geologists benefit humanity by discovering, defining, and analysing geologically based risks or conditions that impact, or might impact, humans as they utilize and interact with their built and natural environment.

Tepel (2004*c*) suggested that practitioners must recognize they had a responsibility for ensuring the long-term health of our profession through improving practice, mentoring new graduates and developing the state of the art. This responsibility is one that Peter understood and took seriously from the outset of his career. Throughout his working life Peter developed the state of the art through his successful consultancy practice and publications (Charman 2025), but perhaps even more importantly he encouraged individuals to progress their own careers in his role as a teacher, mentor and consultant. To illustrate this commitment there follows a series of 'Experiences with Peter Fookes' text boxes provided by his students and colleagues. These start with his teaching on the Queen Mary College MSc in Geomaterials, which, in conjunction with Alan Poole, he established in 1979–80. Box 1 is a contribution by Alan and this is followed by an input from two of the course's alumni, Geoff Pettifer (Box 2) and Adrian Morey (Box 3).

In addition to his teaching at QMC, Peter was the external examiner on undergraduate programmes at the University of Plymouth on a number of occasions and provided the funding for an annual prize for the best undergraduate field project. His contributions to the university resulted in him being awarded an honorary DSc in 2005. His other contributions to higher education teaching can also be recognized by his award of an honorary professorship at Birmingham University in 1999 and receiving the title of Distinguished Research Associate at Oxford University in 2001.

From: Griffiths, J. S., Giles, D. P., Davis, J. A. and Dunlop, J. J. (eds) 2025. *The Peter Fookes Engineering Geological Legacy in Geomodels, Geomaterials and Geomorphology*. Geological Society, London, Engineering Geology Special Publications, **30**, 267–291,
First published online 1 July 2025, https://doi.org/10.1144/egsp30-2024-18

Box 1. Alan Poole: Peter's contribution to the QMC Master's course in Geomaterials

I joined Queen Mary College (QMC) as they moved into their new Geology Building in 1962 and gradually progressed through the ranks, and, with the continuing encouragement from Peter, I eventually became Director of the Geomaterials MSc course. The course was initially inspired through discussions with Peter, who by then had become a leading expert on Engineering Geology and geological materials, working principally in the Middle East. Such was the respect for Peter's expertise and experience in the construction industry generally that in 1975 he managed to obtain funding from a British civil engineering company working in the construction boom that was then ongoing in the Middle East. This allowed Bill French and I to journey out to Bahrain and receive a week-long masterclass of instruction from Peter himself in all the pragmatic aspects of dealing with the construction materials problems encountered in Middle Eastern countries, from salt contamination, high saline water tables to the essential need to make use of less-than-ideal geological materials for construction. Based on this experience, on our return to QMC, Bill and I began to provide specialist short courses for individual postgraduates in the selection and use of natural materials for construction, and in methods of forensic investigation of building failures. This enabled these graduates to apply the useful and practical ideas we had gleaned from Peter within their own professional careers in the industry.

These first tentative steps in specialist postgraduate training by Bill and I were strongly supported by Peter, and eventually they led to the instigation in 1979–80 of a formal 1 year taught master's course, initially entitled MSc in Industrial Petrology and later renamed MSc in Geomaterials. In organizing the MSc, we managed to persuade many senior specialists from the worlds of quarrying, construction, materials investigation and testing to visit QMC and contribute their expertise to the course, thus providing the graduates a quite unique and practical insight into the profession they were about to enter. Perhaps most important among these generous experts was Peter himself, who contributed two courses of formal lectures, together with much informal mentoring and appropriate introductions to companies. Peter also led two week-long and very intensive field/site courses for the MSc students. These field courses took place in February and September each year, which Peter would refer to as our spring (it would usually rain!), and the days' work and evening presentations that followed often continued until 10pm or later. Peter's depth and breadth of geological knowledge, coupled with his practical expertise and wide experience, made these field study courses perhaps the most valuable training experience of the whole of the MSc course (Figs 1–3). In addition to these week-long intensive field courses each year, we separately arranged between eight and 12 technical 1 day visits to construction sites, pits and quarries around London that were hosted by directors and senior technical experts from the companies. Peter often provided the necessary introductions for accessing the companies concerned and joined several of the visits himself over the years. The visits to the Thames Barrier and the Channel Tunnel during their construction were particularly memorable.

Fig. 1. Peter, centre left with his wife Edna next to him, and MSc students in the Watts Blake and Bearne kaolin pit, Devon.

Fig. 2. Peter, on left, with the QMC MSc students in 1991–92.

Fig. 3. Peter (standing) on Dartmoor overseeing and answering questions.

To save money, QMC closed the Department of Geology, moving the staff into the Geography Department. This did not work very well, so after a year or so the Applied Geologists and the MSc were moved to the Civil Engineering Department. This arrangement proved a happier collaboration but, flushed with their perceived success with their first cutbacks and in spite of earlier promises of permanence from the Pro-Principal, the QMC decided that to close the Civil

Fig. 4. Paula Carey and Peter (both standing) with the Greenwich MSc in Geomaterials students.

Engineering Department was an even better cost-saving measure, although it also involved several staff redundancies. Consequently, in 2000 the MSc was moved to the University of Greenwich, and Paula Carey became its new director (Fig. 4). However, Peter, the specialist experts from industry and I continued to provide our essential contributions to the course, often providing our time and expertise for free.

This 1 year MSc course ran for 23 years before the withdrawal of the post-graduate studentships (grants) by government caused its eventual closure. In that time well over 130 people had graduated. It is undoubtedly true that Peter, with his breath of knowledge, real interest and care of individuals, together with his unique sense of humour, will certainly have made an enduring impression on every graduate. It is a given that once someone had met Peter, he or she would never forget the encounter, and so many of the course graduates owe their subsequent careers to his influence and post-graduation mentoring skills.

Writing from a more personal standpoint, I feel privileged and honoured to have known Peter as friend, advisor and most valued colleague. His pragmatism, humour and kindness to people never failed him through all of the years I knew him, and indeed continued in spite of the serious medical problems he suffered in his last years. He has provided a model of academic excellence and true humanity for the rest of us to aspire towards.

Beyond his teaching and support for various Earth Science degree programmes, it is the memories of individuals that best identify the contribution that Peter made to both the discipline and people's careers. This is recognized in the personal reflections of Alan Clark (Box 4), and in the following series of boxes written by individuals who knew and worked with Peter during his career. The writers of these reflections identify the work that they were involved with, and the influence that Peter had on both them and the project.

A major project that Peter Fookes was involved with in the mid–late 1970s was the Dharan–Dankuta Road (DDR) in Nepal. This project was one of the earliest examples to demonstrate the value of engineering geomorphological mapping for the construction of a highway in a very challenging mountainous environment where geotechnical data were difficult to obtain (Brunsden *et al.* 1976). Two people who worked on this project provide recollections of their experience of working with Peter out on-site: Dick Martin (Box 5) and Adrian Marsh (Box 6).

Box 7 is a recollection from one of the co-authors of this chapter of a project in West Africa with Peter as the consultant. Like the times with Adrian Marsh in Nepal (Box 6), this was a project involving a lot of walking in often difficult terrain.

Apart from the MSc teaching, Peter was also an experienced PhD examiner, and Gareth Hearn has recollections of his own viva voce in 1987 (Box 8). As he describes, Gareth

Box 2. Geoff Pettifer: QMC MSc Geomaterials Alumnus

I was one of Peter's students on the MSc Geomaterials course at Queen Mary's College (QMC), University of London in 1981. Peter's lectures were packed full of interesting information, and he always stressed the importance of understanding how geology combined with past and present hydrology and geomorphology to produce the ground conditions that were the basis of excavation and construction engineering. He was keen to help us advance our careers using this knowledge.

Fig. 5. Peter and chapter co-author David Shilston in a hotel site office in eastern Java planning a campaign to gather field evidence of the character and stability of cut slopes in young sediments and even younger volcanic rocks.

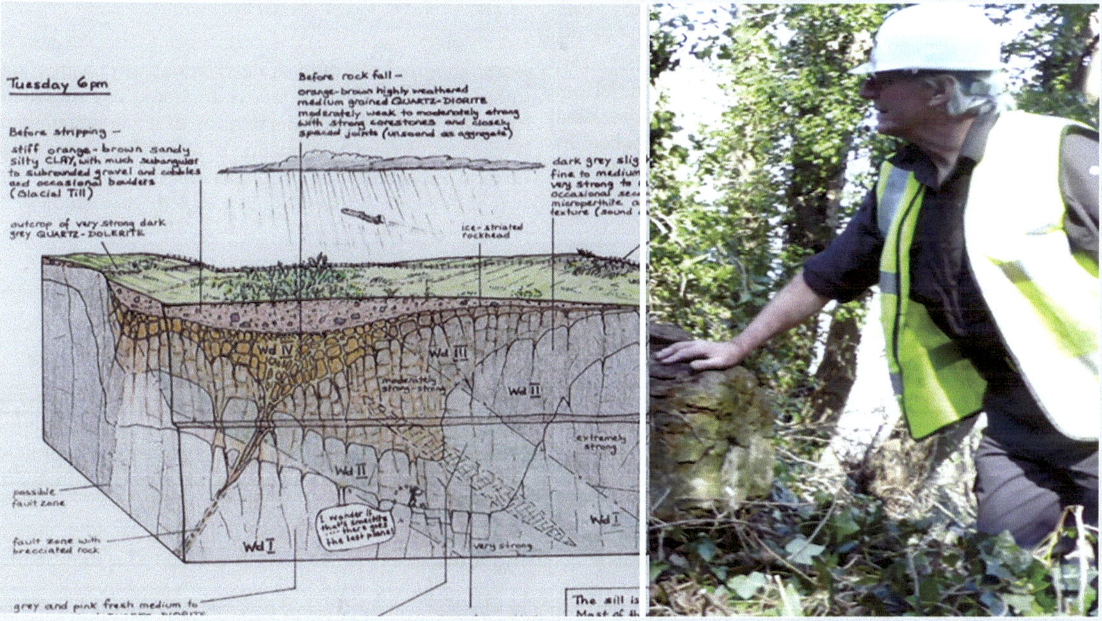

Fig. 6. Geoff Pettifer posing at a new Regionally Important Geodiversity Site (RIGS) in Dorset Jurassic strata, trying to see the bigger picture. The block diagram seems to fit but has different rock types from the exposure, so it represents an initial geomodel that would be modified following more detailed studies and re-examination of data. Geoconservation often forms part of a multidisciplinary environmental project in the UK.

In the 1990s, I worked with Peter on post-construction arbitration cases for Sandberg, Atkins and other companies, investigating excavation conditions for road, tunnel and power-station projects in Lesotho, Hong Kong, North Wales and Indonesia (Fig. 5). He persuaded me to construct a painted cardboard model to illustrate ground conditions for a road in Lesotho, in a way that that could be easilybe understood by non-specialists, and it may still be in the Roads Authority headquarters in Maseru. It became clear that considerable cost-savings could be made by more comprehensive investigations at the pre-construction stage.

In 1994 we collaborated on a paper dealing with rock excavation methods (Pettifer and Fookes 1994). Peter thought I had a knack for presenting relevant ground information in block diagrams (Fig. 6), and I did several for the first Glossop Lecture in 1997 (Fookes 1997*b*). We last worked together overseas during construction of an onshore oil pipeline in Sakhalin, Russia in 2007, where he and Gareth Hearn got me to draw a sequence of block diagrams depicting geology and landscape evolution along the route. In 2014, Peter, Tony Waltham (see Box 9) and I compiled a book on the use of geomodels in engineering geology (Fookes *et al.* 2015). Peter was an inspiring and steadfast teacher, friend, and colleague.

then went on to work with Peter on many projects throughout his career (see Hearn and Martin, this volume).

Whilst Peter's expertise in ground models and geomaterials is well documented, he was also skilled in the understanding of karst landscapes, notably those associated with gypsum. In developing his expertise, he worked closely with one of the world experts in the subject, Tony Waltham (Waltham and Fookes 2003). In Box 9 Tony describes working with Peter on a pipeline project in Turkey that had to cross some truly spectacular gypsum karst landscapes.

Throughout the 2000s, Peter worked closely with David Shilston and Mark Lee (co-authors of this chapter) both on publications (e.g. Fookes and Shilston 2001; Fookes *et al.* 2005, 2007) and on consultancy projects, notably in his favourite environment, hot deserts (e.g. Fookes *et al.* 2001; Lee and Fookes 2005). In Box 10 David Shilston describes his first experience of working with Peter, on an arbitration in Bahrain, and in Box 11 Mark Lee describes a typical desert project working with Peter in central Algeria when Peter was already 76 years old.

It is fitting that practically all the images in this wonderfully eclectic mix of memories show Peter out on site. Hatheway (2002), when referring to the core skills required of an engineering geologist, said, 'we become real when we leave the office or laboratory and enter the field'. Peter recognized that, above all, engineering geology is a field science and, as discussed in the next section, that is why so much of his published work was centred around case studies.

The use of case studies

Authors aiming to publish descriptions of construction and other projects commonly face restrictions due to confidentiality and the ever-present prospect of contractual disputes. Papers that have a combination of research and project information are usually less problematic. Such combined papers, which for ease we have simply called 'case studies', are a

Box 3. Adrian Morey – QMC MSc Geomaterials Alumnus

As a member of the 1984–85 MSc Geomaterials cohort, I reflect fondly on a year marked by the presence of a charismatic guest lecturer we affectionately dubbed 'The Prof'. Upon entering the realm of engineering geology, the magnitude of Peter's influence wasn't immediately apparent to us; yet, the joy of having him as our lecturer remains unparalleled. Peter emanated an infectious energy, coupled with an unwavering passion for the profession, knowledge, fun and his wife – the delightful Edna.

Peter's unique charm extended to assigning endearing nicknames to each member of our cohort – I proudly wore the moniker 'Flattop'. His memorable phrase, often accompanied by the distinctive sound of him sucking on his tooth, echoes in my mind: 'OK Flattop, you're on!'.

Beyond the lecture halls, Peter's impact on my career trajectory was profound. He graciously referred me for my first two professional roles in the field of professional services and remained a steadfast supporter throughout. Until I relocated overseas, I would enjoy the occasional rendezvous with Peter and Edna in Winchester or aboard their narrowboat *Buffoonery*. Undoubtedly, my experiences echo those of countless others who were touched by Peter's mentorship and the enduring legacy he carved within our profession.

Box 4. Alan Clark – personal reflections

Peter first arrived in my life in 1971 when I was working for Wimpey Laboratories, under Ian Higginbottom. Peter at this time was a specialist advisor to Rendel Palmer & Tritton (RPT), who were the consultants for the design and supervision of the Taff Valley Trunk Road (TVTR) from Cardiff to Merthyr Tydfil. In late 1971, during construction of Stage 2 of the road at Upper Boat, Nantgarw, there was a series of failures of the cut slopes in fluvio-glacial materials of the partially constructed road (Fig. 7). Wimpey Laboratories carried out a borehole investigation of the area and Peter recommended that geomorphological and geotechnical field mapping was carried out on the cut slopes and the upslope areas to assist in the design of the remedial measures. A colleague and I carried out this work in early 1972 under Peter's guidance. Peter was keen for us to use the new techniques of field mapping outlined in the Geological Society's Working Party Report 'The Preparation of Maps and Plans in terms of Engineering Geology', which was in draft form at this time. Peter would turn up at the end of the day at our site office where we were drawing up the details of the day's mapping to enthusiastically review the work. The Working Party Report was eventually published in 1972 (Anon.) and Peter actively encouraged us to publish the details of our work (Clark and Johnson 1975).

Peter's subsequent long-term involvement with the promotion and use of engineering geomorphology to assist in route selection and design of the TVTR is well documented. I was fortunate to continue my association with Peter and engineering geomorphology on the northern stages of the road when, then as an employee of RPT, I worked with Jim Griffiths on the investigations and route selection of the stages 4 and 5 of the road. The many other RPT projects that had his involvement included the Thames Barrier site investigations (Fookes and Martin 1978) and rock face stabilization of old quarries used as tank farms in Plymouth (Fookes and Sweeney 1976).

The early 1970s saw the start of the development boom in the Middle East that continues today. Peter was involved in various RPT projects in Bahrain, Kuwait and Abu Dhabi. At the same time, he was also working with Queen Mary College, London to develop his seminal works on Middle Eastern rocks as aggregates and their classification and description. His paper ((Fookes and Higginbottom 1975) was our bible at the time for logging cores. Further afield and up to the 1990s, Peter was a guiding light on the geotechnical investigations in Bangladesh, Ecuador, the Bahamas and, in particular, Libya. RPT at the time (1970s and the 1980s) were designing major new roads in Tripoli and Benghazi, and all the major ports in the country. His great experience with the geotechnics of desert environments was invaluable. Peter would come to the office full of ideas, which to us as the listeners was like standing under a waterfall as the suggestions flowed. Our task was to find the gems and put them into practice.

Peter was a force of nature, sometimes controversial, and was someone who brought knowledge, insight, inspiration and encouragement to all who worked with him. It was an honour to have known him and worked with him.

Fig. 7. Mudslide at Nangarw, A470 trunk road, South Wales.

Box 5. Dick Martin – Nepal 1975–80

I first met Peter on a plane to Delhi and Kathmandu in January 1975. I was a greenhorn research student, just a few months into a 3 year studentship sponsored by Rendel Palmer & Tritton (RPT), and was accompanying Denys Brunsden and David K.C. Jones to carry out a geomorphological reconnaissance survey of a new RPT alignment for the Dharan–Dhankuta Road (DDR) in east Nepal. By then I had read some of Peter's early papers (e.g. on Mangla Dam, and the mapping

Fig. 8. Peter during the Dharan–Dankuta Road (DDR) construction phase giving a mini lecture to local labourers and onlookers about gabion stone packing.

Fig. 9. Peter (with cravat and hand lens) and the then RPT site staff (left to right: Adrian Marsh, Dick Martin and Chris Manby) at the Dharan–Dankuta Road (DDR) base camp, daubed with red pigment on Nepal Holi (Spring Festival) Day.

review with Bill Dearman) and realized that he was already at the forefront of engineering geology. Being introduced to him that day was a daunting experience, but he quickly set me at ease with his open and friendly manner.

The first day's fieldwork with Peter and first night in our tented camp left an unforgettable memory, as he prompted intensive discussions on what he'd observed and how this might affect cross-section designs along the trace, as well as telling jokes and tales of derring-do from his work on other projects. Camp conditions were pretty basic at that time, but I never heard him complain about them.

After finishing the studentship, I spent nearly 3 years from 1977 to 1980 working for RPT on-site during construction of the DDR. Peter made week-long visits twice a year during this period and I quickly came to regard these as rich learning experiences. His method of working typically combined quiet observation on works progress, occasional detailed inspections (e.g. of excavations and foundations), quickfire questions to us site staff on seemingly every aspect of our work (Fig. 8), and every few hours he would stop and spend 5–10 min summarizing his findings and making suggestions. By then we had well-established fixed camps. Despite some long and tiring days out on the road, Peter was usually very sociable in the evenings. He mixed easily and often humorously with everyone involved, from the local cleaners and kitchen staff to the managing contractor's works supervisors and engineers (Fig. 9). He was an important mentor and inspiration in my early career, and much-admired friend, colleague and raconteur.

Box 6. Adrian Marsh – Nepal 1977–79

In November 1977, I joined the resident team responsible for the detailed design and construction of the Dharan–Dhankuta road as Materials Engineer with a site laboratory in the Himalayan foothills of eastern Nepal. This was my first such appointment and a daunting task for a young engineering geologist. Peter was already involved in the project, having been a guiding mind in the route selection through the mountainous and often unstable terrain. For me, his visits to site equated to intense, advanced crash courses in everything concerning the sourcing, testing and approval of the construction materials used. Vehicle access to most of the route alignment was impossible until the earthworks (Fig. 10), slope stabilization and drainage measures were well advanced. Hence, quarrying and rock crushing were mainly carried out by hand labour stretched out at points along the 51 km route that intersected various low to moderately metamorphosed phyllite, quartzite, schistose and gneissic strata (Fookes and Marsh 1981*a*). Peter surmised that the metamorphic mineralogy and fabric, faulting, and unpredictable weathering profiles could result, *inter alia*, in potential deleterious alkali-aggregate reactivity, flakiness/elongation and unsoundness of the crushed aggregates (Fookes and Marsh 1981*b*). Hand crushing trials, as

Fig. 10. Peter tackling the steep mountain terrain with quartzite debris underfoot in eastern Nepal with a little help from his new local friends.

Fig. 11. Peter was concerned about the availability and quality of locally sourced aggregates for concrete and roadstone. He suggested that trials like this were carried out along the route to assess the output of hand-crushed stone to provide suitable materials.

pictured, and associated laboratory testing were carried out as a key tool in determining the suitability of each prospective quarry and their derived aggregates (Fig. 11).

Box 7. Jim Griffiths – Liberia 1983

In April 1983 I travelled with Peter to Monrovia, the capital of Liberia. I say travelled with, but in fact Peter flew first class whilst I was back in economy, although he did arrange for the stewardess to send me a complimentary drink. In Liberia we were looking to identify a suitable alignment for a railway linking the iron ore mine at Bong Town to new processing facilities some 150 km distant. This rail link was to be cut through a primary equatorial rain forest. Like colleagues before

Fig. 12. Peter pointing out the development of ferricrete gravel nodules.

Fig. 13. Peter, under the white handkerchief, with railway engineer Rex Edwards, taking a breather whilst hiking along the cut trace of the railway alignment.

Fig. 14. Peter, with his back to the camera, describing what he has seen to the track maintenance crew during a stopover in a railway siding.

and since, I was Peter's amanuensis on this project and whilst I had a reasonable amount of overseas experience it had mostly been in desert terrain and this was my first time in the jungle. Peter was with us for my first 2 weeks on-site, and I treated this time with him as a personal tutorial in topical soils and weathering (Fig. 12). Despite the wet and humid conditions, Peter hiked along the alignment trace (Fig. 13) whilst I took copious notes. We also spent time in a track maintenance train and whenever we stopped Peter was always happy to talk to the railway men about their work (Fig. 14). Apart from the technical knowledge I acquired, what I remember most was how Peter was able to draw upon such a wide range of personal experience and apply it to the project. He was then able to communicate that knowledge in an effective way to everyone.

large part of Peter's corpus of published works and were an important component of his approach to consultancy and teaching. Typically, in a single paper, Peter and his co-authors would advance the research understanding of a remote area or topic and provide one of more examples of its practical application in engineering geology, geomaterials and so on.

One might ask: 'Why did Peter publish case studies?'. Indeed, why does anyone publish case studies? The answer is multifaceted. For Peter, one motivation was his wish to advance and promulgate his and his co-authors' growing understanding of a technical topic, an innovative method of working, a lesson learnt, etc. Building collaborations with like-minded geoprofessionals was another motivation.

Box 8. Gareth Hearn – career reflections from 1985

One of my earliest memories of Peter is when he gave his presentation at the First International Symposium on Geomorphology in Manchester in 1985. In his summing up, he advised that the engineer 'should put his money where the geomorphologist's mouth is.' Having only met Peter fleetingly before, I was shocked by the matter-of-fact way in which he made this statement, yet tremendously buoyed by its implications for me as a PhD student of engineering geomorphology. My next memorable encounter with Peter was when he and Professor Ron Cooke sat opposite me at my PhD viva in 1987. I entered the room with much trepidation but, as it turned out, my supervisor (Professor David Jones) and I sat quietly for most of the interview as Ron and Peter proceeded to overrule one another's questions on the grounds of them being unfair or ambiguous. I got off very lightly under the circumstances and, with the help of a bottle of correction fluid, managed to get the thesis over the line.

I have since had the benefit of working with Peter on projects in Indonesia, Guinea, Papua New Guinea, Romania and Russia (Fig. 15). Peter's ability to make immediate sense of complex geological and geomorphological settings was remarkable, and he was adept in conveying this complexity in simple terms to his clients. As a peer reviewer, Peter would 'keep his powder dry' as he interviewed geoteam and client members. He would then reveal his findings during a final meeting, with all contributors assembled, rather like Hercule Poirot with his room of suspects. Peter, Andrew Hart (Peter's trusted geomorphologist and report-writing scribe) and I would usually be the first to arrive followed, one-by-one, by the geoteam and client members. Once all were sitting comfortably, Peter would open the meeting noting that, in his experience, the degree of lateness tended to be positively correlated with seniority. This had the immediate effect of levelling the playing field and ensuring that he had the attention of everyone concerned from start to finish.

Fig. 15. Peter with Geoff Pettifer and others about to board a Thunderbird 3 all-terrain personnel carrier to Sakhalin Island, Russia, for the SEICL oil and gas pipelines project. Source: photograph reproduced with permission of Andrew Hart.

Peter encouraged and inspired me immensely throughout my career, both as a 'boots-on-the-ground' practitioner and, later, as a multidisciplinary geoteam manager. He would keep me on my toes at all times; posing questions that I knew were designed to test my knowledge and powers of observation. Whatever his views on my performance (he kept these to himself), Peter was always amusingly accommodating. On a flight from Paris to Conakry, the client and I missed the flight while sat in the business lounge. The plane was delayed while our bags were offloaded. Peter and the rest of the geoteam were on board. Peter must have wondered what on earth had happened to us, especially as I was the person responsible for the entire field visit programme (at least that is what was going through my mind). We were lucky enough to be able to catch an alternative flight via Casablanca, arriving in Conakry only a few hours later than Peter and being on time for his arrival in the office the next morning. Peter didn't say a word, not even when the client was out of earshot. He simply gave me a wry smile and his familiar chuckle that said it all. He was a great man and, along with Profs David Jones and Denys Brunsden, has contributed immeasurably to my career.

Box 9. Tony Waltham – Turkey 2001

Fig. 16. Peter, with his back to the camera, discussing karstic collapse with two Turkish colleagues on the rim of one of the large old sinkholes that we had just seen for the first time.

Fig. 17. Peter and his team stop to look at a closed depression in the polygonal karst, this one with a small stream sinking at the edge of its soil fill.

Fig. 18. The finest of the great sinkholes that showed what could develop in the Sivas gypsum, but fortunately only over a long period of time. The rim diameter is about 350 m. Note the person on the rim top right.

New oil and gas pipelines from the Caspian Sea to Europe traverse the length of Turkey where a variety of ground conditions provide some interesting challenges. Peter was supremo for the geological team that had all the appropriate specialists, and he had roped me in to look at the potential sinkhole hazard in a block of spectacular gypsum karst near Sivas in central Turkey (Figs 16–18). Soil settlement by suffosional losses could be recognized and resolved, but a scatter of very large sinkholes got Peter very excited. Was there potential that ground collapse on a scale larger than a pipeline (Fig. 19) could survive intact when suddenly suspended in space? A bit of fieldwork indicated that giant cavities would not develop in the gypsum, and the existing landforms had developed piecemeal in small stages that were pipeline-survivable. But it was a delight to be out in the field with Peter as we discovered more and more of these grand collapse features (Fig. 20), and to see his approach to assessment of a geohazard that was likely to be rare but could be catastrophic. After a few days at Sivas, he got me drawing models for the different types of karst terrain that we had distinguished on the gypsum outcrop. Constructing these geomodels was a new experience for me, but it was Peter's big thing at the time, and really did prove its value in focusing the mind on the various hazards that the karst could present. Understanding the big

Fig. 19. Burying a pipeline in a chaotic mixture of clay and gypsum best described as a pinnacled rockhead.

Fig. 20. Polygonal karst on the Sivas gypsum, with dark soils on the floor of each of numerous closed depressions.

picture was Peter's greatest strength, and while my time with him on the Sivas karst was, I hope, productive, it was also instructive and most enjoyable.

Box 10. David Shilston – learning in the rocky desert of Bahrain, 1994

My first experience of working with Peter was on an arbitration for a construction project in Bahrain (Figs 21 & 22). One of the questions at the heart of the arbitration was deceptively simple: 'How should a contractor identify, excavate and process fill materials for a large earthworks project in the desert of Bahrain?'

Fig. 21. Peter, wearing a white hat and nearest the camera, viewing a limestone quarry and material processing in Bahrain.

Fig. 22. Peter contemplating the ground surface at the project site.

Peter's approach to the work was also simple: review the project data and obtain published geological and geomorphological maps; make an engineering geological and geomorphological reconnaissance of the project site; dig some trial pits; open and operate a small quarry; test the quarried materials in a crushing and screening plant; and bring together all the work into an overall understanding that would be suitable for forming a well-argued and evidence expert opinion. I was appointed as Atkins' project manager and assistant to Peter.

This was my first opportunity to work in a desert with Peter; indeed, it was my first desert project and my first big arbitration project. For the trial pitting, I needed to provide a detailed record and understanding of the near-surface geology. I had done numerous trial pit investigations in the past, and the forensic detail and recording that Peter required enabled me to draw on my experience of archaeological investigation methods. Peter knew all this, of course, and, without letting me know, he turned the trial pitting into practical lessons in the understanding of desert terrain and geology. He said almost nothing as he looked down into the trial pits – he just observed me wrestling with unfamiliar ground conditions; he watched and listened to me drawing on my limited experience to identify and interpret what was there; and he seemed to make a mental note when I got it right, but seldom actually said so.

Fig. 23. A rare in-door photograph: a pensive Peter in the arbitration hearing room in Bahrain, with some posters made from figures in his expert report on the wall and rock specimens on the desk. Note also the presence of Peter's well-travelled suitcase, which was used to carry his paperwork, etc., from our hotel to the hearing.

One notable difficulty with the earthworks was the presence of gypsum that had formed as layers between the beds of the rock, thereby jacking them apart. It took me an embarrassingly long time to comprehend the situation – and all the while Peter just let me get on with it. This was one of numerous occasions, over many years, that I experienced Peter in the role of teacher and mentor.

Peter was a testifying expert witness at the subsequent arbitration hearings, which were held in Bahrain. I assisted him in preparing for the hearings, including the creation of the illustrations and physical exhibits that he devised (Fig. 23), such as the set of large cardboard rocks that he used to demonstrate to the arbitrators the size of blocks from the quarry trials. Peter's personal preparation for the hearing and presentational style were my introduction to the rigors of expert witness testimony. I enjoyed it and learnt much and, as with the trial pitting, I have put the experience to good use during my post-Bahrain career – in both the investigation of desert terrain and as an expert witness.

Box 11. Mark Lee – Algeria 2010

In January 2010, Peter (aged 76 at that time) and I were part of a field team undertaking a reconnaissance of a proposed pipeline route through the central Algerian Sahara, from near Aougrout to Hassi R'Mel, a distance of more than 600 km. We stayed in a simple, white-walled guest house in Timimoun, overlooking the date trees that surrounded the oasis. Each

Fig. 24. The Geoteam.

Fig. 25. The Grand Erg Occidental.

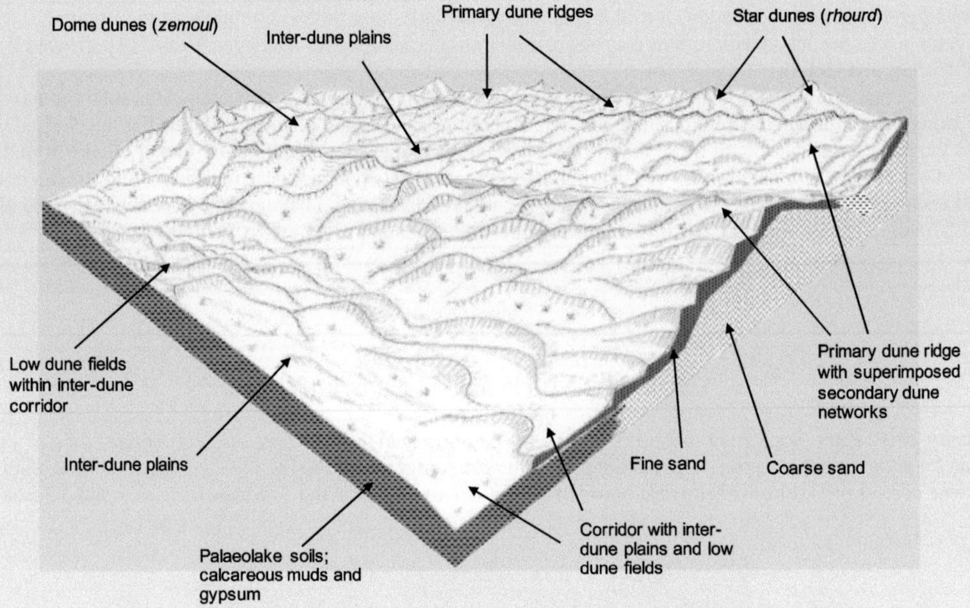

Fig. 26. A conceptual block model of the dune fields. Source: from the author's own files.

day we would overfly sections of the alignment in a five-seater Pilatus PC-6 light aircraft (Fig. 24), flown by our friend Pascal Campion, a desert pilot in the spirit of Antoine de Saint-Exupéry. With views reminiscent of the opening scenes of Anthony Minghella's film, *The English Patient*, we would fly low across the Great Western Sand Sea (Grand Erg Occidental, covering 78 000 km^2 and more than 150 km wide) with its parallel 100–200 m-high linear dunes and superimposed star, dome and barchanoid dunes (Fig. 25). From the air we 'read the ground' and made judgements about the terrain and geohazard challenges facing the project, focusing on a possible routing strategy through the sand sea that maximized the use of inter-dune flats (old lake beds; gassi) and sand plains (feidj). Away from the dune fields we observed strong duricrusts and cap rock on the Tademait Plateau and the Hamada Bet Touadine to the north of the sand sea, and made predictions about pipeline trench excavatability. At night we drank Algerian red wine and ate couscous, whilst Pascal shared stories about the life of a desert pilot and I drew block models (Fig. 26).

However, Peter's publications were not entirely altruistic as a further important motivation was that they raised and maintained his personal profile with practitioners and with universities, and such links were important for his consultancy business. On reflection, the 'why?' question could be reversed. It would surely have been surprising if someone with Peter's intellectual and enquiring character, wide-ranging technical expertise, communication skills, and engaging persona did not place great emphasis on publishing his work and the work of others, commonly through the medium of case studies.

The paragraphs below provide an overview of Peter's case study publications. In doing so, they track the phases of development of his technical interests and consultancy work. Each phase usually included one or more review publications in journals, conference proceedings or industry magazines where they would reach an appropriate audience. These review publications are not, strictly speaking, case studies but were derived from work published in case studies, such as 'Middle East – inherent ground problems' (Fookes 1978), *A Handbook of Engineering Geomorphology* (Fookes and Vaughan 1986) and 'Desert environments: landscapes and stratigraphy' (Fookes and Lee 2009). In addition, Peter was often the behind-the-scenes driving force for publications, but not one of the named authors. An excellent example of this is 'The Bahrain surface materials resources survey and its application to regional planning' (Brunsden *et al.* 1979).

The list of Peter's publications compiled by John Charman (Charman 2025) encompasses some 200 technical papers, book chapters and magazine articles, including his early career work at Mangla Dam in West Pakistan, aggregates and concrete, the engineering geology of deserts and of an enormous range of other geological environments, and his 1997 Glossop Lecture on geological models. In only a small proportion of his publications was Peter the sole author. More usually, he was a co-author with people involved in the project or research that was being described. Indeed, some of Peter's most well-known publications are case studies reported by members of a project team in which he played a role. The first example from the list of his publications is Little *et al.* (1963) 'Bedrock grouting tests at Mangla Dam, West Pakistan'. Peter submitted his PhD in 1967, on the Siwalik Clay bedrock at Mangla Dam. Five years later he was the lead author of a paper about the site investigations for the Taff Vale Trunk Road in South Wales (Fookes *et al.* 1972) and joint author in a paper about site investigations for a highway in tropical conditions in Fiji (Lovegrove and Fookes 1972). A paper about the use of aerial photography to detect slope instability in Kent followed a few years later (Norman *et al.* 1975).

It was in the 1970s that Peter began his long involvement with concrete and aggregates in the Middle East (notably Fookes and Collis 1975*a*, *b*, 1976; Fookes 1978; Fookes and Higginbottom 1980*a*, *b*; Venečanin 1981; Lee *et al.* 2025). His work grew to include North Africa, where the Great Man-Made River project in Libya was the subject of a series of papers such as Stoner and Fookes (1991), Stoner *et al.* (1991), Woodford and Ray (1992) and Fookes *et al.* (1993).

Although Peter's case studies and more general papers about projects in the 1980s and early 1990s are predominantly about geomaterials in the Middle East, he was also active in other geological environments. There are, for example, papers on instability of cut slopes in Malaysia (Fookes 1986), limestone weathering (Fookes and Hawkins 1988), an earthfill dam in Papua New Guinea (Fookes and Dale 1992) and tropical residual soils (Fookes 1994).

Peter's Glossop Lecture in 1997 is one of, if not the most important of, his published works. It is a long paper that intertwines, in a typically engaging manner, the fundamentals of the subject with practical examples and data from projects (i.e. case studies). It is sad that there is no recording of Peter delivering his lecture – a tour de force that no one who was in the Royal Institution's famous lecture theatre that evening will forget.

Soon after presenting his Glossop Lecture, Peter became heavily involved in the investigation and assessment of major long-distance oil and gas pipeline routes. The individual projects were presented as a series of case studies at a conference at the Institution of Civil Engineers in London in 2004: 'Terrain and Geohazard Challenges Facing Onshore Oil and Gas Pipelines'. One of the conference papers illustrates particularly well the collaborative effort that underpinned such projects and Peter's involvement in them: 'Terrain, ground conditions and geohazards: evaluation and implications for pipelines' (Charman *et al.* 2005).

Perhaps the finest examples of Peter's ability to bring people together for a collaborative technical objective were his chairmanship, contributing authorship and behind-the-scenes encouragement of working parties on behalf of the Geological Society of London's Engineering Group, all of which make extensive use of case studies:

- Preparation of maps and plans in terms of engineering geology (Anon. 1972).
- Aggregates: sand, gravel and crushed rock aggregates for construction purposes (Sims 1986; Smith and Collis 2001 – 3rd edition revised by Fookes *et al.*).
- Tropical residual soils (Fookes 1997*a*).
- Clay material in construction (Reeves *et al.* 2006)
- Hot deserts (Walker 2012); chaired by John Charman.

Throughout his long career, Peter's writing, use and encouragement of published case studies were not only important parts of his publication endeavours but were also essential to how he undertook his work. He drew on them and his wide-ranging personal experience to help him understand geological (and geomorphological) conditions and processes, and to erect the classification schemes and models that are a key part of many of his contributions to the practice of engineering geology and geomaterials.

To conclude, it seems most appropriate that Peter's last two publications took him and us back to the deserts that he loved (Fookes *et al.* 2019) and to the engineering geology of concrete (Fookes and Lee 2019).

Beyond site ground models

In his Glossop Lecture, Peter's focus was on the interaction between the total geological history and site conditions. He wrote that (Fookes 1997*b*, p. 395):

> [A]t any location the geology is the sum of its history:
>
> - from the genesis of the bedrock and any subsequent alteration by tectonic and other forces capable of folding, faulting or physically disturbing the bedrock
> - by diagenesis, hydrothermal, igneous (intrusive and extrusive), metamorphic, and other geological activities which can alter or modify the original rock or, in the case of igneous activity for example, add to it
> - subaerial weathering and other surface or near-surface influences, including groundwater and fluvial, marine, aeolian, glacial and periglacial processes; erosion and deposition with the formation of superficial deposits.

This reflects a basic concept that the conditions and geotechnical characteristics of the ground are the 'inheritance' of the geological and geomorphological history of the site, including past and present climatic conditions. However, this inheritance rarely reflects any one climate or period of geomorphological change; a landscape often comprises an assemblage of landforms of different ages and origins, some of which may be of considerable antiquity.

The total geology concept was a major contribution to engineering geology. Over the last 25 years and more, it has led to the geological model approach with its focus on current ground conditions with respect to foundation design, earthworks, etc. (e.g. IAEG C25: Baynes and Parry 2022) or the almost ubiquitous use of software to create digital models by correlating ground investigation data in three dimensions (e.g. Culshaw 2005). These approaches tend to promote the view that the current ground conditions at a site are the inevitable product of history. However, it seems likely that, in many cases, the materials and processes observed today are conditional on a non-deterministic sequence of past events and weathering of materials. This may explain why sites in similar settings can have different and, sometimes, unexpected conditions.

In the ground model approach, often the implicit assumption is that site conditions will be static over the life of a project. However, the Earth's surface is dynamic, not static, and landforms and materials change through time as a result of weathering and surface processes such as landsliding, erosion, sediment transport and deposition. In many places, change will be slow and imperceptible over engineering time. But this is not always the case, and these changes have the potential to cause significant harm to civil engineering projects (i.e. hazards or geohazards). For example, as part of the Millenium celebrations, Higgitt and Lee (2001) provided an important insight into the surprising extent and scale of geomorphological process activity and landscape change in Britain over the last 1000 years, notably around the coast and estuaries, along river channels, and on some hillslopes.

Although less widely known than his Glossop Lecture, Peter's book *Engineering Geomorphology: Theory and Practice* (Fookes *et al.* 2007) was probably the first attempt to explain the importance of understanding landscape change in a ground engineering context. Written with engineering geomorphologists Mark Lee and Jim Griffiths, this was an attempt to set out the key principles of the subject. They viewed engineering geomorphology as being directed towards understanding the way landforms or Earth surface systems respond to relatively short- to medium-term (<1–1000 years) changes in energy inputs (e.g. resulting from climatic variability, changes in sediment supply, land-use change, neotectonics and anthropogenic effects). In his prologue to the book, Peter wrote of his desire to see greater appreciation of how the subject could be:

> [O]f value to a wide spectrum of engineering situations, particularly in providing a new set of judgemental skills to enable a better and more rigorous evaluation of situations than hitherto. Regrettably, in the last couple of decades, engineering geology has increasingly tended to be driven by the commercial needs of site investigation and, as with soil mechanics and geotechnics as a whole, bedevilled by clients who on occasion, short-sightedly, want the ostensibly cheapest version of everything
>
> (Fookes *et al.* 2007).

The book describes how evidence of past change can be organized into a conceptual model to explain the historical evolution of a landform system. Such models tend to be qualitative and subjective but can provide a framework for understanding and predicting future change. A simple evolutionary model of the West Dorset coast, UK, is presented in Figure 27. It was developed to help determine the potential impacts of proposed coast protection works at West Bay (east of Lyme Regis) on the nearby Chesil Beach (a landform of unique scientific importance, both nationally and internationally: Lee and Brunsden 2001). Historical chart/map and photograph evidence suggests that the four discrete shingle transport cells at the present day are probably the remnants of what once may have been a continuous shingle beach extending some 45 km eastwards from Lyme Regis to the Isle of Portland.

Prior to the construction of The Cobb breakwater at Lyme Regis in 1756, the shingle beach may have even extended westwards into East Devon (Lee 2001; Brunsden and Lee 2004). However, the long-term effects of differential erosion of the more resistant headlands (Golden Cap and Doghouse Hill–Thorncombe Beacon) and the softer rocks in the intervening bays has gradually created the present-day sequence of pocket beaches, which are linked episodically by shingle transfers. The fragmentation of the continuous beach into discrete pocket beaches may have been a relatively recent phenomenon, occurring over the last 250 years or so. As a result, Chesil Beach has been cut off from potential sources of shingle in East Devon and West Dorset, and can be regarded as a relict feature. It was considered unlikely, therefore, that coastal defence improvement works at West Bay, Lyme Regis would have a significant impact on the sediment budget and morphology of Chesil Beach.

This example was based on an understanding of the site (Chesil Beach) and the situation (the East Devon–West Dorset coasts). This 'broader perspective' is important for identifying and evaluating geohazards on many projects. Hillside sites need to be seen in the context of the slope or terrain unit or land system, river floodplain sites as part of the whole catchment, and coastal sites as part of a sediment transport cell. Each of these broader units represents an 'Earth surface system'.

Although Peter was well aware of the importance of 'site and situation' (e.g. Fookes *et al.* 2005; Lee and Fookes 2015), he didn't pursue this in his Glossop Lecture and related ground model publications. His explicit focus was on 'site', a focus that has continued through the development of Eurocode 7 (see Griffiths 2025). However, his post-Glossop experiences working on remote region pipelines reinforced his life-long appreciation of engineering geomorphology: 'I have spent much of my life endeavouring to improve my geomorphology' (Fookes *et al.* 2007, prologue).

With this in mind, it is worth speculating on how Peter might have seen his ground model approach developing to take account of the future challenges posed by climate change and sea-level rise. In our view, the Earth surface systems concept would be central to the approach as it forms a useful spatial framework for evaluating:

- How hazards and risks to a particular site can arise as a result of processes operating elsewhere within a system.

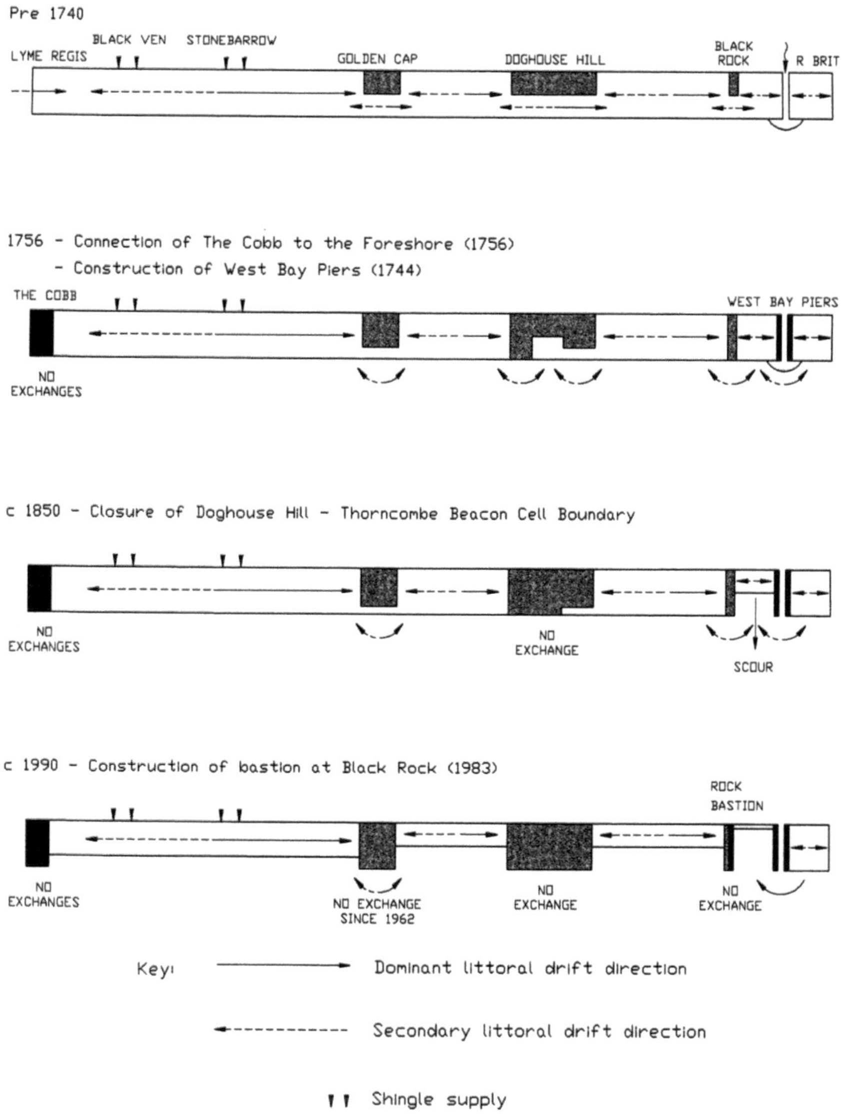

Fig. 27. A schematic model of the fragmentation of coastal cells on the West Dorset coast. Source: Lee and Brunsden (2001).

For example, changes in land use in the headwaters of a catchment can lead to changes in flood frequency and river-channel erosion or deposition downstream.

- The potential impacts of an engineering project on other landforms within a system. For example, construction of a harbour breakwater can disrupt sediment transport and cause accelerated erosion downdrift.

In the context of engineering geology, the systems concept can be defined as: the geological risk profile of a site is a function of both the site conditions (the ground model) and the processes operating within the relevant Earth surface system. However, the response of a system to the energy inputs (e.g. rainfall, seismicity, wind and waves) can be complex and non-linear, with multiple possible outcomes. Many systems can be viewed as having a 'memory' insofar as the current and future behaviour is influenced by the effects of past events on the system (Lee 2008) (Fig. 27). For example, the response of a coastal cliff system to wave-energy inputs is not simply a reflection of the beach levels, but also depends on the volume of debris supplied onto the foreshore by the recession process (i.e.

system outputs) and the length of time that this debris has been exposed to wave attack; the presence and volume of cliff foot debris is conditional on the recent cliff behaviour. This debris provides protection from wave run-up; cliff foot erosion only occurs when the debris has been removed. Future behaviour is, thus, critically dependent on recent cliff behaviour and on processes operating elsewhere within the coastal cell, especially fluctuations in the longshore transport of beach-building material. This concept of site change can be expressed in a way that we are sure would find favour in Peter's eyes as: at any location, the future conditions are the sum of its history and the potential responses to forcing events.

Conclusions

Hatheway (2002) set out a comprehensive statement on the engineering geological activities that are of value to society (Table 1) and are at the core of being an engineering geologist. When we reflect on Peter's career through his writing and support for other practitioners, it is apparent that he covered almost all the aspects of engineering geology described in this table.

Peter's legacy, therefore, is that his commitment to engineering geology (and engineering geomorphology) is reflected in his contribution to the core values of the subject and its value to society (Table 1) through his teaching,

Table 1. Engineering geological activities of value to society (based on Hatheway 2002)

Activities	Value to society	Basic impact
Identifying materials: Soil Weak rock Rock Surface water Groundwater	Defining and describing geological conditions affecting the design, construction, operation, and maintenance of engineered works	Material and engineering properties Discontinuities Geological contacts, boundaries and lateral extent
Defining site conditions	No other profession has or maintains an ability to explore and define conditions relating to the success of the works of mankind located on or in the ground	Provides the basis for engineers to optimize the way in which projects are designed to achieve the owner's/ client's objectives
Relative assurance against the effects of unknown geological conditions or constraints	Identification of conditions and uncertainties affecting design, construction, operation and maintenance of engineered works	Static and dynamic foundation conditions Identification of geological constraints (geohazards) Magnitude and frequency of damaging phenomena
Materials for construction (aggregates, cladding stone, rip rap)	Exploration Characterization Planning for recovery Post-operational remediation	Site development plan Site closure plan
Environmental assessments and management of associated geoenvironmental hazards	Property valuations and property transfers	Detection of conditions inimical to the owner's development plans; assessing the environmental impact
Mitigation of the effects of the extremes of nature	Choice of alternatives Avoidance of difficult sites Application of safety buffer zones Attempt to defuse the cause	Prediction of events
Protection from extremes of nature	Quantification of magnitude and frequency of damaging geohazards, including floods, earthquakes, tsunamis, volcanism, mass movement, subsidence, soil erosion	Recurrence interval Prediction Duration Magnitude Spatial extent
Protection from pollution	Siting and design contribution to waste-management facilities Site and waste characterization of dump and spill sites	Characterizing site and nature of waste Developing pollution protection measures Detection of conditions inimical to operation of the facility
General resources assessment, including groundwater, geothermal and surface water	Exploration Characterization Planning for recovery	Resource assessment

publications and training of the next generation. Peter's achievements are even more noteworthy because few people realized that he was dyslexic and never learnt to drive, and relied on his wife, Edna, to send his faxes and emails.

However, we must ask whether the present working environment in today's civil engineering industry could accommodate an independent consultant like Peter? Was Peter a unique man of his times? One thing is certain, all who worked with Peter George Fookes know the debt they owe him and will no doubt wish to raise a glass to his memory.

Acknowledgements The authors thank the reviewers for constructive comments that improved the article. Box 1 by Alan Poole is a revised version of an article that appeared in the 2020 Kirkaldy Society Magazine for Queen Mary College geology alumni edited by Dr Ian Sims.

Competing interests The authors declare that they have no known competing financial interests or personal relationships that could have appeared to influence the work reported in this paper.

Author contributions **DTS**: writing – original draft (equal); **EML**: writing – original draft (equal); **JSG**: writing – original draft (equal).

Funding This research received no specific grant from any funding agency in the public, commercial, or not-for-profit sectors.

Data availability Data sharing is not applicable to this article as no datasets were generated or analysed during the current study.

References

Anon. 1972. The preparation maps and plans in terms of engineering geology: Report by the Geological Society Engineering Group Working Party. *Quarterly Journal of Engineering Geology*, **4**, 293–382, https://doi.org/10.1144/GSL.QJEG.1972.005.04.01

Baynes, F.J. and Parry, S. 2022. *Guidelines for the Development and Application of Engineering Geological Models on Projects*. IAEG Commission 25 Publication No.1. International Association for Engineering Geology and the Environment (IAEG), https://iaeg.info/wp-content/uploads/2022/12/IAEG-C25-EGM-Guidelines-v1.0-14-December-2022.pdf

Brunsden, D. and Lee, E.M. 2004. Behaviour of coastal landslide systems: an inter-disciplinary view. *Zeitschrift für Geomorphologie*, **134**, 1–112.

Brunsden, D., Doornkamp, J.C., Fookes, P.G., Jones, D.K.C. and Kelly, J.M.H. 1976. Large-scale geomorphological mapping. *Quarterly Journal of Engineering Geology*, **8**, 227–253, https://doi.org/10.1144/GSL.QJEG.1975.008.04.01

Brunsden, D., Doornkamp, J.C. and Jones, D.K.C. 1979. The Bahrain surface materials resources survey and its application to regional planning. *The Geographical Journal*, **145**, 1–35, https://doi.org/10.2307/633073

Charman, J. 2025. The life and times of Peter George Fookes: 1933–2020. *Geological Society, London, Engineering Geology Special Publications*, **30**, https://doi.org/10.1144/egsp30-2024-4

Charman, J.H., Fookes, P.G., Hengesh, J., Lee, E.M., Pollos-Pirallo, S., Shilston, D.T. and Sweeney, M. 2005. Terrain, ground conditions and geohazards: evaluation and implications for pipelines. *In*: Sweeney, M. (ed.) *Terrain and Geohazard Challenges Facing Onshore Oil and Gas Pipelines. Proceedings of the Conference in 2004*. Institution of Civil Engineers, London, 78–94.

Clark, A.R. and Johnson, D.K. 1975. geotechnical mapping as an integral part of site investigation – two case histories. *Quarterly Journal of Engineering Geology*, **8**, 211–224, https://doi.org/10.1144/GSL.QJEG.1975.008.03.04

Culshaw, M.G. 2005. From concept towards reality: developing the attributed 3D geological model of the shallow subsurface. *Quarterly Journal of Engineering Geology and Hydrogeology*, **38**, 231–284, https://doi.org/10.1144/1470-9236/04-072

Fookes, P.G. 1978. Middle East – inherent ground problems. *Quarterly Journal of Engineering Geology*, **11**, 33–49, https://doi.org/10.1144/GSL.QJEG.1978.011.01.04

Fookes, P.G. 1986. *A Preliminary View on Cut Slope Instability of Roads in Peninsular Malaysia*. Public Works Department, Kuala Lumpur.

Fookes, P.G. 1994. A review: Genesis and classification of tropical residual soils for engineers. *In*: Blight, G.E., Fourie, A.B., Luker, I., Mouton, D.J. and Scheurenberg, R.J. (tech. eds) *Geotechnics in the African Environment: Proceedings of 10th Regional Conference for Africa on Soil Mechanics & Foundation Engineering, and the 3rd International Conference on Tropical & Residual Soils*, Maseru, 23–27 September 1991, Volume 2, 423–442.

Fookes, P.G. (ed.) 1997*a*. *Tropical Residual Soils*. Geological Society Professional Handbooks. Geological Society, London.

Fookes, P.G. 1997*b*. Geology for engineers: the geological model, prediction and performance. The First Glossop Lecture. *Quarterly Journal of Engineering Geology*, **30**, 293–424, https://doi.org/10.1144/GSL.QJEG.1997.030.P4.02

Fookes, P.G. and Collis, L. 1975*a*. Problems in the Middle East. *Concrete*, **9**, 12–17.

Fookes, P.G. and Collis, L. 1975*b*. Aggregates and the Middle East. *Concrete*, **9**, 14–19.

Fookes, P.G. and Collis, L. 1976. Cracking and the Middle East. *Concrete*, **10**, 14–19.

Fookes, P.G. and Dale, S.G. 1992. Comparison of interpretations of a major landslide at an earthfill dam site in Papua New Guinea. *Quarterly Journal of Engineering Geology*, **25**, 313–330, https://doi.org/10.1144/GSL.QJEG.1992.025.04.06

Fookes, P.G. and Hawkins, A.B. 1988. Limestone weathering: its engineering significance and a proposed classification scheme. *Quarterly Journal of Engineering Geology*, **21**, 7–31, https://doi.org/10.1144/GSL.QJEG.1988.021.01.02

Fookes, P.G. and Higginbottom, I.E. 1975. The classification and description of near shore carbonate sediments for engineering purposes. *Géotechnique*, **25**, 406–411, https://doi.org/10.1680/geot.1975.25.2.406

Fookes, P.G. and Higginbottom, I.E. 1980*a*. Some problems of construction aggregates in desert areas, with particular reference to the Arabian peninsula, 1: Occurrence and special characteristics. *Proceedings of the Institution of Civil Engineers, Part 1*, **68**, 39–67, https://doi.org/10.1680/iicep.1980.2504

Fookes, P.G. and Higginbottom, I.E. 1980*b*. Some problems of construction aggregates in desert areas, with particular reference to the Arabian peninsula, 2: Investigation, production and quality control. *Proceedings of the Institution of Civil Engineers, Part 1*, **68**, 69–90, https://doi.org/10.1680/iicep.1980.2505

Fookes, P.G. and Lee, E.M. 2009. Desert environments: landscapes and stratigraphy. *Geology Today*, **25**, 172–180, https://doi.org/10.1111/j.1365-2451.2009.00722.x

Fookes, P.G. and Lee, E.M. 2019. The engineering geology of concrete in hot drylands. *Quarterly Journal of Engineering Geology*, **52**, 257–259, https://doi.org/10.1144/qjegh2018-185

Fookes, P.G. and Marsh, A.H. 1981*a*. Some characteristics of construction materials in the low to moderate grade metamorphic rocks of the Lower Himalayas of east Nepal, 1: occurrence and geological features. *Proceedings of the Institution of Civil Engineers Part 1*, **70**, 123–138, https://doi.org/10.1680/iicep.1981.1961

Fookes, P.G. and Marsh, A.H. 1981*b*. Some characteristics of construction materials in the low to moderate grade metamorphic rocks of the Lower Himalayas of east Nepal, 2: engineering characteristics. *Proceedings of the Institution of Civil Engineers Part 1*, **70**, 139–162, https://doi.org/10.1680/iicep.1981.1962

Fookes, P.G. and Martin, P.L. 1978. Site investigations and geotechnical considerations. *In*: *Thames Barrier Design: Proceedings of the October 1977 Conference*. Institution of Civil Engineers, London, 29–50, https://doi.org/10.1680/tbd.00575.0006

Fookes, P.G. and Shilston, D.T. 2001. Building the ground model: case study of a rock tunnel in SW England. *Geological Society, London, Engineering Geology Special Publications*, **18**, 123–128, https://doi.org/10.1144/GSL.ENG.2001.018.01.18

Fookes, P.G. and Sweeney, M. 1976. Stabilization and control of local rock falls and degrading rock slopes. *Quarterly Journal of Engineering Geology*, **9**, 37–55, https://doi.org/10.1144/GSL.QJEG.1976.009.01.03

Fookes, P.G. and Vaughan, P.R. (eds) 1986. *A Handbook of Engineering Geomorphology*. Surrey University Press, Guildford, UK.

Fookes, P.G., Hinch, L.W. and Dixon, J. 1972. Geotechnical considerations of the site investigation for Stage IV of the Taff Vale Trunk Road in South Wales. *In*: *Permanent International Association of Road Congresses Second British Regional Congress, Cardiff*. Permanent International Association of Road Congresses (PIARC), 1–25.

Fookes, P.G., Stoner, J.R. and Mackintosh, J.H. 1993. The Libyan Great Man-Made River Project, Phase 1: a case study on the influence of climate and geology on concrete technology. *Quarterly Journal of Engineering Geology*, **26**, 25–60, https://doi.org/10.1144/GSL.QJEG.1993.026.01.04

Fookes, P.G., Lee, E.M. and Sweeney, M. 2001. Pipeline route selection and ground characterization, Algeria. *Geological Society, London, Engineering Geology Special Publications*, **18**, 115–121, https://doi.org/10.1144/GSL.ENG.2001.018.01.17

Fookes, P.G., Lee, E.M. and Milligan, G. (eds) 2005. *Geomorphology for Engineers*. Whittles Publishing, Dunbeath, UK.

Fookes, P.G., Lee, E.M. and Griffiths, J.S. 2007. *Foundations of Engineering Geomorphology*. Whittles Publishing, Dunbeath, UK.

Fookes, P.G., Pettifer, G. and Waltham, A. 2015. *Geomodels in Engineering Geology – An Introduction*. Whittles Publishing, Dunbeath, UK.

Fookes, P.G., Waring, D. and Lee, E.M. 2019. Wilfred Thesiger and the Umm as Samim, Oman. *Geology Today*, **35**, 221–227, https://doi.org/10.1111/gto.12290

Griffiths, J.S. 2025. What do we mean by a ground model – the Fookes geomodel legacy. *Geological Society, London, Engineering Geology Special Publications*, **30**, https://doi.org/10.1144/egsp30-2023-11

Hatheway, A.W. 2002. Site characterization, raison d'ětre of engineering geology. *In*: Tepel, R.E. (ed.) *Visioning the Future of Engineering Geology: Sustainability and Stewardship*. Association of Engineering Geologists Special Publications, **19**, Vol. 1 (CD-ROM).

Higgitt, D.L. and Lee, E.M. (eds) 2001. *Geomorphological Processes and Landscape Change: Britain in the Last 1000 Years*. Blackwell, Oxford, UK, https://doi.org/10.1002/9780470712832.

Lee, E.M. 2001. Estuaries and coasts: morphological adjustment and process domains. *In*: Higgett, D. and Lee, E.M. (eds) *Geomorphological Processes and Landscape Change: Britain in the Last 1000 Years*. Blackwell, Oxford, UK, 237–268.

Lee, E.M. 2008. Coastal cliff behaviour: observations on the relationship between beach levels and recession rates. *Geomorphology*, **101**, 558–571, https://doi.org/10.1016/j.geomorph.2008.02.010

Lee, E.M. and Brunsden, D. 2001. Sediment budget analysis for coastal management, west Dorset. *Geological Society, London, Engineering Geology Special Publications*, **18**, 181–189, https://doi.org/10.1144/GSL.ENG.2001.018.01.25

Lee, E.M. and Fookes, P.G. 2005. Hot deserts. *In*: Fookes, P.G., Lee, E.M. and Milligan, G. (eds) *Geomorphology for Engineers*. Whittles Publishing, Dunbeath, UK, 419–453.

Lee, E.M. and Fookes, P.G. 2015. A note on the origins of engineering geomorphology in the UK. *Quarterly Journal of Engineering Geology and Hydrogeology*, **48**, 147–156, https://doi.org/10.1144/qjegh2014-048

Lee, E.M., Sweeney, M. and Clarke, J. 2025. Arabia Felix: Fookes in the desert. *Geological Society, London, Engineering Geology Special Publications*, **30**, https://doi.org/10.1144/egsp30-2023-33

Little, A.L., Stewart, J.C. and Fookes, P.G. 1963. Bedrock grouting tests at Mangla Dam, West Pakistan. *In*: *Grouts & Drilling Muds in Engineering Practice*. Institution of Civil Engineers, London, 91–97.

Lovegrove, G.W. and Fookes, P.G. 1972. The planning and implementation of a site investigation for a highway in tropical conditions in Fiji. *Quarterly Journal of Engineering Geology*, **5**, 43–68, https://doi.org/10.1144/GSL.QJEG.1972.005.01.05

Norman, J.W., Leibowitcz, T.H. and Fookes, P.G. 1975. Factors affecting the detection of slope instability with air photographs in an area near Sevenoaks, Kent. *Quarterly Journal of Engineering Geology*, **8**, 159–176, https://doi.org/10.1144/GSL.QJEG.1975.008.03.01

Pettifer, G.S. and Fookes, P.G. 1994. A revision of the graphical method for assessing the excavatability of rock. *Quarterly Journal of Engineering Geology*, **27**, 145–164, https://doi.org/10.1144/GSL.QJEGH.1994.027.P2.05

Reeves, G.M., Sims, I. and Cripps, J.C. (eds) 2006. *Clay Materials Used in Construction*. Geological Society Engineering Geology Special Publications, **21**, https://doi.org/10.1144/GSL.ENG.2006.021.01.21

Sims, I. 1986. Sand, gravel and crushed rock aggregates for construction purposes. *Quarterly Journal of Engineering Geology*, **19**, 325–338, https://doi.org/10.1144/GSL.QJEG.1986.019.03.13

Smith, M.R. and Collis, L. (eds) 2001. *Aggregates: Sand, Gravel and Crushed Rock Aggregates for Construction Purposes*, 3rd edn (revised by Fookes, P.G., Lay, J., Sims, I., Smith, M.R. and West, G.). Geological Society Engineering Geology, London,

Special Publications, **17**, https://doi.org/10.1144/GSL.ENG.2001.017.01.00

Stoner, J.R. and Fookes, P.G. 1991. The Libyan Great Man-Made River Project Phase 1. Part 1: manufacture of prestressed concrete cylinder pipe. *Proceedings of the Institution of Civil Engineers, Part 1*, **90**, 853–879, https://doi.org/10.1680/iicep.1991.15386

Stoner, J.R., Fookes, P.G. and Mackintosh, J.H. 1991. The Libyan Great Man-Made River Project Phase 1. Part 2: concrete technology of prestressed concrete pipe manufacture. *Proceedings of the Institution of Civil Engineers, Part 1*, **90**, 881–919, https://doi.org/10.1680/iicep.1991.15387

Tepel, R.E. (ed.). 2002*a*. *Visioning the Future of Engineering Geology: Sustainability and Stewardship*. Association of Engineering Geologists Special Publications, **19**, Vol. 1 (CD-ROM).

Tepel, R.E. 2002*b*. Mission-capable and mission ready? The role of engineering geology professional associations as stewards of, and standard bearers, for the profession. *In*: *Association of Engineering Geologists Special Publications*, **19**, Vol. 1 (CD-ROM).

Tepel, R.E. (ed.). 2003. *Visioning the Future of Engineering Geology: Degree Program Quality and Accreditation*. Association of Engineering Geologists Special Publications, **19**, Vol. 2 (CD-ROM).

Tepel, R.E. (ed.). 2004*a*. *Visioning the Future of Engineering Geology: Understanding conflict and co-operation in the practice of engineering geology*. Association of Engineering Geologists Special Publications, **19**, Vol. 4 (CD-ROM).

Tepel, R.E. 2004*b*. Confounded confusion: implications and consequences of the great definitional obfuscation between engineering geology and environmental geology. *Association of Engineering Geologists Special Publications*, **19**, Vol. 4 (CD-ROM).

Tepel, R.E. 2004*c*. Towards a philosophy of engineering geology. *Association of Engineering Geologists Special Publications*, **19**, Vol. 4 (CD-ROM).

Tepel, R.E. 2004*d*. The engineering geology site characterization as the core function of engineering geology practice. *Association of Engineering Geologists Special Publications*, **19**, Vol. 3 (CD-ROM).

Venečanin, D.S.D. 1981. Some problems of construction aggregates in desert areas, with particular reference to the Arabian peninsula. *Discussion of Papers 8284 and 8302 by P.G. Fookes and I.E. Higginbottom Proceedings of the Institution of Civil Engineers, Part 1*, **70**, 519–522, https://doi.org/10.1680/iicep.1981.1787

Walker, M.J. (ed.). 2012. *Hot Deserts: Engineering Geology and Geomorphology*. Geological Society, London, Engineering Geology Special Publications, **25**, https://doi.org/10.1144/EGSP25.00

Waltham, A.C. and Fookes, P.G. 2003. Engineering classification of karst ground conditions. *Quarterly Journal of Engineering Geology and Hydrogeology*, **36**, 101–118, https://doi.org/10.1144/1470-9236/2002-33

Woodford, A.J. and Ray, W.J.F. 1992. Discussion: The Libyan Great Man-Made River Project Phase 1; parts 1 and 2. Discussion of Papers 9749 and 9750 by J.R. Stoner, P.G. Fookes and J.H. Mackintosh. *Proceedings of the Institution of Civil Engineers, Part 1*, **94**, 501–502, https://doi.org/10.1680/istbu.1992.21510

Index

Page numbers in *italics* denote figures. Page numbers in **bold** denote tables.